Michael Quinten

**Optical Properties of
Nanoparticle Systems**

Related Titles

Schmid, G. (ed.)

Nanoparticles

From Theory to Application

2010
ISBN: 978-3-527-32589-4

Gubin, S. P. (ed.)

Magnetic Nanoparticles

2009
ISBN: 978-3-527-40790-3

Fendler, J. H. (ed.)

Nanoparticles and Nanostructured Films

Preparation, Characterization and Applications

1998
ISBN: 978-3-527-29443-5

Bohren, C. F., Huffman, D. R.

Absorption and Scattering of Light by Small Particles

1998
ISBN: 978-0-471-29340-8

Michael Quinten

Optical Properties of Nanoparticle Systems

Mie and Beyond

WILEY-VCH Verlag GmbH & Co. KGaA

The Author

Dr. Michael Quinten
Mauritiusstr. 7
52457 Aldenhoven
ulmi.quinten@t-online.de

Cover

Lycurgus Cup, a dichroic glass cup with a mythological scene; Late Roman, 4th century AD; probably made in Rome. The illustration on the right shows the same glass cup when held up to the light. © Trustees of the British Museum.

■ All books published by **Wiley-VCH** are carefully produced. Nevertheless, authors, editors, and publisher do not warrant the information contained in these books, including this book, to be free of errors. Readers are advised to keep in mind that statements, data, illustrations, procedural details or other items may inadvertently be inaccurate.

Library of Congress Card No.: applied for

British Library Cataloguing-in-Publication Data
A catalogue record for this book is available from the British Library.

Bibliographic information published by the Deutsche Nationalbibliothek
The Deutsche Nationalbibliothek lists this publication in the Deutsche Nationalbibliografie; detailed bibliographic data are available on the Internet at <http://dnb.d-nb.de>.

© 2011 Wiley-VCH Verlag & Co. KGaA, Boschstr. 12, 69469 Weinheim, Germany

All rights reserved (including those of translation into other languages). No part of this book may be reproduced in any form – by photoprinting, microfilm, or any other means – nor transmitted or translated into a machine language without written permission from the publishers. Registered names, trademarks, etc. used in this book, even when not specifically marked as such, are not to be considered unprotected by law.

Composition Toppan Best-set Premedia Limited, Hong Kong
Printing and Binding Fabulous Printers Pte. Ltd., Singapore
Cover Design Adam Design, Weinheim

Printed in Singapore
Printed on acid-free paper

ISBN: 978-3-527-41043-9

To Ulrike, Eva and Christoph

Contents

Preface *XIII*

1 **Introduction** *1*

2 **Nanoparticle Systems and Experimental Optical Observables** *9*
2.1 Classification of Nanoparticle Systems *10*
2.2 Stability of Nanoparticle Systems *14*
2.3 Extinction, Optical Density, and Scattering *21*
2.3.1 The Role of the Particle Material Data *25*
2.3.2 The Role of the Particle Size *26*
2.3.3 The Role of the Particle Shape *29*
2.3.4 The Role of the Particle Concentration *33*
2.3.4.1 Dilute Systems *33*
2.3.4.2 Closely Packed Systems *34*

3 **Interaction of Light with Matter – The Optical Material Function** *37*
3.1 Classical Description *37*
3.1.1 The Harmonic Oscillator Model *38*
3.1.2 Extensions of the Harmonic Oscillator Model *40*
3.1.3 The Drude Dielectric Function *41*
3.2 Quantum Mechanical Concepts *42*
3.2.1 The Hubbard Dielectric Function *43*
3.2.2 Interband Transitions *47*
3.3 Tauc–Lorentz and OJL Models *50*
3.4 Kramers–Kronig Relations and Penetration Depth *52*

4 **Fundamentals of Light Scattering by an Obstacle** *55*
4.1 Maxwell's Equations and the Helmholtz Equation *56*
4.2 Electromagnetic Fields *59*
4.3 Boundary Conditions *61*
4.4 Poynting's Law and Cross-sections *62*
4.5 Far-Field and Near-Field *65*

4.6	The Incident Electromagnetic Wave	66
4.7	Rayleigh's Approximation for Small Particles – The Dipole Approximation	69
4.8	Rayleigh–Debye–Gans Approximation for Vanishing Optical Contrast	71

5 Mie's Theory for Single Spherical Particles 75
5.1 Electromagnetic Fields and Boundary Conditions 76
5.2 Cross-sections, Scattering Intensities, and Related Quantities 83
5.3 Resonances 87
5.3.1 Geometric Resonances 88
5.3.2 Electronic Resonances and Surface Plasmon Polaritons 91
5.3.2.1 Electronic Resonances 92
5.3.2.2 Surface Plasmon Polariton Resonances 94
5.3.2.3 Multiple Resonances 101
5.3.3 Longitudinal Plasmon Resonances 104
5.4 Optical Contrast 108
5.5 Near-Field 112
5.5.1 Some Further Details 122

6 Application of Mie's Theory 123
6.1 Drude Metal Particles (Al, Na, K) 124
6.2 Noble Metal Particles (Cu, Ag, Au) 127
6.2.1 Calculations 127
6.2.2 Experimental Examples 129
6.2.2.1 Colloidal Au and Ag Suspensions 129
6.2.2.2 Gold and Silver Nanoparticles in Glass 131
6.2.2.3 Copper Nanoparticles in Glass and Silica 132
6.2.2.4 Ag_xAu_{1-x} Alloy Nanoparticles in Photosensitive Glass 134
6.2.2.5 Silver Aerosols 135
6.2.2.6 Further Experiments 137
6.3 Catalyst Metal Particles (Pt, Pd, Rh) 139
6.4 Magnetic Metal Particles (Fe, Ni, Co) 141
6.5 Rare Earth Metal Particles (Sc, Y, Er) 142
6.6 Transition Metal Particles (V, Nb, Ta) 145
6.7 Summary of Metal Particles 147
6.8 Semimetal Particles (TiN, ZrN) 148
6.9 Semiconductor Particles (Si, SiC, CdTe, ZnSe) 151
6.9.1 Calculations 151
6.9.2 Experimental Examples 154
6.9.2.1 Si Nanoparticles in Polyacrylene 154
6.9.2.2 Quantum Confinement in CdSe Nanoparticles 154
6.10 Carbonaceous Particles 156
6.11 Absorbing Oxide Particles (Fe_2O_3, Cr_2O_3, Cu_2O, CuO) 162
6.11.1 Calculations 162

6.11.2	Experimental Examples	163
6.11.2.1	Aerosols of Fe_2O_3	163
6.11.2.2	Aerosols of Cu_2O and CuO	165
6.11.2.3	Colloidal Fe_2O_3 nanoparticles	167
6.12	Transparent Oxide Particles (SiO_2, Al_2O_3, CeO_2, TiO_2)	168
6.13	Particles with Phonon Polaritons (MgO, NaCl, CaF_2)	170
6.14	Miscellaneous Nanoparticles (ITO, LaB_6, EuS)	172

7 Extensions of Mie's Theory 177

7.1	Coated Spheres	177
7.1.1	Calculations	177
7.1.1.1	Metallic Shells on a Transparent Core	180
7.1.1.2	Oxide Shells on Metal and Semiconducting Core Particles	184
7.1.2	Experimental Examples	187
7.1.2.1	Ag–Au and Au–Ag Core–Shell Particles	187
7.1.2.2	Multishell Nanoparticles of Ag and Au	189
7.1.2.3	Optical Bistability in Silver-Coated CdS Nanoparticles	190
7.1.2.4	Ag and Au Aerosols with Salt Shells	193
7.1.2.5	Further Experiments	196
7.2	Supported Nanoparticles	198
7.3	Charged Nanoparticles	206
7.4	Anisotropic Materials	210
7.4.1	Dichroism	210
7.4.2	Field-Induced Anisotropy	211
7.4.3	Gradient-Index Materials	211
7.4.4	Optically Active Materials	213
7.5	Absorbing Embedding Media	214
7.5.1	Calculations	214
7.5.2	Experimental Examples	219
7.5.2.1	Absorption of Scattered Light in Ag and Au Colloids	219
7.5.2.2	Ag and Fe Nanoparticles in Fullerene Film	220
7.6	Inhomogeneous Incident Waves	223
7.6.1	Gaussian Beam Illumination	223
7.6.2	Evanescent Waves from Total Internal Reflection	226

8 Limitations of Mie's Theory – Size and Quantum Size Effects in Very Small Nanoparticles 233

8.1	Boundary Conditions – the Spill-Out Effect	233
8.2	Free Path Effect in Nanoparticles	234
8.3	Chemical Interface Damping – Dynamic Charge Transfer	240

9 Beyond Mie's Theory I – Nonspherical Particles 245

9.1	Spheroids and Ellipsoids	247
9.1.1	Spheroids (Ellipsoids of Revolution)	247
9.1.1.1	Electromagnetic Fields	248

9.1.1.2	Scattering Coefficients	251
9.1.1.3	Cross-sections	252
9.1.1.4	Resonances	252
9.1.1.5	Numerical Examples	254
9.1.1.6	Extensions	254
9.1.2	Ellipsoids (Rayleigh Approximation)	255
9.1.3	Numerical Examples for Ellipsoids	259
9.1.3.1	Metal Particles	259
9.1.3.2	Semimetal and Semiconductor Particles	265
9.1.3.3	Carbonaceous Particles	266
9.1.3.4	Particles with Phonon Polaritons	267
9.1.3.5	Miscellaneous Particles	267
9.1.4	Experimental Results	268
9.1.4.1	Prolate Spheroidal Silver Particles in Fourcault Glass	268
9.1.4.2	Plasma Polymer Films with Nonspherical Silver Particles	269
9.1.4.3	Further Experiments	272
9.2	Cylinders	273
9.2.1	Electromagnetic Fields and Scattering Coefficients	273
9.2.2	Efficiencies and Scattering Intensities	277
9.2.3	Resonances	279
9.2.4	Extensions	281
9.2.5	Numerical Examples	282
9.2.5.1	Metal Particles	283
9.2.5.2	Semimetal and Semiconductor Particles	288
9.2.5.3	Carbonaceous Particles	291
9.2.5.4	Oxide Particles	292
9.2.5.5	Particles with Phonon Polaritons	293
9.2.5.6	Miscellaneous Particles	294
9.3	Cubic Particles	296
9.3.1	Theoretical Considerations	296
9.3.2	Numerical Examples	298
9.3.2.1	Metal Particles	299
9.3.2.2	Semimetal and Semiconductor Particles	299
9.3.2.3	Particles with Phonon Polaritons	300
9.3.2.4	Miscellaneous Particles	301
9.4	Numerical Methods	302
9.4.1	Discrete Dipole Approximation	302
9.4.2	T-Matrix Method or Extended Boundary Condition Method	305
9.4.3	Other Numerical Methods	307
9.4.3.1	Point Matching Method	307
9.4.3.2	Discretized Mie Formalism	307
9.4.3.3	Generalized Multipole Technique	307
9.4.3.4	Finite Difference Time Domain Technique	307
9.5	Application of Numerical Methods to Nonspherical Nanoparticles	308

9.5.1	Nonmetallic Nanoparticles	308
9.5.2	Metallic Nanoparticles	310

10 Beyond Mie's Theory II – The Generalized Mie Theory 317

10.1	Derivation of the Generalized Mie Theory	318
10.2	Resonances	321
10.3	Common Results	325
10.3.1	Influence of Shape	325
10.3.2	Influence of Length	327
10.3.3	Influence of Interparticle Distance	327
10.3.4	Enhancement of Scattering and Extinction	329
10.3.5	The Problem of Convergence	331
10.4	Extensions of the Generalized Mie Theory	335
10.4.1	Incident Beam	335
10.4.2	Nonspherical Particles	336

11 The Generalized Mie Theory Applied to Different Systems 341

11.1	Metal Particles	342
11.1.1	Calculations	342
11.1.2	Experimental Results	346
11.1.2.1	Extinction of Light in Colloidal Gold and Silver Systems	346
11.1.2.2	Total Scattering of Light by Aggregates	353
11.1.2.3	Angle-Resolved Light Scattering by Nanoparticle Aggregates	355
11.1.2.4	PTOBD on Aggregated Gold and Silver Nanocomposites	358
11.1.2.5	Light-Induced van der Waals Attraction	360
11.1.2.6	Coalescence of Nanoparticles	361
11.1.2.7	Further Experiments with Gold and Silver Nanoparticles	363
11.2	Semimetal and Semiconductor Particles	364
11.3	Nonabsorbing Dielectrics	367
11.4	Carbonaceous Particles	369
11.5	Particles with Phonon Polaritons	372
11.6	Miscellaneous Particles	375
11.7	Aggregates of Nanoparticles of Different Materials	376
11.8	Optical Particle Sizing	379
11.9	Stochastically Distributed Spheres	382
11.10	Aggregates of Spheres and Numerical Methods	387
11.10.1	Applications of the Discrete Dipole Approximation	387
11.10.2	Applications of the T-Matrix approach	389
11.10.3	Other Methods	389

12 Densely Packed Systems 393

12.1	The Two-Flux Theory of Kubelka and Munk	394
12.2	Applications of the Kubelka–Munk Theory	397
12.2.1	Dense Systems of Color Pigments: Cr_2O_3, Fe_2O_3, and Cu_2O	398
12.2.2	Dense Systems of White Pigments: SiO_2 and TiO_2	399

12.2.3	Dense Systems of ZrN and TiN Nanoparticles	400
12.2.4	Dense Systems of Silicon Nanoparticles	401
12.2.5	Dense Systems of IR Absorbers: ITO and LaB_6	403
12.2.6	Dense Systems of Noble Metals: Ag and Au	404
12.2.7	The Lycurgus Cup	406
12.3	Improvements of the Kubelka–Munk Theory	407

13 Near-Field and SERS *411*
13.1 Waveguiding Along Particle Chains *412*
13.2 Scanning Near-Field Optical Microscopy *416*
13.3 SERS with Aggregates *420*

14 Effective Medium Theories *427*
14.1 Theoretical Results for Dielectric Nanoparticle Composites *431*
14.2 Theoretical Results for Metal Nanoparticle Composites *433*
14.3 Experimental Examples *437*

References *441*
Color Plates *479*
Index *485*

Preface

The optical response of heterogeneous matter consisting of nanoparticles in a surrounding matrix material is often easy to measure, but rather difficult to describe, yet it is almost impossible to give a complete description.

I met this problem first when I joined Professor Kreibig's group in Saarbrücken, Germany, in 1983. I soon learned that classical electrodynamics is helpful, but solid-state physics is also indispensable when dealing with the optical properties of nanoparticles and nanoparticle systems. In my subsequent studies in Saarbrücken and Aachen electrodynamics became dominant, without losing sight of solid-state physics. This was enabled by the long-lasting cooperation with Uwe Kreibig, who is still engaged in the physics of interfaces and surfaces of very small particles.

During the long period of research on the optical properties of nanoparticle matter from 1983 to 2000 at the Universities of Saarbrücken, Aachen, Graz, Chemnitz, and Bochum, I became acquainted with several aspects of light scattering and absorption by small particles in combination with solid-state physics. I learned that a great variety of scientific and engineering disciplines have significant interest in the optical properties of such inhomogeneous nanoparticle matter. Even today, after ten successful years in industry, inhomogeneous nanoparticle matter hits me again via optical metrology solutions for photovoltaics, thin films, organic LEDs, and some further applications of nanoparticle systems. Hence the most important motivation for this book came from applications of nanoparticle matter.

The purpose of this book is to give first an overview of analytical and numerical models for the optical response of nanoparticles and nanoparticle systems. It may, therefore, appear to have a more theoretical character, but many experimental results complement the various calculations. Second, and in the main, this book provides many calculations on the *spectral behavior* of light scattering and absorption by nanoparticles and nanoparticle systems. Here, electrodynamics again meets solid-state physics. The initial interaction of light with matter, expressed by the frequency-dependent dielectric function of the particle material, enters the electrodynamic scattering model and yields characteristic spectra that are determined by the material-specific properties, the particle-specific topological properties and statistics.

To write this book required reading and evaluating of many monographs and an even larger number of other publications on this subject. However, the amount of published work is too immense to consider them all in such a book. I hope to have included the most relevant and up-to-date literature, and apologize for all the contributions not considered here.

Last but not least, I want to acknowledge all the inspiration and encouragement from many people during the writing of this book. Special thanks are due to Uwe Kreibig for his continuous interest and so many fruitful discussions on several aspects of nanoparticles and nanoparticle matter. In addition, I want to thank all the people who gave me the chance to increase and to improve my knowledge on this subject with a stay in their groups: A. Heilmann (Halle), R. Wannemacher (Leipzig), F. R. Aussenegg and A. Leitner (Graz), R. Hempelmann (Saarbrücken), Th. Henning (Jena), and G. Schweiger (Bochum). I also gratefully acknowledge all the people who supported me by providing relevant data, namely U. Kreibig, H. Eckstein, G. Reuter, M. Gartz, K.-J. Berg, J. Porstendorfer, H. Hofmeister, W. Hoheisel, H. Mutschke, H. Amekura, Y. Takeda, L. M. Liz-Marzan, G. C. Schatz, R. Jin, and R. M. Magruder III. Finally, many thanks go to my family for their support and patience with me during the writing of this book.

Aldenhoven, October 2010 *Michael Quinten*

1
Introduction

About 40 years ago, research on nanoparticles and nanoparticle matter restarted. The development of new techniques (laser, ESCA, STM, AFM, SNOM, etc.) and the continuous improvement of existing techniques (vacuum technology, electron microscopy, etc.) has allowed new insights into an old subject – the transition from the atom or molecule to the bulk solid matter.

The results obtained at the beginning of this period showed that the *size* of the object plays the key role: almost all physical properties show a significant size dependence on going to sizes less than 100 nm. This is, however, not new at all. For more than 150 years chemists and physicists have been concerned with the colloidal state of matter, being called by Ostwald *Welt der vernachlässigten Dimensionen* (world of the disregarded dimensions) [1]. Well known are, for example, the colloidal gold dispersions of Faraday in the nineteenth century. Colloidal systems can even be traced back to the Romans, who empirically developed methods to stain glass with small inclusions of gold or silver. One famous example is the Lycurgus cup (fourth century AD; see book cover), which can be viewed at the British Museum. First descriptions of how to obtain ruby gold glass can be found as early as in the bibliography of Ashurbanipal in Nineve (seventh century BC). Industrial manufacturing of stained glass with colloidal particles was established in the seventeenth century, for example by Kunckel [2]. In the nineteenth and twentieth centuries, colloidal nanoparticles found applications mainly in photography based on silver halide nanocrystals, as color pigments, and in catalysis, with their size often exceeding the magic threshold of 100 nm used in nanotechnology. The basic structures in nanotechnology are one-dimensional structures which are less than 100 nm in size in all three dimensions (nanoparticles, nanopores), line-shaped structures with two dimensions less than 100 nm (nanowires, nanocages, nanogrooves), and layers which are nanoscaled in only one dimension. To avoid further semantic problems, I will use the term *nanoparticle* for all particles that are at least in one dimension less than 999 nm. Hence I include a larger range of particles relevant for applications beyond the nanotechnology horizon, such as organic and inorganic color pigments.

The newly recognized size and quantum size effects of the last 40 years not only led to a better insight into the transitions from molecule to bulk matter, but also to a new, rapidly growing market of possible applications. Although we usually do

Optical Properties of Nanoparticle Systems: Mie and Beyond. Michael Quinten
Copyright © 2011 WILEY-VCH Verlag GmbH & Co. KGaA, Weinheim
ISBN: 978-3-527-41043-9

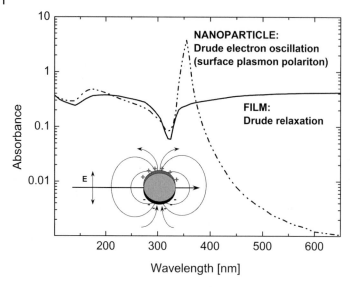

Figure 1.1 The conduction electron oscillator in a silver nanoparticle compared with the conduction electron relaxator in a silver film.

not realize it, nanoparticles belong to our everyday life: cosmetics, medicines, alternative energy, communication, and displays are examples which benefit from the basic science of nanotechnology. As in various applications the magic threshold of 100 nm is exceeded, this book also treats nanoparticles up to 999 nm in size so as to include these cases.

A reduction in size down to a few nanometers often leads to size- or material-specific peculiarities which can be used in new applications of the materials. These peculiarities cannot be observed with macroscopic pieces of the same material. For example, nanoparticulate matter exhibits increased hardness, fracture strength, additional electronic states, increased chemical selectivity, and increased surface energy. One of the most interesting peculiarities is the resonant absorption of light by nanometer-sized gold or silver particles. Unlike in bulk gold and silver, the collective excitation of the conduction electrons by an external electromagnetic field does not result in a relaxator but is transformed into an oscillator type of behavior with a distinct resonance, called surface plasmon polariton. Figure 1.1 illustrates this behavior for a silver nanoparticle (oscillator) compared with a bulk silver film (relaxator).

Hence metal nanoparticles play a particularly pronounced role in nanomaterial science. Nature even seems to have a preference for metals. More than two-thirds of all elements are metals. When looking at the optical properties of nanoparticles, not all metals exhibit such striking properties as gold and silver, nor do most of the metal particles remain unchanged under ambient conditions. Rather, oxides, sulfides, nitrides, and so on are formed with their own specific peculiarities. Therefore, this book considers also the optical properties of nonmetallic nanoparticles and provides many examples of various nanoparticle systems of metallic, semiconducting, carbonaceous, and dielectric particles.

During the last 40 years, a huge number of papers, reviews, and books have appeared which are concerned with nanotechnology, nanoelectronics, nanooptoelectronics, information technology with nanomaterials, self-assembly, nanostructured magnetic materials, nanocomposites, nanowires, and nanobelts. They mainly cover mechanical, electronic, quantum mechanical, and medical aspects, with little attention to optical properties. The most perceptible to humans, however, are the optical properties, for example, the color of paintings and the colors due to interference.

Optical properties of nanomaterials include elastic light scattering, absorption, reflectance and transmittance, second harmonic generation, third-order nonlinear optical properties, surface-enhanced Raman scattering, and others. This book concentrates on the *linear* optical properties: elastic light scattering and absorption of single nanoparticles and on reflectance and transmittance of nanoparticle matter.

Elastic light scattering has turned out to be a powerful tool for examination of the properties of small particles. Scientists and engineers from a large variety of disciplines – physics, electrical engineering, meteorology, chemistry, biophysics, and astronomy – are concerned with this field.

Hence light scattering and absorption by small particles has already been treated in numerous textbooks [3–5] and monographs [6–16], so it would appear almost unnecessary to write a further book on this topic. However, these monographs either deal mainly with particles larger than 1000 nm that are important in geophysics, planetary science, and astrophysics; nanoparticles appear here almost only for species that are relevant for the Earth's radiation budget and in astrophysics, for example, carbon or silicates; or, they mainly consider in detail the spatial (angular) distribution of scattered light. Only Kreibig and Vollmer [11] have given a comprehensive overview of the optical properties of metallic nanoparticles, clusters, and cluster matter, including a discussion of size and quantum size effects. Their book is restricted to particle sizes less than approximately 100 nm and to metals for which nanoparticles exhibit characteristic resonances in absorption and scattering. It gives a good overview of the developments in nanoparticle science from the beginning in the 1970s until 1995 and includes also an overview of preparation techniques. The present book is intended to fill the gap in the description of the optical properties of small particles with sizes less than 1000 nm and to provide a comprehensive overview of the *spectral behavior* of nanoparticulate matter of metallic, semiconducting, carbonaceous, and dielectric particles.

From the physical point of view, the spectral behavior must be a function of the photon energy $\hbar\omega$. On the other hand, the optical properties of small particles strongly depend on the size compared with the size of the electromagnetic radiation, that is, the wavelength λ. Moreover, in commonly used spectrometers for the ultraviolet, visible, and near-infrared spectral ranges, the output is usually given versus the wavelength. In this respect, the wavelength seems to be the appropriate quantity which permits direct comparison with experimental results, for which I

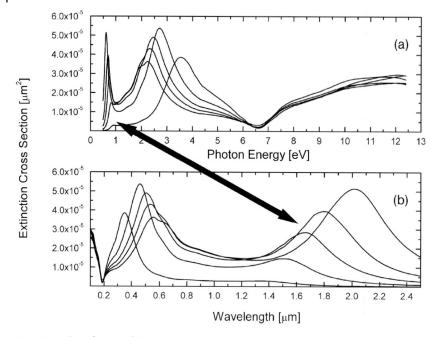

Figure 1.2 The influence of abscissa scaling on the appearance of optical data: extinction cross-section spectra of yttrium nanoparticles (a) versus photon energy and (b) versus wavelength.

preferably use the wavelength as abscissa to allow comparison with measured optical properties. Note that in the mid- and far-infrared regions the wavenumber in cm^{-1} is often used. This quantity is proportional to the energy $\hbar\omega$, but has the advantage of being measured in simple natural numbers instead of floating point numbers. Figure 1.2 demonstrates the difference between wavelength and photon energy as abscissa in the optical absorption spectra of nanoparticles. Using the photon energy, the UV region becomes spread and the IR region squeezed, and vice versa when using the (vacuum) wavelength.

In general, there are two steps on the way to the optical properties of a nanoparticle system. The first step (Figure 1.3) considers the geometry of a single isolated nanoparticle and its intrinsic optical properties, that is, its dielectric function. This information enters a suitable classical electrodynamic model, derived either rigorously or approximately, or formulated as a numerical method. The enormous practical advantage of any electrodynamic scattering model is that it enables one to compute *numerically* the optical response for arbitrary *realistic* particle materials. However, the classical electrodynamics used, being a phenomenological theory to describe light propagation, does not yield information about optical material properties. They enter via the dielectric functions inserted into the Maxwell boundary conditions, which must be taken from elsewhere, for example, from experiments or from quantum solid-state theory model calculations. The results of the first step are the optical properties of a single particle.

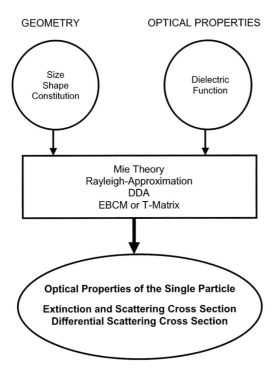

Figure 1.3 Step 1 on the way to the optical properties of a nanoparticle system: optical properties of the single particle.

The second step (Figure 1.4) considers the optical properties of the single particle from the first step and combines it with statistical information on size, shape, and the spatial distribution of many particles. New models that take into consideration all this information allow the calculation of, finally, the macroscopic optical properties of the nanomaterial, such as reflectance, transmittance, absorbance, and color.

The present book continues in Chapter 2 with an attempt to classify nanoparticle systems according to different properties and parameters. Then, it is emphasized how the main parameters of nanoparticle matter – size, shape, and constitution of the individual particle, optical constants of the particle material, concentration and spatial distribution of the particles – influence in principle the optical properties of a nanoparticle system.

One of the most important parameters – the dielectric function of the individual particle material – is discussed in Chapter 3 before any electromagnetic scattering model, because the understanding of intrinsic material-dependent interaction of light with matter is essential for the understanding of light scattering processes by an arbitrary obstacle. Once having established optical constants for the particle material, either from a table in a published work or by well-known and well-defined models, they can be used in an electromagnetic scattering model. Each of these

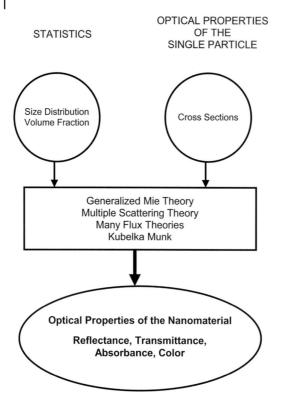

Figure 1.4 Step 2 on the way to the optical properties of a nanoparticle system: optical properties of the nanomaterial.

scattering models, rigorously established as numerical method for solving Maxwell's equations, has the same basis, described in Chapter 4.

In the subsequent chapters, various electrodynamic models are introduced which consider either the shape of the particle or the concentration of the particles in the nanoparticle matter. The models are supplemented by a considerable number of calculations of the spectral dependence of scattering and absorption, and by selected experimental results.

The most evolved model on light scattering and absorption by small particles stems from Gustav Mie [17] and describes the scattering by a spherical particle. Mie's theory, presented in Chapter 5, is fundamental and has found many applications in physics, astrophysics, chemistry, and engineering whenever light scattering by particles is important. Resonances in scattering and absorption play an important role for the spectrally resolved optical response of the nanoparticles and are discussed in detail in this chapter.

The importance of Mie's theory is emphasized in Chapter 6, where it is applied to calculate the size-dependent spectra of various nanoparticles. Metallic nanoparticles of different metals play a pronounced role also in these calculations. Further relevant material groups are semiconductors and oxides, with corresponding calculations presented here. The calculations are supplemented by many experimental examples.

Mie's theory has often been extended to permit calculations on particles taking into account certain boundary conditions. Chapter 7 considers the most important extensions on coated spheres, spheres on substrates, and inhomogeneous incident waves and other extensions. Again, appropriate experimental examples are given.

So far, Mie's theory and its extensions appear to be almost the ideal model to describe the optical properties of nanoparticles. However, particularly for metal particles where electrons form a quasi-free carrier gas with generic properties, the limitations of Mie's theory become obvious on going to sizes less than about 30 nm. They are discussed in Chapter 8.

The strongest limitation of Mie's theory, however, follows from its applicability to only spherical particles. Beyond Mie's theory, further electrodynamic models have been developed for particles of nonspherical shape. They are summarized and discussed in Chapter 9. Models for light scattering by ellipsoids, spheroids, cylinders, and cubes are available, partly as a rigorous solution of Maxwell's equations, and partly in the approximation of particles that are small compared with the wavelength of light. All these models are again supplemented by representative calculations and experimental results. Finally, numerical models for arbitrarily shaped particles have been developed. Among them, the *discrete dipole approximation* and the *T-matrix method* are the mostly commonly used and best known models.

Beyond Mie's theory, not only models for light scattering by nonspherical particles are available, but also models that take into account electromagnetic interactions among neighboring particles in nanoparticulate matter. With the *generalized Mie theory* (GMT) we present in Chapter 10 a rigorous solution for interacting spherical particles. Interaction introduces new parameters on which the optical properties now depend, in addition to the parameters of the isolated particle. The main new parameters are the size and shape of so-called aggregates – more or less densely lumped particles – and the interparticle distance. Their influence on the optical properties are discussed in detail in this chapter.

Similarly to single isolated particles, Chapter 11 gives an overview of the optical properties of aggregates and on nanoparticulate matter where the concentration of particles is high as the electromagnetic interaction among the nanoparticles can never be neglected. This is done with numerous calculations for various particle materials and certain aggregate topologies, and also with calculations on higher concentration nanoparticle systems. The calculations are supplemented by some experimental data.

So far, the GMT seems to be almost the most comprehensive model to describe the linear optical properties of clustered nanoparticles, similar to the Mie-theory for isolated particles. However, the computational burden increases rapidly with increasing number of particles and requires more and more computer capabilities with correspondingly longer computation times. Therefore, for densely packed nanoparticle systems such as paintings and lacquers the GMT must be replaced by faster models.

Faster and even much simpler models for dense nanoparticle matter are introduced and discussed in Chapter 12. They are based on the calculation of energy fluxes through the samples from left to right and vice versa. Individual particle properties enter these calculations by absorption and scattering rates. The results of such calculations are the reflectance and transmittance of the nanoparticulate matter. For the simplest flux model by Kubelka and Munk [18] improvements are discussed using the GMT for the calculation of the absorption and scattering rates of individual blocks of nanoparticle matter.

Special emphasis is given to the electromagnetic near-field in Chapter 13. This plays a certain role in applications such as surface-enhanced Raman spectroscopy (SERS) and also for near-field microscopy techniques and for waveguides with sizes below the wavelength.

Chapter 14, finally, is dedicated to the optical properties of bulk inhomogeneous matter which consists of a matrix material with nanoparticles as inclusions. The optical response of the inhomogeneous matter is described here as the response of homogeneous matter with modified dielectric function. Various *effective medium theories* are available in which the size of the inclusions is meaningless and scattering by the particles is neglected. The result of all these models is a so-called effective dielectric function that enters the macroscopic optical response of the bulk inhomogeneous matter, that is, the reflectance, transmittance, and intrinsic absorption. We discuss in this final chapter both the strengths and weaknesses of these models.

2
Nanoparticle Systems and Experimental Optical Observables

The single isolated nanoparticle is certainly the ideal subject to study the transition between the molecular state and the bulk solid or liquid state. Unfortunately, however, due to their large surface-to-volume-ratio, the stored surface energy renders them thermodynamically unstable. Hence they need extra stabilization, by charging them uniformly, by depositing them on a substrate, or by embedding them in some matrix material. In all cases, the pure intrinsic properties of the single nanoparticle are influenced more or less due to electrical charge or the surrounding media. Only if the nanoparticles are prepared under high-vacuum conditions in special setups in a free particle jet are single isolated particles available for a rather short investigation time at the price of high costs of preparation and detection. Therefore, single nanoparticles are of prime interest for fundamental research.

However, even if such more complex systems containing only single stabilized nanoparticles are available, they are of limited interest for technical and practical applications. Additionally, the detection of a single nanoparticle is almost impossible, because its traces are too weak. For example, with common optical methods it is not possible to detect a single nanoparticle due to its small cross-sections for scattering and extinction. Only in optical near-field microscopy and related methods are experiments on single (but deposited) nanoparticles successful. Hence, instead of systems with single nanoparticles we often meet systems with many particles. Also in nature, nanoparticles are also mostly found in macroscopic ensembles.

The large diversity of possible sample structures resulting from this has not yet allowed the successful derivation of a general theoretical description for nanoparticle systems such as is known, for example, for crystalline matter. As a consequence, to obtain information that can be generalized, we must classify nanoparticle systems according to certain parameters. This is attempted in Section 2.1, which is followed by a brief discussion of the stability of nanoparticle systems in Section 2.2. In Section 2.3 the linear optical properties of nanoparticle systems accessible by measurement are presented and discussed in dependence on particle size and shape, optical material data, and concentration.

2.1
Classification of Nanoparticle Systems

A nanoparticle system is a system of two or more homogeneous phases (solid, liquid, gaseous), where the properties of the boundary between two phases relevantly determines the properties of the nanoparticle system. This simple picture, however, neglects all parameters and properties of the individual nanoparticles and all parameters that lead to the specific properties of the nanoparticle system when putting the single particles in the system. A nanoparticle system is a particular case of inhomogeneous granular or composite matter, because the structural elements – the nanoparticles – are large compared with the atomic scale. Therefore, for a classification, it seems appropriate first to classify the nanoparticles and then to turn to the nanoparticle system with additional parameters.

In the following, we attempt first a classification of nanoparticles according to different parameters.

A) Origin
- undesired by-products
  - diesel exhaust (soot)
  - sulfuric acid droplets
- natural
  - aerosols (dust, sea salt, volcanic ash, fire ash, …)
  - hydrosols
  - geo-colloids (inclusions in minerals, opals, …)
- synthetic
  - various technical and medical "vacuosols", aerosols, and hydrosols
  - heterogeneous catalysts
  - sol–gel systems
  - ferrofluids
  - nanoceramics
  - electrical devices (cermets, varistors, …)
  - island films
  - photographic systems
  - gas sensors
  - solar absorbers
  - recording tapes, data storage systems
  - color pigments, lacquers
  - cosmetics (sun protection creams, toothpaste, …)
  - medical drugs
  - organic systems (proteins, viruses, biomolecules, …).

It becomes obvious from this list that synthetic nanoparticles are widespread and determine many parts of our life, partly without us knowing that we use nanoparticle systems. They are used as filling materials, coatings, and paintings, and have applications in electronics, medicine, cosmetics, food and drugs, and housekeeping. Colloidal systems (hydrosols) are not only esthetically appealing but also serve as model systems for understanding optical, magnetic, electrokinetic, and absorptive properties of bulk matter. Aerosols (both natural and from pollution) on the one hand affect the radiative energy balance of the Earth–atmosphere system, but on the other hand can also be helpful in medical applications concerning the respiratory tract.

B) **Preparation methods**
- gas-phase processes
  - thermal evaporation and condensation
  - atom deposition and condensation
  - supersonic beam expansion
  - sputtering
  - laser ablation
  - exploding wire
  - pyrolysis
  - diffusion (also in liquid and solid matrices)
- liquid-phase processes
  - chemical reduction
  - hydrolysis
  - photolysis
  - radiolysis
  - sol–gel methods
- mechanical processes
  - milling
  - ultrasound
  - spraying.

The various production methods result in rather different production volumes per year. Whereas for laboratory use the production is far below 1 ton per year, in industry many kilotons are produced per year (e.g., carbon black, color pigments, heterogeneous catalysts, Aerosil). Except for spraying, the mechanical processes have a relatively low efficiency because of the increasing efforts and costs with decreasing size and the tendency of the small particles to coagulate again to form larger particles.

C) **Dispersion in**
- gases
- liquids
- solids (composites).

Dispersion in gases, liquids, and solids strongly influences the stability of the nanoparticle system because different interactions among the particles may become relevant, affecting the mobility of the particles or their spatial distribution.

D) **Form and structure**
- spheres
- needles
- plates
- fibers
- tubes.

Form and structure strongly influence the mechanical, magnetic, and optical properties of the single particle.

E) **Chemical composition**
- metals
- semiconductors
- dielectrics, oxides
- polymers
- carbon
- biomolecules.

The chemical composition of the single particle affects the stability of the nanoparticle system and also optical and magnetic properties.

F) **Surface modification**
- untreated
- coated (core–shell particles)
- passivated
- functionalized for catalysis, medical applications, biological applications.

The surface modification is comparable to the chemical composition in its effects.

G) **Applications**
- photothermal heat source
- fluorescence labels of biological structures
- magnetic labels
- biochemical sensors
- alternative energy, solar cells
- communications, data storage
- displays, LEDs
- near-field optics
- UV absorbers, IR absorbers.

On going from the nanoparticles to nanoparticle systems we have to consider further parameters. Various interactions among the particles lead to new properties of the nanoparticle system which are not available for the single particles. For example, the beauty of opals is the result of *multiple light scattering* by silica (SiO_2) spheres of about 250 nm diameter closely packed in a cubic lattice in the opal.

The main parameters which have to be considered are the *concentration* and the *aggregation state*. For the opal, optical interactions among densely packed nanoparticles lead to multiple scattering. Van der Waals forces among the particles lead to a cubic lattice.

In low-concentration disordered nanoparticle systems the particles are well separated, but statistically distributed. The topology of the system can then be described by the average volume fraction of nanoparticles in the sample, often also called the *filling factor*, f:

$$f = \frac{\sum_i N_i V_{p,i}}{V} \tag{2.1}$$

where N_i is the number of particles i with volume $V_{p,i}$. In general, the particles may not only differ in size, but differ also in their shape and chemical composition. Hence, another practical definition of the filling factor is

$$f = \sum_i \frac{1}{\rho_i} \frac{m_i}{V} \tag{2.2}$$

where ρ_i is the mass density of component i with mass concentration m_i/V.

The filling factor is well suited for systems where the mean distance among particles is large.

If in low-concentration systems neighboring nanoparticles are more or less strongly pinned to each other by interaction forces (e.g., van der Waals forces) or chemical bonds, they form nanoparticle aggregates. Provided that these aggregates keep well separated in the system, the filling factor in Equations 2.1 and 2.2 is also applicable, because the portion of aggregates is already contained as component i with mass concentration m_i/V.

We can distinguish between *coagulation aggregates* and *coalescence aggregates*.

In coagulation aggregates or agglomerates, the individuality of the particles is maintained. Neither spontaneous atomic migration takes place nor does charge transfer from one to the other particle by exchanging electrons occur.

Coagulation aggregates are usually treated with statistical methods. They are described by system properties characterizing the whole aggregate:

- mean number N_A of nanoparticles per aggregate
- mean aggregate size d and the local properties characterizing the surrounding of the single nanoparticle within an aggregate
- nanoparticle pair correlation functions
- higher order correlation functions
- next neighbor distances
- coordination numbers
- the *Hausdorff dimension* $H = \partial(\log N_A)/\partial(\log d)$, which is equivalent to the fractal dimension of extended structures.

Schönauer and Kreibig [19] introduced the additional parameter

- *compactness* $S = B/(N_A - 1)$, where B is the number of all close connections between all neighboring particles within the aggregate. The larger the value of

S, the more compact is an aggregate. For example, $S = 1$ for the linear chain aggregate, $S = 3$ or 6 for the close-packed extended two- or three-dimensional aggregate, and, by extra definition, $S = 0$ for the single particle.

Coalescence aggregates are created when strong chemical bonding causes neighboring particles to grow together by building common extended grain boundaries. This process usually induces the formation of new, larger units by reduction of surface area and by exothermic reduction of surface energy. Then, the individuality of the nanoparticles involved is lost, and a reduced number of larger units appear with changed electronic and optical properties and often irregular shapes. If the particles are metallic or semiconducting, percolation paths for electrical and heat conduction are formed and charge conservation for the individual parts no longer holds. Incomplete coalescence of spherical nanoparticles can lead to irregular rod- or chain-like structures.

A certain class of coalescence aggregates are *fractal aggregates* or *fractals*. These aggregates contain a large number of coalesced primary particles. The peculiarity of fractals is their self-similarity. They form a matter neither two-dimensional (2D) nor three-dimensional (3D) because of many empty sites within the complete fractal. Overviews of fractals can be found elsewhere [20–22].

2.2
Stability of Nanoparticle Systems

Spontaneous coagulation of nanoparticles mainly appears in aerosols and mechanically generated nanoparticle systems, but rarely in hydrosols. The reason for this different behavior of these colloidal systems is due to the charging of the nanoparticles in both systems.

In fact, hydrosols are always electrolytes containing various positively and negatively charged ions. Bringing a metal electrode with excess charge into this electrolyte, this excess charge will be screened by the ions in the electrolyte. This is achieved by a thin ion layer in direct contact to the electrode – the Stern layer – followed by a diffuse ion cloud extending in the electrolyte. Both form the so-called double layer. The extension of the cloud is given by the Debye–Hückel screening length $1/\kappa$ with the Debye-Hückel parameter given by

$$\kappa = \frac{e^2}{\varepsilon \varepsilon_0 k_B T} \sum_j n_j z_j^2 \qquad (2.3)$$

where n_j is the concentration and z_j the ion valence of ions of kind j.

Nanoparticles diluted in electrolytes also exhibit this phenomenon of a screening double layer. Hence the particles appear to be uniformly charged all with the same sign of charge. This is not so in aerosols: in aerosols, one finds uncharged particles and also positively and negatively charged particles for which coagulation is accelerated in aerosols. In hydrosols, the charging of the particles prevents them from coagulation due to a screened Coulomb potential V_C resulting from the double layer. The Coulomb potential counteracts the attractive van der Waals interaction V_{vdW} among the particles which is always present.

For two neighboring spherical particles with diameter $2R$, V_C and V_{vdW} are given by

$$V_C(r) = 4\pi\varepsilon\varepsilon_0 R \psi_0^2 \frac{R-r}{r}\log\left\{1+\frac{R}{r-R}\exp[-\kappa(r-2R)]\right\} \quad (2.4)$$

$$V_{vdW}(r) = -\frac{H}{6}\left[\frac{2R^2}{r^2-4R^2}+\frac{2R^2}{r^2}+\log\left(\frac{r^2-4R^2}{r^2}\right)\right] \quad (2.5)$$

Equation 2.4 is the generalization of the Coulomb potential of Derjaguin [23] and Verwey and Overbeek [24] derived by McCartney and Levine [25]. The attraction in Equation 2.5 has its origin in the van der Waals–London attraction between two atoms which was integrated over all atoms in both spherical particles; H is the Hamaker constant.

Obviously, only the repulsive Coulomb potential depends on the ion strength $\sum_j n_j z_j^2$ and can be strong or weak, depending on the Debye–Hückel parameter κ.

As an example, Figure 2.1 shows the total interaction potential $V(r) = V_C(r) + V_{vdW}(r)$ as a function of κ. The curves are calculated for a 2–2 electrolyte with $\psi_0 = 25\,\text{mV}$ and $H = 125\,\text{meV}$.

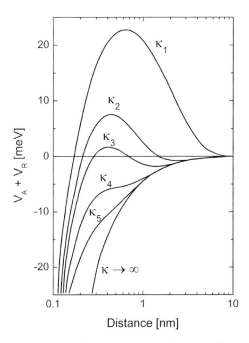

Figure 2.1 Total interaction potential $V(r)$ as a function of the Debye–Hückel parameter κ: $\kappa_1 = 0.01$, $\kappa_2 = 0.1$, $\kappa_3 = 0.2$, $\kappa_4 = 0.5$, and $\kappa_5 = 1\,\text{nm}^{-1}$.

For low values of κ, the potential $V(r)$ is repulsive. The particles must cross a relatively high potential wall before sticking together due to the van der Waals attraction. In this case they keep distant and cannot coagulate. With increasing κ, however, the potential wall decreases and when exceeding a certain value – in the example for $\kappa \geq 0.2\,\mathrm{nm}^{-1}$ – the repulsion can never prevent coagulation. At values between this threshold value and low values there exists a rather interesting secondary minimum of the potential (in Figure 2.1 the curves with κ_2 and κ_3). In these cases the particles can approach each other to a certain extent and are fixed at this distance, forming regularly arranged coagulation aggregates like liquid crystals. The stability of these aggregates is not very high; it only needs a small amount of energy to destroy the regular pattern.

The influence of the ions on the stability can be demonstrated when ions become exchanged in colloidal solutions by dialysis. Figure 2.2 shows the optical extinction spectra obtained for silver colloids with mean sizes $2R = 40$ and $60\,\mathrm{nm}$ before and after dialysis for 12 h. With dialysis the concentration of ions becomes reduced and the pH of the silver colloid changes from usually 8–9 to 7. This means that the suspension becomes more *acidic* after dialysis than the original suspension. In comparison with the spectra before dialysis (squares), the spectra of the colloids after dialysis (solid lines) show a decrease in extinction in the wavelength range where the extinction is peaked due to a resonant absorption in the silver particles (the reason for this absorption peak will be explained later in Chapter 5), and an increase in extinction at longer wavelengths. Also, the peak position is red shifted. The differences in the spectra come from reversible aggregation of the particles in the colloids. As the ion concentration of the particles in the suspension is reduced due to dialysis, van der Waals attraction among the particles comes increasingly into play, leading to the shallow minimum in a short distance from the particle surfaces as in Figure 2.1. As this distance is small, electromagnetic interactions become important, increasing the extinction at longer wavelengths. This *aggregation* is reversible because already Brownian motion of the particles can destroy these *aggregates*.

As the repulsive potential $V_C(r)$ depends on the ion strength, controlled aggregation into aggregates of various mean sizes and topologies can also be induced by addition of a variable amount of a salt solution to the colloidal suspension over a threshold concentration. The added cations (anions) differ in their effect. According to Hofmeister [26], they can be arranged in a line in which the effect increases and the corresponding threshold concentration decreases on going from right to left:

Cations:	Th	Al	Cu	Sr	Ca	Mg	H	Cs	Rb	K	Na	Li
Anions:	citrate	tartrate	SO_4	acetate	Cl	NO_3	Br	I	CNS			

For a constant amount of added salt solution larger than the threshold concentration, aggregation increases with increasing time. The aggregation process is diffusion limited and can be stopped by addition of a solution containing

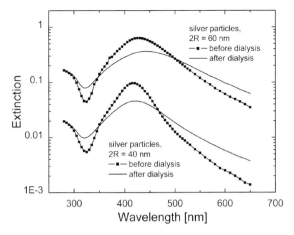

Figure 2.2 Changes in optical extinction spectra of silver particles due to dialysis of the samples.

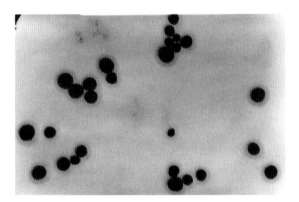

Figure 2.3 Colloidal particles and aggregates with a protecting gelatin shell.

macromolecules, for example, gelatin solution. Gelatin forms mechanically protecting shells around the aggregates and particles. An example is shown in the micrograph in Figure 2.3.

The number N of excess charges on nanoparticles in colloidal suspensions can be estimated. The excess charge q_s on the surface of a metal particle in an electrolyte is compensated by a countercharge $-q_s$ from ions in the electrolyte, leading to a capacitor with capacitance C. Whereas q_s is concentrated in a surface layer, $-q_s$ is randomly distributed in the solution as a diffuse ion cloud around the particle. From various experimental results in electrochemistry [27], it is known that for metal nanoparticles the capacitance C consists of two parts:

$$\frac{1}{C} = \frac{1}{C_{el}} + \frac{1}{C_{met}} \tag{2.6}$$

from the electrolyte and the metal. The capacitance C_{el} follows from DLVO theory as

$$C_{el} = \varepsilon_0 \varepsilon_{el}/\kappa \tag{2.7}$$

The capacitance C_{met} results from the *jellium model* [28]. Typically, the values for C_{met} range from 30 to $50\,\mu\text{F}$ per cm² of the metal surface. For a 1–1 electrolyte (e.g., NaCl solution), $C_{met} = 30\,\mu\text{F}\,\text{cm}^{-2}$, and a surface potential of $U = 100\,\text{mV}\,\text{cm}^{-2}$, one obtains the number N depending on particle size $2R$ in nanometers as

$$N(R) = 2.9 \cdot \text{nm}^{-1} \cdot R + 0.5 \cdot \text{nm}^{-2} \cdot R^2 \tag{2.8}$$

If this electrostatic protection is not given, as in aerosols or when forming particles by evaporation and condensation, the mobile particles continue to grow by coagulation until they finally form macroscopic precipitates. The final sizes are determined by the available nanoparticle material. The continuing growth is supported not only by the addition of atoms but also by coagulation of already existing particles. As a consequence, special efforts must be made if stable, long-living nanoparticles of a given size are required. The most common ways of stabilization are the following:

- to fix the nanoparticles on a proper substrate (island films)
- to freeze the nanoparticles inside a rigid matrix, preventing their mutual motion
- to protect the nanoparticle surfaces by adsorbed or chemically bound nonmetallic interlayers, or
- electric surface charges, thus preventing the nanoparticles from approaching.

The stabilization of nanoparticles on supports or in matrices unfortunately goes along with another complication. The particle shape as a whole may differ from the shape energetically favored in the free nanoparticle due to adhesion and chemical binding processes which contribute to the equilibrium total energy. Typical effects are the deformation of particles when deposited on substrates.

The coagulation process can be described as a series of particle collisions, each of which destroys pairwise particles and forms new, larger particles. Figure 2.4 illustrates this process schematically.

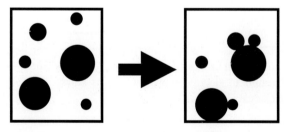

Figure 2.4 Schematics of forming new particles by collision of existing particles.

If $Z_n = N_n/V$ is the concentration of particles of kind n, which consists of n primary particles, then the time evolution of Z_n is given by

$$\frac{\partial Z_n}{\partial t} = -Z_n \sum_m K_{nm} Z_m + \frac{1}{2} \sum_{k=n-m}^{n-1} K_{km} Z_k Z_m \tag{2.9}$$

where K_{nm} is the probability of coagulation of particle n with particle m, with

$$K_{nm} = A \exp\left(-\frac{E_a}{k_B T}\right) \tag{2.10}$$

E_a being the activation energy for this coagulation step. K_{nm} can be interpreted as the mean number of collisions of type $(n) + (m) - (n + m)$. In general, A and E_a are different for each coagulation step. A simplification was given by von Smoluchowski [29]: all A and E_a are identical. In that case, the probabilities K_{nm} become all identical and constant $K_{nm} = K_D$ = constant.

With this simplification, the coagulation process is described by the equation

$$\frac{\partial Z_n}{\partial t} = -\frac{1}{2} K_D Z^2 \tag{2.11}$$

with the solution

$$Z(t) = \frac{Z_0}{1 + 0.5 K_D Z_0 t} \tag{2.12}$$

Coagulation happens after the coagulation time

$$t_C = \frac{2}{K_D Z_0} \tag{2.13}$$

For the value of K_D, different models are available. If, for example, coagulation is driven by diffusion, a particle with radius R_1 and diffusion constant D_1 collects a particle with radius R_2 and diffusion constant D_2. Then, the probability of coagulation is

$$K_D = 4\pi (R_1 + R_2)(D_1 + D_2) \tag{2.14}$$

For a moving particle with velocity $u(R_1)$ which collects another moving particle with $u(R_2)$ it is

$$K_D = \pi (R_1 + R_2)^2 [u(R_2) - u(R_1)] \tag{2.15}$$

Particle growth can also be induced by *Ostwald ripening* [30]. Ostwald ripening is the effect that larger particles grow in time at the cost of smaller particles which become even smaller and even may vanish. Figure 2.5 illustrates this process schematically.

The reason is that the condensed phase of matter (here the solid or liquid particle) is in a thermodynamic balance with the gas phase. At each temperature a certain number of atoms or molecules of the same material as the particle material surround the particle as a cloud. From this cloud, atoms condense on the particle

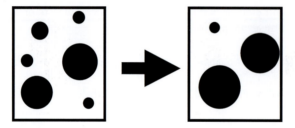

Figure 2.5 Schematics of Ostwald ripening.

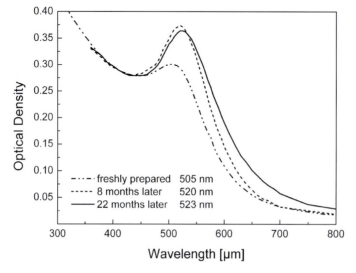

Figure 2.6 Experimental proof of Ostwald ripening with a colloidal gold solution.

but, vice versa, also atoms from the particle enter this cloud. In addition to the temperature, the number of atoms in the cloud depends strongly on the particle size: small particles have a highly numbered cloud, large particles have a poor cloud. If, at least, diffusion comes into play, even through the embedding medium, this diffusion leads to redistribution of atoms in the corresponding clouds. In the time average, however, this redistribution yields an increase in the number of atoms surrounding the larger particles and, therefore, by condensation to the growth of the larger particles at the cost of the smaller particles.

In an experimental study on colloidal gold particles over 22 months, this Ostwald ripening was observed. The gold particles exhibit an extinction peak in the visible spectral region with its peak position depending on the particle size: with increasing size the peak position shifts to longer wavelengths. The reason for this extinction peak will be explained later in Chapter 5. Comparing the spectra measured directly after preparation of the gold colloid with those recorded on the same colloid after 8 and 22 months, a clear shift to longer wavelengths can be recognized in Figure 2.6. This is caused by an increasing particle size due to Ostwald ripening.

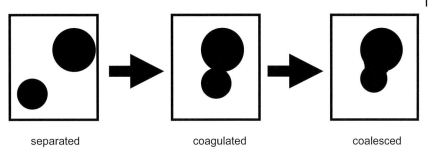

Figure 2.7 Schematics of coalescence.

It can also be recognized that in the first 8 months the particle growth is faster than in the subsequent 14 months.

Finally, we consider the *coalescence* of particles. Coalescence occurs when strong chemical or physical bonding causes coagulated particles to grow together by building common extended grain boundaries. Figure 2.7 illustrates the process schematically.

The reason for the atomic migration is the same as for Ostwald ripening. Unlike Ostwald ripening, however, the particles already stick together due to an earlier coagulation process. Therefore, the formation of common grain boundaries preferably occurs where the coagulated particles are in contact.

2.3
Extinction, Optical Density, and Scattering

When turning to the linear optical properties of a nanoparticle system, the classical quantities reflectance and transmittance of a macroscopic body must be replaced by extinction and scattering by the particles. Reflectance and transmittance are meaningful again when the nanoparticle system forms a macroscopic body with an almost infinite number of nanoparticles. The corresponding linear response of a nanoparticle system is, however, in all cases the result of many factors that must be taken into account: particle sizes, shapes, and constitution, optical material constants of the particles and the surrounding, and the structural parameters of the nanoparticle system, as for example the distances among the particles. An example is given in Figure 2.8 for suspensions of colloidal gold. The changes in size and shape of more or less densely lumped gold nanoparticles (aggregates) lead to characteristic changes of the optical properties of the colloidal suspension. The reference is the spectrum of the single isolated nanoparticles at the top of the graph.

The results prove that the sizes, shapes, and distances between particles are usually governed by complex statistical laws which, in general, vary from sample

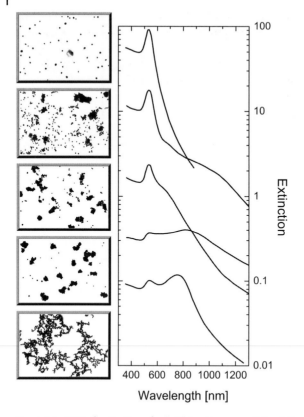

Figure 2.8 Optical extinction of colloidal gold nanoparticles: influence of size and shape of aggregates on the optical properties.

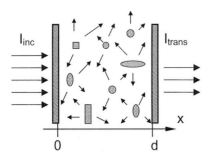

Figure 2.9 Scattering and absorption in an ensemble of nanoparticles.

to sample. There are only a few very special kinds of samples (e.g., photonic crystals or self-organized systems) where all constituents have almost identical parameters.

Consider an assembly of nanoparticles illuminated by a plane wave (Figure 2.9). The incident light is scattered and absorbed by each particle or aggregate in the volume to a certain extent, for the transmitted light is diminished along the propagation direction of the incident light by these absorption and scattering processes.

If the assembly is contained in a transparent cell, as for example for colloidal suspensions, reflection losses at the cell walls additionally diminish the transmitted light. The intensity of the transmitted light is then

$$I_{trans}(\lambda) = [1 - R_1(\lambda)]^2 [1 - R_2(\lambda)]^2 A(\lambda) I_{inc}(\lambda) \tag{2.16}$$

for each wavelength λ, neglecting multiple reflections at the cell walls. The factor $A(\lambda)$ contains all absorption and scattering losses in the measuring volume. The reflection losses $R_1(\lambda)$ at the front side and $R_2(\lambda)$ at the rear side seriously influence the measurement and must be eliminated with a reference measurement on a particulate-free system:

$$I_{ref}(\lambda) = [1 - R_1(\lambda)]^2 [1 - R_2(\lambda)]^2 I_{inc}(\lambda) \tag{2.17}$$

For particles in free space, for example in aerosols or free particle beams, the reflection losses are missing and the reference measurement at least yields the spectral distribution $I_{inc}(\lambda)$ of the incident light. We note that occasionally it is difficult to measure the reflection losses correctly. In some special cases of particle assemblies, for example, interstellar dust or atmospheric aerosols the reference measurement is also critical.

The ratio of transmitted to incident intensity defines the *(internal) transmittance*, $T(\lambda)$

$$T(\lambda) = \frac{I_{trans}(\lambda)}{I_{ref}(\lambda)} = A(\lambda) \tag{2.18}$$

which is the factor $A(\lambda)$ in Equation 2.16.

Note that for a correct determination of the transmittance it is important that the transmitted light is collected only in a small aperture around the forward direction $\Omega = 0°$.

The transmittance $T(\lambda)$ is connected with the *optical density* $\tau(\lambda)$ (often also called *extinction* or *absorbance*, the latter term often being used in commonly used spectrometers) via the decadic logarithm:

$$\tau(\lambda) = -\log_{10}[T(\lambda)] \tag{2.19}$$

The definitions of $T(\lambda)$ and $\tau(\lambda)$ in Equations 2.18 and 2.19 are valid for all particle assemblies, also if higher extinction processes such as multiple scattering are dominant. The major task is to determine $T(\lambda)$ or $\tau(\lambda)$ as a function of the optical properties and the topologies of the nanoparticle composite and the concentration of particulate matter.

The straightforward way to calculate the dielectric (optical) response would be to sum all contributions to the electrical polarization of the whole sample, including retarded electrodynamic multipole interactions of neighboring particles and the size, shape, and interparticle distance distributions in the sample. This approach was investigated by Bedeaux and Vlieger [31–33].

If the filling factors are smaller than 10^{-2}, that is, $f \leq 10^{-2}$, and the sample topology is isotropic and statistically disordered, the optical properties of the composite follow directly from the sum over the optical properties of all individual particles. Then, the phenomenological *Lambert–Beer law* or *Bouguer's law* may be applied and yields a simple relation between $T(\lambda)$ or $\tau(\lambda)$ and the single particle optical properties:

$$T(\lambda) = \exp\left[-\sigma_{ext}(\lambda)\frac{N}{V}d\right] \qquad (2.20)$$

$$\tau(\lambda) = \sigma_{ext}(\lambda)\frac{N}{V}\log(e)d \qquad (2.21)$$

In these equations we assumed for simplicity identical particles with extinction cross-section σ_{ext} and particle number concentration N/V. The quantity d is the lateral size of the measuring volume in the propagation direction (e.g., the thickness of the sample). In this case it is also simple to relate N/V with the filling factor f and the single particle volume V_T: $N/V = f/V_T$. Equation 2.21 also defines the extinction coefficient $e(\lambda)$ as

$$e(\lambda) = \sigma_{ext}(\lambda)\frac{N}{V} \qquad (2.22)$$

We already assumed that scattering plays a certain role in the optical properties of nanoparticle composites. Indeed, scattering is the essential parameter to distinguish between various concepts on different levels of approximation. The key parameter for classification is the parameter d/λ, where d is the linear dimension of typical structural elements and λ is the wavelength of the radiation in the sample.

If $d/\lambda \leq 10^{-2}$, nanoparticle systems behave like homogeneous materials, that is, the laws of geometric optics for the beam directions hold (rectilinear propagation of light beams; existence of the regularly refracted, transmitted, and reflected beams, uniform light propagation velocity) without light scattering at inhomogeneities, and for planar sample geometry the Fresnel equations for the intensities are sufficient for description. As a consequence, the electrodynamic response of the macroscopic sample can be described by a proper, in general tensorial, complex-valued dielectric function, and the task of many theoretical models comprehensively called *effective-medium theories* is to determine this *effective dielectric function* $\varepsilon_{eff}(\lambda)$ from the optical properties of the sample constituents. A typical feature of such models is that the particle size does not appear explicitly, a direct consequence of the condition $d/\lambda \leq 10^{-2}$. We will introduce the effective-medium theories in Chapter 14.

The optical response of a nanoparticle composite changes substantially if d/λ increases above $\sim 10^{-2}$ because then the optical effect of light scattering becomes effective. The propagation of light in the sample depends explicitly on the size of the contained structural elements and neither follows any longer the laws of geometric optics, nor can it be described by a dielectric function and the Fresnel equations.

Typical for light scattering is its angular spread. Hence it contributes to the dissipation of radiation energy of an incident parallel light beam in the forward direction except for a narrow aperture around $\Omega = 0°$. This includes all forward scattering and all backward scattering.

Assuming negligible multiple scattering, the diffuse forward scattering by an ensemble of particles called *haze*, H, can be calculated from integration of the angular distribution of the scattered light $i_{sca}(\Omega)$ of a single particle in the forward direction ($\Omega = 0$–$90°$). For example, for an ensemble of spherical nanoparticles the haze is given by

$$H = \frac{d \dfrac{f}{V_P} \dfrac{\lambda^2}{4\pi} \int\limits_{\Omega_0}^{\pi/2} i_{sca}(\Omega) d\Omega}{1 + d \dfrac{f}{V_P} \dfrac{\lambda^2}{4\pi} \int\limits_{0}^{\pi/2} i_{sca}(\Omega) d\Omega} \qquad (2.23)$$

where V_P is the single particle volume and d the thickness of the sample. The integration in the numerator excludes a small aperture Ω_0 around $\Omega = 0°$.

More complicated to describe is the optical response of the nanoparticle system in the backward direction. Backscattering may lead to diffuse reflectance of the incident light. This plays a role especially in high-concentration composites where multiple scattering occurs. In less concentrated systems the backscattering into a certain angle is used, for example, in LIDAR (light detection and ranging) remote sensing it is $\Omega = 180°$. However, a similar quantity like haze in forward scattering is missing in backward scattering.

Roughly, for regular particle arrangements geometric optical beams (reflected light) follow from the coherent superposition of the backscattered light. This has been demonstrated [34] and is the reason for the weak localization of light, which was observed in systems of nonabsorbing polystyrene particles [35]).

2.3.1
The Role of the Particle Material Data

The dielectric function ε of the particle material is commonly well defined. Problems arise when it appears to be dependent on the experimental methods or theoretical models when determining it. This is almost always the case for metals and often for dielectrics in regions where strong resonances occur and for semiconducting materials. Then, the prediction of the optical properties of the nanoparticle system also depends on the dielectric functions used for the particles and the surrounding. We demonstrate this with computed extinction efficiency spectra of silver nanoparticles. The optical constants were taken from various sources [36–40]. Figure 2.10 summarizes computed extinction efficiency spectra of silver nanoparticles with $2R = 20$ nm in vacuum, using these different Ag optical constant sets.

The nanoparticles exhibit a narrow extinction band peaked around a wavelength of 0.36 μm, which is associated with the surface plasmon polariton resonance in

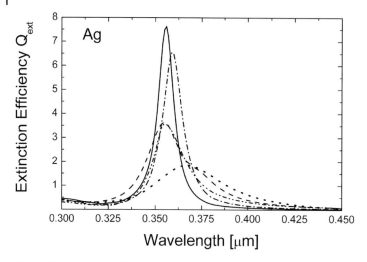

Figure 2.10 Extinction efficiency spectra of Ag nanoparticles with $2R = 20$ nm, computed with different sets of optical constants of Ag: —— [36], – – – [37], ····· [38], ···· [39], ······ [40].

small silver particles. An explanation of the surface plasmon polariton resonance is given later in Chapter 5.

It can clearly be recognized from Figure 2.10 that the peak position and also the half-width of the resonance depend on the data set used for the optical constants. Although the narrowest and highest peak is obtained with the data from Johnson and Christy [37], the minimum wavelength (the highest photon energy) at which the surface plasmon occurs is obtained with the data from Paquet [36] and Hagemann et al. [39]. The less pronounced and broadest peak with the largest shift to longer wavelengths is obtained with the data from Quinten [40]. The peak positions vary from 355 to 367 nm, or from 3.378 to 3.493 eV photon energy.

2.3.2
The Role of the Particle Size

Equation 2.21 holds true for uniformly sized particles in the ensemble. In general, however, already the particle size is statistically distributed. Then, the optical density becomes a sum over all different contributions:

$$\tau(\lambda) = \sum_k \sigma_{ext}(\lambda, R_k) \frac{n_k(R_k)}{V} \log(e) d \qquad (2.24)$$

where $n_k(R_k)/V$ is the particle number concentration of particles of size R_k.

As an example, Figure 2.11 shows a typical transmission electron microscopy (TEM) image and the corresponding particle size histogram for a gold and silver nanoparticle system. The particles appear to be almost spherical in shape, although their intrinsic structure is dominantly a face-centered cubic (fcc) lattice.

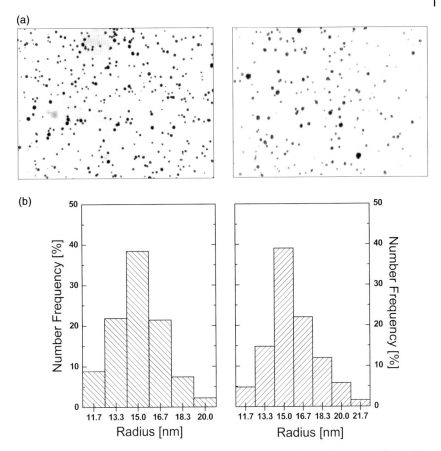

Figure 2.11 Typical transmission electron micrograph and particle size histogram of (a) gold particles ($2R_{mean} = 30.2\,nm$) and (b) silver particles ($2R_{mean} = 31.6\,nm$), prepared in a colloidal solution.

Often, the experimentally determined size distribution $n_k(R_k)$ can be modeled with mathematical distribution functions. The most important is the *log-normal distribution*:

$$\frac{\partial N}{\partial \log R} = \frac{N_0}{\log \sigma \sqrt{2\pi}} \exp\left[-\frac{1}{2}\left(\frac{\log R - \log R_{50}}{\log \sigma}\right)^2\right] \quad (2.25)$$

where R_{50} is the median radius, which corresponds to the radius at which 50% of all N_0 particles are accumulated, and σ is the geometric standard deviation with $\log \sigma = \log(R_{84}/R_{50}) = \log(R_{50}/R_{16})$, R_{16} being the radius at which 16% of all particles are accumulated and R_{84} the radius at which 84% of all particles are accumulated. A standard deviation of $\sigma = 1.75$ means that the turning points of the distribution function are at the radii $R = 1.75 R_{50}$ and $R = R_{50}/1.75$; 68% of all N_0 particles lie in the interval $[R_{50}/\sigma, \sigma R_{50}]$.

Another common distribution is the *normal distribution*:

$$\frac{\partial N}{\partial R} = \frac{N_0}{\sigma_n \sqrt{2\pi}} \exp\left[-\frac{1}{2}\left(\frac{R-R_{50}}{\sigma_n}\right)^2\right] \tag{2.26}$$

The standard deviation σ_n has the same unit as the particle size and satisfies the relations $\sigma_n = R_{84} - R_{50} = R_{50} - R_{16}$; 68% of all N_0 particles lie in the interval $[R_{50} - \sigma, R_{50} + \sigma]$.

Note that a problem may arise with the normal distribution when used for modeling an experimentally determined size distribution: in some cases the normal distribution may lead to negative particle sizes, which are not meaningful. In contrast, the log-normal distribution never leads to negative particle sizes.

In some cases, a good approximation to the size distribution is given by the *Gamma distribution*:

$$\frac{\partial N}{\partial R} = \frac{N_0 z^z}{R_{50} \Gamma(z)} \left(\frac{R}{R_{50}}\right)^z \exp\left(-z\frac{R}{R_{50}}\right) \tag{2.27}$$

where z is defined as $z = 1/\sigma^2$, with σ being standard deviation in percent, and $\Gamma(z)$ is the Gamma function.

For particles generated by milling processes, often the *Weibull distribution* is appropriate to describe the size distribution:

$$\frac{\partial N}{\partial R} = \frac{N_0 w \ln 2}{R_{50}} \left(\frac{R}{R_{50}}\right)^w \exp\left[-\ln 2 \left(\frac{R}{R_{50}}\right)^w\right] \tag{2.28}$$

where w is the Weibull parameter, which proved to be material dependent in milling processes.

For all these theoretical distributions, the mean size R_{mean} equals the median size R_{50}, which need not be the same in size distribution histograms.

For a continuous size distribution, the extinction coefficient $e(\lambda)$ at wavelength λ is given as

$$e(\lambda) = \int_0^\infty \sigma_{\text{ext}}(\lambda, R) \frac{\partial N}{\partial R} dR = \int_0^\infty \sigma_{\text{ext}}(\lambda, R) \frac{\partial N}{\partial \log R} d(\log R) \tag{2.29}$$

Granqvist and Buhrman [41] found that the size of ultrafine metal particles generated by inert gas evaporation, in supported catalysts or in colloids, often is log-normally distributed. In colloids, alternatively, a normal distribution can be found. From our own evaluation of TEM images of more than 60 various gold and silver colloids, a representative number of particle size distribution histograms was obtained. Some distributions are displayed in Figure 2.12. Compared with log-normal size distributions, the standard deviations lie in the range $1.11 \leq \sigma \leq 1.26$, which is rather narrow. Note that these values are significantly smaller than those obtained in gas condensation methods.

At the end of this section, we present numerical results for the influence of the size distribution of nanoparticles on the extinction spectra of low-concentration

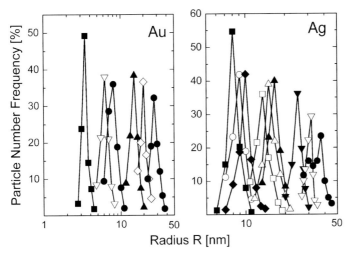

Figure 2.12 Size distributions of various gold and silver particles in aqueous suspension.

particle systems. The calculations were made for silver nanoparticles and hematite (α-Fe$_2$O$_3$) nanoparticles with median size $2R_{50} = 200$ nm and a standard deviation varying from $\sigma = 1.1$ to 1.5 in a log-normal distribution. The spectra in Figure 2.13 are compared with the corresponding single particle spectrum (dashed lines). For better comparison, the spectra of the silver particles are shifted along the ordinate.

In both figures pronounced changes in the spectra can be recognized when the standard deviation is increased. With increasing σ, particularly the particles with sizes larger than the median size contribute to the spectrum and smear out the resonance-like features in the single particle spectra which are clearly resolved in the single particle spectrum.

2.3.3
The Role of the Particle Shape

The shape of nanoparticles is not unique for all particles. Often, dielectric particles from polymerization are spherical, whereas for other materials, such as oxides and ionic crystals, the shape deviates from spherical. One reason is that the atoms and molecules that form a nanoparticle by condensation interact and complex surface topologies including edges, corners, and statistical roughness are formed. If preferably the same atom lattice as in the bulk is obtained in the nanoparticle, cubic shapes and also needles result. Gold and silver nanoparticles from chemical reduction in aqueous solutions are often almost spherical in shape, although their atom lattice is fcc. However, depending on the nucleation rate or the reduction velocity, it may happen that they grow into wires or pentagonal particles. Examples of various shapes are shown with the TEM images in Figure 2.14.

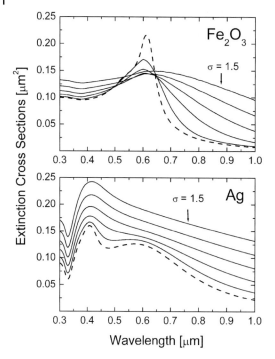

Figure 2.13 Influence of the particle size distribution on sodium and silver nanoparticle extinction spectra. Median size $2R_{50} = 200\,\text{nm}$, standard deviation varying from $\sigma = 1.1$ to 1.5.

Figure 2.14 Exemplary transmission electron micrographs of nanoparticles of various shapes.

Numerous, mostly theoretical, studies have been published considering the effect of the particle shape on light scattering by exact solutions for single spheres, cylinders, and spheroids or by means of various models [aggregates of spheres, discrete dipole approximation (DDA), T-matrix, and extended boundary condition method (EBCM)]. The best evolved and best understood model for light scattering and absorption by nanoparticles is the light scattering by a spherical particle. Details of this model and its extensions are given in Chapters 5 and 7.

A first rough approximation for describing light scattering and absorption by nonspherical particles is to replace the particle by either the surface area-equivalent sphere or the volume-equivalent sphere. Chýlek and Ramaswamy [42] showed that

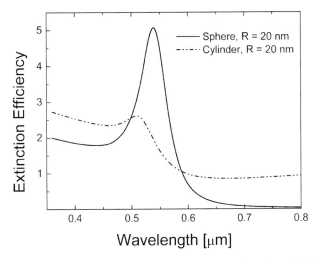

Figure 2.15 Computed extinction efficiency spectra of a spherical gold particle and an infinitely long cylinder with same diameter $2R = 40$ nm, embedded in glass.

for nonspherical particles with particle size parameters $x = 2\pi R/\lambda < 0.6$ the sphere with the equivalent surface area yields a better approach to the extinction cross-section of a nonspherical particle than the volume-equivalent sphere. However, the difference is small. The extinction cross-section of the sphere with the same surface area deviates by less than 2% from the extinction cross-section of the nonspherical particle, whereas the volume-equivalent sphere deviates by less than 6%. For larger size parameters they stated that the volume-equivalent sphere is a fairly good approximation. This approximation works fairly well for dielectric particles as long as resonances can be excluded in these particles (these resonances occur as size-dependent resonances, so-called morphology-dependent resonances). For absorbing particles the approximation often fails, particularly if resonances such as the already mentioned surface plasmon resonance occur.

In the following, we demonstrate in two examples the influence of the particle shape on the optical extinction by nanoparticles.

In the first example in Figure 2.15, the computed extinction efficiency spectrum of a gold sphere with $2R = 20$ nm is compared with that of an infinitely long cylinder (a wire) of gold with diameter $2R = 20$ nm. These nanoparticles exhibit a narrow extinction band, which is associated with the surface plasmon polariton resonance in those nanoparticles (a definition of the surface plasmon polariton is given in Chapter 5). One obvious effect of the particle shape is that the surface plasmon polariton of the wire is shifted to shorter wavelengths compared with the spherical particle. Another effect is the increased extinction at longer wavelengths. The reasons become clear in Section 9.2, where the scattering by infinitely long cylinders is discussed.

The second example is from experiments on aqueous gold colloids. It was observed that while preparing colloidal gold particles with wanted mean diameter

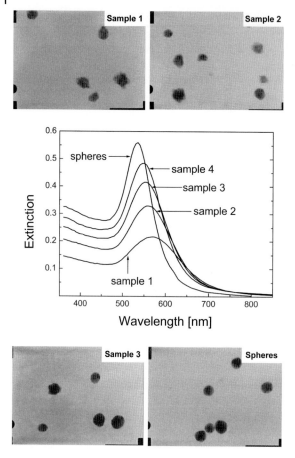

Figure 2.16 Extinction spectra and transmission electron micrographs of irregularly shaped gold particles.

$2R = 50$ nm by growing smaller particles with mean diameter $2R_{nucleus} = 8$ nm, the color of the colloidal suspension passed through different stages from blue to violet to red. At different time points samples were withdrawn from the mixture and stabilized with a small amount of a gelatin solution. The samples were analyzed with TEM and from each sample an optical extinction spectrum was recorded in the wavelength range 360–1000 nm. The results are presented in Figure 2.16.

All spectra exhibit an extinction band which can be associated with the surface plasmon polariton resonance for the spherical gold particles. The peak position and half width of the extinction bands are different in all samples. The micrographs show that the particles in the samples are irregularly shaped but their sizes do not exceed the size of the spherical particles.

Comparing the measured spectra with spectra computed for spherical particles, a mean particle size of $2R = 52$ nm is obtained for the spherical particles, which is in agreement with the size $2R = 46 \pm 6$ nm obtained from evaluation of the TEM images. For the nonspherical particles, larger sizes of volume-equivalent spheres are obtained: $2R = 76$ nm (sample 4), 80 nm (sample 3), 88 nm (sample 2), and 104 nm (sample 1).

2.3.4
The Role of the Particle Concentration

2.3.4.1 Dilute Systems

Coming back to the connection between the optical properties of nanoparticle systems and the optical properties of the embedded nanoparticles, we show the influence of increasing filling factor f on the optical absorption and scattering by the assembly in the case of low filling factors, when the Lambert–Beer law is applicable.

For convenience we assume identical particles. Then, the optical density of the assembly of nanoparticles as a function of the (mass) concentration m/V of the particulate matter is

$$\tau(\lambda) = \frac{\sigma_{\text{ext}}(\lambda, R)}{V_p} \frac{m}{V} \frac{\log(e)}{\rho} d \qquad (2.30)$$

where V_p and ρ are the single particle volume and the mass density of the particulate matter, respectively. The filling factor f in terms of m/V and ρ is

$$f = \frac{m}{V} \frac{1}{\rho} \qquad (2.31)$$

As the optical density in Equation 2.30 is proportional to the mass concentration or the filling factor, the optical density spectra of particle assemblies do not differ in shape from the spectra of single particles when the filling factor or the mass concentration is increased. In contrast, the transmittance $T(\lambda)$ must change considerably on increasing the concentration m/V, since it is

$$T(\lambda) = 10^{-\tau(\lambda)} \qquad (2.32)$$

This is demonstrated in Figure 2.17 with computed transmittance spectra for silver nanoparticles with $2R = 20$ and 50 nm dispersed in glass. The sample thickness is $d = 1$ mm. The mass concentration m/V is 0.075, 0.2, 0.75, 2.0, 7.5, 20, and 75 mg cm^{-3} for the small particles (filling factors $f = 7.14 \times 10^{-5}$, 1.9×10^{-4}, 7.14×10^{-4}, 1.9×10^{-3}, 7.14×10^{-3}, 1.9×10^{-2}, 7.14×10^{-2}), and $m/V = 0.075, 0.125$, 0.2, 0.75, 1.25, 2.0 and 7.5 mg cm^{-3} (filling factor $f = 7.14 \times 10^{-5}$, 1.19×10^{-4}, 1.9×10^{-4}, 7.14×10^{-4}, 1.19×10^{-3}, 1.9×10^{-3}, 7.14×10^{-3}) for the larger particles. On increasing the concentration m/V the minimum in the transmittance at $\lambda = 422$ nm for $2R = 20$ nm and at $\lambda = 446$ nm for $2R = 50$ nm becomes deeper. On exceeding a certain concentration the transmittance in the minimum is almost

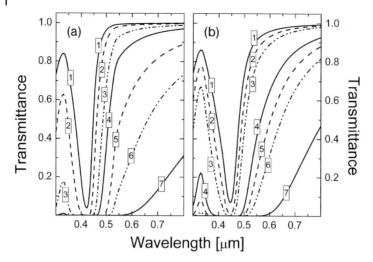

Figure 2.17 Transmittance spectra of assemblies of silver nanoparticles with (a) $2R = 20$ nm and (b) $2R = 50$ nm. The spectra are numbered according to the increasing concentration m/V, with $m/V = 0.075, 0.2, 0.75, 2.0, 7.5, 20$, and $75\,\text{mg cm}^{-3}$ in (a), and $m/V = 0.075, 0.125, 0.2, 0.75, 1.25, 2.0$, and $7.5\,\text{mg cm}^{-3}$ in (b). The sample thickness is $d = 1$ mm.

$T = 0$. With further increase in m/V the transmittance is almost $T = 0$ in a small spectral region. The width of this spectral region increases with increasing concentration, so that the transmittance first for violet and blue visible light vanishes and then also for green visible light. In consequence, the corresponding color changes. We determined tristimulus values and chromaticity coordinates from these spectra and they are summarized in a chromaticity diagram in Figure 2.18. The effect of an increase in the mass concentration on the color of the disperse Ag system is obvious. The apparent color changes from yellow to red with increasing concentration, independently of the particle size.

This example demonstrates that even in low-concentration particle assemblies with negligible multiple scattering the optical properties of the nanoparticle system can vary strongly with the filling factor. Indeed, this effect was used in the coloring of glass in several technical applications where nowadays they are replaced by organic pigments. An example is red bulbs for medical applications. Figure 2.19a shows a photograph of such a red bulb, and in Figure 2.19b the corresponding measured transmittance spectrum of the bulb is depicted. The spectrum is similar to the spectra in Figure 2.17 numbered "7." This means that the mean particle size could be, for example, $2R = 50$ nm with a mass concentration $m/V = 7.5\,\text{mg cm}^{-3}$.

2.3.4.2 Closely Packed Systems

In turning to dense packing of nanoparticles, not only may the increase in the filling factor result in changes, but additionally now electromagnetic interaction

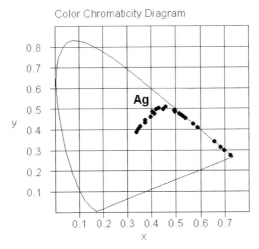

Figure 2.18 Chromaticity diagram showing the color change with increasing concentration m/V of nanoparticles.

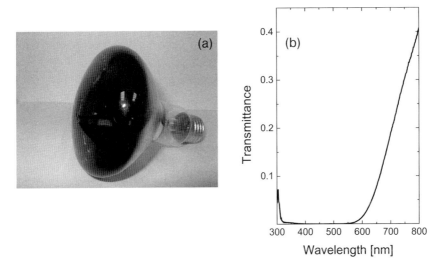

Figure 2.19 (a) Photograph of a red bulb for medical applications and (b) transmittance spectrum of such a red bulb. A color version of this figure can be found in the color plates at the end of the book.

between the particles occurs, and also mechanical, magnetic (if the particles are magnetic), and electric interactions are of increasing importance. These interactions come into play when the filling factor f becomes larger than 10^{-2}. Then, the Lambert–Beer law starts to fail to give a correct description of the observed optical properties of nanoparticle composites and Equations 2.21, 2.24, and 2.30 must be corrected. Mechanical interaction may be examined with special light scattering methods, for example, the Brownian motion of colloidal particles becomes disturbed at high volume fractions, which is examined with photon correlation spectroscopy. In this book, we only consider electromagnetic interaction due to superposition of scattered fields of neighboring particles at the site of a certain particle.

The electromagnetic response of the whole densely packed particle assembly may be obtained either from *radiative transfer* models or from coherent superposition of electromagnetic fields and determining optical cross-sections for the whole system. The latter approach is solved for spherical nanoparticles and is called the *generalized Mie theory* (GMT). An extensive discussion of the GMT is given in Chapters 10 and 11. The main advantage of the GMT compared with radiative transfer models is the coherent superposition of electromagnetic fields, instead of the incoherent superposition of intensities in radiative transfer models. Then, interferences are fully taken into account. The radiative transfer considers the stationary distribution of intensities in space caused by incoherent interaction of radiation with matter, in which the geometry, the radiation source distribution, and boundary conditions are well known. The radiative transfer models take into account the energy conservation law, and also principles of thermodynamics, electrodynamics, and geometric optics.

In fact, radiative transfer is applicable only if the phases of the scattered light are statistically distributed. Therefore, it is applicable only for diluted systems, or for systems where the particles are larger than the wavelength. Its applicability to submicron- and nanometer-sized particles must be considered to be critical. However, some *many-flux theories* have been successfully established, which assume certain numbers of photon fluxes in the (+x) and (−x) directions as an approach to the radiative transfer through a particle assembly. Among the various flux models (six-, four-, two-flux), the two-flux model of Kubelka and Munk [18] is the most popular and attained distinct importance, since it is simple and yields reasonable results that compare well with experimental results. It is described in Chapter 12.

3
Interaction of Light with Matter – The Optical Material Function

The physically relevant quantities in describing fundamentals of the interaction of light with matter are the photon energy $\hbar\omega$ and the complex dielectric function $\varepsilon(\omega) = \varepsilon_1(\omega) + i\varepsilon_2(\omega)$. Nevertheless, when discussing wave propagation in and through media, the wavelength λ and the complex refractive index $n + i\kappa$ are involved. They are combined by two relations:

$$n + i\kappa = \sqrt{\varepsilon_1 + i\varepsilon_2} \quad \text{Maxwell's relation} \tag{3.1}$$

$$\frac{\omega}{c} = \frac{2\pi}{\lambda}(n + i\kappa) \quad \text{Dispersion relation} \tag{3.2}$$

In this chapter, we introduce the interaction of light with bulk matter described by the dielectric function $\varepsilon(\omega)$. We begin with a classical description of $\varepsilon(\omega)$ from a macroscopic view in Section 3.1. Then, in Section 3.2, we briefly introduce quantum mechanical concepts for the dielectric function of condensed matter. Special solutions for amorphous semiconducting materials or oxides are presented in Section 3.3. Finally, we give in Section 3.4 with the Kramers–Kronig relations a tool for the calculation of the imaginary part of the dielectric function from the real part and vice versa and discuss the penetration of light into different materials.

3.1
Classical Description

The interaction of electromagnetic fields with matter is dominated by the forces exerted by the incident electric (and magnetic) field on the electric charges in the matter. At high frequencies, the electric field E inside the body of condensed matter usually displaces the electrons in the atoms of condensed matter while the ions are too inert as to follow the electric field with the same frequency. Thereby each atom becomes an electric dipole with dipole moment p. At low frequencies in the far-infrared region, the incident light can also couple to the ions via TO phonons and induces dipole moments by displacing the negatively and positively charged ions in different directions.

Optical Properties of Nanoparticle Systems: Mie and Beyond. Michael Quinten
Copyright © 2011 WILEY-VCH Verlag GmbH & Co. KGaA, Weinheim
ISBN: 978-3-527-41043-9

3 Interaction of Light with Matter – The Optical Material Function

The dipole moments add up to a macroscopic net polarization P of the sample. The connection between P and E can be described by the general equation

$$P = \varepsilon_0 \chi E \tag{3.3}$$

defining the macroscopic *susceptibility* χ of the matter. Physics enters this formal relation by interpreting χ. In general, χ itself depends on E, that is, $\chi(E)$. This field dependence can be described by a series expansion of P in powers of E, where susceptibilities $\chi_{ijk...}$ are introduced which now are a constant in relation to E:

$$P_i = \varepsilon_0 \underbrace{\sum_j \chi_{ij} E_j}_{\text{linear term } \chi^{(1)} \text{ (anisotropic medium)}} + \varepsilon_0 \underbrace{\sum_j \sum_k \chi_{ijk} E_j E_k}_{\text{quadratic term } \chi^{(2)}} + \varepsilon_0 \underbrace{\sum_j \sum_k \sum_l \chi_{ijkl} E_j E_k E_l}_{\text{cubic term } \chi^{(3)}} + \ldots \tag{3.4}$$

for each component $i = x, y, z$ of P. For sufficiently small electric fields ($|E| \leq 100\,\text{V m}^{-1}$), which is fulfilled in most optical applications, the relation Equation 3.4 is linear. Then, the susceptibility χ is a tensor of second rank, if the material is optically anisotropic. At large intensities the response of condensed matter on electric (and magnetic) fields becomes dependent on higher powers of the fields. This phenomenon is called *optical nonlinearity*, in contrast to the linear response at moderate intensities.

3.1.1
The Harmonic Oscillator Model

Restricting to sufficiently small electric fields ($|E| \leq 100\,\text{V m}^{-1}$), Equation 3.3 is a linear relation between P and E. The linearity means that charges which are displaced from their position of equilibrium are retreated by forces which are proportional to the distance from the position of equilibrium. In consequence, a charge q_j with mass m_j executes forced oscillations in a time-periodic electric field $E = E_0 \exp(-i\omega t)$. This is the *harmonic oscillator model*, because the retreating force is linear in the displacement following from a harmonic (quadratic) potential:

$$V(r) = \sum_j \frac{1}{2} m_j \omega^2 r_j^2 \tag{3.5}$$

This condition leads to harmonic oscillations of the charges q_j which are most easily described. From the force balance:

$$F_{\text{inertia}} + F_{\text{damping}} + F_{\text{repulsive}} = F_{\text{electrical}} \tag{3.6}$$

we obtain

$$\sum_j m_j \frac{\partial^2 r_j}{\partial t^2} + m_j \gamma_j \frac{\partial r_j}{\partial t} + D_j r_j = \sum_j q_j E(t) \tag{3.7}$$

for the displacement r_j. The second term on the left-hand side accounts for the perturbation of the movement by interactions with other charges and lattice defects. Their contributions are assumed to lead to damping of the oscillation with damping constant γ_j. In general there is no reason to assume a linear dependence on $\partial r_j/\partial t$. Different relaxation processes can lead to other dependences. However, this assumption is most commonly applied and often justified.

A time-harmonic solution of Equation 3.7, that is, a solution for a time harmonic field $E(t) = E_0 \exp(-i\omega t)$, is given as

$$r_j = \frac{q_j}{m_j} \frac{E_0}{\omega_j^2 - \omega^2 - i\omega\gamma_j} \qquad (3.8)$$

with the resonance frequency $\omega_j = D_j/m_j$. The macroscopic net polarizability P follows from all dipole moments $p_j = q_j r_j$ in the sample volume V via

$$P = \frac{1}{V}\sum_j N_j p_j = \frac{1}{V}\sum_j N_j q_j r_j \qquad (3.9)$$

from which we obtain the susceptibility χ_j for all N_j charges q_j as

$$\chi_j(\omega) = \frac{1}{V}\sum_j \frac{N_j q_j^2}{\varepsilon_0 m_j} \frac{1}{\omega_j^2 - \omega^2 - i\omega\gamma_j} \qquad (3.10)$$

N_j/V is the total number density of charges q_j with mass m_j and eigenfrequency ω_j. In this respect, χ in Equation 3.10 is also valid for ionic crystals (e.g., NaCl, MgO), with the corresponding ionic charges and masses. The assumption of linearity of Equation 3.3 allows to the summation of all various contributions χ_j to give a total susceptibility χ.

The polarization P in Equation 3.3 is a contribution to the current displacement D:

$$D = \varepsilon_0 E + P = \varepsilon_0 (1 + \chi) E \qquad (3.11)$$

This equation is also a definition for the *dielectric function* $\varepsilon(\omega) = \varepsilon_1(\omega) + i\varepsilon_2(\omega)$ as

$$\varepsilon(\omega) = 1 + \chi = 1 + \sum_j \frac{\omega_{pj}^2}{\omega_j^2 - \omega^2 - i\omega\gamma_j} \qquad (3.12)$$

This *ansatz* is well suited to describe the optical constants if only discrete electronic excitations have to be considered, which are harmonic oscillators. Deviations occur for special cases, e.g., the bandgap of semiconducting materials or for the continuous band of interband transitions. They will be described later in Section 3.2. This harmonic oscillator model was developed by H. A. Lorentz around the beginning of the twentieth century.

Note that the ansatz $\exp(-i\omega t)$ for the time dependence of the electric field leads to the definition of $\varepsilon(\omega)$ as $\varepsilon(\omega) = \varepsilon_1(\omega) + i\varepsilon_2(\omega)$ and to $\tilde{n}(\omega) = n(\omega) + i\kappa(\omega)$. Alternatively, the ansatz $\exp(i\omega t)$ leads to the definition of $\varepsilon(\omega)$ as $\varepsilon(\omega) = \varepsilon_1(\omega) - i\varepsilon_2$ and to $\tilde{n}(\omega) = n(\omega) - i\kappa(\omega)$, which can also be found in the common literature.

At low frequencies, that is, for $\omega \to 0$, the real part $\varepsilon_1(0)$ of the dielectric function becomes constant:

$$\varepsilon_1(0) = 1 + \sum_j \frac{\omega_{Pj}^2}{\omega_j^2} \qquad (3.13)$$

while the imaginary part $\varepsilon_2(0)$ vanishes. $\varepsilon_1(0)$ represents the static dielectric constant of the material. The constant ratio ω_{Pj}^2/ω_j^2 defines a new quantity, the *oscillator strength* S_j of the jth harmonic oscillator, so that ω_{Pj}^2 in Equation 3.12 can be replaced by $S_j\omega_j^2$.

3.1.2
Extensions of the Harmonic Oscillator Model

For statistically pertubated or amorphous materials, it seems appropriate to use an extension of the harmonic oscillator model on so-called Brendel oscillators. A Brendel oscillator is a harmonic oscillator with eigenfrequency ω_k and width γ_k that is inhomogeneously broadened by an infinite sum over sharp harmonic oscillators with eigenfrequency x and width γ_k [43]. These oscillators are Gaussian distributed around the harmonic oscillator with eigenfrequency ω_k with a standard deviation σ_k. With additional contributions of Brendel oscillators the dielectric function becomes

$$\varepsilon(\omega) = 1 + \underbrace{\sum_j \frac{\omega_{Pj}^2}{\omega_j^2 - \omega^2 - i\omega\gamma_j}}_{\text{harmonic oscillators}} + \underbrace{\sum_k \frac{\omega_{Pk}^2}{\sigma_k\sqrt{2\pi}} \int_0^\infty dx \frac{\exp\left[-0.5\left(\frac{x-\omega_k}{\sigma_k}\right)^2\right]}{x^2 - \omega^2 - i\omega\gamma_k}}_{\text{Brendel oscillators}} \qquad (3.14)$$

The advantage of a Brendel oscillator compared with a harmonic oscillator broadened by increased γ is that the contours in the real and imaginary part of the dielectric function become smoother due to the Gaussian distribution. In Figure 3.1 we give an example of the dielectric functions of a single harmonic oscillator with $\omega_p = 6 \times 10^{15}\,\text{s}^{-1}$, resonance frequency $\omega_{\text{res}} = 3 \times 10^{15}\,\text{s}^{-1}$, and damping constant $\gamma = 2 \times 10^{14}\,\text{s}^{-1}$. The dashed lines show the dielectric functions of the same oscillator assumed to be a Brendel oscillator with a standard deviation of 10%.

In the simpler model of Kim et al. [44], the damping γ_k of a harmonic oscillator ω_k is assumed to be frequency dependent:

$$\varepsilon(\omega) = 1 + \chi_\infty + \underbrace{\sum_j \frac{\omega_{Pj}^2}{\omega_j^2 - \omega^2 - i\omega\gamma_j}}_{\text{harmonic oscillators}} + \underbrace{\sum_k \frac{\omega_{Pk}^2}{\omega_k^2 - \omega^2 - i\omega\gamma_k(\omega)}}_{\text{Kim oscillators}} \qquad (3.15)$$

with

$$\gamma_k(\omega) = \gamma_k \exp\left[-\frac{1}{1+\sigma_k^2}\left(\frac{\omega-\omega_k}{\gamma_k}\right)^2\right] \qquad (3.16)$$

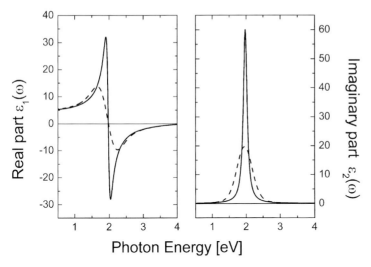

Figure 3.1 Inhomogeneous line broadening: harmonic oscillator with $\omega_{res} = 3 \times 10^{15}\,s^{-1}$ and damping constant $\gamma = 2 \times 10^{14}\,s^{-1}$ (full line), and Brendel oscillator with the same parameters and a standard deviation $\sigma = 10\%$ (dashed line).

The parameter $\sigma_k \geq 0$ is used to switch between a Gaussian or a Lorentzian shape of $\chi_k(\omega)$. For $\sigma_k = 0$ a pure Gaussian shape is obtained and for $\sigma_k > 5$ a Lorentzian shape is obtained.

3.1.3
The Drude Dielectric Function

In metals, semimetals and semiconductors, an important contribution to the dielectric function stems from unbound charge carriers, the free electrons. Within the harmonic oscillator model their contribution is obtained when assuming the eigenfrequency $\omega_{fe} = 0$, corresponding to moving in a potential $V(r) = 0$. Then, the susceptibility of the free electrons reads

$$\chi_{fe}(\omega) = -\frac{\omega_p^2}{\omega^2 + i\omega\gamma_{fe}} \tag{3.17}$$

with the abbreviation

$$\omega_p^2 = \frac{Ne_0^2}{Vm_{eff}\varepsilon_0} \tag{3.18}$$

being the plasma frequency of the electrons assuming them as a plasma. In a parabolic band structure the effective mass m_{eff} of the electrons is identical with the electron mass m_e, but in nonparabolic band structures m_{eff} may differ from m_e.

With the contribution of free electrons but without Brendel or Kim oscillators, the dielectric function becomes

$$\varepsilon(\omega) = 1 - \underbrace{\frac{\omega_p^2}{\omega^2 + i\omega\gamma_{fe}}}_{\substack{\text{Drude susceptibility}\\\text{(free electrons)}}} + \underbrace{\sum_j \frac{\omega_{Pj}^2}{\omega_j^2 - \omega^2 - i\omega\gamma_j}}_{\substack{\text{harmonic oscillators}\\\text{(bound electrons)}}} \quad (3.19)$$

At $\omega = 0$, $\varepsilon_1(0)$ still becomes constant but takes now large negative values depending on ω_p and γ_{fe} due to the contribution of the free electrons. The imaginary part $\varepsilon_2(\omega)$ always increases as $1/\omega$ with decreasing ω for small ω and diverges at $\omega = 0$. However, the DC conductivity remains finite, approaching the value

$$\sigma_{DC} = \varepsilon_0 \frac{\omega_p^2}{\gamma_{fe}} \quad (3.20)$$

The damping constant γ_{fe} of the free electrons is closely related to the relaxation time τ, after which the common drift motion of the free electrons is relaxed by interactions with other charges and lattice defects:

$$\gamma_{fe} = \tau^{-1} = \sum_i \tau_i^{-1} = \tau^{-1}_{\text{point defects}} + \tau^{-1}_{\text{dislocations}} + \tau^{-1}_{\text{grain boundaries}}$$

$$+ \tau^{-1}_{\text{surface/interface}} + \tau^{-1}_{\text{e-phonon}} + \tau^{-1}_{\text{e-e}} \quad (3.21)$$

This *Mathiessen rule* only holds for independent relaxation processes. Experimental values for τ are usually determined from the DC conductivity (Equation 3.20). For example, for sodium τ amounts to $\tau(\text{Na}) = 3.2 \times 10^{-14}$ s [45]. The relaxation time τ proves to be almost frequency dependent; for instance, $\tau(\text{Ag}) = 3.7 \times 10^{-14}$ s for $\omega = 0$ changes to 1.5×10^{-14} s in the optical region [46].

3.2
Quantum Mechanical Concepts

In Section 3.1 we presented classical models for the dielectric function from a macroscopic viewpoint. In this section we briefly introduce quantum mechanical concepts for the dielectric function of condensed matter.

The general structure of the dielectric function $\varepsilon(\omega)$ of the bulk material is based on the nonlocality of the response, meaning that the dielectric response $\mathbf{D}$ due to the incident electromagnetic wave at a volume element $dV(\mathbf{r})$ and at a time t is influenced by the excitations, polarizations, and fields at all other volume elements $dV(\mathbf{r}')$, occurring at properly retarded times t'.

Nonlocal response is usually incorporated into electrodynamics [47] by replacing the material relation

$$\mathbf{D}(\mathbf{r},t) = \varepsilon_0 \varepsilon \mathbf{E}(\mathbf{r},t) \quad (3.22)$$

with the local dielectric function by

3.2 Quantum Mechanical Concepts

$$D_i(r,t) = \varepsilon_0 \sum_j \int dt' \int d^3r' \varepsilon_{ij}(r,r',t-t') E_j(r,t) \tag{3.23}$$

For homogeneous materials, the Fourier transform of this equation is

$$D_i(K,\omega) = \varepsilon_0 \sum_j \varepsilon_{ij}(K,\omega) E_j(K,\omega) \tag{3.24}$$

For isotropic materials, ε_{ij} is a symmetric tensor containing only diagonal elements $\varepsilon_\perp$ and $\varepsilon_\parallel$:

$$\underline{\underline{\varepsilon}}(K,\omega) = \begin{pmatrix} \varepsilon_\perp & 0 & 0 \\ 0 & \varepsilon_\parallel & 0 \\ 0 & 0 & \varepsilon_\parallel \end{pmatrix} \tag{3.25}$$

$\varepsilon_\parallel$ is used for longitudinal fields D and E (K is parallel to D and E), for example, for a moving electron, while $\varepsilon_\perp$ is used for transverse fields (K is perpendicular to D and E). Hence $\varepsilon_\perp(K, \omega)$ is the linear response function for the interaction of light with extended matter. However, so far, it is not possible to calculate $\varepsilon_\perp(K, \omega)$ from first principles. On the other hand, it can be shown that for $K \to 0$ – this is valid for electromagnetic radiation[1] – $\varepsilon_\parallel(K \to 0, \omega)$ fully describes the interaction of light with matter. This was proved in the random phase approximation (RPA), and also in general using field-theoretical methods. In the following, the microscopic theory for $\varepsilon_\parallel(K, \omega)$, abbreviated as $\varepsilon(K, \omega)$, will briefly be recalled [48], and then the case $K \to 0$ will be discussed.

3.2.1
The Hubbard Dielectric Function

We consider the linear response of the electron system on a weak pertubation given by an external scalar electrical potential $\varphi_{\text{ext}}(r, t)$. As a consequence of the pertubation, the electron density $n(r)$ will locally fluctuate, described by the density fluctuation $\langle \delta n(r, t) \rangle$. This is the *local density approximation* (LDA) of the density functional theory. $\langle \delta n(r, t) \rangle$ can be calculated from the fluctuation-dissipation theorem as

$$\langle \delta n(r,t) \rangle = (-e) \int d^3r' \int dt' \cdot \eta(r,r',t-t') \varphi_{\text{ext}}(r',t) \tag{3.26}$$

with the density autocorrelation function

$$\eta(r,r',t-t') = -\frac{i}{\hbar} \theta(t-t') \langle [n(r), n(r')] \rangle \tag{3.27}$$

---

1) For electromagnetic radiation $|K| < 1/a$, with a being the lattice constant. This holds even true in the X-ray region, where the lattice constant has to be replaced with the extension of the electronic core levels involved.

In the following, we use the Greek letter η for the autocorrelation function instead of the commonly used Greek letter χ, to avoid confusion with the susceptibility. The step function $\theta(t - t')$ makes the density autocorrelation causal. The density fluctuation causes an induced electrical potential φ_{ind}, for which the electron system is now located in the potential

$$\varphi_{tot} = \varphi_{ext} + \varphi_{ind} \tag{3.28}$$

The current displacement D results from the external potential only:

$$D = -\varepsilon_0 \text{grad}\,\varphi_{ext} \tag{3.29}$$

whereas the total electric field E results as a gradient of the whole potential:

$$E = -\text{grad}\,\varphi_{tot} \tag{3.30}$$

For the following it is more convenient to express all relations in the Fourier space. The corresponding Fourier transforms of Equations 3.29 and 3.30 are

$$D(K,\omega) = \varepsilon_0 K \varphi_{ext}(K,\omega) \tag{3.31}$$

$$E(K,\omega) = K \varphi_{tot}(K,\omega) \tag{3.32}$$

Comparison with

$$D(K,\omega) = \varepsilon(K,\omega)\varepsilon_0 E(K,\omega) \tag{3.33}$$

yields for the dielectric function

$$\frac{1}{\varepsilon(K,\omega)} = \frac{\varphi_{tot}}{\varphi_{ext}} = 1 + \frac{\varphi_{ind}}{\varphi_{ext}} \tag{3.34}$$

The next step is to find an expression for φ_{ind} as a function of φ_{ext}. φ_{ind} is the solution of Poisson's equation (as the Fourier transform):

$$\varphi_{ind}(K,\omega) = -\frac{e}{\varepsilon_0 K^2}\langle \delta n(K,\omega)\rangle \tag{3.35}$$

With the Fourier-transform of Equation 3.26:

$$\langle \delta n(K,\omega)\rangle = -e\eta(K,\omega)\varphi_{ext}(K,\omega) \tag{3.36}$$

it follows that

$$\frac{1}{\varepsilon(K,\omega)} = 1 + \frac{e^2}{\varepsilon_0 K^2}\eta(K,\omega) \tag{3.37}$$

The remaining task is to derive the autocorrelation function $\eta(K,\omega)$. This is an elaborate task, so we refer the reader to common textbooks of quantum mechanics for the solution to this problem. We adopt the results from Sturm [48] and discuss $\varepsilon(K,\omega)$ for some special cases.

The simplest case is the *homogeneous electron gas*, in which the ions appear as the constant positive charge background (*jellium model*) with the task of compensating the integrated negative charge of the electrons. However, exact solutions of the many-body Schrödinger equation are even in this case not available due to the long-range acting Coulomb forces. An approximation that has proven to be very successful in quantum mechanics is the *self-consistent field* (SCF) *approximation* [49]. The starting point is the linear response of *independent* particles (remember that at realistic densities the electrons are strongly coupled, in fact). For the homogeneous electron gas the eigenfunctions are plane waves and the energy eigenvalues form a parabolic band. The autocorrelation function of the homogeneous electron gas is

$$\eta^0(K,\omega) = \frac{2}{V} \sum_{K'} \frac{f(E_K) - f(E_{K+K'})}{\hbar\omega + i\Delta + E_K - E_{K+K'}} \quad (3.38)$$

with

$$E_K = \frac{\hbar^2 K^2}{2m} \quad (3.39)$$

and $f(E_K)$ being the Fermi function.

As we cannot expect that η^0 is a good approximation for the *interacting* system of electrons, it is more appropriate to use

$$\langle \delta n(K,\omega) \rangle = -e\eta(K,\omega)\varphi_{tot}(K,\omega) \quad (3.40)$$

instead of Equation 3.36, because the electron system is located in φ_{tot}. Moreover, Equation 3.35 must be modified to take into account exchange and correlation terms:

$$(-e)\varphi_{ind}(K,\omega) = \frac{e^2}{\varepsilon_0 K^2} \langle \delta n(K,\omega) \rangle [1 - G^0(K)] \quad (3.41)$$

Within the LDA, the additional term $G^0(K)$ is proportional to K^2 for small K and becomes constant for large K [50–52]. Resolving Equations 3.28, 3.40, and 3.41 for $\langle \delta n(K,\omega) \rangle$ leads to

$$\langle \delta n(K,\omega) \rangle = (-e) \frac{\eta^0}{1 - \frac{e^2 \eta^0}{\varepsilon_0 K^2}(1 - G^0)} \varphi_{ext}(K,\omega) \quad (3.42)$$

Comparison with Equation 3.38 shows that we obtain a more realistic response function:

$$\eta^{SCF} = \frac{\eta^0}{1 - \frac{e^2}{\varepsilon_0 K^2}(1 - G^0)\eta^0} \quad (3.43)$$

With this response function inserted in Equation 3.37, the dielectric function becomes

$$\varepsilon(K,\omega) = 1 - \frac{e^2}{\varepsilon_0 K^2} \frac{\eta^0}{1 + \frac{e^2}{\varepsilon_0 K^2} G^0 \eta^0} \quad (3.44)$$

This relation is known as the *Hubbard dielectric function* [53].

In the random phase approximation (RPA), Equation 3.44 can be simplified because the correlation and exchange terms (G^0) are neglected. The corresponding dielectric function in this approximation is the *Lindhard dielectric function*:

$$\varepsilon^L(K,\omega) = 1 - \frac{e^2}{\varepsilon_0 K^2} \eta^0(K,\omega) \tag{3.45}$$

A further simplification results for $K \to 0$, namely the *Drude dielectric function*:

$$\varepsilon^D(\omega) = \lim_{K \to 0} \varepsilon^L(K,\omega) = 1 - \frac{e^2}{m\varepsilon_0 V} \frac{1}{\omega^2} \sum f(E_K) = 1 - \frac{\omega_p^2}{\omega^2} \tag{3.46}$$

For further discussion, it seems appropriate to introduce the energy loss function α. For an electron passing through a medium, the loss of kinetic energy due to electrical interaction with the electrons of the medium is proportional to α. In this case it is

$$\alpha = \operatorname{Im}\left[-\frac{1}{\varepsilon(K,\omega)}\right] \tag{3.47}$$

On the other hand, when an electromagnetic wave passes through an absorbing medium, the loss function is

$$\alpha = \operatorname{Im}[\varepsilon(K \to 0, \omega)] \tag{3.48}$$

From Equation 3.47 we have directly

$$\alpha = -\frac{e^2}{\varepsilon_0 K^2} \operatorname{Im}[\eta(K,\omega)] \tag{3.49}$$

Using Equation 3.38 as a first attempt for η we obtain

$$\alpha = \operatorname{Im}[\varepsilon^L(K,\omega)] = \frac{e^2}{\varepsilon_0 K^2} \frac{1}{V} \sum_{K'}[f(E_K) - f(E_{K+K'})]\delta(\hbar\omega + E_K - E_{K+K'}) \tag{3.50}$$

There are electron–hole transitions in the parabolic energy band of the homogeneous electron gas, as illustrated in Figure 3.2. Using instead directly the Lindhard dielectric function $\varepsilon^L(K,\omega)$ in Equation 3.47, we have

$$\alpha = \frac{\operatorname{Im}[\varepsilon^L(K,\omega)]}{|\varepsilon^L(K,\omega)|^2} \tag{3.51}$$

We can now recognize additional excitations of the electron gas for $\varepsilon^L(K, \omega) = 0$. They are collective excitations of the whole electron gas, called plasmons, which can also appear for $K = 0$ at the *plasma frequency* ω_p. Plasmon excitations are not restricted to $K = 0$ but can also occur for $K \neq 0$. The dispersion relation for the plasmons is

$$\omega_p(K) = \omega_p + \frac{\hbar}{m}\left(\frac{3E_F}{5\hbar\omega_p}\right)K^2 + \ldots \tag{3.52}$$

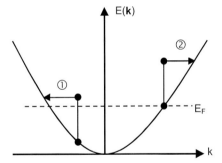

Figure 3.2 Electron–hole transitions with energy $\hbar\omega$ and momentum transfer K_1 (①) and K_2 (②) in the parabolic energy band of the homogeneous electron gas.

3.2.2
Interband Transitions

Up to now we have introduced the dielectric function of the homogeneous electron gas. In condensed matter, however, the electrons are located in a periodic lattice, leading to a *periodically inhomogeneous electron gas*. The main changes are as follows:

1) $n(K,\omega)$ and $\langle \delta n(K,\omega) \rangle$ are periodic with $K + G$, G being a reciprocal lattice vector.
2) The electrons can be described with Bloch wavefunctions, since the energy band deviates from the parabolic band.

Calculations of the dielectric function in the periodic crystal were made by Adler [54] and Wiser [55] in the SCF approach. These calculations are very extensive and we refer the reader to the original literature for further reading. The main effect of the periodic potential of the lattice on the parabolic band of the homogeneous free electron gas is that it is split into separate bands in the reduced Brillouin scheme with a bandgap at the Brillouin zone edge (Figure 3.3).

From Figure 3.3, it is obvious that now direct transitions, so-called *interband transitions*, with $K = 0$ are also possible which contribute substantially to the loss function. This is an interesting case because evidently they can be excited also with electromagnetic radiation. Unfortunately, the calculation of $\varepsilon(K, \omega)$ is very time consuming, so we refer the reader to quantum mechanics textbooks and monographs.

In Figure 3.4 we show representatively a part of the band structure of Au with direct and indirect (intraband) transitions.

A basic expression for the susceptibility of direct interband transitions χ^{IB} was given by Bassani and Parravicini [57]. They examined first the probability per unit time P_{if} for a direct transition from the initial state i into the unoccupied final state f, when an electromagnetic field $A = A_0 \hat{e} \exp(-i\omega t)$ is applied to the electronic system with momentum $p = -i\hbar \nabla$. The quantity $\hat{e}$ is the polarization vector of the incident field. The probability P_{if} then is given as

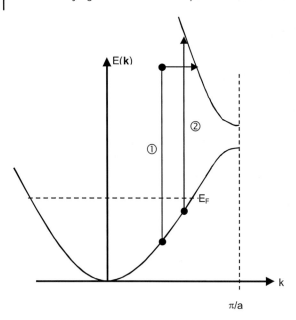

Figure 3.3 Interband transitions with energy $\hbar\omega$ in the reduced energy band scheme of the periodically inhomogeneous electron gas: ① with momentum transfer K and ② without momentum transfer ($K = 0$, direct transition).

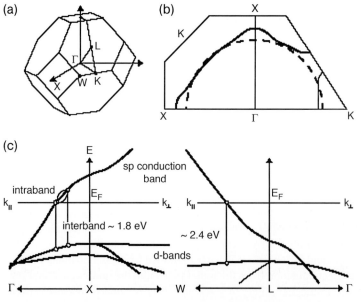

Figure 3.4 Part of the band structure of Au in the first Brillouin zone considering direct transitions from the 5d band to the hybridized 6sp band. After [56].

$$P_{if} = \frac{2\pi}{\hbar} \left| \left\langle f \left| \frac{e_0}{m} \mathbf{A} \cdot \mathbf{p} \right| i \right\rangle \right|^2 \delta(E_f - E_i - \hbar\omega)$$

$$= \frac{2\pi}{\hbar} \left(\frac{e_0 A_0}{m} \right)^2 |\langle f | \hat{e} \cdot \mathbf{p} | i \rangle|^2 \delta(E_f - E_i - \hbar\omega)$$

$$= \frac{2\pi}{\hbar} \left(\frac{e_0 A_0}{m} \right)^2 |\hat{e} M_{if}|^2 \delta(E_f - E_i - \hbar\omega) \tag{3.53}$$

where $M_{if} = \langle f | \mathbf{p} | i \rangle$ is the momentum transition matrix element.

The number of transitions $W(\omega)$ per unit time follows from summation over all possible electronic states in the Brillouin zone BZ:

$$W(\omega) = \sum_{i,f} \int_{BZ} \frac{2 d^3 K}{(2\pi)^3} P_{if} \tag{3.54}$$

From this number of transitions the absorption constant $\alpha(\omega)$ is calculated by dividing the energy of absorbed photons $\hbar\omega W(\omega)$ by the total energy flux $\varepsilon_0 A_0^2 \omega^2 c$ at frequency ω in the electromagnetic field:

$$\alpha(\omega) = \frac{\hbar\omega W(\omega)}{\varepsilon_0 A_0^2 \omega^2 c} \tag{3.55}$$

On the other hand, the absorption constant is connected with the imaginary part of the dielectric function ε_2 and the imaginary part of the susceptibility χ_2, via

$$\alpha(\omega) = \frac{\omega}{c} \varepsilon_2 = \frac{\omega}{c} \chi_2 \tag{3.56}$$

From Equations 3.55 and 3.56, a relation for the imaginary part χ_2^{IB} of the susceptibility of interband transitions follows, which can be used in Kramers–Kronig relations (see Section 3.4) to obtain the corresponding real part χ_1^{IB}. Then, the complex susceptibility of the interband transitions reads

$$\chi^{IB} = \frac{2\hbar^2 e_0^2}{\varepsilon_0 m_{eff}^2} \sum_{i,j} \int_{BZ} \frac{2 d^3 K}{(2\pi)^3} |\hat{e} M_{if}(K)|^2 \left(\frac{1}{[E_f(K) - E_i(K)]\{[E_f(K) - E_i(K)]^2 - (\hbar\omega)^2\}} \right.$$

$$\left. + i \frac{\pi}{2\hbar^2 \omega^2} \delta[E_f(K) - E_i(K) - \hbar\omega] \right) \tag{3.57}$$

With this formalism, the linear optical response regarding all relevant electronic transitions is traced back to the band structure model of the solid state. The resulting dielectric function is determined by the band structure $E(K)$ and transition matrix elements M_{if}, describing interband transitions between initial (i) and final (f) states.

Assuming that the matrix elements are independent of the wavevector, that is, constant throughout the Brillouin zone, one finds for χ_2^{IB} that

$$\omega^2 \chi^{IB}(\omega) \propto \sum_{i,j} |M_{if}|^2 J_{if}(\omega) \tag{3.58}$$

50 | *3 Interaction of Light with Matter – The Optical Material Function*

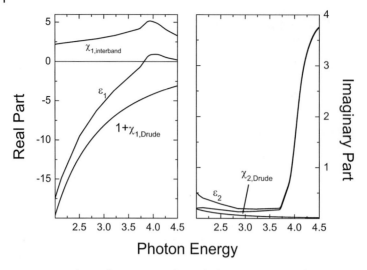

Figure 3.5 Dielectric function $\varepsilon(\omega)$ of Ag with decomposition into free electron contribution and interband contribution.

where $J_{if}(\omega)$ is the joint density of states given by the integral over the Brillouin zone:

$$J_{if}(\hbar\omega) = \frac{2}{(2\pi)^3} \int_{BZ} d^3K \delta[E_f(K) - E_i(K) - \hbar\omega] \quad (3.59)$$

$J_{if}(\omega)$ is called the joint density of states because it gives the density of pairs of states – one occupied and one empty – separated by the energy $\hbar\omega$. It reflects the shapes of the electronic energy bands involved in the transition, but information about the influence of the individual density of states on the transition probability is obscured.

As an example, Figure 3.5 depicts the real part $\varepsilon_1(\omega)$ and the imaginary part $\varepsilon_2(\omega)$ of the dielectric function of silver metal with decomposition into the Drude free electron contribution and the interband contribution.

3.3
Tauc–Lorentz and OJL Models

Amorphous semiconductor and oxide materials often have optical functions that depend on the deposition conditions and do not have so sharp features like a harmonic oscillator. To model them, Jellison and Modine [58] derived a model based on a combination of the Tauc band edge and the Lorentz oscillator formulation. With this *Tauc–Lorentz model* the imaginary part of the complex dielectric function of amorphous materials with bandgap (mainly semiconductor materials) can be modeled as

3.3 Tauc–Lorentz and OJL Models

$$\varepsilon_{2\text{TL}}(\omega) = \frac{\omega_P^2}{\omega} \frac{\omega_{\text{res}}\gamma(\omega - \omega_{\text{gap}})^2}{(\omega^2 - \omega_{\text{res}}^2)^2 + \omega^2\gamma^2} \quad (3.60)$$

where the oscillator has a resonance frequency ω_{res} and a damping constant γ, ω_{gap} is the frequency corresponding to the bandgap energy $E_{\text{gap}} = \hbar\omega_{\text{gap}}$. The plasma frequency is proportional to the momentum transition matrix element. Note that in the original paper, the equation is expressed in terms of photon energies.

The real part $\varepsilon_{1,\text{TL}}$ is obtained from the imaginary part by the Kramers–Kronig relation (see Equation 3.65 in the next section).

Another model for the dielectric function of amorphous semiconductor materials stems from O'Leary, Johnson, and Lim [59] and is well known as the *OJL-model*. In a defect-free crystalline semiconductor, the absorption spectrum terminates abruptly at the energy gap. In contrast, in an amorphous semiconductor the absorption spectrum reaches into the gap region. The reason is that the electronic states in the valence band and conduction band can be divided into localized states and states which are randomly distributed through these amorphous semiconductors. Whereas the distribution of localized states follow a square-root functional dependence in the band region, the distribution shows an exponential functional dependence in the tail region. Introducing breadths γ_C and γ_V (both energies) for the conduction and valence band tails, respectively, O'Leary, Johnson, and Lim modeled the density of states for the conduction and valence band of an amorphous semiconductor. With these relations they derived the optical absorption coefficient $\alpha_{\text{OJL}}(E)$ with $E = \hbar\omega$ as

$$\alpha_{\text{OJL}}(E) = D^2(E) \frac{\sqrt{2}}{\pi^2\hbar^3} m_C^{\frac{3}{2}} \frac{\sqrt{2}}{\pi^2\hbar^3} m_V^{\frac{3}{2}} J(E) \quad (3.61)$$

where m_C and m_V are the effective mass of electrons in the conduction and valence band, respectively. The optical transition matrix element $D^2(E)$ is proportional to $1/E$ but its exact functional dependence remains unknown and must be adjusted accordingly.

The normalized joint density-of-states (JDOS) $J(E)$ in the OJL model differs for the two cases (1) $E \leq E_{\text{gap}} + (\gamma_C + \gamma_V)/2$ and (2) $E \geq E_{\text{gap}} + (\gamma_C + \gamma_V)/2$:

1) $E \leq E_{\text{gap}} + (\gamma_C + \gamma_V)/2$:

$$J(E) = \frac{\gamma_C^2}{\sqrt{2e}} \exp\left(\frac{E - E_{\text{gap}}}{\gamma_C}\right) Y\left(\frac{\gamma_V}{2\gamma_C}\right) + \frac{\gamma_V^2}{\sqrt{2e}} \exp\left(\frac{E - E_{\text{gap}}}{\gamma_V}\right) Y\left(\frac{\gamma_C}{2\gamma_V}\right)$$

$$+ \frac{1}{2\sqrt{e}} \frac{(\gamma_C\gamma_V)^{\frac{3}{2}}}{\gamma_V - \gamma_C} \left[\exp\left(\frac{E - E_{\text{gap}} - \frac{\gamma_C}{2}}{\gamma_V}\right) - \exp\left(\frac{E - E_{\text{gap}} - \frac{\gamma_V}{2}}{\gamma_C}\right)\right] \quad (3.62)$$

with $Y(z) = \sqrt{z}\exp(-z) + 0.5\sqrt{\pi}\operatorname{erfc}(\sqrt{z})$ and erfc = error function.

2) $E \geq E_{\text{gap}} + (\gamma_C + \gamma_V)/2$:

$$J(E) = \frac{\gamma_C^2}{\sqrt{2e}} \exp\left(\frac{E-E_{gap}}{\gamma_C}\right) Y\left(\frac{E-E_{gap}}{\gamma_C} - \frac{1}{2}\right)$$
$$+ \frac{\gamma_V^2}{\sqrt{2e}} \exp\left(\frac{E-E_{gap}}{\gamma_V}\right) Y\left(\frac{E-E_{gap}}{\gamma_C} - \frac{1}{2}\right)$$
$$+ \frac{1}{2\sqrt{e}} (E-E_{gap})^2 L\left(\frac{\gamma_C}{2(E-E_{gap})}, \frac{\gamma_V}{2(E-E_{gap})}\right) \quad (3.63)$$

with $L(x,y) = \int_x^{1-y} \sqrt{z}\sqrt{1-z}\,dz$.

The imaginary part of the complex dielectric function of amorphous semiconductor materials with bandgap E_{gap} finally follows from the optical absorption coefficient $\alpha_{OJL}(E)$ as

$$\varepsilon_{2,OJL}(E) = \frac{\hbar c \alpha_{OJL}(E)}{E} \quad (3.64)$$

The real part $\varepsilon_{1,OJL}$ is obtained from the imaginary part by the Kramers–Kronig relation (see Equation 3.65 in the next section).

3.4
Kramers–Kronig Relations and Penetration Depth

The real and imaginary parts of the dielectric function, describing the electric polarization and the optical energy dissipation in matter, respectively, are dependent on each other and related by the *Kramers–Kronig* relations, sometimes called *dispersion integrals*. These very general relations hold for any frequency-dependent function that connects an output with an input in a linear causal way.

Assuming $\lim_{\omega \to \infty} \chi(\omega) = 0$, one finds

$$\varepsilon_1(\omega) = 1 + \frac{2}{\pi} \wp \int_0^\infty \frac{\Omega \varepsilon_2(\Omega)}{\Omega^2 - \omega^2} d\Omega \quad (3.65)$$

$$\varepsilon_2(\omega) = -\frac{2}{\pi} \wp \int_0^\infty \frac{\varepsilon_1(\Omega) - 1}{\Omega^2 - \omega^2} d\Omega \quad (3.66)$$

where $\wp$ is the principal value of the integral. In principle, the measurement of $\varepsilon_1(\omega)$ and $\varepsilon_2(\omega)$ requires two independent experiments. The Kramers–Kronig relations can replace one of them, making their determination easier. Many published optical functions result from such a Kramers–Kronig analysis of the reflectance or the absorption coefficient or from combination of electron energy-loss experiments with Kramers–Kronig relations. Problems usually arise from the fact that the integrals in Equations 3.65 and 3.66 extend from 0 to ∞. Experimental values of $\varepsilon_1(\omega)$ and $\varepsilon_2(\omega)$ are, however, only available for restricted regions.

Table 3.1 Skin depth δ (nm) for selected materials as a function of wavelength.

	Al	Ag	Au	Cu	Si	Fe$_2$O$_3$	CuO
δ(300 nm)	13	48	25	28	12	30	32
δ(500 nm)	13	25	43	31	1093	199	117
δ(700 nm)	13	23	27	27	8868	4456	359
δ(900 nm)	17	22	25	25	38206	28648	4775

Optical constants from [36, 37, 60].

An electromagnetic wave impinging on a surface of absorbing material [i.e., $\varepsilon_2(\omega) \neq 0$ in the considered spectral range] has only a certain penetration depth due to energy dissipation which can be estimated from the optical functions. For simplicity, we assume here a plane wave propagating along the z-axis, that is, normal to the sample surface. Then, the skin depth δ determining attenuation of the field results from the imaginary part of the complex refractive index $\tilde{n}(\omega) = n(\omega) + i\kappa(\omega)$:

$$\delta = \frac{c}{\omega \kappa(\omega)} = \frac{\lambda}{2\pi \kappa(\lambda)} \quad (3.67)$$

where the skin depth δ refers to the electric field. The corresponding depth for the light intensity is smaller by a factor of 2. The intensity dependence is usually expressed as

$$I(\omega, z) = I_0(\omega, z = 0) \exp(-\gamma z) \quad (3.68)$$

where $\gamma = 2/\delta$ is the extinction constant, that is, the field is limited to a skin of thickness of order of δ. In Table 3.1 we list skin depths for several materials at vacuum wavelengths of the incident light of 300 nm (4.133 eV), 500 nm (2.480 eV), 700 nm (1.771 eV), and 900 nm (1.378 eV).

It becomes obvious from Table 3.1 that for the metals the skin depth remains approximately constant on going to longer wavelengths. The reason is an increase in κ with increase in wavelength, due to the contribution of the free electrons to the susceptibility. In contrast, silicon and the selected absorbing dielectrics become more and more transparent with increasing wavelength. The reason is that the interband transitions close to the UV spectral region do not contribute further to the imaginary part of the susceptibility at longer wavelengths, in agreement with the harmonic oscillator model. As no free electrons can contribute to the susceptibility, κ decreases with increasing wavelength, approaching $\kappa = 0$ for still longer wavelengths. Then, the dielectrics will be completely transparent with skin depths $\delta = \infty$.

4
Fundamentals of Light Scattering by an Obstacle

The scientific study of light scattering began in 1869 with experiments on aerosols by Tyndall [61]. They excited the interest of Lord Rayleigh (John William Strutt, the third Baron Rayleigh) in the problem of light scattering [62, 63]. His most inventive paper [64], published in 1881, utilized a boundary value solution for the circular cylinder as well as a volume integral obtained from Green's theorem, which provided a completely general solution. By relaxing some conditions on size and the dielectric constant, he was able to derive various closed expressions. His analysis of light scattering by small particles and studies on the scattering of an electromagnetic wave by an isotropic sphere by the Dane Ludvig Valentin Lorenz [65] form the base for many following examinations on light scattering by small particles. The general solution of Maxwell's equations for scattering from a sphere of arbitrary size was obtained in 1908 by Gustav Mie [17], who tried to explain the color of gold colloids, that is, nanoparticles of gold in aqueous suspensions. Previously, von Ignatowski in 1905 [66] and Seitz in 1906 [67] had developed a general solution for infinitely long cylinders (wires).

Elastic light scattering phenomena with particles can be roughly divided into two categories: the *Rayleigh scattering regime* and the *Mie scattering regime*. Thereby, the scattering particles are classified according to the ratio of the particle size to the wavelength of the incident light. At wavelengths of visible light, a reasonable but not very distinct limit for Rayleigh scattering is a size of about 300 nm, depending on the particle material. Below that, the particles satisfy the conditions for Rayleigh scattering better the smaller the particles are. Vice versa, Mie scattering fits better the larger the particles become.

In the Rayleigh regime, a simpler theoretical approach to light scattering allows the treatment of spherical and even nonspherical particles. However, most nanometer-sized particles scatter the light rather similarly, and for their distinction special measurement techniques (small-angle scattering, depolarization measurements, etc.) are required.

Exact analytical solutions of the light scattering by small particles that extend to the Mie scattering regime are available only for a few highly symmetric bodies, such as the sphere, the spheroid, and the infinitely long cylinder. The most evolved is the Mie theory for spherical particles. For almost all other particles, light scattering and absorption can be treated by various mostly numerical models. A short

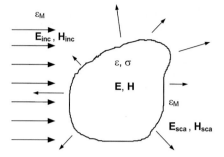

Figure 4.1 Scattering of light by an arbitrarily shaped particle.

review on analytical and numerical methods was given by Wriedt [68]. However, before introducing distinct analytical or numerical solutions in forthcoming chapters, we explain here the step-by-step concept of all analytical particle scattering models.

4.1
Maxwell's Equations and the Helmholtz Equation

The interaction of electromagnetic radiation with matter can be described macroscopically with Maxwell's equations derived from Oerstedt's, Gauss's, and Ampère's laws. Hence we begin the description of light absorption and elastic scattering by an arbitrary obstacle with Maxwell's equations.

Assume an arbitrary particle illuminated by an electromagnetic wave represented by the fields (E_{inc}, H_{inc}). The wave is partially refracted (*E*, *H*) and partially reflected at the surface of the obstacle (E_{sca}, H_{sca}). Since this reflection is not directed, but is different in various directions, it is called *scattering*. The refracted part may be partially absorbed when passing through the obstacle (Figure 4.1).

The response of the particle material to the electromagnetic fields *E* and *H* of the refracted wave generally consists of four contributions: a polarization *P*, a magnetization *M*, a current *J*, and a displacement current $\partial D/\partial t$.

The relation between *P* and *E* is given by the equation

$$P = \varepsilon_0 \chi E \tag{4.1}$$

where χ is the macroscopic susceptibility of the matter. In general, χ itself depends on *E*, that is, $\chi(E)$. This field dependence can be described by a series expansion of *P* in powers of *E*, where susceptibilities $\chi_{ijk...}$ are introduced which are now a constant in relation to *E*:

$$P_i = \underbrace{\varepsilon_0 \sum_j \chi_{ij} E_j}_{\text{linear term } \chi^{(1)} \text{ (anisotropic medium)}} + \underbrace{\varepsilon_0 \sum_j \sum_k \chi_{ijk} E_j E_k}_{\text{quadratic term } \chi^{(2)}} + \underbrace{\varepsilon_0 \sum_j \sum_k \sum_l \chi_{ijkl} E_j E_k E_l}_{\text{cubic term } \chi^{(3)}} + \ldots \tag{4.2}$$

for each component $i = x, y, z$ of P. For sufficiently small electric fields ($|E| \leq 100\,\mathrm{V\,m^{-1}}$), which is fulfilled in most optical applications, the relation in Equation 4.2 is linear. Then, the susceptibility χ is a tensor of second rank, if the material is optically anisotropic. At high intensities the response of condensed matter to electric fields becomes dependent on higher powers of the fields. This phenomenon is called *optical nonlinearity*, in contrast to the linear response at moderate intensities. The various forms of optical nonlinearity (e.g., four-wave mixing, second and higher harmonic generation, optical bistability, optical phase conjugation) occur depending on the matter under consideration with different amplitudes. If the linear term $\chi^{(1)}$ is of the order of 1, $\chi^{(2)}$ is about $10^{-12}\,\mathrm{m\,V^{-1}}$ and $\chi^{(3)}$ is about $10^{-21}\,\mathrm{m^2\,V^{-2}}$ [69]. These are typical values for bulk matter; however, special crystals exceed them by far.

The polarization P in Equation 4.1 contributes to the electrical flux density or displacement D:

$$D = \varepsilon_0 E + P = \varepsilon_0(1+\chi)E \tag{4.3}$$

defining the dielectric constant or permittivity ε as

$$\varepsilon = 1 + \chi \tag{4.4}$$

The susceptibility χ and dielectric constant ε are optical material functions. In the framework of Maxwell's theory, they enter the field relations as constants which are valid for the bulk material under consideration.

In the presence of unbound charges in the bulk matter, for example, free electrons in metals or semiconductors, the electric field E additionally causes an electric current with current density J:

$$J = \sigma E \tag{4.5}$$

where σ is the *(optical) conductivity*. The energy contained in $J \cdot E$ is transformed into oscillations of the crystal lattice (phonons) and is therefore missing in the energy transport between the incident, refracted, and scattered wave. It is *absorbed* in the particle.

Similarly to the electric field E, also the magnetic field H causes two reactions of the material on the applied field, a magnetization M and a current displacement $\partial D/\partial t$. The latter is the result of the Lorentz force on bound and free electrons. The magnetization M contributes to the magnetic induction B in the particle:

$$B = \mu_0 H + \mu_0 M = \mu\mu_0 H \tag{4.6}$$

due to reorientation of permanent magnetic dipoles in the applied field. At frequencies ranging from the far-infrared to infinity, permanent magnetic dipoles are, however, too inert to follow the oscillating magnetic field of the incident electromagnetic wave. Therefore, the *relative permeability* μ can be assumed to be $\mu = 1$, even for magnetic materials throughout the above spectral region. To our knowledge, no material exists with $\mu \neq 1$ within the above frequency range.

So far, we have introduced optical constants ε, σ, and μ as macroscopic quantities for the reactions of the applied electromagnetic fields on condensed matter.

The next step is to determine the electromagnetic fields inside and outside the particle. A transverse electromagnetic field cannot, in general, be derived from a purely scalar function of space and time, in contrast to longitudinal heat flow or acoustic vibrations. Instead, we have to look for vectorial fields **E** and **H** that must satisfy Maxwell's equations and the vector wave equation. For the sake of simplification we restrict the considerations to only harmonic waves, hence the time dependence of the fields can be separated with the ansatz

$$E = E(r)\exp(-i\omega t)$$
$$H = H(r)\exp(-i\omega t) \tag{4.7}$$

The task is now to determine the unknown parts **E(r)** and **H(r)** of the electromagnetic fields of the incident wave, the scattered wave, and the internal wave with the help of Maxwell's equations:

$$\text{div}(\varepsilon E) = 0 \tag{4.8}$$

$$\text{div}(B) = 0 \tag{4.9}$$

$$\text{curl} E = i\omega\mu_0 H \tag{4.10}$$

$$\text{curl} H = (-i\varepsilon\varepsilon_0\omega + \sigma)E \tag{4.11}$$

using Equations 4.1–4.6 and assuming no static charges ($\rho = 0$) in Equation 4.8. Since we always assume homogeneity in time, it is usual, and is done here and in the following, to omit the time dependence $\exp(-i\omega t)$ in the equations, it being assumed to be unaffected by matter (this is not always the case; we recall inelastic scattering).

Equation 4.8 can be further simplified for isotropic and homogeneous matter, for which ε is only a scalar, to

$$\text{div} E = 0 \tag{4.12}$$

This simplification is, however, not always allowed. For example, hematite (α-Fe_2O_3) is anisotropic with two principal directions of crystal symmetry, in which the dielectric constants differ. Nevertheless, an approximate solution for scattering by a hematite particle can be obtained by taking the solution for each direction separately and averaging them accordingly. Another example is materials with a nonvanishing gradient of ε, for example, gradient-index materials commonly used in lightwave technology. In the following, we assume for simplicity homogeneous and isotropic materials, for which Equation 4.12 holds.

For solving these Maxwell equations simultaneously for **E** and **H**, Equations 4.10 and 4.11 must be decoupled. This can be achieved by taking the curl of one equation and inserting the result into the other equation, resulting in the *vector wave equation* or *Helmholtz equation*:

$$\text{curl curl} Z - k^2 Z = 0 \tag{4.13}$$

where **Z** represents either **E** or **H**. The wavenumber k satisfies the dispersion relation

$$k^2 = \frac{\omega^2}{c^2}\left[\varepsilon(\omega) + i\frac{\sigma(\omega)}{\omega\varepsilon_0}\right] = \frac{\omega^2}{c^2}\tilde{n}^2(\omega) \tag{4.14}$$

where $\tilde{n}(\omega)$ is the complex refractive index with $\tilde{n}(\omega) = n(\omega) + i\kappa(\omega)$. The term in parentheses combines the permittivity ε (polarization) with the conductivity σ (absorption) to give the complex dielectric function[1] $\varepsilon(\omega) = \varepsilon_1(\omega) + i\varepsilon_2(\omega)$. Refractive index and dielectric function are related via Maxwell's relation

$$\tilde{n}(\omega) = \sqrt{\varepsilon(\omega)} \tag{4.15}$$

$\tilde{n}(\omega)$ is usually used for the description of wave propagation, its real part determining the phase propagation constant, whereas $\varepsilon(\omega)$ contains most clearly the physical description of the interaction of light with matter since susceptibilities of independent polarization and absorption processes are usually assumed to be additive.

4.2 Electromagnetic Fields

The vector wave equation (Equation 4.13) must have independent solutions for E and H because of Equations 4.10 and 4.11. Furthermore, the divergence of Z has to vanish because of Equations 4.12 and 4.9. In Cartesian coordinates an arbitrary transverse electromagnetic field can be resolved into two partial fields, each derivable from a purely scalar function Φ satisfying the scalar wave equation

$$(\nabla^2 + k^2)\Phi = 0 \tag{4.16}$$

This becomes obvious if we use the so-called Hertz vectors Π_e and Π_h. It was shown by Hertz [70] and more generally by Righi [71] that it is possible under ordinary conditions to define an electromagnetic field in terms of a properly chosen single vector function Π.

Assuming $\Pi_e = (0, 0, \Pi_e)^T \exp(-i\omega t)$ to be a solution, the electric field E and the magnetic field H can be derived by the operations

$$E = \text{curl curl}\Pi_e \tag{4.17a}$$

$$H = (-i\omega\varepsilon_0\varepsilon + \sigma)\text{curl}\Pi_e \tag{4.17b}$$

It is simple to show that the z-component H_z is zero, that is, its axial or longitudinal component is absent. Thus we have derived from a scalar function Π_e an electromagnetic field characterized by the absence of the longitudinal component of the magnetic vector. This field is called electric or more properly transverse magnetic (TM) mode.

---

1) To take into account dependences on either the frequency ω or on the wavevector, the term *dielectric function* is used instead of *dielectric constant*.

In the same way, a second, independent solution for the fields can be derived from a Hertz vector $\Pi_h = (0, 0, \Pi_h)^T \exp(-i\omega t)$ by the operations

$$E = i\omega\mu_0 \operatorname{curl} \Pi_h \tag{4.18a}$$

$$H = \operatorname{curl}\operatorname{curl} \Pi_h \tag{4.18b}$$

It is simple to show that in this case the longitudinal component E_z of the electric vector is zero, defining the magnetic or more properly transverse electric (TE) mode.

In curvilinear coordinate systems, complete solutions of the vector wave equation in a form directly applicable to the solution of boundary-value problems are at present known only for the separable systems of cylindrical, spheroidal, and spherical coordinates.

In curvilinear coordinate systems, it is also common practice to introduce vector harmonics M and N instead of using Hertz vectors. The vector harmonics are defined by

$$M = \operatorname{curl}(\Phi a) \tag{4.19a}$$

$$N = k^{-1}\operatorname{curl} M \tag{4.19b}$$

where the quantity Φ is a scalar potential satisfying Equation 4.16 and a is an arbitrary constant vector. Comparing Equation 4.19 with Equations 4.17 and 4.18, the corresponding Hertz vectors are given by

$$k\Pi_e = \Phi a \tag{4.20a}$$

$$i\omega\mu_0 \Pi_h = \Phi a \tag{4.20b}$$

With the vector harmonics M and N the TM mode or electric mode is given as

$$E^{TM} = E_0 N \tag{4.21a}$$

$$H^{TM} = \frac{kE_0}{i\omega\mu_0} M \tag{4.21b}$$

and the TE mode or magnetic mode as

$$E^{TE} = E_0 M \tag{4.22a}$$

$$H^{TE} = \frac{kE_0}{i\omega\mu_0} N \tag{4.22b}$$

The quantity E_0 is a dimensionless constant that represents the magnitude of the electric field. If the electric *and* the magnetic field vector have no longitudinal component, the solution is called TEM wave. Generally, E and H are given as a linear superposition of TM and TE modes. Note that in several electrodynamic problems it is not possible to obtain pure TE or TM solutions.

With the introduction of vector harmonics M and N, the problem of finding electromagnetic fields E and H that correspond to electromagnetic waves is transformed to the problem of finding a solution Φ of the scalar wave equation (Equation 4.16) and a constant vector *a*. Often, it is easier to solve this problem instead of finding Hertz vectors.

It is worth mentioning that a third, irrotational solution of Maxwell's equations and the vector wave equation exists as

$$L = -\text{grad}\,\phi_L \tag{4.23}$$

that satifies the Laplace equation

$$\text{div}\,L = -\nabla^2 \phi_L = 0 \tag{4.24}$$

where *L* describes longitudinal waves with wavenumber k_L. These waves cannot be described with the dielectric function defined in Equation 4.14, but are described with the dielectric function $\varepsilon_L(\omega, k_L)$ that satisfies

$$\varepsilon_L(\omega, k_L) = 0 \tag{4.25}$$

If longitudinal waves can propagate in the material in a certain frequency range, the function *L* describes longitudinal electron density fluctuations in the material, the *longitudinal bulk plasmon modes*. At these frequencies, the interior TM fields are coupled to the bulk plasmons, whereas the TE fields are not affected.

4.3 Boundary Conditions

After solving Maxwell's equations, the next step is to relate the electromagnetic fields inside the particle (*E*, *H*) with the electromagnetic fields (E_{inc}, H_{inc}, E_{sca}, H_{sca}) outside the particle. This is usually attained with Maxwell's boundary conditions at the surface of the particle:

$$(E_{\text{inc}} + E_{\text{sca}}) \times n\big|_{\text{surface}} = E \times n\big|_{\text{surface}} \tag{4.26a}$$

$$(H_{\text{inc}} + H_{\text{sca}}) \times n\big|_{\text{surface}} = H \times n\big|_{\text{surface}} \tag{4.26b}$$

where *n* is the vector normal to the surface of the particle. For an arbitrarily shaped particle, it is not appropriately defined for the whole surface. This is the main reason why in general analytical solutions of the scattering problem are available only for a few geometric bodies: the sphere, the spheroid, and the infinite cylinder. For other geometries, either approximate solutions exist, or numerical models (e.g., aggregates of spheres, discrete dipole approximation, T-matrix or others) are used. We will introduce the exact solutions, the approximations, and the numerical techniques in the following chapters.

With the additional longitudinal field *L*, these boundary conditions must be supplemented by an *additional boundary condition* including the continuity of the normal component of *E*:

$$(E_{\text{inc}} + E_{\text{sca}}) \cdot n\big|_{\text{surface}} = E \cdot n\big|_{\text{surface}} \tag{4.27}$$

a condition primarily introduced for plane surfaces by Sauter [72], Forstmann [73] and Melnyk and Harrison [74]. Later Clanget [75] and Ruppin [76–78] applied it to spherical geometry.

4.4
Poynting's Law and Cross-sections

Under experimental conditions, the rates at which energy from the incoming light is scattered and absorbed are of more practical interest than the electromagnetic fields. They are obtained from the energy flux balance from and to the particle.

The total rate at which energy changes in time inside a closed volume V containing the particle is given by

$$\frac{\partial}{\partial t} w = \frac{\partial}{\partial t}(D \cdot E + B \cdot H) = J \cdot E + \text{div} S \qquad (4.28)$$

where w is the energy density. The quantity $S = E \times H$ is the Poynting vector or radiant flux density of the electromagnetic wave outside the particle. In the stationary case ($\partial w/\partial t = 0$), the total loss of electromagnetic energy in the particle by absorption ($J \cdot E$) equals the radiant flux outside the particle (div$S$) through a volume V in which the particle is completely contained (see Figure 4.2), according to the law of conservation of energy.

Integrating Equation 4.28 in the stationary case over the closed volume V, the rate W_{abs} at which energy is absorbed in the particle follows as

$$W_{\text{abs}} = \iiint J \cdot E \, dV = -\iiint \text{div} S \, dV = -\frac{1}{2} \text{Re} \oiint (E \times H^*) n_A \, dA \qquad (4.29)$$

Here, we used the time-averaged Poynting vector $S = \frac{1}{2}\text{Re}(E \times H^*)$ outside the particle because at optical frequencies detectors are not able to measure the rapid

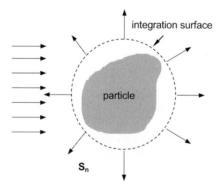

Figure 4.2 Sketch of the radiant flux through a volume in which the particle is completely contained.

oscillations of the electromagnetic fields directly, but yield time-averaged quantities. Re means the real part and the asterisk denotes the complex conjugate. The volume integration can be replaced by a surface integration over the closed surface around the volume V, defined by the normal vector n_A. As the electromagnetic fields outside the particle are assumed to result from the superposition of the incident fields and the fields of the scattered wave, the absorption rate contains three terms:

$$W_0 = -\frac{1}{2}\text{Re}\oiint (E_{inc} \times H_{inc}^*) \tag{4.30}$$

$$W_{ext} = -\frac{1}{2}\text{Re}\oiint (E_{inc} \times H_{sca}^* + E_{sca} \times H_{inc}^*) \tag{4.31}$$

$$W_{sca} = -\frac{1}{2}\text{Re}\oiint (E_{sca} \times H_{sca}^*) \tag{4.32}$$

where W_0 represents the rate at which energy is absorbed from the incident wave in the integrating volume in absence of the particle, and vanishes for nonabsorbing embedding media. Later, in Section 7.5, we will also discuss absorbing embedding media. W_{ext} represents the extinction rate, that is, consumptive (absorptive) losses and scattering losses out of the direction of the incident wave, and W_{sca} the scattering rate of the particle.

Dividing the rates W_{ext}, W_{sca}, and W_{abs} by the intensity I_0, that is, the energy flux density of the incident light, expressions for the *optical cross-sections* σ_{ext}, σ_{sca}, and σ_{abs} are found:

$$\sigma_{abs} = \sigma_{ext} - \sigma_{sca} \tag{4.33}$$

These quantities represent particle-specific quantities which are independent of the strength of the incident electromagnetic fields. Remember that we assumed a linear response of the particle material. In some cases, it is useful to normalize the cross-sections σ on the geometric cross-section G of the particle, to obtain the dimensionless *efficiencies* Q:

$$Q_{abs} = Q_{ext} - Q_{sca} \tag{4.34}$$

The geometric cross-section G is the projection of the particle on the plane perpendicular to the direction of incidence. For example, for a sphere it is $G = \pi R^2$, where R is the radius of the sphere.

We note that the use of cross-sections or efficiencies is helpful for comparing the size, shape and wavelength dependences of the scattering and absorption by various nanoparticles. Nevertheless, the experimentally observable quantity is the scattered intensity, which may be very small for small particles. For example, for a single spherical particle with radius $R = 1$, 10, or 100 nm, the scattering cross-section at wavelengths in the visible spectral region are approximately 10^{-22}, 10^{-16}, and 10^{-10} cm^2, respectively. Hence it is almost always necessary and also the usual case for nanoparticles to measure light scattering and absorption by many particles in the measuring volume.

INFO – Comparison with a Slab

Comparing light scattering and absorption by an arbitrarily shaped particle with the reflection, transmission, and absorption of a slab, one recognizes some similarities (Figure 4.3).

1) We have three kinds of fields and waves:
   - The incident wave.
   - The wave reflected at the top surface plus the wave coming out the slab at the rear side of the slab. Both together correspond to the scattered wave in the case of the particle.
   - The wave passing through the slab. Here, also absorption can diminish the magnitude of the fields inside the slab.
2) The absorption inside the slab cannot be directly measured but can be determined from reflectance and transmittance. For the particle, absorption in the particle can be followed from the scattering.
3) The *coefficients* of the reflected wave and transmitted wave follow from applying Maxwell's boundary conditions at the surfaces of the slab (Fresnel equations).

So far, both systems appear to be almost identical. However, whereas the particle response contains an explicit wavelength dependence [e.g., the scattering by a small particle in the Rayleigh approximation (see Section 4.7) is proportional to $1/\lambda^4$], the reflectance and transmittance by a slab are independent of the wavelength, except for the wavelength dependence of the optical constants. On the other hand, this means that for a large particle that is comparable to a slab, the light scattering models must result in the same wavelength-independent reflectance and transmittance as for a slab.

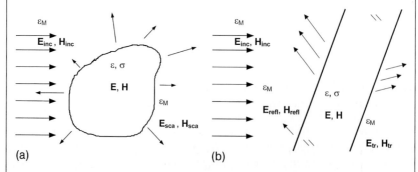

Figure 4.3 (a) Scattering by a particle. (b) Reflection and transmission for a slab.

With the extinction and scattering cross-sections, integral properties of the single particle are defined which are a measure of the ability to scatter and absorb light. They do not provide information about the angular distribution of the scattered light. This information is contained in the integrand of the scattering rate in Equation 4.32: $-\frac{1}{2}\text{Re}[(E_{sca} \times H^*_{sca}) \cdot n_A]$. It defines the *differential scattering cross-section* $\partial\sigma_{sca}/\partial\Omega$, which depends on the solid angle Ω. The differential cross-section can be divided into two additive terms, the *scattering intensities* $i_{per}(\Omega)$ and $i_{par}(\Omega)$ perpendicular and parallel to the plane subtended by the incident and scattered directions (scattering plane).

4.5
Far-Field and Near-Field

σ_{ext} and σ_{sca} are measures of the ability to extinguish and to scatter an electromagnetic wave at incident wavelength λ. They do not provide information about the actual strength of the electromagnetic fields in the vicinity of or far away from the particle, because they are integral properties independent of the distance from the particle surface. The local fields, however, are of great interest in several applications, for example, SERS (surface-enhanced Raman scattering) or near-field optical microscopy, or for nonlinear optical effects.

Here, we give a brief introduction to the *near-field zone* and the *far-field zone* in terms of the energy flux given by the time-averaged Poynting vector. A detailed discussion of the near-field of a spherical particle is given in Section 5.5. For simplicity, we assume spherical coordinates. Then, it is convenient to use a spherical shell of diameter $2R_{cs}$ as integration surface in Equations 4.30 and 4.32 with the normal vector n_A equal to the radial unit vector e_r. It follows that only the radial component S_r of the Poynting vector contributes to the extinction and scattering rate and, hence, to the optical cross-sections. The Poynting vector, however, also has components S_θ and S_φ:

$$S = \begin{pmatrix} S_r \\ S_\theta \\ S_\varphi \end{pmatrix} = \frac{1}{2}\text{Re}\begin{pmatrix} E_\theta H^*_\varphi - E_\varphi H^*_\theta \\ E_\varphi H^*_r - E_r H^*_\varphi \\ E_r H^*_\theta - E_\theta H^*_r \end{pmatrix} \quad (4.35)$$

The scattered wave can be described in spherical coordinates by outgoing spherical waves for which the exact analytical solution yields that the radial components E_r and H_r of the scattered wave decrease with R_{cs}^{-2} faster than the tangential components E_θ, H_θ, E_φ and H_φ, which decrease with R_{cs}^{-1} with increasing R_{cs}. From this radial component dependence it follows that in the far-field, that is, for large $R_{cs} \gg R$, the radial component S_r of the Poynting vector is much larger than the tangential components, $S_r \gg S_\theta, S_\varphi$. Then, the Poynting vector is almost

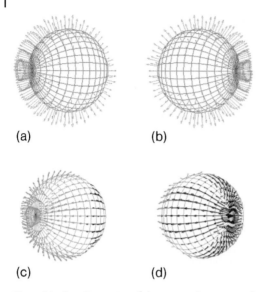

Figure 4.4 Poynting vector of the scattered wave on a fictitious shell around a spherical particle with $2R = 20\,\text{nm}$: (a) and (b) in the far-field, (c) and (d) at $R_{cs} = R$ (particle surface).

perpendicular to the surface used in integration and the electromagnetic fields become transverse.

This is shown in Figure 4.4a and b for a dipole scatterer. The arrow indicates the direction of the incident light. On going in the near-field zone, that is, $R_{cs} \approx R$, the tangential components now determine the Poynting vector. This means that a significant energy flux is set up around the spherical particle with components also directed towards the particle. This is illustrated in Figure 4.4c and d. The net flux through the integration sphere, given by S_r, however, is the same as in the far-field.

4.6
The Incident Electromagnetic Wave

Since in experimental set-ups and in applications lasers are often used, the incident electromagnetic wave has become increasingly important also in light scattering by small particles. The fundamental theories, however, always assumed a plane electromagnetic wave impinging the particle. Therefore, a refinement of the scattering theory is necessary, extending to inhomogeneous incident waves, as for example the ground mode of a laser, the Gaussian mode.

A plane wave is characterized by the fact that the surfaces of constant phase and constant amplitude coincide and are planes normal to the direction of propagation. Mathematically a monochromatic plane wave is described by the equation

4.6 The Incident Electromagnetic Wave

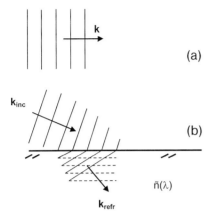

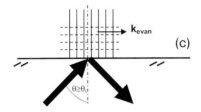

Figure 4.5 Illustration of the planes of constant phase (solid lines) and constant amplitude (dashed lines) for (a) a plane wave, (b) a wave in an absorbing medium (oblique incidence), and (c) an evanescent wave from total internal reflection.

$$A(r,t) = A_0 \exp[i(kr - \omega t)] \qquad (4.36)$$

The surfaces of constant phase satisfy the equation

$$kr - \omega t = \text{constant} \qquad (4.37)$$

which is the vectorial representation of planes perpendicular to k. This presentation assumes that k is a real-valued vector (Figure 4.5a). If, however, absorption in the medium comes into play, k is a complex-valued vector $k = k_1 + ik_2$. Then, the wave becomes inhomogeneous because its amplitude decreases with $A_0 \exp(-k_2 r)$. If k_1 and k_2 are collinear, for example, for perpendicular incidence at the boundary to an absorbing medium, the wave evanesces *along* the propagation direction. The surfaces of constant phase and constant amplitude still coincide in this special case. If k_1 and k_2 include an angle Ω, the surfaces of constant phase and constant amplitude never coincide but are oblique (Figure 4.5b).

For total internal reflection there is no energy flux perpendicular to the surface at which the total reflection occurs, and the corresponding refracted wave must be restricted in its amplitude to a region close to the surface. This is described mathematically with a vector $k = k_1 + ik_2$ with $k_1 \perp k_2$. While the wave propagates with

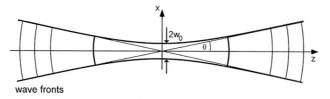

Figure 4.6 Illustration of a Gaussian beam.

$k_1 = (k_x, k_y, 0)^T$ along the surface, it evanesces with $A_0\exp(-k_2 r) = A_0\exp(-k_z z)$ perpendicular to this surface. In consequence, the surfaces of constant phase and constant amplitude are perpendicular (Figure 4.5c) for this evanescent wave from total internal reflection.

The first treatment of inhomogeneous incident waves as converging beams stems from Möglich [79] in 1933. Möglich gave explicit expressions for beam expansion coefficients as a function of the size, wavelength, and beam shape.

An inhomogeneous wave of technical importance is the *Gaussian beam*, which is the ground mode of each gas or solid-state laser. The mathematical description is more complicated than that of plane waves or spherical waves. For a Gaussian beam propagating in the direction of the z-axis, it is

$$E(x,y,z,t) = E_0 \exp\left[-\frac{(x^2+y^2)}{w^2(z)}\right]\exp\left(-i\left\{\omega t - k\left[z + \frac{(x^2+y^2)}{2R(z)}\right] + \phi(z)\right\}\right) \quad (4.38)$$

in which the beam in the direction of propagation z is described by a Gaussian function, justifying the name of the beam (Figure 4.6).

The beam radius $w(z)$ is defined by the distance from the beam axis at which the field amplitude has decayed by a factor $1/e$:

$$w(z) = w_0\sqrt{1+\left(\frac{z}{z_R}\right)^2},\ z_R = \frac{\pi w_0^2}{\lambda},\ R(z) = z\left[1+\left(\frac{z_R}{z}\right)^2\right],\ \text{and}\ \phi(z) = \tan^{-1}\left(\frac{z}{z_R}\right)$$

(4.39)

where z_R is the Rayleigh length. The beam diameter increases with increasing distance from the origin, the laser, and at sufficiently large distances the beam radius increases linearly with distance. The surfaces of constant phase are planes perpendicular to the direction of propagation near the origin and are spherical surfaces far from it. The function $\phi(z)$ accounts for the stretching of the spherical wave fronts to form plane wave fronts as the beam approaches the coordinate origin and the bending of the plane fronts to spherical wave fronts with increasing distance from the origin. The smallest beam diameter is called the beam waist w_0.

For inhomogeneous waves, it is reasonable to redefine the intensity incident on a particle, since it is not constant over the cross-sectional area of the particle, or decreases along with the particle axis in absorbing surrounding media. A first attempt, but not necessarily satisfactory approach, is to average over the cross-sectional area of the particle perpendicular to the Poynting vector of the incident wave. This allows comparison with results for plane waves. We discuss this in examples in Section 7.6.

4.7
Rayleigh's Approximation for Small Particles – The Dipole Approximation

For a description of elastic light scattering and absorption by a nanoparticle we can consider the particle also as a source of electromagnetic radiation. The simplest radiator is an oscillating electric dipole with dipole moment p. Its properties can be found in any standard textbook on electrodynamics, for example, those by Born and Wolf [3] and Jackson [5]. For a dipole small compared with the wavelength λ, the emitted electric and the magnetic fields in the direction of r are given in the far-field zone (distance $r = |r| \gg \lambda$)[2)] by

$$E = \frac{k^2}{4\pi\varepsilon_0\varepsilon_M}[r\times(r\times p)]\frac{e^{i(kr-\omega t)}}{r^3} \tag{4.40}$$

and

$$H = \frac{1}{Z_0}\frac{k^2}{4\pi\varepsilon_0\varepsilon_M}(r\times p)\frac{e^{i(kr-\omega t)}}{r^2} \tag{4.41}$$

where $Z_0 = \sqrt{\mu_0/\varepsilon_0}$ is the vacuum impedance and ε_M is the dielectric constant of the surrounding medium.

These fields are also obtained for a small particle if the particle fulfills certain conditions, so that the particle can be replaced by an oscillating dipole. The conditions are

$$x \ll 1 \text{ and } |m|x \ll 1 \tag{4.42}$$

where m is the relative index of refraction, defined as

$$m = \frac{\tilde{n}(\lambda)}{n_M(\lambda)} = \sqrt{\frac{\varepsilon(\lambda)}{\varepsilon_M(\lambda)}} \tag{4.43}$$

The size parameter x is generally defined as

$$x = \frac{\text{circumference}}{\text{wavelength}} \tag{4.44}$$

The above conditions (4.42) are called the *Rayleigh approximation*, in honor of Lord Rayleigh, who first analyzed in 1871 the elastic light scattering by small particles [62, 63].

As the dipole in the particle is induced by the incident electric field E_0, the dipole moment p is connected with E_0 via

$$p = \varepsilon_0\varepsilon_M\underline{\alpha}E_0 \tag{4.45}$$

---

2) We restrict ourselves here to the far-field because measurements on absorption and scattering are usually made far away from the particle.

where $\underline{\underline{\alpha}}$ is a polarizability tensor of second rank with its elements being proportional to the particle volume V_p, similar to the atomic polarizability. For simplicity we assume in the following that $\underline{\underline{\alpha}}$ is a diagonal tensor.

We can then calculate the rates at which energy is absorbed and scattered by the particle, using the above fields, and can express them in terms of cross-sections for extinction (= absorption plus scattering) and scattering:

$$\sigma_{ext} = k \cdot \mathrm{Im} \sum_{i=1}^{3} \alpha_i + \frac{k^4}{6\pi} \sum_{i=1}^{3} |\alpha_i|^2 \tag{4.46}$$

$$\sigma_{sca} = \frac{k^4}{6\pi} \sum_{i=1}^{3} |\alpha_i|^2 \tag{4.47}$$

where Im means the imaginary part.

A notable property of the outgoing dipole radiation (scattering) is that it is proportional to V_p^2/λ^4, a finding of Lord Rayleigh. For absorbing particles an additional term proportional to V_p/λ contributes to the extinction cross-section. This term dominates the extinction of light for strongly absorbing particles, for example, carbonaceous particles and metal particles. In addition to this size dependence, the shape of the nanoparticle and the dielectric properties of the nanoparticle and the surrounding enter the scattering and absorption, for the polarizability elements α_i are in general

$$\alpha_i = V_p \frac{\varepsilon - \varepsilon_M}{\varepsilon_M + G_i(\varepsilon - \varepsilon_M)} = V_p \frac{m^2 - 1}{1 + G_i(m^2 - 1)} \tag{4.48}$$

The polarizability is determined on the one hand by the optical contrast $\varepsilon - \varepsilon_M$, and on the other by a factor G_i in the denominator that takes into account the shape of the particle. The latter can be determined for several particle shapes. This is discussed in detail later in Chapter 9. For example, for a sphere it is $G_i = 1/3$, $i = 1, 2, 3$.

The geometry factors G_j satisfy the conditions

$$\sum_j G_j = 1 \text{ and } G_j \leq 1 \, \forall j \tag{4.49}$$

The angular distribution of the dipole radiation – the differential cross-section – is given by

$$\frac{\partial \sigma_{sca}}{\partial \Omega} = \frac{k^4}{24\pi^2} \frac{1 + \cos^2 \theta}{2} \sum_{i=1}^{3} |\alpha_i|^2 \tag{4.50}$$

which is composed of a part with the incident light polarized parallel to the scattering plane ($\cos^2 \theta$), and a part with the incident light polarized perpendicular to the scattering plane (1).

If the particle size grows, this simple approach of a single dipole fails. The reason is that the homogeneous distribution of all microscopic dipoles in the volume is disturbed by the introduction of any curvilinear surface. As long as the particle is very small, the pertubation is negligible and the particle as a whole can be described by a single dipole. For larger particles, however, the microscopic dipoles rearrange close to the surface, leading to higher moments in the macroscopic polarization *P*, the multipole moments. Then, rigorous solutions for the light scattering are needed, the most evolved being the Mie theory [17] for a spherical particle of arbitrary size.

4.8 Rayleigh–Debye–Gans Approximation for Vanishing Optical Contrast

In addition to the simplification for very small particles compared with the wavelength treated in the preceding section, a simplification can also be made for vanishing optical contrast $\varepsilon - \varepsilon_M$. Vanishing contrast can be achieved twice. The first method is the *method of index matching*, which is applied preferably in static and dynamic light scattering to examine static and dynamic mechanical next-neighbor interactions in densely packed colloids. The second method is to examine the scattering of X-rays by small particles in powders to obtain particle sizes and shapes. In this case the wavelength is very small, for example, for Cu Kα radiation it is $\lambda = 0.154$ nm, but the relative refractive index m is $m = 1 - \delta + i\gamma$, with $\delta \approx 10^{-5}$ and $\gamma \approx 10^{-7}$ outside resonances and, therefore, close to unity. Then, the Rayleigh–Debye–Gans approximation holds true also for larger particles with mean size $<\lambda/|m-1|$. As the scattering angles θ are less than ~3°, this light scattering is also called *small-angle scattering*.

For vanishing optical contrast, each volume element d*V* of an arbitrary particle acts as a radiating dipole with dipole moment (see Figure 4.7):

$$\alpha = 3\varepsilon_0 \frac{m^2 - 1}{m^2 + 2} dV = 2\varepsilon_0 (m-1) dV \quad (4.51)$$

The simplification to the right-hand side of this equation follows from $m \approx 1$.

To obtain the scattering by the whole particle, the contributions of each volume element must be summed, considering that the scattered fields of all volume elements differ by a phase factor $\exp[i\delta(\theta, \varphi)]$, with the polar angle θ and the azimuthal angle φ. The totally scattered intensity is then

$$I_{sca} = I_0 \frac{1 + \cos^2\theta}{2} \frac{k^4 V^2}{r^2} (2\varepsilon_0 |m-1|)^2 |R(\theta, \varphi)|^2 \quad (4.52)$$

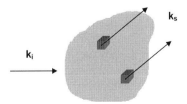

Figure 4.7 Scattering by a volume element d*V* of an arbitrary particle.

4 Fundamentals of Light Scattering by an Obstacle

Table 4.1 Form factor $P(Q)$ of various particles.

Particle	$P(Q)$
Ellipsoid, axes a, b, c Spheroid, $a = b$, $c = \varepsilon a$ $\varepsilon > 1$ prolate $\varepsilon < 1$ oblate Sphere, $a = b = c = R$	$\left[3\dfrac{\sin(Q\tilde{R}) - Q\tilde{R}\cos(Q\tilde{R})}{(Q\tilde{R})^3}\right]^2$ with $\tilde{R} = \left[(a^2 \sin^2\beta + b^2 \cos^2\beta)\sin^2\alpha + c^2\cos^2\alpha\right]^{0.5}$
Disc with radius R	$\dfrac{2}{(QR)^2}\left[1 - \dfrac{J_1(2QR)}{QR}\right]$, with J_1 the Bessel function of order 1
Rod with length L	$2\dfrac{Si(QL)}{QL} - \left[\dfrac{\sin\left(\dfrac{QL}{2}\right)}{\dfrac{QL}{2}}\right]^2$, with $Si(x) = \displaystyle\int_0^x \dfrac{\sin(t)}{t}dt$
Cylinder with radius R and length L	$\left[2\dfrac{J_1(QR\sin\alpha)}{QR\sin\alpha}\dfrac{\sin\left(\dfrac{QL}{2}\sin\alpha\right)}{\dfrac{QL}{2}\sin\alpha}\right]^2$, with J_1 the Bessel function of order 1
Parallelepiped, a, b, c Cube $a = b = c = L$	$\left[\dfrac{\sin(Qa\sin\alpha\cos\beta)}{Qa\sin\alpha\cos\beta}\dfrac{\sin(Qb\sin\alpha\cos\beta)}{Qb\sin\alpha\cos\beta}\dfrac{\sin(Qc\cos\alpha)}{Qc\cos\alpha}\right]^2$

The term $R(\theta,\varphi)$ can be calculated from

$$R(\theta,\varphi) = \frac{1}{V}\int \exp[i\delta(\theta,\varphi)]dV \tag{4.53}$$

Although it is not obvious, often the φ dependence is omitted, and the phase term (4.53) is written as a function of the quasi-momentum Q between the incident light with momentum k_0 and the scattered light with momentum k_s:

$$Q = k_0 - k_s, \quad Q = \frac{4\pi}{\lambda}\sin\left(\frac{\theta}{2}\right) \tag{4.54}$$

The square of $R(\theta,\varphi)$ is then abbreviated as $P(Q)$, and is called the *form factor*, because it contains information about the shape of the particle.

The calculation of $P(Q)$ is not easy. The simplest case is the sphere for which the symmetry of the sphere allows the omission of the φ dependence. This was already done in 1881 by Lord Rayleigh [64] and independently in 1925 by Gans [80]. Kerker [7] argued that Debye, not Gans, should share honors with Rayleigh. We follow the arguments of van de Hulst [6] and call the theory the Rayleigh–Debye–Gans theory. In Table 4.1 we give some examples for $P(Q)$, following Pedersen [81].

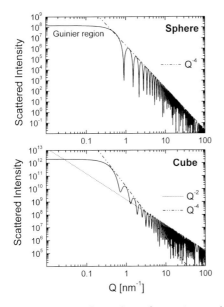

Figure 4.8 SAXS by a sphere of $2R = 10\,\text{nm}$ and a cube of length $L = 5\,\text{nm}$.

Table 4.2 Envelope of the form factor $P(Q)$ of various particles.

Particle	Envelope of $P(Q)$
Ellipsoid, spheroid, sphere	Large $Q\tilde{R}$: Q^{-4}
Disc	Large QR: Q^{-2}
Rod	Large QL: $Q^{-1} + Q^{-2}$
Cylinder	Large QR, QL: Q^{-4}
	Still larger QR, QL: $Q^{-3} + Q^{-5}$
Parallelepiped, cube	Large Qa, Qb, Qc: Q^{-4}
	Still larger Qa, Qb, Qc: Q^{-2}

The discussion of the form factors is not simple. For $x \to 0$ ($x = QR$, QL, Qa, etc.) it follows for all $P(Q)$ that $P(Q) = $ constant. This region is called the Guinier region. For large $x \gg 1$, the envelope of $P(Q)$ is for a sphere $P(Q) \propto Q^{-4}$. This is called the Porod region. However, this Q dependence is not universal for all particle shapes, as can be seen from Figure 4.8 and Table 4.2. As can be recognized, for a cube the slope of the envelope changes from Q^{-4} to Q^{-2} at about $3\,\text{nm}^{-1}$, as discussed in Table 4.2.

In the usual set-ups for small-angle X-ray scattering (SAXS) the available Q extends from $Q \approx 0.03\,\text{nm}^{-1}$ (minimum Q because the detector is protected by a beam stop for $Q < 0.03\,\text{nm}^{-1}$) to $Q \approx 3\,\text{nm}^{-1}$. Then, the particle size range which

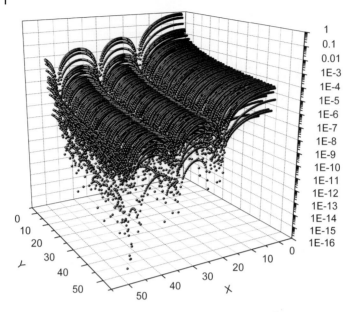

Figure 4.9 Two-dimensional plot of the form factor $|R(\theta,\varphi)|^2$ of a cylinder.

can be examined is between approximately 1 and 150 nm. These values are valid for Cu Kα radiation.

At the end of this section, we will consider the limits of the above $P(Q)$ due to neglect of the φ dependence. Often, either the detector is one-dimensional or the particle shapes are uniformly distributed in the examined sample. Then, the φ dependence is averaged and the above $P(Q)$ apply. However, when using two-dimensional detectors, a well-defined orientation of the particles will lead to different behavior of $|R(\theta,\varphi)|^2$ along the x-axis and the y-axis of the detector, except for highly symmetric particles such as a sphere. We demonstrate this effect for a cylinder, for which the form factor $|R(\theta,\varphi)|^2$ is

$$|R(\theta,\varphi)|^2 = \left[2\frac{J_1(QR\sin\theta)}{QR\sin\theta} \frac{\sin\left(\frac{QL}{2}\sin\theta\cos\varphi\right)}{\frac{QL}{2}\sin\theta\cos\varphi} \right]^2 \qquad (4.55)$$

Figure 4.9 shows the distribution of scattered light according to this form factor. It is clear that the scattering is different along the x-axis and the y-axis, for which an asymmetric contour can be seen at the detector area.

5
Mie's Theory for Single Spherical Particles

Light scattering and absorption by a sphere surely constitute the most evolved model for light scattering by obstacles. The scattering of a plane electromagnetic wave by a spherical particle was analyzed already before the turn of the last century by the Dane Ludvig Valentin Lorenz [65] in 1890. Lorenz did not use Maxwell's equations to derive electrical and magnetic fields of the scattered wave, but based his own theory of light scattering based on the concept of a light vector rather than on the concept of electromagnetic fields. This made the physical interpretation more complicated and ultimately led to the neglect of Lorenz's contribution. In 1908, Gustav Mie [17] proposed his general solution for scattering and absorption from single, isolated spheres based on Maxwell's equations. His intention was to explain the color of colloidal gold nanoparticles and its size dependence. It is worth noting that besides his theoretical work, Mie's great achievement was to calculate for the first time the spectra of gold nanoparticles of various sizes. This is of importance because in 1908 neither electron microscopy for size determination nor computers for calculation of spectra were available. With the size-dependent spectra presented in Mie's contribution, however, it was possible to compare measured spectra with the calculated spectra and to determine the size of gold nanoparticles to a certain extent. The first application of Mie's theory was therefore the interpretation of the experiments of Steubing [82] on colloidal gold particles.

The great importance of the theoretical work of both, Lorenz and Mie can be best recognized from the fact that it has often been treated later in many books on classical electrodynamics, for example, by Born and Wolf [3], Stratton [4], and Jackson [5], and also in several monographs on light scattering, for example, by van de Hulst [6], Kerker [7], and Bohren and Huffman [8]. In honor of Lorenz the theory is often also called the *Lorenz–Mie theory*, but is referred to as the *Mie theory* in the following.

The original Mie theory is restricted to plane wave scattering by a homogeneous, isotropic sphere in a nonabsorbing embedding medium. As it is not limited to any sphere size, it has found many applications from nanoparticles to millimeter-sized droplets in the atmosphere and is used in solid-state physics, physical chemistry, engineering, and astrophysics. With the development of computers, even fast simulation of scattering and extinction data for comparison with experimental data or for prediction of optical properties is available.

Optical Properties of Nanoparticle Systems: Mie and Beyond. Michael Quinten
Copyright © 2011 WILEY-VCH Verlag GmbH & Co. KGaA, Weinheim
ISBN: 978-3-527-41043-9

5.1
Electromagnetic Fields and Boundary Conditions

The symmetry of a sphere always allows one to choose the incident plane wave as propagating along the positive z-axis as depicted in Figure 5.1. Then, in spherical coordinates r, θ, φ the scalar wave equation (4.16) in Chapter 4 has an infinite number of independent solutions Φ_{nm}:

$$\Phi_{{}^e_o nm}(r,\theta,\varphi) = z_n(kr) P_{nm}(\cos\theta) \begin{matrix} \cos(m\varphi) \\ \sin(m\varphi) \end{matrix} \qquad (5.1)$$

with $n = 1, 2, ..., \infty$ and $m = 0, \pm 1, \pm 2, ..., \pm n$, obtained with a separation of variables method. The quantity z_n is any of the four spherical Bessel functions, either spherical Bessel functions j_n, or spherical Neumann functions y_n, or spherical Hankel functions of the first kind $h_n^{(1)}$, or of the second kind $h_n^{(2)}$. $P_{nm}(\cos\theta)$ are the associated Legendre polynomials. The subscripts "o" for odd and "e" for even account for the symmetry of the functions when changing the azimuthal angle φ to $-\varphi$. For more detailed information about spherical Bessel functions and associated Legendre polynomials, we refer to the *Handbook of Mathematical Functions* by Abramovitz and Stegun [83].

Stratton [4] pointed out that in spherical coordinates – and only in this case – also the radial vector $\mathbf{r} = r\mathbf{e}_r$ can be chosen for the constant vector $\mathbf{a}$ when creating solutions of the vector wave equation. With this choice, a set of linear independent and orthogonal *vector spherical harmonics* $\mathbf{M}_{{}^e_o nm}$ and $\mathbf{N}_{{}^e_o nm}$ is created from[1])

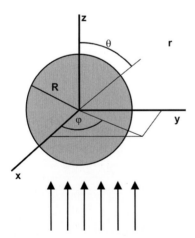

Figure 5.1 Sketch of the geometry in light scattering by a spherical particle.

---

1) We point out that Born and Wolf [3] used Hertz potentials Π_e and Π_m instead of vector spherical harmonics $\mathbf{M}$ and $\mathbf{N}$.

$$\boldsymbol{M}_{\substack{e\\o}nm} = \operatorname{curl}(\Phi_{\substack{e\\o}nm}\boldsymbol{r}) \tag{5.2}$$

$$\boldsymbol{N}_{\substack{e\\o}nm} = k^{-1}\operatorname{curl}\boldsymbol{M}_{\substack{e\\o}nm} \tag{5.3}$$

The explicit expressions are

$$\boldsymbol{M}_{enm} = \begin{pmatrix} 0 \\ -\sin(m\varphi)\dfrac{mP_{nm}(\cos\theta)}{\sin\theta}z_n(\rho) \\ -\cos(m\varphi)\dfrac{dP_{nm}(\cos\theta)}{d\theta}z_n(\rho) \end{pmatrix} \tag{5.4}$$

$$\boldsymbol{N}_{onm} = \begin{pmatrix} 0 \\ \cos(m\varphi)\dfrac{mP_{nm}(\cos\theta)}{\sin\theta}z_n(\rho) \\ -\sin(m\varphi)\dfrac{dP_{nm}(\cos\theta)}{d\theta}z_n(\rho) \end{pmatrix} \tag{5.5}$$

$$\boldsymbol{N}_{enm} = \begin{pmatrix} \cos(m\varphi)\cdot n(n+1)\cdot P_{nm}(\cos\theta)\dfrac{z_n(\rho)}{\rho} \\ \cos(m\varphi)\dfrac{dP_{nm}(\cos\theta)}{d\theta}\dfrac{1}{\rho}\dfrac{d}{d\rho}[\rho z_n(\rho)] \\ -\sin(m\varphi)\dfrac{mP_{nm}(\cos\theta)}{\sin\theta}\dfrac{1}{\rho}\dfrac{d}{d\rho}[\rho z_n(\rho)] \end{pmatrix} \tag{5.6}$$

$$\boldsymbol{N}_{onm} = \begin{pmatrix} \sin(m\varphi)\cdot n(n+1)\cdot P_{nm}(\cos\theta)\dfrac{z_n(\rho)}{\rho} \\ \sin(m\varphi)\dfrac{dP_{nm}(\cos\theta)}{d\theta}\dfrac{1}{\rho}\dfrac{d}{d\rho}[\rho z_n(\rho)] \\ \cos(m\varphi)\dfrac{mP_{nm}(\cos\theta)}{\sin\theta}\dfrac{1}{\rho}\dfrac{d}{d\rho}[\rho z_n(\rho)] \end{pmatrix} \tag{5.7}$$

with the abbreviation $\rho = kr$. Any solution for the electromagnetic fields can now be expanded in an infinite series of these functions. For the transverse magnetic (TM) mode the magnetic field has no radial component ($H_r = 0$), and vice versa for the transverse electric (TE) mode the electric field has no radial component ($E_r = 0$). The symmetry of the sphere allows m to be restricted to $m = \pm 1$, for the reason that the reference frame always can be chosen with the z-axis being the direction of propagation for the incident plane wave. Later, in the extensions of the Mie theory, we will show that for inhomogeneous incident waves, for example, Gaussian beams and evanescent waves, this symmetry is broken.

In the following, the electromagnetic fields of the incident wave, the scattered wave, and the interior wave are expressed in terms of the vector spherical harmonics.

The refractive index n_p of the particle may be a complex number, which includes absorption in the sphere. The particle is assumed to be embedded in a nonabsorbing medium with refractive index n_M and to be illuminated by an incident plane wave with wavenumber $k_M = (2\pi n_M)/\lambda$. The parameter $m = n_p/n_M$ represents the relative refractive index of the particle. The size parameter x is defined as $x = k_M R$, so that $mx = k_1 R$.

Then, the electromagnetic fields for a plane wave propagating along the z-axis, being polarized along the x-axis, are as follows:

incident wave:

$$\mathbf{E}_{\text{inc}} = E_0 \sum_{n=1}^{\infty} \frac{i^n (2n+1)}{n(n+1)} \left[\mathbf{M}_{\text{on1}}^{(1)}(k_M) - i\mathbf{N}_{\text{en1}}^{(1)}(k_M) \right] \tag{5.8a}$$

$$\mathbf{H}_{\text{inc}} = \frac{-k_M E_0}{\omega \mu_0} \sum_{n=1}^{\infty} \frac{i^n (2n+1)}{n(n+1)} \left[\mathbf{M}_{\text{en1}}^{(1)}(k_M) + i\mathbf{N}_{\text{on1}}^{(1)}(k_M) \right] \tag{5.8b}$$

scattered wave:

$$\mathbf{E}_{\text{sca}} = -E_0 \sum_{n=1}^{\infty} \frac{i^n (2n+1)}{n(n+1)} \left[b_n \mathbf{M}_{\text{on1}}^{(3)}(k_M) - ia_n \mathbf{N}_{\text{en1}}^{(3)}(k_M) \right] \tag{5.9a}$$

$$\mathbf{H}_{\text{sca}} = \frac{k_M E_0}{\omega \mu_0} \sum_{n=1}^{\infty} \frac{i^n (2n+1)}{n(n+1)} \left[a_n \mathbf{M}_{\text{en1}}^{(3)}(k_M) + ib_n \mathbf{N}_{\text{on1}}^{(3)}(k_M) \right] \tag{5.9b}$$

wave inside the particle:

$$\mathbf{E}_1 = E_0 \sum_{n=1}^{\infty} \frac{i^n (2n+1)}{n(n+1)} \left[\beta_n^1 \mathbf{M}_{\text{on1}}^{(1)}(k_1) - i\alpha_n^1 \mathbf{N}_{\text{en1}}^{(1)}(k_1) \right] \tag{5.10a}$$

$$\mathbf{H}_1 = \frac{-k_1 E_0}{\omega \mu_0} \sum_{n=1}^{\infty} \frac{i^n (2n+1)}{n(n+1)} \left[\alpha_n^1 \mathbf{M}_{\text{en1}}^{(1)}(k_1) + i\beta_n^1 \mathbf{N}_{\text{on1}}^{(1)}(k_1) \right] \tag{5.10b}$$

The abbreviations a_n and b_n are the expansion coefficients of the TM and TE modes of the scattered wave.

The use of spherical Bessel functions j_n (superscript 1 of the vector harmonics) in the expansions of the electromagnetic fields inside the particle and of the incident wave guarantees the continuity of the fields at the origin of the reference frame centered in the spherical particle. The use of spherical Hankel functions $h_n^{(1)}$ (superscript 3 of the vector harmonics) in the expansion of the scattered wave takes into consideration outgoing spherical waves. Note that for the time-harmonic ansatz $\exp(i\omega t)$ for the fields, the refractive index would be $\tilde{n}(\omega) = n(\omega) - i\kappa(\omega)$. Then, an outgoing spherical wave would be represented by $h_n^{(2)}$ instead of $h_n^{(1)}$. The physical meaning of the sum over n in the above equations is that the fields are expanded to linear independent *radiating multipoles of order n*, starting with the radiating dipole with $n = 1$. For a more detailed discussion of the fields as radiating multipoles, see the information box **Radiating multipoles of a sphere**.

The next step is to apply Maxwell's boundary conditions at the surface of the particle:

$$(\mathbf{E}_{inc} + \mathbf{E}_{sca}) \times \mathbf{e}_r|_{r=R} = \mathbf{E}_1 \times \mathbf{e}_r|_{r=R} \tag{5.11}$$

$$(\mathbf{H}_{inc} + \mathbf{H}_{sca}) \times \mathbf{e}_r|_{r=R} = \mathbf{H}_1 \times \mathbf{e}_r|_{r=R} \tag{5.12}$$

Maxwell's boundary conditions can be resolved to obtain all unknown expansion coefficients α_n, β_n of the waves inside the particle and the expansion coefficients a_n and b_n of the scattered wave. We only give the scattering coefficients a_n and b_n because we are interested in scattering and absorption.

$$a_n = \frac{\psi_n(x)\psi'_n(mx) - m\psi'_n(x)\psi_n(mx)}{\xi_n(x)\psi'_n(mx) - m\xi'_n(x)\psi_n(mx)} \tag{5.13}$$

$$b_n = \frac{m\psi_n(x)\psi'_n(mx) - \psi'_n(x)\psi_n(mx)}{m\xi_n(x)\psi'_n(mx) - \xi'_n(x)\psi_n(mx)} \tag{5.14}$$

$\Psi_n(x) = xj_n(x)$, $\chi_n(x) = xy_n(x)$, and $\xi_n(x) = xh_n^{(1)}(x)$ are Riccati–Bessel, Riccati–Neumann and Riccati–Hankel functions, respectively. The prime denotes derivation with respect to the argument.

It is worth mentioning that the task of resolving Maxwell's boundary conditions to obtain these coefficients is not as trivial as it seems at first sight. The reason is that inserting the above fields in Maxwell's boundary conditions results in four equations where an *infinite* sum over the multipole n still remains. To solve this problem, one must use the following trick:

1) Multiply each of the four equations with the function $P_{q1}/\sin(\theta)$.
2) Integrate each equation from -1 to $+1$ over $x = \cos(\theta)$.
3) Multiply each of the four equations with the derivative $dP_{q1}/d\theta$.
4) Integrate each equation from -1 to $+1$ over $x = \cos(\theta)$.

The reason for this procedure is that one can then use the orthogonality relations of the associated Legendre polynomials to obtain four equations for α_n, β_n, a_n, and b_n for each order n separately.

INFO – Radiating Multipoles of a Sphere

We illustrate here the multipole expansion of the electromagnetic fields in spherical coordinates. Without loss of generality we can restrict considerations to TM partial waves in the plane of incidence ($\varphi = 0$). The discussion for TE partial waves is similar.

In this case the incident, scattered, and interior fields for a compact sphere read as follows:

$$\mathbf{E}^{(j)} = E_0 \sum_{n=1}^{\infty} \frac{i^n(2n+1)}{n(n+1)}(-iA_n) \begin{pmatrix} n(n+1) \cdot P_{n1} \dfrac{z_n[x^{(j)}]}{x^{(j)}} \\ \dfrac{dP_{n1}}{d\theta}\left\{z_{n-1}[x^{(j)}] - n\dfrac{z_n[x^{(j)}]}{x^{(j)}}\right\} \\ 0 \end{pmatrix} \tag{5.15}$$

$$H^{(j)} = -\frac{k^{(j)}E_0}{\mu_0\omega}\sum_{n=1}^{\infty}\frac{i^n(2n+1)}{n(n+1)}(A_n)\begin{pmatrix} 0 \\ 0 \\ -\frac{dP_{n1}}{d\theta}z_n[x^{(j)}] \end{pmatrix} \quad (5.16)$$

with $x^{(j)} = k^{(j)}R$ and

j	A_n	$k^{(j)}$	z_n
Incident	1	k_M	j_n
Scattered	a_n	k_M	$h_n^{(1)}$
Interior	α_n	k	j_n

For further discussion of the field distribution we must discuss the functions $z_n(x)$ and $z_n(x)/x$, as well as $P_{n1}(\theta)$ and $\partial P_{n1}(\theta)/\partial\theta$. Since the scattered fields are only defined for $r \geq R$, we can restrict the discussion to $z_n = j_n$ in the particle and for the incident light.

Bessel functions $j_n(x)$ and $j_n(x)/x$

In Figure 5.2 we show the functions $j_n(x)$ and $j_n(x)/x$ for $n = 0, 1, 2, 3, 4, 10$, and 20. In addition, comparison is made between the functions with $x = k^{\text{incident}}r$ and $mx = k^{\text{interior}}r$, with $m = 1.5$.

All functions are finite at $x = 0$, except for $j_0(x)/x$. However, this function does not play a role since it is not contained in the above field equations. It can be recognized that for $n = 1$ the electric field components are maximum at $x = 0$, while the magnetic field components and all higher order contributions $n > 1$ are 0 at $x = 0$. They become maximum at larger x, that is, for larger particle sizes.

For each order n, the functions are oscillating with x. That means that along with the radial direction any field component keeps its sign between two zeros of the corresponding Bessel function.

Comparing Bessel functions with argument x with those of argument mx, it turns out that they match at certain values of x. For example, $j_{20}(x)$ matches $j_{20}(1.5x)$ around the points $x = 6.2, 10, 15, 18$, and so on.

Associated Legendre polynomials P_{n1} and derivatives $\partial P_{n1}(\theta)/\partial\theta$

The behavior of the Legendre polynomials is illustrated in Figure 5.3 for $n = 1$, 2, 3, and 20. They are oscillating functions with the polar angle θ and symmetric for angles between 0° and −180°, for which it is sufficient to display them only in the range 0–180°. Hence they divide the plane of incidence (and the plane of scattering) into regions with different signs. For example, for $n = 1$,

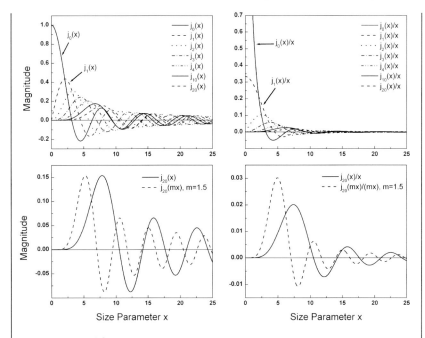

Figure 5.2 Bessel functions versus size parameter.

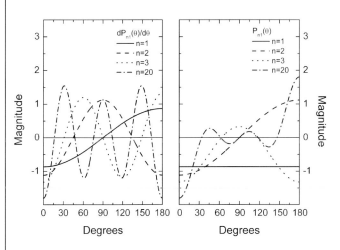

Figure 5.3 Legendre polynomials versus angle.

the zeros of $\partial P_{11}(\theta)/\partial \theta$ are at 90° and −90°. They are positive between 0° and 90° and between 0° and −90°. In the other regions they are negative. Therefore, the plane is divided into two regions which differ in sign. For the higher orders, more zeros occur, so that the plane will be divided in more regions with alternating sign.

Electromagnetic Fields

Combining now Bessel functions with Legendre polynomials, we can discuss the behavior of the field components. As an example, we discuss the component E_θ:

$$E_\theta = E_0 \sum_{n=1}^{\infty} \frac{i^n(2n+1)}{n(n+1)}(-iA_n)\frac{dP_{n1}}{d\theta}\left[j_{n-1}(x) - n\frac{j_n(x)}{x}\right] \quad (5.17)$$

and show $\dfrac{dP_{n1}}{d\theta}\left[j_{n-1}(x) - n\dfrac{j_n(x)}{x}\right]$ for the orders $n = 1, 2, 3,$ and 20 (Figure 5.4).

It becomes obvious that actually the field components behave like multipoles. Moreover, with increasing multipole order n the regions of high magnitude become very small, as can be best recognized for $n = 20$.

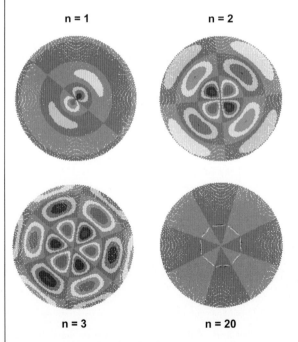

Figure 5.4 False color plot of E_θ for $n = 1, 2, 3,$ and 20. A color version of this figure can be found in the color plates at the end of the book.

5.2
Cross-sections, Scattering Intensities, and Related Quantities

Under experimental conditions, the rates at which energy from the incoming light is scattered and absorbed are of more interest than the electromagnetic fields. They are obtained from the energy flux balance from and to the particle. According to the general discussion in Chapter 4, the rates are obtained from surface integration of the Poynting vector outside the particle. In spherical coordinates it is convenient to use the surface of a conceptual sphere of radius $R_{cs} \geq R$ concentric with the particle with diameter $2R$ as integration surface. Then, the rates are

$$W_0 = -\frac{1}{2}\text{Re}\int_0^{2\pi}\int_0^{\pi}(E_{\text{inc}} \times H_{\text{inc}}^*)e_R R_{cs}^2 \sin\theta\,d\theta\,d\varphi \tag{5.18}$$

$$W_{\text{ext}} = -\frac{1}{2}\text{Re}\int_0^{2\pi}\int_0^{\pi}(E_{\text{inc}} \times H_{\text{sca}}^* + E_{\text{sca}} \times H_{\text{inc}}^*)e_R R_{cs}^2 \sin\theta\,d\theta\,d\varphi \tag{5.19}$$

$$W_{\text{sca}} = -\frac{1}{2}\text{Re}\int_0^{2\pi}\int_0^{\pi}(E_{\text{sca}} \times H_{\text{sca}}^*)e_R R_{cs}^2 \sin\theta\,d\theta\,d\varphi \tag{5.20}$$

where W_0 represents the rate at which energy is absorbed from the incident wave in the integrating volume in the absence of the sphere, and vanishes for nonabsorbing embedding media; W_{ext} represents the extinction rate and W_{sca} the scattering rate of the embedded sphere. Dividing these rates by the intensity I_0 of the incident light, the *optical cross-sections* σ of a sphere are obtained. For their explicit calculation we insert the electromagnetic fields of the incident wave (Equation 5.8) and of the scattered wave (Equation 5.9) in Equations 5.19 and 5.20 and make use of the orthogonality relations of the vector spherical harmonics:

$$\int_0^{2\pi}\int_0^{\pi}(M_{nm} \times M_{qp}^*) \cdot e_r \sin\theta\,d\theta\,d\varphi = \int_0^{2\pi}\int_0^{\pi}(N_{nm} \times N_{qp}^*) \cdot e_r \sin\theta\,d\theta\,d\varphi = 0 \tag{5.21}$$

$$\int_0^{2\pi}\int_0^{\pi}(M_{nm} \times N_{qp}^*) \cdot e_r \sin\theta\,d\theta\,d\varphi = \frac{z_n(kr)}{kr}[kr \cdot z_q^*(kr)]' \cdot \delta_{nq}\delta_{mp} \tag{5.22}$$

and analogously for all other combinations of vector spherical harmonics. The quantities δ_{nq} and δ_{mp} are Kronecker symbols. After some manipulations, the results for the cross-sections are as follows:

$$\sigma_{\text{ext}} = -\frac{2\pi}{k_M^2}\sum_{n=1}^{\infty}(2n+1)\begin{Bmatrix}\text{Re}(a_n+b_n)\cdot\text{Im}[\xi_n(k_M R_{cs})\psi_n^*(k_M R_{cs}) \\ -\xi_n'(k_M R_{cs})\psi_n^*(k_M R_{cs})] + \text{Im}(a_n+b_n) \\ \cdot\text{Re}[\xi_n(k_M R_{cs})\psi_n^*(k_M R_{cs}) - \xi_n'(k_M R_{cs})\psi_n^*(k_M R_{cs})]\end{Bmatrix} \tag{5.23}$$

$$\sigma_{\text{sca}} = \frac{2\pi}{k_M^2}\sum_{n=1}^{\infty}(2n+1)(|a_n|^2+|b_n|^2)\cdot\text{Im}[\xi_n(k_M R_{cs})\xi_n^*(k_M R_{cs})] \tag{5.24}$$

Obviously, the cross-sections still depend upon the size $2R_{cs}$ of the integrating sphere because the Riccati–Bessel functions $\psi_n(kR_{cs})$ and the Riccati–Hankel functions $\xi_n(kR_{cs})$ and their derivatives depend on kR_{cs}. Using the identity $\xi_n(kR_{cs}) = \psi_n(kR_{cs}) + i\chi_n(kR_{cs})$, the R_{cs}-dependent terms in Equations 5.23 and 5.24 can be rewritten. For real arguments kR_{cs}, for which the Bessel and Neumann functions are also real, it is

$$\xi_n \psi_n'^* - \xi_n' \psi_n^* = -i(\chi_n \psi_n' - \psi_n \chi_n') \tag{5.25}$$

The value of the Wronskian of real functions in parentheses is always 1, independently of the argument kR_{cs}! For that reason, also $\text{Im}(\xi_n \xi_n'^*) = 1$ for all arguments kR_{cs}. In consequence, Equations 5.23 and 5.24 reduce to

$$\sigma_{ext} = \frac{2\pi}{k_M^2} \sum_{n=1}^{\infty} (2n+1) \text{Re}(a_n + b_n) \tag{5.26}$$

$$\sigma_{sca} = \frac{2\pi}{k_M^2} \sum_{n=1}^{\infty} (2n+1)\left(|a_n|^2 + |b_n|^2\right) \tag{5.27}$$

These well-known results are independent of R_{cs} and, hence, valid in both the *far-field zone* and the *near-field zone*.

For numerical evaluation of the cross-sections, the infinite sum over the multipole order n must be restricted to a finite number $NMAX$. As found empirically by Wiscombe [84], a sufficient number $NMAX$ is given by

$$NMAX = \begin{cases} \text{integer}\left(x + 4x^{\frac{1}{3}} + 1\right) & x \leq 8 \\ \text{integer}\left(x + 4.05x^{\frac{1}{3}} + 2\right) & 8 < x < 4200 \\ \text{integer}\left(x + 4x^{\frac{1}{3}} + 2\right) & 4200 \leq x \end{cases} \tag{5.28}$$

With the extinction and scattering cross-sections, integral properties of the single particle are defined which are a measure of the ability to scatter and to absorb light. They do not provide information about the angular distribution of the scattered light. This information is contained in the integrand of the scattering rate in Equation 5.20:

$$E_{sca,\theta} H^*_{sca,\varphi} - E_{sca,\varphi} H^*_{sca,\theta} = \frac{I_0}{k_M^2 R^2}\left[i_{per}(\theta)\sin^2\varphi + i_{par}(\theta)\cos^2\varphi\right] \tag{5.29}$$

where $i_{per}(\theta)$ and $i_{par}(\theta)$ are the *scattering intensities* perpendicular and parallel to the plane subtended by the incident and scattered directions (scattering plane) with

$$i_{per}(\theta) = \left|\sum_{n=1}^{\infty} \frac{2n+1}{n(n+1)}[a_n \pi_n(\theta) + b_n \tau_n(\theta)]\right|^2 = |S_1(\theta)|^2 \tag{5.30}$$

$$i_{par}(\theta) = \left| \sum_{n=1}^{\infty} \frac{2n+1}{n(n+1)} [a_n \tau_n(\theta) + b_n \pi_n(\theta)] \right|^2 = |S_2(\theta)|^2 \qquad (5.31)$$

where $S_1(\theta)$ and $S_2(\theta)$ are known as (complex) *phase* functions. The angular dependent functions $\tau_n(\theta)$ and $\pi_n(\theta)$ are defined as $\tau_n(\theta) = \partial P_{n1}/\partial \theta$ and $\pi_n(\theta) = P_{n1}/\sin\theta$. $P_{n1}(\cos\theta)$ are associated Legendre polynomials for $m = 1$.

In Figure 5.5 we give examples of the scattering intensities i_{par} in the scattering plane and i_{per} in the plane perpendicular to the scattering plane for a small particle

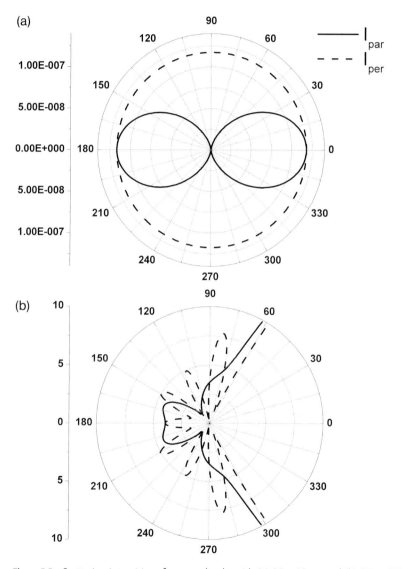

Figure 5.5 Scattering intensities of a water droplet with (a) $2R = 20$ nm and (b) $2R = 1000$ nm.

with $2R = 20$ nm and a large particle with $2R = 1000$ nm. The small particle acts as a dipole ($n = 1$) with i_{par} proportional to $(\cos\theta)^2$, and a constant scattering intensity i_{per}. If the particle grows, the light is dominantly scattered in the forward direction. The scattering intensities become very similar close to the forward direction, but differ for larger angles due to the different zeros of the functions $\tau_n(\theta)$ and $\pi_n(\theta)$. This can clearly be recognized from Figure 5.5b. In this figure, the scattering intensity in the forward direction is not plotted completely because it amounts to 1209 at $\theta = 0°$ for both scattering intensities.

The scattering intensities $i_{per}(\theta)$ and $i_{par}(\theta)$ define the *differential scattering cross-section* $\partial\sigma_{sca}/\partial\Omega$ as

$$\frac{\partial\sigma_{sca}}{\partial\Omega} = \frac{1}{k_M^2}[i_{par}(\theta) + i_{per}(\theta)] \tag{5.32}$$

from which the scattering cross-section σ_{sca} follows as

$$\sigma_{sca} = \frac{1}{k_M^2}\int_0^{2\pi}\int_0^{\pi}[i_{par}(\theta) + i_{per}(\theta)]\sin\theta\, d\theta\, d\varphi \tag{5.33}$$

The light scattered by a nanoparticle is polarized to a certain amount P. The *degree of polarization* P is related to $i_{per}(\theta)$ and $i_{par}(\theta)$:

$$P = \frac{i_{par}(\theta) - i_{per}(\theta)}{i_{par}(\theta) + i_{per}(\theta)} \tag{5.34}$$

For a dipole scatterer as in Figure 5.5a, it follows that the scattered light is completely unpolarized in the forward and backward directions [$P(0°) = P(180°) = 0$], and is completely polarized at 90° [$P(90°) = 1$], a finding of Lord Rayleigh for the light scattered by air molecules.

Another important quantity connected with $i_{per}(\theta)$ and $i_{par}(\theta)$ is the *asymmetry parameter* or weighted cosine of the scattering angle:

$$g = \langle\cos\theta\rangle = \frac{\frac{1}{k_M^2}\int_0^{2\pi}\int_0^{\pi}[i_{par}(\theta) + i_{per}(\theta)]\cos\theta\sin\theta\, d\theta\, d\varphi}{\frac{1}{k_M^2}\int_0^{2\pi}\int_0^{\pi}[i_{par}(\theta) + i_{per}(\theta)]\sin\theta\, d\theta\, d\varphi}$$

$$= \frac{4}{k_M^2 R^2 Q_{sca}}\mathrm{Re}\sum_{n=1}^{\infty}\left\{\frac{n(n+2)}{n+1}(a_n a_{n+1}^* + b_n b_{n+1}^*) + \frac{2n+1}{n(n+1)}a_n b_n^*\right\} \tag{5.35}$$

The asymmetry parameter enters the calculation of the *radiation pressure*, a force acting on the spherical particle, as the incident electromagnetic wave carries momentum in addition to energy. This force was proposed as early as 1619 by Johannes Kepler for the sunlight blowing back a comet's tail so that it always points away from the Sun. The first experimental proof of the radiation pressure, however, was done just in 1903 by Nichols and Hull [85]. Although in 1873 James Clerk Maxwell established theoretically that waves do exert pressure, the radiation pressure was first analyzed theoretically by Debye [86] in 1909, six years after the first experimental proof of the radiation pressure by Nichols and Hull.

If W is the total energy flux of the incident wave, its total momentum flux is W/c. Absorption and scattering of the incident wave lead to a change of the component of momentum in the propagation direction. The rate of change of momentum is the radiation pressure. Debye's analysis yielded the following for the radiation pressure on a spherical particle:

$$F_{\text{pr}} = \frac{I_0}{c}(\sigma_{\text{ext}} - g\sigma_{\text{sca}}) \tag{5.36}$$

Note that this analysis is based on a plane incident wave, and therefore must be appropriately extended for focused Gaussian beams as used to illuminate particles in *optical levitation*.

Also, the amount scattered back to the light source is of interest. It is called the *radar backscattering cross-section* although we act with light. It is of importance for LIDAR (light detection and ranging) and is calculated from the differential scattering cross-section for the angle $\theta = 180°$. For a spherical particle it is

$$\sigma_{\text{back}} = \frac{\partial \sigma_{\text{sca}}(\theta = 180°)}{\partial \Omega} = \frac{\pi}{k_M^2}\left|\sum_{n=1}^{\infty}(2n+1)(-1)^n(a_n - b_n)\right|^2 \tag{5.37}$$

5.3 Resonances

A notable property of the scattering coefficients a_n and b_n (Equations 5.13 and 5.14) and also the coefficients α_n, β_n of the interior wave is that their denominator can become very small. The minima are complex-numbered poles. At these poles the corresponding partial wave exhibits a resonance with different magnitudes for the scattered and interior fields due to the different numerator. The resonances are given for

$$m\frac{\xi_n'(x)}{\xi_n(x)} = \frac{\psi_n'(mx)}{\psi_n(mx)} \tag{5.38}$$

for the nth TM mode with a_n, and for

$$\frac{1}{m}\frac{\xi_n'(x)}{\xi_n(x)} = \frac{\psi_n'(mx)}{\psi_n(mx)} \tag{5.39}$$

for the nth TE mode with b_n. Again, $x = k_M R$ and $mx = k_1 R$.

As ψ_n and ξ_n and their derivatives in Equations 5.38 and 5.39 are oscillating functions, these conditions can be fulfilled multiple times for the same multipole order n at different size parameters $x_1 \leq x_r$. These resonances are distinguished by the *order number l* or *radial mode number*. The resonance of a particular mode with order number $l = 1$ is the resonance with the lowest size parameter and, hence, the innermost resonance of the multipole n between $0 \leq r \leq R$.

Equations 5.38 and 5.39 may be fulfilled for geometric reasons *and* for electronic reasons. For the latter, we can further distinguish two cases: (i) bound electrons which exhibit a harmonic oscillator behavior and (ii) free electrons. The resonances caused by bound electrons are denoted in the following *electronic resonances* and the resonances by free electrons are denoted *surface plasmon polaritons*. All three resonances are discussed in the following, starting with the *geometric resonances*.

5.3.1
Geometric Resonances

Geometric resonances occur if the particle size approaches a certain relation with the wavelength of the incident wave. The corresponding resonances are called *geometric resonances* or *morphology-dependent resonances* (MDRs) or *whispering gallery modes* (WGMs). The latter term was first used by Lord Rayleigh for sound waves guided along a cathedral ceiling [87]. The MDRs of spheres were first explored experimentally by Ashkin and Dziedzic [88], and they observed resonances in the radiation pressure on dielectric spheres [89] and in the scattered light from such spheres [90, 91].

For a first understanding of MDRs, we assume a sphere with refractive index n_P as large as geometric optics can approximately be applied. A resonance occurs if a beam refracted into the particle is multiple totally reflected at the boundary particle surrounding and is superposed with the incoming refracted beam. If the phase of the reflected beam matches exactly the phase of refracted beam modulus 2π, constructive interference occurs. Otherwise, interference diminishes the field amplitudes rapidly. A more detailed analysis yields a condition for constructive interference as a function of the radius r_i of the inner caustics and the wavelength in the particle (see Figure 5.6):

$$2\pi r_i = q \frac{\lambda}{n_P}, \quad q = \text{integer number} \tag{5.40}$$

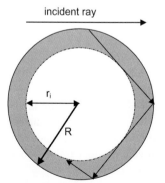

Figure 5.6 Schematic diagram of the constructive interference of a totally reflected beam with the incoming ray–a geometric optical model for the explanation of MDRs.

For the radius r_i of the inner caustics, the following inequality can be deduced from geometric optics that connects it with the radius R of the sphere and the refractive indices n_P of the sphere and n_M of the surrounding matrix:

$$r_i < \frac{n_M}{n_P} R \tag{5.41}$$

Although helpful for understanding the MDRs, geometric optics is not well suited to explain MDRs in nanoparticles. A more precise picture is obtained when looking at the resonance conditions in Equations 5.38 and 5.39. For that purpose, we calculated the functions $f(mx) = \psi'_n(mx)/\psi_n(mx)$, $F_{TM}(x) = m[\xi'_n(x)/\xi_n(x)]$, and $F_{TE}(x) = (1/m)[\xi'_n(x)/\xi_n(x)]$ for the multipole order $n = 20$ and for $m = 1.5$ in steps of $\Delta x = 0.01$. As in this case $f(mx)$ is real valued, but $F_{TM}(x)$ and $F_{TE}(x)$ are complex numbers, the resonance condition is approximately fulfilled when the real parts of F_{TM} and F_{TE} are equal to $f(mx)$ and the imaginary parts almost vanish. The calculations showed that resonances can occur for $x \approx 16.235$ for the TE_{20} mode, where the imaginary part of F_{TE} amounts to 0.008. The next TE_{20} resonance occurs at $x \approx 19.11$ where the imaginary part of F_{TE} amounts to 0.1096. The TM_{20} mode shows resonances for $x \approx 16.65$, where the imaginary part of F_{TM} amounts to 0.0303, and for $x \approx 19.33$, where the imaginary part of F_{TM} amounts to 0.27595. Further resonances for larger x are not further investigated since the imaginary parts of F_{TE} and F_{TM} still increase.

Morphology-dependent resonances appear in extinction and scattering as very narrow peaks in the corresponding cross-section when varying the size parameter by either the size or the wavelength of light. As an example, the extinction efficiency Q_{ext} is plotted versus size for water droplets in air in Figure 5.7. In this example the wavelength was kept constant ($\lambda = 514.5$ nm, $m = n_P/n_M = 1.334$) while the particle size was varied from $2R = 2$ to 10 000 nm in steps of 2 nm. The corresponding size parameter varied between $x = 0.01$ and 62.1. In Figure 5.7b the part between $2R = 3500$ and 5000 nm is zoomed out to show the morphology-dependent resonances.

A similar dependence of Q_{ext} is obtained when keeping the particle size constant and varying the wavelength. Then, however, it must be considered that the refractive index of the particle is also wavelength dependent. For the above water droplets that means that at short wavelengths (large size parameters) the strong UV absorption of water, and at long wavelengths (small size parameters) the strong IR absorption of water, influence the resulting spectrum considerably. In detail, the MDRs are damped away in the UV and sharp resonances occur in the IR for small size parameters caused by phonon polaritons, which will be discussed later.

MDRs are well known mainly for liquid droplets (water, sulfuric acid, nitric acid, ammonia) or solid polymer particles or glass beads. The low refractive index of these materials, ranging from $n_P = 1.3$ to 1.7, usually requires micron-sized particles of these materials that are outside the scope of this book. On the other hand, it follows from Equation 5.40 that with increasing particle refractive index n_P the radius of the inner caustics decreases. That means that in highly refractive particle materials MDRs can be obtained for nanoparticles. For example, we show in Figure 5.8 the extinction efficiency for a spherical diamond particle with $2R = 600$ nm and

90 | *5 Mie's Theory for Single Spherical Particles*

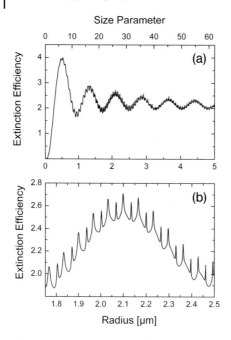

Figure 5.7 Extinction efficiency of water droplets in air showing morphology-dependent resonances. (a) Size range $2R = 2–10000$ nm, $m = 1.334$, $\lambda = 514.5$ nm; (b) size range $2R = 3500–5000$ nm.

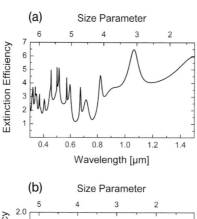

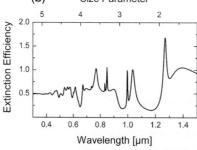

Figure 5.8 Extinction efficiency of (a) a diamond particle, $2R = 600$ nm, in air and (b) a silicon particle, $2R = 500$ nm, in air, showing morphology-dependent resonances.

a silicon particle with $2R = 500$ nm in air, plotted versus the wavelength, that is, the size parameter was varied via the wavelength in the range between 300 and 1500 nm, and the dispersion of the refractive index of diamond and silicon was fully taken into account, using optical constants from the literature[36]. For the diamond particle the size parameter x amounts to $x < 7$ and for the silicon particle $x < 4$. In this wavelength range the refractive indices vary between 2.55 and 2.4 for diamond and between 5 and 3.5 for silicon. The latter is strongly absorbing at wavelengths lower than 400 nm. Obviously, in both cases MDRs are obtained which are comparable to those in Figure 5.7b. For silicon they are strongly damped away at wavelengths lower than 400 nm (size parameters larger than 4.9).

5.3.2
Electronic Resonances and Surface Plasmon Polaritons

Polarization of materials by electromagnetic waves is based on

- displacement of bound charges (both electrons and ions)
- electronic transitions between occupied and unoccupied electronic levels
- displacement of free electrons, if free electrons are present.

For small particles especially the electronic transitions between occupied and unoccupied electronic levels can lead to resonances. For metal particles additionally the free electrons can cause resonances in the metal particle.

For understanding these *electronic resonances* we assume very small particles so that the Riccati–Bessel functions in Equations 5.38 and 5.39 can be approximated by the first term of their series expansion. In this *quasi-static approach*, these equations simplify to

$$\varepsilon = -\frac{n+1}{n}\varepsilon_M \tag{5.42}$$

for the TM modes, and

$$\varepsilon = \varepsilon_M \tag{5.43}$$

for the TE modes of multipole order n. The quantities ε and ε_M are the dielectric functions of the particle material and the surrounding medium, respectively. It follows that

1) Electronic resonances appear independently of the particle size. However, we point to the fact that with increasing particle size this rough approximation for the Riccati–Bessel functions no longer holds true, so that Equations 5.42 and 5.43 must be modified with a size-dependent term.

2) Resonances of the TE modes can only occur when the above approximation of the Riccati–Bessel functions not longer holds true.

3) Sharp resonances of TM modes requires that the real part ε_1 of the dielectric function $\varepsilon(\omega)$ is negative and approaches $-[n + 1/n]\varepsilon_M$ at a certain frequency ω_n, while the imaginary part ε_2 almost vanishes at this frequency.

Negative values in $\varepsilon_1(\omega)$ can be achieved twice:

- close to the resonance frequency of a harmonic oscillator
- for free electrons.

This can be seen from the classical Drude–Lorentz–Sommerfeld equation for the dielectric function:

$$\varepsilon(\omega) = 1 \underbrace{- \frac{\omega_P^2}{\omega^2 + i\omega\gamma_{fe}}}_{\substack{\text{Drude susceptibility} \\ \text{(free electrons)}}} + \underbrace{\sum_j \frac{S_j \omega_j^2}{\omega_j^2 - \omega^2 - i\omega\gamma_j}}_{\substack{\text{harmonic oscillators} \\ \text{(bound electrons)}}} \qquad (5.44)$$

In spectral regions where the free electron contribution is negligible, the harmonic oscillator contribution results in negative values for the real part of the dielectric function, since the corresponding term in Equation 5.44 alternates in sign on going from low to high frequencies close to the resonance frequency ω_j. Then, small particles can exhibit electronic resonances. Far away from such a resonance frequency, the contribution of bound electrons is almost constant and resonances can only occur by particle size.

If free electrons are present, as for metal particles, the dielectric function of the metal becomes negative even for a constant contribution χ of the bound electrons at frequencies with

$$\omega^2 \leq \frac{\omega_P^2}{1 + \chi} \qquad (5.45)$$

due to the free electron contribution. The corresponding particle resonances will be discussed in the following as *surface plasmon polaritons*.

5.3.2.1 Electronic Resonances

As discussed before, electronic resonances can appear in the wavelength region where the dielectric function of the particle material exhibits a harmonic oscillator behavior. Good candidates for electronic resonances are semiconducting nanoparticles and nanoparticles of ionic crystals such as NaCl.

For demonstration, Figure 5.9 shows the dielectric function of NaCl [36] in the far-infrared (FIR) region and the spectrum of an NaCl sphere with $2R = 200\,\text{nm}$. Such a sphere is still small compared with the wavelength of the light in the FIR region.

The dielectric function $\varepsilon_{\text{NaCl}}$ clearly shows harmonic oscillator behavior. Assuming that the damping constant γ_0 of this harmonic oscillator is small compared with the frequency ω in this spectral region, we obtain for the resonance position of the TM multipoles

$$\omega_n^2 = \omega_0^2 \left(1 + \frac{S_0}{1 + \frac{n+1}{n}\varepsilon_M} \right) \qquad (5.46)$$

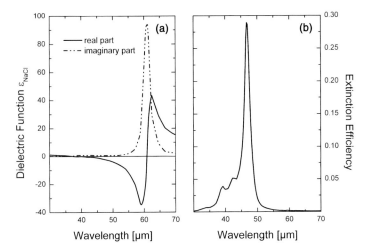

Figure 5.9 (a) Dielectric function ε_{NaCl} of sodium chloride in the wavelength range 5–30 μm. (b) Extinction efficiency of an NaCl sphere with $2R = 200$ nm in the wavelength range 5–30 μm.

where S_0 is the oscillator strength and ω_0 the resonance frequency of the harmonic oscillator.

As the NaCl particle is very small compared with the wavelength, only a TM dipole resonance can be expected in the extinction efficiency. The resonance position at the wavelength $\lambda = 46.9$ μm corresponds exactly to the position where the real part of the dielectric function takes the value $\varepsilon_1 = -2$.

Similar behavior of the dielectric function and the extinction cross-section of a spherical particle can be found for almost all ionic crystal materials, such as KBr, LiF, CaF_2, MgO, and many others. Some examples are shown and discussed later in Section 6.5.

Another example is shown in Figure 5.10 for silicon. Here, the dielectric function [36] clearly shows the behavior resulting from two consecutive harmonic oscillators in the UV spectral region. However, only for the short-wavelength oscillator does ε_1 becomes negative and takes the value $\varepsilon_1 = -2$ at wavelength $\lambda = 127$ nm. The second oscillator is still influenced by the first oscillator, so that the real part ε_1 never becomes negative at longer wavelengths. Looking at the maximum of the extinction efficiency of an Si sphere with $2R = 20$ nm, its maximum appears at the wavelength $\lambda = 172$ nm. This position is far away from the position where $\varepsilon_1 = -2$. Calculations with smaller particles (down to $2R = 2$ nm) indicated that the wavelength of the extinction efficiency maximum shows a strong wavelength dependence. The reasons for the difference compared with the previous example NaCl are as follows:

- a stronger influence of the oscillator strength S of this harmonic oscillator
- a stronger influence of the imaginary part ε_2 on the resonance position of the particle.

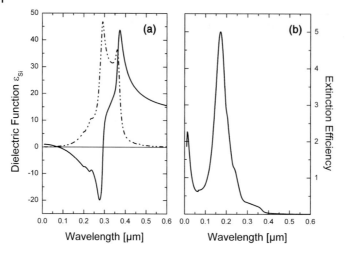

Figure 5.10 (a) Dielectric function ε_{Si} of silicon in the wavelength range 0.01–0.6 μm. (b) Extinction efficiency of an Si sphere with $2R = 20$ nm in the wavelength range 0.01–0.6 μm.

The imaginary part results not only in damping but also in a red shift, that is, a shift of the resonance position to longer wavelengths.

5.3.2.2 Surface Plasmon Polariton Resonances

The most important condition for excitation of *surface plasmon polaritons* (SPPs) is the existence of free charge carriers, which overall act as a polarizable medium. Then, the entirety of free charge carriers becomes excited by the incident light. Free charge carriers can be found in metal particles, but also in (doped) semiconductors and semimetals.

A pronounced SPP resonance can be found for aluminum, alkali metal, and noble metal nanoparticles. An example is given in Figure 5.11 for Na, Ag, Au, and Al nanospheres with $2R = 10$ nm. For discussion of Figure 5.11 we look at the dielectric functions of bulk Al, Na, Ag, and Au in Figure 5.12. The horizontal dotted line indicates the condition $\varepsilon_1 = -2$.

For all four metals only one dipole ($n = 1$) resonance is obtained for $\varepsilon_1 = -2\varepsilon_M$. This condition is fulfilled at different wavelengths, indicating that the wavelength position is determined not only by the free carriers but also by the interband transitions in the metal. For a free carrier gas the resonance position ω_n of the nth multipole follows from Equation 5.42 as

$$\omega_n = \frac{\omega_p}{\sqrt{1 + \frac{n+1}{n}\varepsilon_M}} \tag{5.47}$$

with a width Γ of [93]

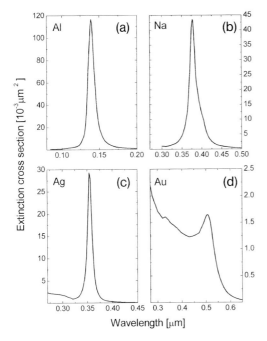

Figure 5.11 Computed extinction cross-section spectra of spherical metal particles with $2R = 10\,\text{nm}$, showing an SPP (resonance of the TM-dipole mode). (a) Na; (b) Ag; (c) Au; (d) Al. Optical constants from [36, 37, 92].

$$\Gamma = \frac{2\varepsilon_2}{\sqrt{\left(\dfrac{\partial \varepsilon_1}{\partial \omega}\right)^2 + \left(\dfrac{\partial \varepsilon_2}{\partial \omega}\right)^2}} \tag{5.48}$$

The derivatives are taken at $\omega = \omega_{\text{max}}$. In most cases it is $(\partial \varepsilon_2/\partial \omega)^2 \ll (\partial \varepsilon_1/\partial \omega)^2$. For example, for $n = 1$ and $\varepsilon_M = 1$ (vacuum) $\omega_1 = \omega_p/\sqrt{3}$.

The interband transitions contribute to this position in a different way. For alkali metals their influence is perceptible only for χ_1^{IB} and is negligible for χ_2^{IB}. The prediction of the simple free electron theory yields a volume plasmon energy of $\hbar\omega_p = 5.95\,\text{eV}$ for Na [94], giving $\hbar\omega_1 = 3.44\,\text{eV}$ for the dipolar SPP. The corrections for the core polarizability and the effective electron mass [95, 96] shift the volume plasmon energy to $5.6\,\text{eV}$ and the spherical SPP to $3.23\,\text{eV}$. This is in quantitative agreement with the value of $3.16\,\text{eV}$ obtained from evaluation of the spectrum for an Na nanoparticle in Figure 5.11b. The dipolar surface plasmon polariton is very pronounced since damping is small in the visible spectral region. In potassium the analog values are $\hbar\omega = 4.29\,\text{eV}$ (SPP: $\hbar\omega = 2.48\,\text{eV}$) for the free electron and $\hbar\omega = 3.86\,\text{eV}$ (SPP: $\hbar\omega = 2.23\,\text{eV}$) including core polarization. The measured dielectric function [92] yields $\hbar\omega = 2.27\,\text{eV}$. Hence no influences of interband absorption χ_2^{IB} are apparent for the alkali metals within the accuracy of the experimental data.

96 | *5 Mie's Theory for Single Spherical Particles*

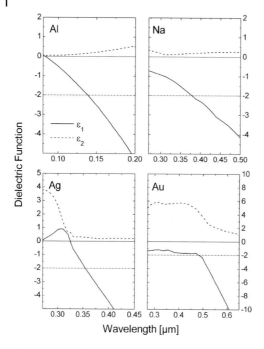

Figure 5.12 Dielectric constants of Na, Ag, Au, and Al [36, 37, 92] in the same wavelength range as in Figure 5.11.

In contrast to the alkali metals, Ag shows pronounced deviations. The contribution χ^{IB} of interband transitions from the 4d to the 5sp band has an enormous influence on the positions of the plasmons. The volume plasmon energy $\hbar\omega_p$ shifts from the free electron value of about 9.2 eV to 3.8 eV and the corresponding SPP resonances from $\hbar\omega \approx 5.6$ eV to ~3.3 eV. The SPP resonances in Ag therefore cannot be regarded as only free electron resonances even as a rough approximation, but rather are hybrid resonances resulting from cooperative behavior of both the 4d and the conduction electrons. This effect is even stronger in the other noble metals. In copper, the free electron Drude volume plasmon energy $\hbar\omega \approx 9.3$ eV almost coincides with that of silver, yet their SPP resonances differ strongly. For Ag, the interplay between the negative ε_1 of the conduction electrons and the positive susceptibility χ_1^{IB} of the interband transitions shifts the SPP value beyond the interband transition edge at 3.9 eV. Below the edge, $\varepsilon_2(\omega)$ is small enough to permit still a sharp resonance (see the spectrum of an Ag nanoparticle in Figure 5.11c). In contrast, the corresponding shift in Cu leads to an SPP at 2.1 eV, where $\varepsilon_1 = -6$. The deviation from the $\varepsilon_1 = -2$ condition is forced by the large and strongly frequency-dependent contributions of ε_2. In fact, the resonance of Cu nanoparticles in vacuum coincides approximately with the minimum of $\varepsilon_2(\omega)$. Summarizing, the noble metals are distinguished from the alkali metals by core influences which are orders of magnitude larger. In Ag, χ_1^{IB} is most effective, for example,

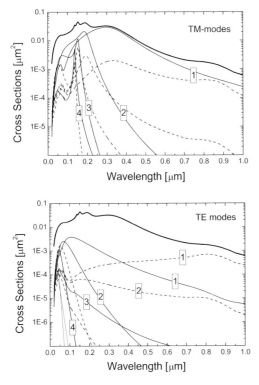

Figure 5.13 Decomposition of the extinction spectrum of an aluminum nanoparticle with $2R = 100$ nm (thick solid line) into TM and TE contributions. The solid lines belong to the scattering and the dashed lines to the absorption.

$|\chi_1^{IB}|/|\chi_1^{Drude}| = 0.54$ at 3.6 eV whereas in Li the same value is smaller than 4% at the resonance position of 3.45 eV [97]. In addition, in Cu and Au, χ_2^{IB} is of comparable importance. An example of the SPP in Au nanoparticles is shown in Figure 5.11d.

Higher order surface plasmon resonances for multipole order $n > 1$ can be resolved for larger nanoparticles They are, however, not further determined only by the electronic properties but also by the size of the particle. As an example, we present spectra of aluminum nanoparticles with $2R = 100$ nm. We decomposed the total extinction spectrum into the contributions of TM and TE partial waves, and moreover into the contribution of scattering and absorption. The results are shown in Figure 5.13.

The total extinction reveals a complex multipeak structure, which results from the various multipole contributions of SPPs (resonances of the TM multipoles) to the absorption and scattering. In particular, structures up to the order $n = 4$ are present. For the large particles, scattering outweighs absorption by far; for very small clusters, however, scattering would be negligible.

The dipole absorption and scattering contributions peak around 310 nm (4 eV), which is far away from the resonance at 155 nm (8 eV) that would be expected from the condition in Equation 5.47 for $n = 1$. This exemplifies the huge shift of the multipole modes due to retardation, that is, the size dependence of the resonance conditions, as given by Equations 5.38 and 5.39 for a compact sphere.

INFO – Plasmonic Excitations in Small Spheres

The resonance condition for the TM dipole mode a_1, Equation 5.40 for $n = 1$, is discussed in almost all relevant publications as the condition for the excitation of the collective excitation of free electrons at the surface of a sphere – the surface plasmon polariton. However, what then is the difference from the TM dipole mode which is also resonant at the same condition?

Following Raether's derivation [98, 98a] for the surface plasmon of a plane surface, the surface plasmon is a solution of Maxwell's equations. For excitation with light the dispersion curve must cross the light straight. At this frequency it is possible to excite the plasmon polariton.

For further discussion, we consider an infinitely long cylinder. Then, we can distinguish three cases, all following from the same electrodynamics, the Maxwell equations in cylindrical coordinates:

1) light scattering and absorption
2) light propagation in the wire
3) surface waves (Lecher wires).

The differences lie in the boundary conditions for the Maxwell differential equations which lead in all three cases to different solutions.

The first case presumes an incident plane wave which generates a wave propagating through the cylinder and generates a scattered wave which propagates outside the cylinder from the cylinder surface to the far-field.

In the second case, how the waves have been generated is neglected. The solutions of the Maxwell equations will be waves that propagate along the z-axis. Inside the cylinder they are periodic in r, θ, and z. Outside the cylinder they must decrease exponentially in the radial direction as $\exp(-\beta r)$. The solutions are the modes of a cylindrical waveguide. The most important difference from case 1 is that the modes are not coupled to waves outside the cylinder.

In the third case, again how the waves have been generated is neglected. The solutions of the Maxwell equations will be waves that propagate along the z-axis, but they should decrease exponentially in the radial direction both inside and outside the cylinder. These solutions are known as surface-guided waves. Also in this case the modes are not coupled to waves outside the cylinder.

Applying the three cases to a sphere, the first case corresponds exactly to Mie's theory in spherical coordinates with TM and TE multimodes. The resonance of the TM dipole mode in the Rayleigh approximation is given by

Equation 5.40. For the second and third cases, we do not yet know of any published solutions in spherical coordinates. The third case means for a sphere that the solution is periodic in θ and φ but decreases exponentially in the radial direction inside and outside the sphere. This solution should correspond to the SPP for metallic spheres. Consequently, a TM solution in addition to a TE solution should be apparent.

Coming back again to the nanoparticles, it is easy to show that Equation 5.40 is fulfilled for particles with diameter $2R < 15$ nm. Comparing this size with the penetration depth of light in metals (see Table 3.1) it follows that such small particles are only surface for the light. Therefore, it is self-evident that for these particles the TM dipole resonance and an SPP resonance cannot be distinguished. For larger particles for which the TM dipole mode becomes size dependent it is not clear whether the SPP also becomes size dependent. The reason is that, so far, obviously a solution of case 3 does not exist that corresponds to an SPP also for metallic spheres of larger size of micrometers or even millimeters in diameter.

At the end of this section, we discuss three further aspects of nanoparticles which can be best recognized from SPPs: (1) the thermodynamic phase of the particles, (2) alloys, and (3) the concentration of free electrons.

Thermodynamic Phase The optical material functions of a metal can be very different for the solid and the liquid phases, hence giving rise to different extinction spectra. This is demonstrated in a series of spectra of Au nanoparticles of $2R = 5$ nm at increasing temperature up to the liquid phase in Figure 5.14a, using optical functions from the literature [99].

Whereas the solid Au nanoparticles show a distinct resonance which becomes broader with increasing temperature, for the free liquid particles it is completely smeared out. However, increasing the ε_M of the surrounding medium causes the peak of the liquid clusters to appear again, but with a large halfwidth.

Another example is given in Figure 5.14b for mercury. Mercury has the extraordinary property of being liquid at room temperature. The bulk optical functions for solid and liquid mercury [100] are well known. Hg nanoparticles with $2R = 5$ nm exhibit a well-defined SPP resonance even in the liquid state, which, however, is blue shifted compared with the solid particle peak, indicating a higher concentration of free electrons in the liquid phase of Hg.

Alloys A further group of nanoparticles which show SPPs are alloy nanoparticles of Ag and Au. Their analysis yields information about particular alloy properties such as the concentrations of the constituents, the miscibility, and so on. As an example, we present extinction efficiency spectra in Figure 5.15 calculated with measured dielectric functions of Ag_xAu_{1-x} nanoparticles with x varying between 0.2 and 0.9 [38]. The spectra clearly show one plasmon polariton peak which shifts

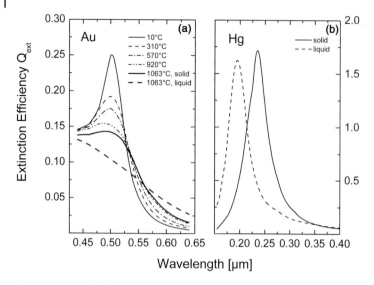

Figure 5.14 (a) Calculated extinction efficiency spectra of Au nanoparticles ($2R = 5$ nm) for the solid phase at different temperatures and the liquid phase. In the liquid phase, the TM dipole resonance (surface plasmon) is damped away. (b) Calculated extinction efficiency spectra of solid and liquid Hg nanoparticles ($2R = 5$ nm).

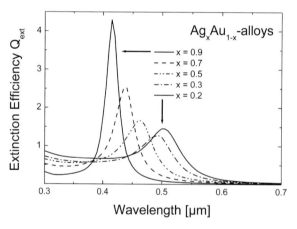

Figure 5.15 Calculated extinction spectra of Ag_xAu_{1-x} alloy nanoparticles with varying concentration x and size $2R$. The values (x, $2R$ [nm]) were chosen to be (0.9, 9), (0.7, 11), (0.5, 12) (0.3, 13), and (0.2, 15) in order to allow comparison with experimental results.

monotonically from the position due to pure Ag to the position due to pure Au nanoparticles when x is varied between 0 and 1. Hence the peak positions directly reflect x and can be used to determine x from experimental spectra. The spectral shape, in particular the high-frequency part, contains information about the alloy band structure.

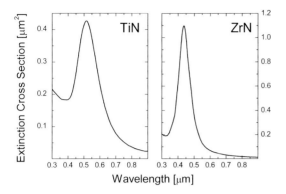

Figure 5.16 Extinction efficiencies of a titanium nitride and a zirconium nitride nanoparticle with $2R = 20$ nm.

Semimetals In the three transition metal nitrides TiN, ZrN, and HfN, the metals form highly directional bondings with the nitrogen by hybridization. A characteristic and unusual result of this intricate bonding is that these compounds are *ceramic* in terms of hardness, inertness, and directed bonds, but are *metallic* with characteristically high electrical and thermal conductivity.

Optical applications of materials from this group are less important, although especially TiN is used as a cosmetic coating to replace gold in situations when a noble metal would not be sufficiently wear resistant. ZrN and HfN have a less yellow but still bright appearance.

Due to their metallic character, nanoparticles of these materials should also show an SPP resonance in extinction and scattering spectra. We show as examples in Figure 5.16 the extinction spectra of TiN and ZrN nanoparticles with $2R = 20$ nm in vacuum (optical constants from [101, 102]). The extinction spectra can be divided into two parts. At short wavelengths, interband transitions from occupied electronic levels into unoccupied electronic levels contribute to the absorption of light by the particles, with an onset wavelength of these interband transitions lying for TiN in the violet–blue spectral region, leading to the characteristic golden color of bulk TiN. The interband transitions are followed at longer wavelengths by an extinction band with the maximum at $\lambda = 516$ nm for TiN and at $\lambda = 436$ nm for ZrN. This extinction band can be interpreted for TiN as SPP [101]. The scattering is negligible compared with the absorption. For ZrN, we assume that the extinction band can also be interpreted as SPP, since the electronic structures of TiN and ZrN are similar.

5.3.2.3 Multiple Resonances

In the following, we discuss the case that multiple resonances occur which may be either electronic resonances or SPP resonances. This may happen if the dielectric function of the particle material exhibits either multiple separated harmonic oscillator contributions or harmonic oscillator contributions plus a Drude

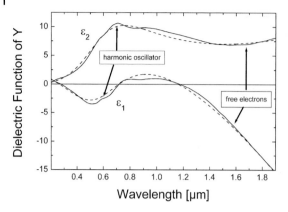

Figure 5.17 Dielectric function of yttrium.

susceptibility from free carriers. An example in which both material resonances contribute to the extinction and scattering of light by nanoparticles is graphitic nanoparticles. Graphite exhibits two sharp resonances in the UV region around wavelength 200 nm which can be assigned to separated electronic interband transitions. Their contribution to the dielectric constant of graphite can be approximated by harmonic oscillators. Graphite, however, is also conductive, with a conductivity much less than that of metals. Hence the plasma frequency ω_p lies in the FIR region. In this spectral range at frequencies $\omega < \omega_p$ the contribution of the transitions in the UV region to the dielectric constant is constant, and the free electrons now determine the dielectric function that becomes negative. In spectra of graphitic nanoparticles, electronic resonances in the UV region and also an SPP in the FIR region can be observed in extinction spectra, as will be demonstrated later in Chapter 6.

Here, we give another example with the metal yttrium. The dielectric function of yttrium in the wavelength range 0.2–1.9 μm can be composed of a harmonic oscillator contribution and a free electron contribution. This can be seen from Figure 5.17, in which the measured dielectric function [103] is compared with a model dielectric function. The real part ε_1 becomes negative between 0.35 and 0.65 μm whereas the imaginary part ε_2 exhibits a broad maximum at $\lambda = 0.65$ μm. In this spectral region the dielectric function is similar to that of a harmonic oscillator. This oscillator can be assigned – but not ambiguously – to a broad interband transition between the parallel electronic levels in the direction from T to T′ extending out of K in the hexagonal close-packed (hcp) lattice of bulk yttrium [103]. At longer wavelengths >1.2 μm the real part ε_1 of the dielectric function of yttrium is negative and decreases with increasing wavelength, whereas the imaginary part ε_2 increases with increasing wavelength. This behavior of ε is characteristic for the Drude susceptibility of free electrons.

The experimental data can be modeled with a dielectric function of a harmonic oscillator and the Drude susceptibility of free electrons:

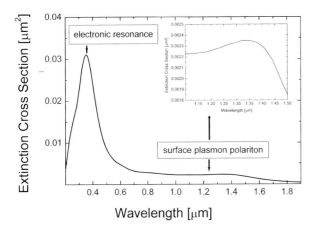

Figure 5.18 Extinction efficiency of an yttrium nanoparticle with $2R = 20\,\text{nm}$.

$$\varepsilon(\omega) = 1 + \chi_\infty + \frac{S_{IB}\omega_{IB}^2}{(\omega_{IB}^2 - \omega^2 - i\omega\gamma_{IB})} - \frac{\omega_P^2}{\omega^2 + i\omega\gamma_{fe}} \quad (5.49)$$

where S_{IB}, ω_{IB}, and γ_{IB} are the oscillator strength, the frequency, and the damping constant of the harmonic oscillator, respectively, and ω_P^2 the plasma frequency and γ_{fe} the damping constant of the free electron contribution, respectively. Their values are $S_{IB} = 9.63$, $\omega_{IB} = 2.9 \times 10^{15}\,\text{s}^{-1}$, $\gamma_{IB} = 3 \times 10^{15}\,\text{s}^{-1}$, $\omega_P = 5.2 \times 10^{15}\,\text{s}^{-1}$ and $\gamma_{fe} = 1.5 \times 10^{14}\,\text{s}^{-1}$. The quantity χ_∞ amounts to $\chi_\infty = 1$ in our approximation of the bulk dielectric function of yttrium in the considered spectral range.

The real part satisfies the condition $\varepsilon_1 = -2$ three times: for the harmonic oscillator at $\lambda = 0.401\,\mu\text{m}$ (3.092 eV) and $\lambda = 0.646\,\mu\text{m}$ (1.919 eV), and for the free electrons at $\lambda = 1.335\,\mu\text{m}$ (0.929 eV). Therefore, there should be three resonances in the extinction cross-section of yttrium nanoparticles, with one of them being an SPP at long wavelengths. To prove this assumption, the calculated extinction efficiency for an yttrium nanoparticle with diameter $2R = 20\,\text{nm}$ is plotted in Figure 5.18.

The spectrum exhibits a sharp resonance-like feature at $\lambda = 0.352\,\mu\text{m}$ (3.522 eV), which can be assigned to the harmonic oscillator, and a very broad feature of low magnitude at about $\lambda = 1.34\,\mu\text{m}$, which can be assigned to the SPP. Obviously, as in the case of silicon in Figure 5.10, the imaginary part of the dielectric function of yttrium is of great importance for the positions and the widths of the resonances. The resonance expected at $\lambda = 0.402\,\mu\text{m}$ is blue shifted to $\lambda = 0.352\,\mu\text{m}$ due to the increasing imaginary part ε_2. Moreover, the expected resonance at $\lambda = 0.646\,\mu\text{m}$ is damped away.

Another example is the metal tantalum. In the wavelength range $0.1–1\,\mu\text{m}$ the dielectric function can be composed of two successive harmonic oscillators and a Drude susceptibility for the free electrons in the metal (see Figure 5.19a). The first harmonic oscillator leads to the condition $\varepsilon_1 = -2$ at two wavelengths $\lambda_1 = 167\,\text{nm}$ and $\lambda_2 = 226\,\text{nm}$ in the UV spectral region. In consequence, the extinction by a

104 | 5 Mie's Theory for Single Spherical Particles

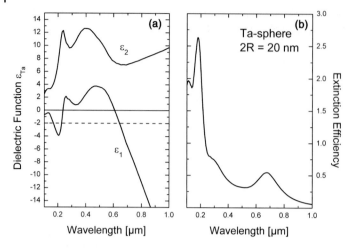

Figure 5.19 (a) Dielectric function ε_{Ta} of tantalum in the wavelength range 0.1–1 µm [36]. (b) Extinction efficiency of a Ta sphere with 2R = 20 nm in the wavelength range 0.1–1 µm.

spherical Ta particle with $2R = 20$ nm exhibits an electronic resonance at $\lambda = 179$ nm, which is close to λ_1 but red shifted due to the high imaginary part ε_2 in this wavelength region. The condition $\varepsilon_1 = -2$ is also fulfilled at a third wavelength $\lambda_3 = 643$ nm, due to the contribution of the free carriers to the dielectric function. The corresponding resonance at $\lambda = 677$ nm is clearly to recognize and can be assigned to the SPP. It is also red shifted because of the high imaginary part ε_2 in this wavelength region.

5.3.3
Longitudinal Plasmon Resonances

We already mentioned in Section 4.2 that the Maxwell equations and the vector wave equation have three solutions **M**, **N**, and **L**, where **M** and **N** represent transverse fields and **L** is the solution for longitudinal fields. Transverse waves propagating in the particle material with wavenumber k are described with a dielectric function $\varepsilon(\omega)$ that satisfies the dispersion relation

$$k^2 = \frac{\omega^2}{c^2}\varepsilon(\omega) \tag{5.50}$$

Longitudinal waves with wavenumber k_L cannot be described with this dielectric function, but are described with the dielectric function $\varepsilon_L(\omega, k_L)$ that satisfies

$$\varepsilon_L(\omega, k_L) = 0 \tag{5.51}$$

If longitudinal waves can propagate in the particle material in a certain frequency range, the function **L** describes longitudinal electron density fluctuations in the material, the *longitudinal bulk plasmon modes*.

5.3 Resonances

At these frequencies, the TM fields interior the particle are coupled to the bulk plasmons, whereas the TE fields are not affected:

$$E = E_0 \sum_{n=1}^{\infty} \frac{i^n(2n+1)}{n(n+1)} \left[\beta_n M_{on1}^{(1)}(k) - i\alpha_n N_{en1}^{(1)}(k) + i\gamma_n L_{en1}^{(1)}(k) \right] \quad (5.52a)$$

$$H = \frac{-kE_0}{\omega\mu_0} \sum_{n=1}^{\infty} \frac{i^n(2n+1)}{n(n+1)} \left[\alpha_n M_{en1}^{(1)}(k) + i\beta_n N_{on1}^{(1)}(k) \right] \quad (5.52b)$$

The incident and scattered waves remain unchanged.

To resolve the unknown scattering coefficients a_n and b_n and the coefficients α_n, β_n, and γ_n of the internal wave, Maxwell's boundary conditions

$$(E_{inc} + E_{sca}) \times n\big|_{surface} = E \times n\big|_{surface} \quad (5.53)$$

$$(H_{inc} + H_{sca}) \times n\big|_{surface} = H \times n\big|_{surface} \quad (5.54)$$

are applied again. However, due to the additional longitudinal field with expansion coefficients γ_n they must be supplemented by the *additional boundary condition* (abc) including the continuity of the normal component of E:

$$(E_{inc} + E_{sca}) \cdot n\big|_{surface} = E_i \cdot n\big|_{surface} \quad (5.55)$$

After resolving for the coefficients, the modified coefficient a_n of the TM scattered field now reads [76–78]

$$a_n = \frac{\psi_n(x)\psi_n'(mx) - m\psi_n'(x)\psi_n(mx) + C_n \psi_n(x)\psi_n(mx)}{\xi_n(x)\psi_n'(mx) - m\xi_n'(x)\psi_n(mx) + C_n \xi_n(x)\psi_n(mx)} \quad (5.56)$$

with

$$C_n = \frac{n(n+1)}{mx} \frac{j_n(m_L x)}{m_L x \cdot j_n'(m_L x)} (m^2 - 1) \quad (5.57)$$

and $m_L = k_L/k_{inc}$.

So far, we have assumed the transverse dielectric function of the particle material to be local, that is, it is a function only of the frequency ω and does not depend on the propagation constant k. This means that energy transport is restricted to electromagnetic waves. Nonlocal dielectric functions include spatial dispersion, that is, $\varepsilon(\omega,k)$. For a free electron-like metal, this dielectric function is as derived by Lindhard:

$$\varepsilon_T(\omega, k_T) = 1 - \frac{\omega_P^2}{\omega^2 + i\omega\gamma_{fe}} \frac{3}{2a^2} \left[\frac{1+a^2}{a} \tan^{-1}(a) - 1 \right] \quad (5.58)$$

while the corresponding longitudinal dielectric function reads

$$\varepsilon_L(\omega, k_L) = 1 - \frac{\omega_P^2}{\omega^2 + i\omega\gamma_{fe}} \frac{3}{a^2} \left(1 - \frac{\tan^{-1}(a)}{a}\right) \cdot \left(1 + i\frac{\gamma_{fe}}{\omega}\left(1 - \frac{\tan^{-1}(a)}{a}\right)\right)^{-1} \quad (5.59)$$

with

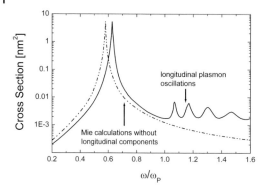

Figure 5.20 Calculated extinction spectrum of a pure Drude metal nanoparticle with consideration of longitudinal plasmon modes.

$$a^2 = -k_L^2 \frac{v_F^2}{(\omega + i\gamma_{fe})^2} \quad (5.60)$$

and v_F = Fermi velocity in both equations.

These modifications of the scattering coefficients were considered by Ruppin [77] using a pure Drude dielectric constant in Mie calculations on nanoparticles. They result in only minor optical effects, as can be seen in Figure 5.20. The spectrum at low frequencies looks more or less the same as the original Mie theory spectrum; however, a slight shift of the surface plasmon peak towards higher frequencies occurs.

A second difference is the onset of – now allowed – longitudinal plasmon oscillations at frequencies above the Drude plasma frequency according to Equations 5.58 and 5.59. According to Ruppin, the blue shift and the spacing of the oscillations increase with decreasing particle size. So far, however, these effects have not yet been identified either experimentally or in calculations with real dielectric functions of metals.

INFO – Magnetic Resonances

For discussion of *magnetic* resonances in metal nanoparticles, we consider nanoparticles in the quasi-static approximation. This means that only the TM dipole mode a_1 and the TE dipole mode b_1 contribute to the extinction cross-section:

$$\sigma_{ext} = \frac{6\pi c^2}{\omega^2} \mathrm{Re}(a_1 + b_1) \quad (5.61)$$

For simplicity, the surrounding medium is assumed to be vacuum ($\varepsilon_M = 1$).

In the quasi-static approximation, the TM dipole mode a_1 and the TE dipole mode b_1 are given as

$$a_1 = -\frac{2}{3}i\left(\frac{\omega R}{c}\right)^3 \frac{\varepsilon-1}{\varepsilon+2} \tag{5.62}$$

$$b_1 = -\frac{1}{45}i\left(\frac{\omega R}{c}\right)^5 (\varepsilon-1) \tag{5.63}$$

with $\varepsilon(\omega) = \varepsilon_1(\omega) + i\varepsilon_2(\omega)$ being the complex dielectric function of the metal. Remember that at frequencies from the FIR to infinity we have $\mu = \mu_M = 1$, since magnetic dipoles are too inert to follow the high-frequency magnetic field of electromagnetic radiation in this spectral range and, hence, no magnetization occurs.

Inserting these relations into Equation 5.61 yields

$$\sigma_{ext} = 12\pi R^3 \underbrace{\frac{\omega}{c}\frac{\varepsilon_2}{(\varepsilon_1+2)^2+\varepsilon_2^2}}_{\text{TM dipole contribution}} + \underbrace{\frac{2\pi}{15}R^5\left(\frac{\omega}{c}\right)^3 \varepsilon_2}_{\text{TE dipole contribution}} \tag{5.64}$$

Without any restriction of the generality, we assume now that the dielectric function $\varepsilon(\omega)$ of the metal particle can be described by a pure Drude dielectric function with plasma frequency ω_p and damping constant γ_{fe}. This holds true for almost all metals at frequencies in the IR region down to $\omega = 0$, except for an additional contribution χ_∞ in the real part ε_1.

We consider now two cases: (1) $\omega \gg \gamma_{fe}$ and (2) $\omega \ll \gamma_{fe}$.

1. $\omega \gg \gamma_{fe}$ (UV–VIS–NIR)

In this case, the dielectric function is approximately

$$\varepsilon_1(\omega) \approx 1 - \frac{\omega_p^2}{\omega^2}, \varepsilon_2(\omega) \approx \frac{\gamma \omega_p^2}{\omega^3} \tag{5.65}$$

Inserting this equation in Equation 5.64, it follows that the TM contribution is proportional to ω^2, $a_1 \propto \omega^2$, because ε_1 dominates the denominator in the TM dipole contribution, and the TE contribution is almost constant, $b_1 \approx$ constant. Note that the TM dipole mode additionally exhibits a resonance at $\omega = \omega_p/\sqrt{3}$, the SPP. Moreover, the frequency-independent prefactor of b_1 is about two orders of magnitude lower than the prefactor of a_1.

2. $\omega \ll \gamma_{fe}$ (FIR)

In this case, the dielectric function is approximately

$$\varepsilon_1(\omega) \approx 1 - \frac{\omega_p^2}{\gamma^2} = \text{constant}, \varepsilon_2(\omega) \approx \frac{\omega_p^2}{\gamma}\frac{1}{\omega} \tag{5.66}$$

Inserting this equation in Equation 5.64 it follows that now both the TM contribution and the TE contribution are proportional to ω^2. Furthermore, the magnitude of b_1 is slightly higher than that of a_1, as the TE dipole contribution now dominates the total spectrum. These results are depicted in Figure 5.21.

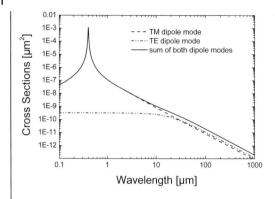

Figure 5.21 Contributions of the TM and TE dipole of a small nanoparticle with Drude dielectric constant.

5.4
Optical Contrast

Resonances of nanoparticles of some materials, such as the SPP resonance of Cu nanoparticles, may be strongly damped away when lying close to interband transitions or other electronic resonances. Then, the theoretical extinction cross-sections show less characteristic features provided that the particles are very small compared with the wavelength. Consequently, optical spectroscopy of nanoparticles in this spectral range as an analytical tool is of limited value if only particles in vacuum are considered. A considerable improvement is possible, however, by embedding the nanoparticles in transparent matrices with $\varepsilon_M > 1$. Then, the optical contrast $\varepsilon - \varepsilon_M$ between particle material and surrounding matrix material is increased. A compilation of actual ε_M values for various matrices is given in Table 5.1.

As a first example, we consider noble metal nanoparticles with $2R = 10$ nm which exhibit an SPP resonance (Figure 5.22). Optical constants were taken from the literature [37]; ε_M amounts to $\varepsilon_M = 1, 3, 5, 7$, or 10 over the whole spectral range.

With increasing ε_M, the SPP resonance shifts to longer wavelengths (red shift), and the magnitude is drastically increased combined with a smaller full width at half-maximum (FWHM) of the peaks. Particularly the weak resonance of Cu nanoparticles in vacuum can be markedly enhanced by shifting it away from the interband transition region using matrices of high ε_M. These results agree with the theoretical results for halfwidth and resonance position of the surface plasmon polariton resonances in Section 5.3.2.2.

The second example is for the metals Y and Ta, where we have both an electronic resonance and an SPP, and the semiconducting material Si, where we have only

Table 5.1 Refractive index n_M and corresponding dielectric constant ε_M of solid transparent matrix materials. All data are at 589 nm.

Surrounding material	n_M	ε_M
Vacuum	1.0	1.0
H$_2$O	1.33	1.78
Various crown glasses	1.5–1.55	2.25–2.4
Various flint glasses	1.6–1.9	2.56–3.61
Fused silica	1.46	2.13
Polycarbonate	1.59	2.53
CaF$_2$	1.43	2.05
MgF$_2$	1.38	1.91
Si$_3$N$_4$	2.03	4.12
Gelatin	1.54	2.37
MgO	1.74	3.03
Al$_2$O$_3$	1.77	3.13
TiO$_2$	2.03	4.12
CeO$_2$	2.47	6.1
ZnO	2.0	4.0

an electronic resonance. The corresponding spectra are shown in Figure 5.23. Optical constants are taken from [36] for Ta and Si and from [103] for Y.

The SPP resonance in Y and Ta at long wavelengths again shows a red shift with increasing ε_M and an increase in its magnitude. The electronic resonance in Y, Ta, and Si also becomes red shifted with increasing ε_M. However, as the condition $\varepsilon = -2\varepsilon_M$ can be reached only in a certain wavelength range, the possible position for the electronic resonance is restricted to this wavelength range. Moreover, its magnitude can decrease again with increasing ε_M. In contrast, the SPP resonance shifts more to longer wavelengths the higher ε_M becomes, because the real part ε_1 decreases continuously with increasing wavelength.

If MDRs contribute to the spectrum, as for highly refractive particle materials, the effect of increasing ε_M is opposite to the effect for SPP resonances: with increasing ε_M first the number of resonances is reduced, and second the resonances become shifted to smaller wavelengths (blue shift). The reason is that for particles of highly refractive materials, the optical contrast $\varepsilon - \varepsilon_M$ between the particle material and the surrounding disappears with increasing ε_M. In this case, the interaction of the particles with the incident light is strongly reduced.

Experimentally, the effect of optical contrast was demonstrated to a certain extent by Underwood and Mulvaney [104] for gold colloids prepared in different liquid media. The change in ε_M from 1.78 for water to 2.505 for CS$_2$ led to a shift of the plasmon polariton peak from wavelength 520 to 545 nm with a corresponding change in the apparent color of the colloid from red to purple.

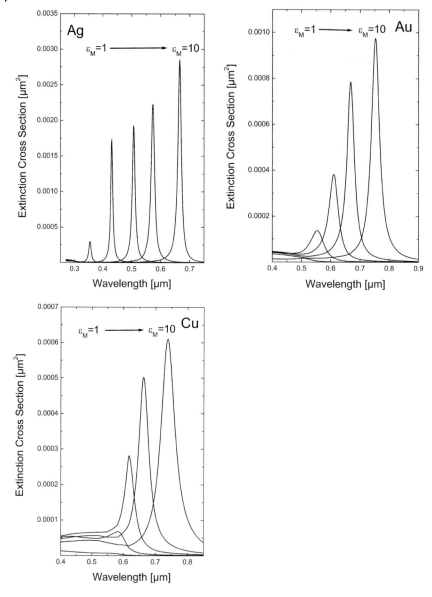

Figure 5.22 Cu, Ag, and Au nanoparticles with $2R = 10$ nm embedded in various dielectric matrices. The dielectric constant ε_M of the transparent matrix material varies from 1 to 10.

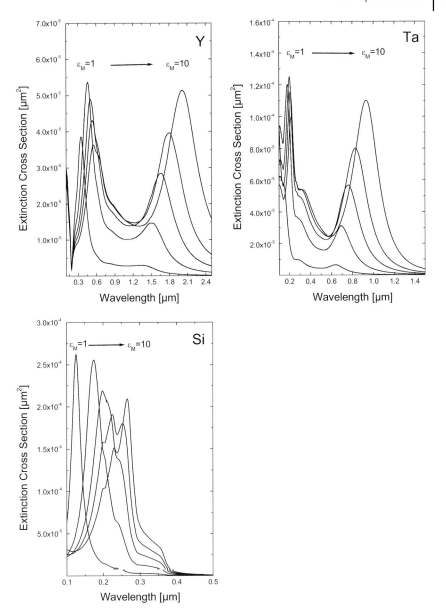

Figure 5.23 Metal and semiconductor nanoparticles ($2R = 10$ nm) embedded in transparent dielectric matrices with ε_M varying from 1 to 10.

5.5
Near-Field

In common experimental set-ups and in technical applications, measurements of absorption of light and elastic light scattering are carried out in the far-field zone, that is, far away from the surface of the particle. The local electromagnetic fields on the surface or in the vicinity of a small particle play an important role for SERS (surface-enhanced Raman scattering), where they are used to enhance Raman scattering of light by adsorbed molecules, and also for nonlinear optical properties of small particles. The optical near-field is also a central topic in scanning near-field optical microscopy (SNOM), where evanescent waves are used.

In Section 5.2, we showed that the cross-sections σ_{ext} and σ_{sca} and the efficiencies Q_{ext} and Q_{sca} for extinction and scattering, respectively, are a measure solely of the ability to extinguish and to scatter an electromagnetic wave at incident wavelength λ to a certain extent. They do not provide information about the strength of the electromagnetic fields nearby or far from the particle at distance $r \geq R$ from the particle center, with $2R$ being the particle diameter. This information is contained in the *near-field efficiencies* Q_{NF} and Q_R introduced in 1981 by Messinger et al. [105]:

$$Q_{NF}(r) = \frac{r^2}{\pi R^2 I_0} \int_0^{2\pi} \int_0^{\pi} E_{sca}(r) \cdot E_{sca}^*(r) \sin\theta \, d\theta \, d\varphi \tag{5.67}$$

$$Q_R(r) = \frac{r^2}{\pi R^2 I_0} \int_0^{2\pi} \int_0^{\pi} E_{r,sca}(r) \cdot E_{r,sca}^*(r) \sin\theta \, d\theta \, d\varphi \tag{5.68}$$

Q_{NF} represents the square of the spatially averaged electric field of the scattered wave, and Q_R represents the square of only the spatially averaged radial component E_r of the electric field of the scattered wave, which is included in Q_{NF}. It is mainly this radial component E_r that determines the difference between the far-field and the near-field. E_r is excluded from the energy flux of the scattered wave through the fictitious shell around the particle when calculating Q_{sca}. The explicit calculation of Q_{NF} and Q_R according to Equations 5.67 and 5.68 yields

$$Q_{NF}(r) = 2\frac{r^2}{R^2} \sum_{n=1}^{\infty} \left\{ |a_n|^2 \left[(n+1)|h_{n-1}^{(1)}(k_M r)|^2 + n|h_{n+1}^{(1)}(k_M r)|^2 \right] \right.$$
$$\left. + (2n+1)|b_n|^2 |h_n^{(1)}(k_M r)|^2 \right\} \tag{5.69}$$

$$Q_R(r) = 2\frac{2}{k_M^2 R^2} \sum_{n=1}^{\infty} (2n+1)(n+1)n|a_n|^2 |h_n^{(1)}(k_M r)|^2 \tag{5.70}$$

where $h_n^{(1)}$ are spherical Hankel functions of first kind and order n and k_M is the wavenumber in the surrounding matrix.

First, we consider the ratio Q_{NF}/Q_{sca} as a function of the particle size. This ratio defines the enhancement of the fields close to the particle surface compared with the far-field. Large enhancement factors are obtained for very small particles, as becomes obvious from Figure 5.24 [106], showing Q_{NF}/Q_{sca} for particles smaller than $2R = 400$ nm at a fixed wavelength $\lambda = 514.5$ nm. Giant electric fields of the scattered wave are obtained for particles with $2R \leq 20$ nm. This result for Q_{NF}/Q_{sca} proved to be *independent of the particle material*, following from many calculations with different materials. For metal particles, however, already Q_{sca} is larger than for dielectric particles of same size, as can be seen from Table 5.2. Then, Q_{NF} also approaches large values for metal particles. A further increase is obtained for nanoparticles exhibiting an SPP. At wavelengths of these resonances Q_{sca} is larger than at off-resonance wavelengths.

Next, we look at the wavelength dependence of Q_{NF}, Q_R, and Q_{sca}. This was done previously by Messinger et al. [105], who discussed the wavelength dependence of Q_{NF} and Q_R at distance $r = R$, that is, at the surface of the spherical particle, for Ag, Au, and Cu nanoparticles of various sizes. We show in Figure 5.25 the $Q_{NF}(\lambda)$, $Q_R(\lambda)$, and $Q_{sca}(\lambda)$ spectra of nanoparticles of Au and Ag with diameter $2R = 10$ nm at distance $r = R$, calculated with optical constants from the literature [40]. For better comparison, Q_{sca} is multiplied with the given factors. $Q_{NF}(\lambda)$ and $Q_R(\lambda)$ exhibit similar spectral features to $Q_{sca}(\lambda)$, interband transitions at short wave-

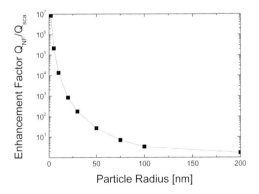

Figure 5.24 Enhancement factor Q_{NF}/Q_{sca} of nanoparticles.

Table 5.2 Values of the dielectric constant and the scattering cross-section of very small particles in the Rayleigh limit of same size and at the same wavelength $\lambda = 514$ nm.

Material	Dielectric constant ε	$Q_{sca} \propto \left|\dfrac{\varepsilon - \varepsilon_M}{\varepsilon + 2\varepsilon_M}\right|^2$
Ag	$-10.3 + i \cdot 0.205$	1.853
GaAs	$17.66 + i \cdot 3.207$	0.725
Si	$17.89 + i \cdot 0.525$	0.72
Si_3N_4	$4.15 + i \cdot 0$	0.263
SiO_2	$2.13 + i \cdot 0$	0.075

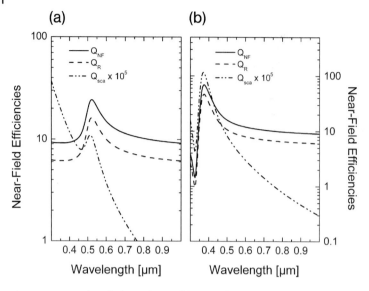

Figure 5.25 Wavelength dependence of the near-field efficiencies of (a) small Au particles and (b) small silver particles with 2R = 10 nm.

lengths, and an SPP resonance. However, there are also clear differences. First, the wavelengths with maximum Q_{NF} and Q_R are slightly red shifted compared with the wavelength at which Q_{sca} is maximum. In detail, Q_{sca} is maximum at $\lambda = 514$ nm for Au and $\lambda = 371$ nm for Ag, whereas Q_{NF} and Q_R are maximum at $\lambda = 522$ nm and $\lambda = 375$ nm, respectively. Moreover, in the interband transition region the near-field efficiencies are almost wavelength independent, in contrast to Q_{sca}. At longer wavelengths, the near-field efficiencies decrease more slowly than Q_{sca}. The reason is that the scattering of light by small particles depends on the particle circumference/wavelength ratio. Then, the effect of an increasing wavelength is similar to the effect of a decreasing particle size at fixed wavelength. Hence, the decrease in Q_{NF} and Q_R with increasing wavelength must be slower than that of Q_{sca}. Indeed, they become approximately constant for large wavelengths. In general, it can be seen that Q_R is the main contributor to Q_{NF}. Q_R amounts to 67% of Q_{NF} for $r = R$. This means that in the near-field the electric field is dominated by the radial component, which is very large.

Finally, the dependence of Q_{NF} and Q_R on the distance $D = r - R$ from the particle surface is shown in Figure 5.26 for a gold nanoparticle with $2R = 100$ nm at two wavelengths $\lambda_1 = 514.5$ nm and $\lambda_2 = 780$ nm. For comparison, Q_{sca} is given as a horizontal line. The behavior of Q_{NF} and Q_R is similar for both wavelengths. Very close to the surface Q_R is the dominant contributor to Q_{NF}. With increasing distance, both quantities decrease. Whereas Q_{NF} approaches Q_{sca}, Q_R decreases more rapidly and amounts to about one-twentieth of Q_{sca} at a distance $D = 0.5\,\mu m$, corresponding in the present example to $D = 10R$. Also for smaller particles many further calculations showed that at $D = 0.5\,\mu m$ Q_R has decreased to approximately one-twentieth of Q_{sca} and, hence, becomes negligible. For the gold particle under

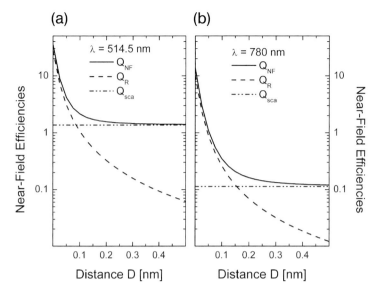

Figure 5.26 Near-field efficiencies of a small Au particle with $2R = 100$ nm as a function of $D = r - R$ at wavelength $\lambda =$ (a) 514 and (b) 780 nm.

consideration, Q_R is in the order of Q_{sca} at $D = 0.076\,\mu\text{m} = 3R$ for $\lambda_1 = 514.5\,\text{nm}$ and $D = 0.15\,\mu\text{m} = 6R$ for $\lambda_2 = 780\,\text{nm}$.

For a quantitative discussion of Figures 5.24–5.26, we assume particles in the Rayleigh approximation. Then, Q_{sca}, Q_{NF}, and Q_R are determined by the TM dipole mode a_1:

$$Q_{sca} = \frac{6|a_1|^2}{(k_M R)^2} \tag{5.71}$$

$$Q_{NF} = \frac{6|a_1|^2}{(k_M R)^2} \frac{(k_M r)^2}{3}\left[2\left|h_0^{(1)}(k_M r)\right|^2 + \left|h_2^{(1)}(k_M r)\right|^2\right] \tag{5.72}$$

$$Q_R = \frac{6|a_1|^2}{(k_M R)^2} 2\left|h_1^{(1)}(k_M r)\right|^2 \tag{5.73}$$

It is obvious that the r dependence of Q_{NF} and Q_R is determined in this approximation by the behavior of the Hankel functions $h_0^{(1)}$, $h_1^{(1)}$, and $h_2^{(1)}$. For small arguments $k_M r$, that is, close to the surface of the particle, the Hankel functions become approximately

$$h_n^{(1)}(k_M r) \approx \frac{\Gamma(0.5)}{2\Gamma(n+1.5)}\left(\frac{k_M r}{2}\right)^n - i\frac{\Gamma(n+0.5)}{2\Gamma(0.5)}\left(\frac{2}{k_M r}\right)^{n+1} \tag{5.74}$$

where the function $\Gamma(x)$ is the Gamma function. For large arguments $k_M r$ they become approximately

$$h_n^{(1)}(k_M r) \approx (-i)^{n+1}\frac{\exp(ik_M r)}{k_M r} \tag{5.75}$$

Within this approximation, it is easy to show that for large arguments $k_M r$ the near-field efficiencies become $Q_{NF} = Q_{sca}$ and $Q_R = 0$. Furthermore, at a fixed wavelength λ the value $Q_R = Q_{sca}/20$ is always obtained for the same distance r, independently of the particle size. On the other hand, the distance r where $Q_R = Q_{sca}/20$ depends on the wavelength. When this value is approached for λ_1 at a distance r_1, it is obtained at r_2 for wavelength λ_2 with $r_1/\lambda_1 = r_2/\lambda_2$.

Small arguments $k_M r$ can be obtained either for distances r close to the particle surface, that is, $r \approx R$, or for long wavelengths, that is, small wavenumbers k_M. For example, for a particle with $2R = 20\,\text{nm}$ and $r = R$, the enhancement Q_{NF}/Q_{sca} amounts to 13 544, which is in very good agreement with the exact value of 13 587 in Figure 5.24. Since for small arguments $k_M r$ due to large wavelengths $Q_{NF} \approx 3 Q_{sca}/(k_M r)^4$ and $Q_R \approx 2 Q_{sca}/(k_M r)^4$, both become approximately constant as already seen in Figure 5.25. The decrease in Q_{sca} proportional to $1/\lambda^4$ is counteracted by the increase in $1/(k_M r)^4$ with λ^4. Moreover, the ratio Q_R/Q_{NF} approaches the value 0.67, as obtained in the evaluation of Figure 5.26.

In addition to the efficiencies, also the scattering intensities (phase functions), that is, the angular distribution of the scattered light, are affected by the radial component of the field. For discussion of the near-field behavior, the far-field scattering intensities i_{per} and i_{par} of a sphere must be redefined as r-dependent scattering intensities.

There are two possibilities to derive r-dependent scattering intensities. The first approach is to discuss the scattering intensities in terms of the time-averaged Poynting vector $\mathbf{S}$. The second is to define them only via the electric field, following the near-field cross-sections according to Messinger et al. [105]. In both cases, the starting points are the electromagnetic fields $\mathbf{E}_{sca}$ and $\mathbf{H}_{sca}$ of the scattered wave:

$$E_r = E_0 \cos\varphi \sin\theta \sum_{n=1}^{\infty} (2n+1) i^{n+1} a_n \frac{h_n(k_M r)}{k_M r} \pi_n(\theta) \tag{5.76a}$$

$$E_\theta = i E_0 \cos\varphi \sum_{n=1}^{\infty} \frac{2n+1}{n(n+1)} i^n \left\{ a_n \left[h'_n(k_M r) + \frac{h_n(k_M r)}{k_M r} \right] \tau_n(\theta) + i b_n h_n(k_M r) \pi_n(\theta) \right\} \tag{5.76b}$$

$$E_\varphi = -i E_0 \sin\varphi \sum_{n=1}^{\infty} \frac{2n+1}{n(n+1)} i^n \left\{ a_n \left[h'_n(k_M r) + \frac{h_n(k_M r)}{k_M r} \right] \pi_n(\theta) \right.$$
$$\left. + i b_n h_n(k_M r) \tau_n(\theta) \right\} \tag{5.76c}$$

$$H_r = \frac{k_M E_0}{\omega \mu_0} \sin\varphi \sin\theta \sum_{n=1}^{\infty} (2n+1) i^{n+1} b_n \frac{h_n(k_M r)}{k_M r} \pi_n(\theta) \tag{5.77a}$$

$$H_\theta = -\frac{k_M E_0}{\omega \mu_0} \sin\varphi \sum_{n=1}^{\infty} \frac{2n+1}{n(n+1)} i^n \left\{ a_n h_n(k_M r) \pi_n(\theta) \right.$$
$$\left. - i b_n \left[h'_n(k_M r) + \frac{h_n(k_M r)}{k_M r} \right] \tau_n(\theta) \right\} \tag{5.77b}$$

$$H_\varphi = -\frac{k_M E_0}{\omega\mu_0}\cos\varphi \sum_{n=1}^{\infty}\frac{2n+1}{n(n+1)}i^n\left\{a_n h_n(k_M r)\tau_n(\theta)\right.$$
$$\left. -ib_n\left[h'_n(k_M r)+\frac{h_n(k_M r)}{k_M r}\right]\pi_n(\theta)\right\} \tag{5.77c}$$

In terms of the time-averaged Poynting vector

$$S = \begin{pmatrix} S_r \\ S_\theta \\ S_\varphi \end{pmatrix} = \frac{1}{2}\text{Re}\begin{pmatrix} E_\theta H_\varphi^* - E_\varphi H_\theta^* \\ E_\varphi H_r^* - E_r H_\varphi^* \\ E_r H_\theta^* - E_\theta H_r^* \end{pmatrix} \tag{5.78}$$

the far-field scattering intensities $i_{per}(\theta)$ and $i_{par}(\theta)$ perpendicular and parallel to the scattering plane are given as

$$i_{per}(\theta) = \lim_{r\gg R}\left[\frac{\text{Re}(-E_\varphi H_\theta^*)}{2k_M^2 r^2 \sin^2\varphi}\right] \tag{5.79}$$

and

$$i_{par}(\theta) = \lim_{r\gg R}\left[\frac{\text{Re}(E_\theta H_\varphi^*)}{2k_M^2 r^2 \cos^2\varphi}\right] \tag{5.80}$$

In the *near-field*, that is, for $r \approx R$, Equations 5.79 and 5.80 define the *angular distribution functions* $s_{per}(\theta)$ and $s_{par}(\theta)$ as

$$s_{per}(\theta) = \frac{\text{Re}(-E_\varphi H_\theta^*)}{2k_M^2 r^2 \sin^2\varphi} \tag{5.81}$$

$$s_{par}(\theta) = \frac{\text{Re}(E_\theta H_\varphi^*)}{2k_M^2 r^2 \cos^2\varphi} \tag{5.82}$$

and we have

$$i_{per}(\theta) = \lim_{r\gg R} s_{per}(\theta) \tag{5.83}$$

$$i_{par}(\theta) = \lim_{r\gg R} s_{par}(\theta) \tag{5.84}$$

As an example, particles in the Rayleigh approximation are discussed in the following. Then, the far-field solution is $i_{per} = \frac{9}{4}|a_1|^2$ and $i_{par} = \frac{9}{4}|a_1|^2\cos^2\theta$. Both scattering intensities are dominated by the dipolar TM mode a_1. These far-field intensities are compared with the angular distribution functions s_{per} and s_{par} for $r = R$, that is, at the particle surface, in Figure 5.27.

Obviously, s_{per} and s_{par} are completely different for $r = R$ from those functions in the far-field. For several scattering angles they even have negative values, which lead to a negative radial component of the Poynting vector of the scattered wave.

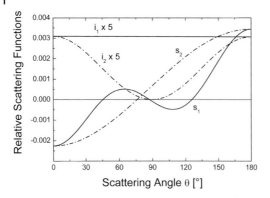

Figure 5.27 Far-field scattering intensities i_{per} and i_{par} of a Rayleigh scatterer in comparison with the angular distribution functions s_{per} and s_{par} for $r = R$.

Then, in these regions the Poynting vector must be directed towards the scattering sphere, indicating an energy flux from the scattered wave to the particle. Furthermore, the magnitudes of s_{per} and s_{par} at $\theta = 180°$ (backward scattering direction) are larger than in the forward scattering direction ($\theta = 0°$).

Negative values of the angular scattering functions result from the spherical Hankel functions and their derivatives, which enter the calculation of the electromagnetic fields as r-dependent complex numbers, together with the complex scattering coefficients a_n and b_n.

In Figure 5.28, it is demonstrated how the angular distribution functions develop with increasing r in a series of four plots. In all plots the far-field scattering intensities i_{per} and i_{par} are additionally given for comparison. Starting with $r = 1.3R$ in Figure 5.28a, it can be seen that the angular regions where the distribution functions are negative are already drastically reduced. Backscattering is still increased compared with the far-field backscattering. The differences between $s_{per}(\theta)$, $s_{par}(\theta)$ and $i_{per}(\theta)$, $i_{par}(\theta)$ decrease with increasing r. In Figure 5.28b at $r = 2R$ $s_{par}(\theta)$ has still negative values between 85 and 100°, but already at $r = 3R$ (Figure 5.28c) both angular distribution functions have positive values. At this distance the curves are rather similar to the far-field scattering intensities. Finally, at a distance of $r = 11R$ (which corresponds to a distance of $D = 10R$ from the particle surface) the differences with respect to the far-field scattering become negligible (see Figure 5.28d).

In Figure 5.29, the corresponding Poynting vector of the scattered wave on a fictitious shell enclosing the particle is plotted at $r = R$ (Figure 5.29a and b) and in the far-field (Figure 5.29c and d). In Figure 5.29a and b, the view is on the north pole of the surrounding fictitious shell and in Figure 5.29c and d the view is on the south pole. Hence one can best recognize the differences in the forward and backward scattering directions. The bold arrows indicate a Poynting vector that is directed towards the spherical particle.

Negative values for the radial component S_r of the Poynting vector are a consequence of energy conservation. As Q_{sca} is independent of the size of the integrating

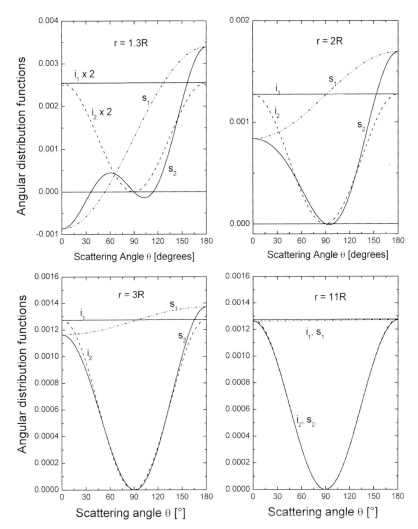

Figure 5.28 Dependence of s_{per} and s_{par} on distance.

sphere, the integral over S_r must also be independent. On the other hand, in the near-field S_r in the backscattering direction is much larger than in the far-field. The increased contribution to the integral from these angular regions is compensated by those regions where S_r is negative.

The above angular distribution functions $s_{per}(\theta)$ and $s_{par}(\theta)$ resulted from the energy flux (Poynting vector) of the scattered wave. Hence they are well founded by physics. Another proposal for angular distribution functions is to define them via only the electric field of the scattered wave, similar to Q_{NF} and Q_R. These *field-based angular distribution functions* $S_{per}(\theta)$ and $S_{par}(\theta)$ are defined via

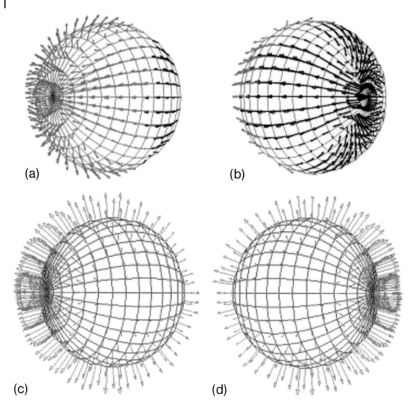

Figure 5.29 Poynting vector of the scattered wave on a fictitious shell around the particle: (a) and (b) at $r = R$ (particle surface), (c) and (d) in the far-field.

$$S_{\text{per}}(\theta,R) = I_0 |e_\varphi(\theta,r)|^2 \tag{5.85}$$

$$S_{\text{par}}(\theta,R) = I_0 \left[|e_\theta(\theta,r)|^2 + |e_r(\theta,r)|^2 \right] \tag{5.86}$$

where e_r, e_θ, and e_φ are the normalized components of the electric field vector in spherical coordinates. The r dependence is included in these field components. For $r \gg R$, that is, in the far-field zone, the component $e_r(\theta, r)$ vanishes and Equations 5.85 and 5.86 become equal to the far-field scattering intensities i_{per} and i_{par}. Due to the definitions in Equations 5.85 and 5.86, S_{per} and S_{par} never become negative, in contrast to s_{per} and s_{par} in Equations 5.81 and 5.82.

The field-based angular distribution functions S_{per} and S_{par} are discussed in detail elsewhere [107]. For example, $S_{\text{per}}(\theta)$ (azimuthal angle $\varphi = 90°$) and $S_{\text{par}}(\theta)$ ($\varphi = 0°$) are plotted in Figure 5.30a and b for a gold particle with diameter $2R = 100$ nm for distances up to $r = 5R$. For comparison i_{per} and i_{par} are plotted in Figure 5.30c and d, calculated using Equations 5.79 and 5.80. The hatched spherical region in the center of each plot corresponds to the particle. The intensities are plotted on a logarithmic scale.

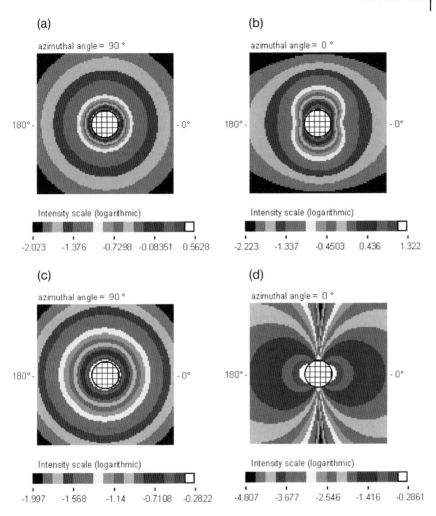

Figure 5.30 Angular distribution of the intensity of the light scattered by a spherical gold particle with diameter $2R = 100$ nm. (a) S_{per}; (b) S_{par}; (c) i_{per}; (d) i_{par}. A color version of this figure can be found in the color plates at the end of the book.

The intensity distribution in Figure 5.30c and d corresponds to that expected for a dipole in the far-field. The near-field intensity S_{per} in Figure 5.30a is rather similar to i_{per} in Figure 5.30c except that the magnitudes are increased by a factor of seven close to the particle surface. At $r = 5R$ the intensities are of comparable magnitude. In contrast, $S_{par}(\theta)$ in the near-field (Figure 5.30b) differs completely from i_{par} (Figure 5.30d). Close to the particle surface the near-field intensity is increased by a factor of 40. At $r = 5R$ the intensities S_{par} and i_{par} are of comparable magnitude. The reason is the contribution of the radial component E_r of the electric field that is excluded in the far-field intensity (Figure 5.30d). More significant than the differences in the magnitudes is the completely different angular distribution of the

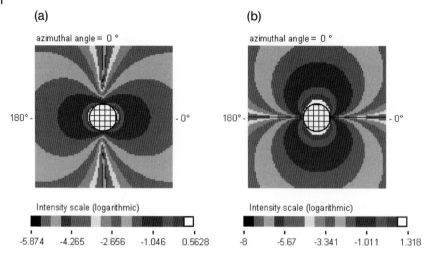

Figure 5.31 Contributions of (a) E_θ and (b) E_r to the modified angular distribution function S_{par} in the plane of incidence for the gold particle in Figure 5.29. A color version of this figure can be found in the color plates at the end of the book.

near-field intensity compared with the far-field intensity. In the far-field, $i_{par}(\theta)$ is proportional to $(\cos\theta)^2$, and hence exhibits minima at $\theta = \pm 90°$. In contrast, in the near-field these minima have vanished and the intensity at $\theta = \pm 90°$ is even larger than that at $\theta = 0°$ and $180°$. These results were confirmed by S. Schelm (personal communications).

For the further discussion, $S_{par}(\theta)$ is divided into the two contributions from the field components e_θ and e_r. Then, it can be seen from Figure 5.31 that the contribution $|e_\theta|^2$ to $S_{par}(\theta)$ is rather similar to $i_{par}(\theta)$ in the far-field, except for the magnitude, which is increased by a factor of seven close to the particle surface. On the other hand, the contribution of the radial field component $|e_r|^2$ is six times larger than that of $|e_\theta|^2$. Therefore, the near-field scattering pattern is dominated by the radial component, as the sum $|e_\theta|^2 + |e_r|^2$ is maximum at $\theta = \pm 90°$ and minimum at $\theta = 0°$ and $180°$, as already shown in Figure 5.30b.

5.5.1
Some Further Details

Xu [108] investigated the electromagnetic (EM) energy flow near single spheres by applying Mie theory. From the patterns of the energy flow, the absorption and the scattering of light can be understood from the microscopic point of view. In the absorption profiles of metallic particles, most absorbed energy is consumed on the surface of the particles, which indicates that the resonance of the surface plasmon is different from that of the bulk plasmon. Two mechanisms to enhance the local EM field are also distinguished: the surface plasmon resonance and the intensified energy flow.

6
Application of Mie's Theory

Mie's theory of light scattering and absorption by a spherical particle has found numerous applications in colloidal chemistry, color science, atmospheric science, and astrophysics, with different main topics. Recently, it has also become a topic for thin-film solar energy conversion.

In atmospheric science, mostly the angular distribution of light scattered by micron-sized particles is of particular interest. Rainbows, glories, and halos or the *blue moon* or the *blue sun* are impressive examples where Mie's theory can explain the characteristic light scattering. Carbonaceous nanometer-sized particles, dust, or ammonia particles become increasingly important because of their contribution to anthropogenic air pollution and their effect on the solar radiation budget.

In astrophysics also nanometer-sized carbonaceous species are of great importance as they are a major part of interstellar dust. In addition, various silicates and silicon based minerals are of interest here. As in astrophysics mostly the infrared spectral wavelength region is used, the particle size/wavelength ratio makes these particles small compared with the wavelength, like nanometer sized particles at visible wavelengths.

Color is not a measurable quantity but rather a sensation triggered by radiation of sufficient intensity. The reason for the color of a certain specimen is light scattering and absorption by nanometer-sized particles. Although the color here is often an effect caused not only by the single particle but also by the scattering and absorption of a particle ensemble, the single particle properties predicted from Mie's theory are the core of any model for paintings based on pigments.

In this chapter, we concentrate on spherical nanoparticles with diameters between 10 and 300 nm. We present numerical results for many particle materials and wavelength regions, obtained by applying Mie's theory rigorously. The restriction to particles with sizes less than 300 nm is not necessary but helpful to keep the amount of information as small as possible to demonstrate all effects. For larger particles, the influence of the particle size on the extinction cross-section spectra increases still further, but a size of $2R = 300$ nm has proven to be sufficient to show the main features which can be expected on increasing the particle size. First numerical results on sizes less than 100 nm have been published [109].

Optical Properties of Nanoparticle Systems: Mie and Beyond. Michael Quinten
Copyright © 2011 WILEY-VCH Verlag GmbH & Co. KGaA, Weinheim
ISBN: 978-3-527-41043-9

The calculations are partly supplemented by experimental results where the filling factor f was as low as electromagnetic interactions among the particles could be neglected, that is, the optical properties of the ensemble were determined by the sum of the optical properties of the individual particles.

We point out that in the following calculated spectra the surrounding is always assumed to have a constant refractive index $n_M = 1$ (vacuum), whereas in experiments it may be larger. Nevertheless, provided that the refractive index n_M remains real valued, that is, the surrounding medium is not absorbing, comparison with the calculations is possible.

To manage the huge amount of possible particle materials, we introduced the following categories:

- Drude metal particles (Al, Na, K)
- noble metal particles (Cu, Ag, Au)
- catalyst metal particles (Pt, Pd, Rh)
- magnetic metal particles (Fe, Ni, Co)
- rare earth metal particles (Sc, Y, Er)
- transition metal particles (V, Nb, Ta)
- semimetal particles (TiN, ZrN)
- semiconductor particles (Si, SiC, CdTe, ZnSe)
- carbonaceous particles (graphite, amorphous carbon, diamond)
- absorbing oxide particles (Fe_2O_3, Cr_2O_3, Cu_2O, CuO)
- transparent oxide particles (SiO_2, Al_2O_3, TiO_2, CeO_2)
- particles with phonon polaritons (MgO, NaCl, CaF_2)
- miscellaneous particles [indium-doped tin oxide (ITO), LaB_6, EuS].

The materials given in parentheses are just representative of the corresponding category, and further materials could be added. The categorization is not unique since, for example, Pt is also a noble metal, TiO_2 is strongly absorbing at wavelengths lower than the bandgap, and MgO is a nonabsorbing oxidic material in the visible spectral range. Nevertheless, it is helpful for discussion of the observed effects with increasing particle size.

For better comparison between spectra of different materials, we plotted the extinction efficiency Q_{ext}, that is, the extinction cross-section normalized to the geometric cross-section. Optical constants are taken from various references, as indicated.

In the examples, the spectral range comprises an extended spectral range including the visible spectral range, to present *genuine* optical properties. Exceptions are materials with phonon polaritons, since these polaritons lie in the far-infrared (FIR) region.

6.1
Drude Metal Particles (Al, Na, K)

We start with spectra of Al, Na, and K nanoparticles. An outstanding property of these metals is that at photon energies lower than the plasma frequency ω_P their dielectric function can be approximated fairly well by the Drude dielectric constant for free electrons:

$$\varepsilon(\omega) = 1 + \chi_1^{IB}(\omega) - \frac{\omega_p^2}{\omega^2} + i\left[\chi_2^{IB}(\omega) + \frac{\gamma_{fe}}{\omega}\frac{\omega_p^2}{\omega^2}\right]. \tag{6.1}$$

The influence of the core electrons on the susceptibility of the interband transitions χ^{IB} is perceptible only for the real part χ_1^{IB} and negligible for χ_2^{IB}, that is, core polarization effects have to be considered. These core polarization effects shift the volume plasmon energy $\hbar\omega_P$ slightly to lower energies (for example, for Na from 5.95 to 5.6 eV). For spherical nanoparticles we therefore can expect surface plasmon polariton (SPP) resonances which are rarely influenced by interband transitions.

In Figure 6.1 we have compiled the calculated spectra for Al, Na, and K nanoparticles. The optical constants used are taken from [36] for Al and from [92] for Na and K. The spectra of these metals exhibit similar features depending on the particle diameter. For small nanoparticles, a single, sharp resonance peak can be recognized that can be assigned to the dipolar TM mode. It corresponds to the SPP in these small particles. Its position shifts to longer wavelengths (lower energies) with increasing diameter. Furthermore, the magnitude increases with increasing diameter, approaching its maximum for a certain size, which is material dependent. The maximum value is $Q_{ext} \approx 16$ for Al and Na and $Q_{ext} \approx 20$ for K. That means that these nanoparticles act similarly to a four-times larger sphere treated within geometric optics.

For larger sizes, the magnitude of the TM dipole peak decreases again and becomes broader. New resonant-like structures appear at short wavelengths. Their number increases with increasing particle size, and also their peak positions shift to longer wavelengths and their halfwidths increase. From the discussion of Figure 5.13 for Al nanoparticles with $2R = 100$ nm, we can assign these structures mainly to TM modes with multipole order $n > 1$. The contribution of the TE modes is still negligible. The high-order TM modes contribute to the extinction mainly by absorption, whereas the low-order TM modes contribute mainly by scattering. The higher order resonances are determined not only by the electronic structure of the metal, that is, its dielectric function, but increasingly also by the ratio of size to wavelength.

As the optical density and the transmittance of an ensemble of these metal particles depend on Q_{ext}, the shift of the peak positions and the varying number of multipole mode contributions also affect the apparent color of particle ensembles.

We should point out that the preparation of ensembles of Na, K, and Al nanoparticles is critical since these metals are chemically very reactive.

Experimentally, Hecht [110] studied the optical properties of K nanoparticles ($2R = 34$–43 nm) in the wavelength range 480–620 nm. He compared the experimental spectra with calculations according to Mie's theory, using three different sets of optical constants. The best agreement was obtained with optical constants from Smith [92], also used in the above calculations.

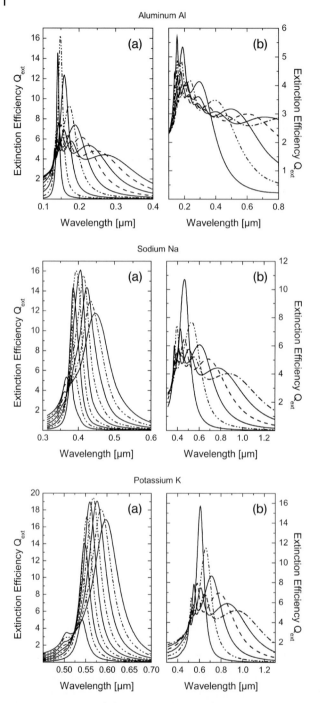

Figure 6.1 Spectra of aluminum, sodium, and potassium nanoparticles. (a) $2R = 10–90$ nm, stepwidth 10 nm; (b) $2R = 100–300$ nm, stepwidth 40 nm.

6.2
Noble Metal Particles (Cu, Ag, Au)

6.2.1
Calculations

Similarly to the Drude metals above, small nanoparticles of noble metals also exhibit an SPP. In contrast to the alkali metals and Al, the noble metals show pronounced deviations, since the contribution χ^{IB} of interband transitions from the 3d (Cu), 4d (Ag), or 5d (Au) electrons to the hybridized 4sp (Cu), 5sp (Ag), or 6sp (Au) band has an enormous influence on the positions of the plasmon polaritons. The SPP resonances in noble metal nanoparticles therefore cannot be regarded as free electron resonances even as a rough approximation; rather, they are hybrid resonances resulting from cooperative behavior of both the d electrons and the conduction electrons. This is demonstrated in Figure 6.2 by representative spectra for Cu, Ag, and Au nanoparticles. Optical constants were taken from Johnson and Christy [37].

The largest influence on the SPP resonance can be recognized in the spectra of Cu nanoparticles. The resonance at wavelengths in the green visible spectral region is strongly quenched by the close-lying interband transitions which also reach to wavelengths in the green visible spectral region. Just if the particle size is larger than $2R = 140$ nm is the resonance clearly separated from the interband transitions. For still larger particles, more TM resonances appear that shift to longer wavelengths and become broader.

For Au nanoparticles, the influence of the interband transitions can also be clearly recognized, but is already strongly reduced because they are more separated from the resonance position of the SPP. Therefore, already for small nanoparticles the resonance of the TM dipole mode can clearly be recognized.

The modest influence of the interband transitions on the SPP is obtained for Ag nanoparticles. Although the interband transitions contribute to the spectra at shorter wavelengths, the peak position of the SPP is well separated from the interband transitions, so a sharp resonance is obtained. The maximum extinction efficiency amounts to $Q_{ext} \approx 14$. This is close to the maximum efficiency of $Q_{ext} \approx 16$ for Al and Na nanoparticles as discussed above. So far, Ag behaves very similarly to the Drude metals.

In general, with increasing particle size the same features as before for the Drude metals can be recognized: increasing number of resonances, shift of peak structures to longer wavelengths, broadening of the resonances. Again, we can assign the resonance structures mainly to TM modes. The contribution of the TE modes is still negligible. The high order TM modes contribute to the extinction mainly by absorption, whereas the low-order TM modes contribute mainly by scattering.

The resonant structures in the extinction lead to characteristic colors for the smaller particles: red for Cu and Au and yellow for Ag. The color changes due the peak position also shift due to the increasing number of TM mode contributions with increasing particle size.

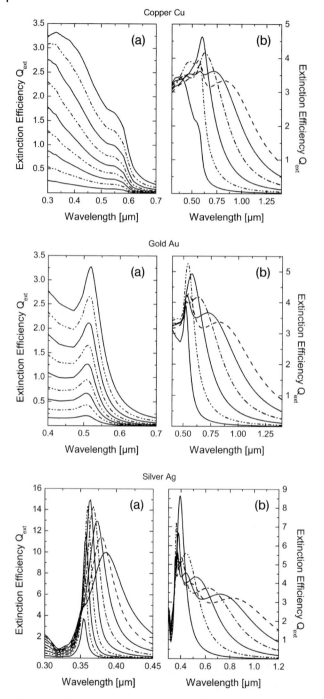

Figure 6.2 Spectra of copper, gold, and silver nanoparticles. (a) $2R = 10–90\,nm$, stepwidth 10 nm; (b) $2R = 100–300\,nm$, stepwidth 40 nm.

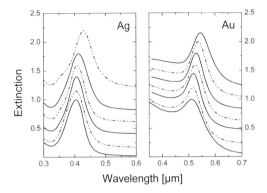

Figure 6.3 Measured optical extinction spectra of several aqueous Ag and Au colloids with different mean particle sizes.

6.2.2
Experimental Examples

6.2.2.1 Colloidal Au and Ag Suspensions

Aqueous suspensions of nanometer-sized Ag and Au particles were prepared by chemical reduction of the metal from a water-soluble salt [111–114]. A certain amount of the reduced atoms spontaneously form nuclei on which the other metal atoms can precipitate, so that the nuclei grow to larger particles. The most critical parameter in this procedure is the nucleation rate. This determines the number of nuclei formed and, hence, the final size of the particles. Still larger particles can be obtained when using the particles from a previous preparation as nuclei for the next reduction process. In this way, most of the colloidal gold and silver particle systems were prepared. However, we note that spherically shaped particles larger than about 60 nm are difficult to obtain by this approach, because for larger sizes increasingly certain crystalline directions grow faster than other directions and the particles become nonspherical.

Figure 6.3 shows selected spectra of several Ag and Au nanoparticle suspensions. Typical filling factors are in the order of $f \approx 10^{-6}$ for gold suspensions and $f \approx 10^{-7}$ for silver suspensions. The spectra are arranged according to the particle size, which increases from bottom to top to show the size dependence of the spectra. For better comparison, the spectra are shifted along the ordinate by an added constant optical density.

The small nanoparticles exhibit an SPP similar to the calculations, which is followed by interband transitions at lower wavelengths. Remember that the peak position depends on the refractive index of the surrounding medium, which is close to $n_M = 1.4$ for gelatin-stabilized aqueous suspensions. Therefore, it is shifted to longer wavelengths (red shift) compared with the calculations in Figure 6.2.

The peak position also depends on the particle size, which is demonstrated in Figure 6.4, where the wavelength at which the SPP is peaked is plotted versus the particle size. The experimental data are taken from various gold and silver colloids

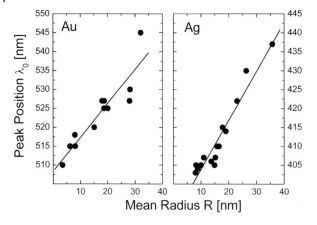

Figure 6.4 Experimental size dependence of the peak position of the surface plasmon polariton of Au and Ag colloids.

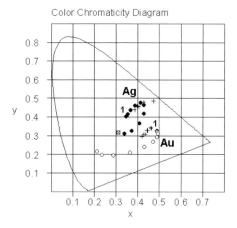

Figure 6.5 Chromaticity diagram showing the color change with increasing particle size for Ag nanoparticles (●) and Au nanoparticles (○). The crosses indicate the chromaticity coordinates for Ag and Au nanoparticles prepared in aqueous colloidal suspensions. For the corresponding mean particle sizes, see text. The "1" indicates the starting point of the curves.

in aqueous suspension, including those in Figure 6.3. The straight lines serve as a guide for the eye. The peak position clearly shifts to longer wavelengths with increasing particle size, confirming the calculation results.

The color of the colloidal suspensions also changes with increasing size. This can be seen from Figure 6.5, where the color coordinates of several aqueous gold and silver nanoparticle suspensions (crosses) are plotted in the chromaticity diagram in comparison with color coordinates from calculated spectra (full and open circles) of gold and silver particles in glass. The norm light source used was D65. The corresponding mean particle sizes in the experiments are $2R = 16.4$,

17.6, 20, 27.4, 31.6, 33, and 52.6 nm for the Ag nanoparticles and $2R = 6.9$, 12.4, 16, 30.2, 37.4, 40.2, and 56.4 nm for the Au nanoparticles. The "1" in the chromaticity diagram indicates the start of the curves.

It is obvious that the apparent color of assemblies containing Ag nanoparticles changes from yellow to red with increasing particle size. The chromaticity coordinates from the calculations lie in a loop in the chromaticity diagram. The loop almost ends at the chromaticity point of the standard illuminant D65. The change in the apparent color of the assemblies containing Au nanoparticles with increasing size is similar to that containing Ag particles. However, the apparent color is red for small particles and changes to blue when the particle diameter exceeds 70 nm.

6.2.2.2 Gold and Silver Nanoparticles in Glass

The art of using colloidal gold and silver for coloring glass is very old and reaches back to the Ancient World. One should keep in mind, however, that it was known how to use colloids for coloring glasses but it was unknown that nanoparticles are the reason for these coloring effects. First descriptions of how to obtain ruby gold glasses can even be found in the bibliography of Ashurbanipal in Ninive (seventh century BC). Another famous example from the Romans is the Lycurgus cup (fourth century AD, British Museum, London). In the Middle Ages, this tradition was continued to prepare wonderfully colored windows of cathedrals [e.g., the gothic cathedral in Halberstadt (fourteenth century) and the windows of Cologne cathedral (1300)]. In 1663, Andreas Cassius described the preparation of red-colored glass with gold by the reaction of gold salt solutions and tin salt solutions. To this day, colloidal gold is called *Cassius Gold Purple*.

Johann Kunckel von Loewenstein is recognized as first person to make the connections between salts, bases, and acids and described recipes for ruby gold in his book *Ars Vitraria Experimentalis oder Vollkommene Glasmacherkunst* [2]. He ran the first glass factory where ruby gold was systematically manufactured.

Finally, it was Michael Faraday who not only described the preparation of gold nanoparticles by chemical reduction of gold salts with phosphorus, but also described the interaction of light with the gold particles using the new theory of waves.

However, not only gold nanoparticles lead to red color in glasses. The red color in some windows of Cologne cathedral is caused by cubo-octahedral copper particles with $2R = 26$ nm, whereas in Kunckel glasses Au particles of $2R = 40$ nm lead to the red color. In addition, Fe_2O_3 nanoparticles also contribute to the red color in Kunckel glasses.

The famous Lycurgus cup is extraordinary since its color depends on whether the light is reflected or transmitted. At places where light is transmitted through the glass it appears red, whereas at places where light is scattered near the surface the scattered light appears opaque green. It has been proven that Ag and Au are contained in the glass in small amounts. As, however, the different colorings in reflectance and transmittance cannot be simply assigned to a certain noble metal and a certain kind of particles, we will treat the Lycurgus cup again later within the discussion of densely packed particle ensembles.

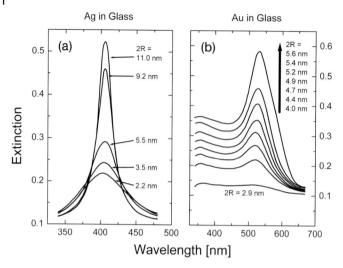

Figure 6.6 Optical extinction spectra of (a) Ag nanoparticles and (b) Au nanoparticles in a glass matrix. Data courtesy of U. Kreibig.

Here, we present experimental results on Ag and Au nanoparticles obtained by Kreibig [115, 116]. Ensembles of nearly spherical, well-separated particles with narrow size distributions were prepared in photosensitive glasses. The particles grow during an annealing process, which can be interrupted at any time to record optical extinction spectra. The advantage of this process is that the number of particles remains constant whereas the particle size increases with increasing annealing time.

Figure 6.6 shows the recorded extinction spectra for the prepared silver particles (a) and gold particles (b). The corresponding particle size was obtained by transmission electron microscopy (TEM) applied to very thin glass layers of the samples. The spectra clearly show that with increasing size the characteristic SPP absorption peak increases in height. A shift of the peak position to longer wavelengths cannot be recognized for the Ag particles because they are still too small. The peak position is at $\lambda \approx 405$ nm, corresponding to the peak position of such small Ag spheres in glass calculated with the Mie theory. For the gold nanoparticles, the peak position shifts from $\lambda \approx 520$ nm to $\lambda \approx 530$ nm with increasing size. Also, the peak positions correspond to the peak positions calculated according to the Mie theory for Au particles in glass. It is notable that the width of the resonance clearly increases with decreasing particle size. For the gold nanoparticles with $2R = 2.9$ nm, the resonance almost has vanished.

6.2.2.3 Copper Nanoparticles in Glass and Silica

As shown earlier, the photosensitive process can be used to create Ag and Au nanoparticles in a glass matrix. In the following, we show that it is also applicable to the third noble metal, copper. Copper, however, plays a special role due to the

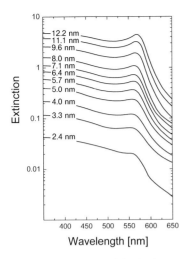

Figure 6.7 Cu nanoparticles in photosensitive glass. Data courtesy of U. Kreibig.

existence of two low-lying oxidation states of the ions: Cu^+ and Cu^{2+}. The photoreduction process only works with Cu^+ and, hence, the production of Cu^{2+} has to be prevented during the melting procedure. Therefore, the glass batches were melted in the presence of high concentrations of additional reducing agents at 1450 °C. The photosensitive process worked well for concentrations of Cu_2O between 0.05 and 0.3 wt%.

Spherical nanoparticles with average diameters between 2.4 and 23.3 nm were produced and their optical transmission spectra were recorded by Kreibig [117]. The spectra are compiled in Figure 6.7. The particle size increases from bottom to top. The spectra look essentially similar to those calculated for Cu nanoparticles but are shifted to longer wavelengths due to the higher refractive index of the glass compared with vacuum. The SPP peak corresponds to the resonance of the TM dipole mode. It shows a red shift from 545 nm at 2.4 nm diameter to 570 nm at 12.2 nm diameter.

Similarly to Kreibig, Uchida *et al.* [118] also prepared copper and silver nanoparticles in glass by means of conventional melt and heat treatment processes. They used the prepared composite glasses for measuring third-order nonlinearities of small metal particles, but also recorded linear absorbance spectra. An increase in the SPP peak intensity and a red shift in the peak position were observed as the particle size increased from $2R = 14.6$ to 95.4 nm.

Another technique often used for the preparation of nanoparticles in a solid matrix is ion implantation with different current densities. Here we show exemplary results from Magruder *et al.* [119, 120] and Takeda *et al.* [121] in Figure 6.8. Magruder *et al.* prepared Cu nanoparticles by ion implantation in fused silica. The spectra in Figure 6.8a represent four different current densities, 0.7, 2.5, 5.0, and 7.5 µA cm^{-2} from bottom to top, when implanting an amount of 6×10^{16} ions cm^{-2}. The corresponding mean sizes are $2R = 5.2, 6.7, 9.9$, and 12.6 nm. The spectra are

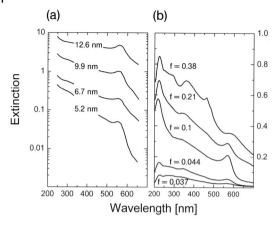

Figure 6.8 Cu nanoparticles in silica by ion implantation at different current densities: (a) Magruder et al. [119], (b) Takeda et al. [121].

comparable to those measured by Kreibig, although the particles were prepared in different ways. Takeda et al. [121] implanted negative Cu ions of 60 keV into amorphous SiO_2 at a fixed flux of $10\,\mu A\,cm^{-2}$ to fluences ranging from 3×10^{16} to $6 \times 10^{17}\,ions\,cm^{-2}$. The optical absorption increased with the Cu concentration but the surface plasmon peak was maximized around a volume fraction of $f = 0.1$ and attenuated beyond that fraction. This work supplements former studies by Kishimoto et al. [122, 123].

6.2.2.4 Ag_xAu_{1-x} Alloy Nanoparticles in Photosensitive Glass

In addition to pure Ag and pure Au nanoparticles, Ag_xAu_{1-x} alloy nanoparticles also show surface plasmon resonances. The position of the TM dipole resonance depends on the concentration x of Ag in the Ag_xAu_{1-x} alloy nanoparticle. This was demonstrated earlier with the calculated spectra in Figure 5.15. In 1987, Radtke [124] studied experimentally the optical properties of such alloy nanoparticles which were produced by reduction of Au and Ag in glasses containing corresponding metal salts. By UV irradiation of the glasses to form nuclei and controlled thermal heating, the different compositions with concentration x varying between 20 and 90% were prepared.

The measured optical extinction spectra are summarized in Figure 6.9a for one series. With the obtained values for x and the optical constants from [38], Ag_xAu_{1-x} spectra of alloy nanoparticles were computed and are displayed in Figure 6.9b. Quantitative correspondence within the limits of accuracy is obtained.

Pal et al. [125] developed a reverse micelle system for the preparation of Ag and Au–Ag alloy nanoparticles at room temperature. By changing the metal ion concentrations at a certain water to surfactant molar ratio, there was no distinguishable change in particle size for Ag and Au–Ag alloy nanoparticles, but agglomeration of particles was observed in the case of gold particles, with a clearly recognizable change of the color from red to blue. Optical absorption spectra showed a continu-

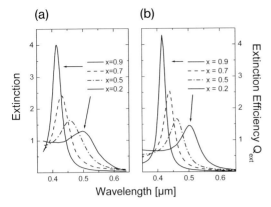

Figure 6.9 (a) Measured extinction spectra of Ag_xAu_{1-x} alloy nanoparticles in glass. (b) Calculated extinction spectra of Ag_xAu_{1-x} alloy nanoparticles with concentration x and size $2R$ as in the experimental spectra and from TEM. The sizes are $2R = 9, 11, 12$, and 15 nm for $x = 0.9, 0.7, 0.5$, and 0.2, respectively.

ous red shift with increasing amount of Au in the Au–Ag alloy nanoparticles, similar to the results of Radtke above.

Torigoe et al. [126] prepared colloidal Ag–Pt alloys by chemical reduction from an Ag–Pt-salt. Optical absorption spectra were recorded. They showed an SPP band at 354 nm for a certain Ag : Pt ratio. The peak wavelength shifts in the range 330–410 nm depending on the Ag : Pt ratio.

6.2.2.5 Silver Aerosols

Unlike colloids with filling factors of $f \approx 10^{-6}$, the particle number concentration in aerosols is low as measurement of the optical density becomes difficult if the aerosol contains nanoparticles. The reason is the low extinction cross-section of single nanoparticles. This problem can be avoided by increasing the optical pathlength through the aerosol to obtain sufficiently high optical density values. Thus, for measuring the extinction of light by aerosols containing nanometer-sized ultrafine particles, a multi-pass cell that works in principle like a White cell [127, 128] was used in experiments [129–132].

The procedure for preparation of Ag aerosols was as follows.

- preparation of an aqueous solution of silver nitrate ($AgNO_3$)
- spraying the solution
- drying the droplets in a diffusion drier
- heating the resulting $AgNO_3$ salt particles in a furnace at different temperatures
- measurement of the extinction of light by the aerosol
- preparation of samples for size determination with TEM.

Silver particles were generated by spray pyrolysis of 0.016, 0.05 and 0.1 M aqueous $AgNO_3$ solution. The extinction spectra were recorded with a multi-pass cell and

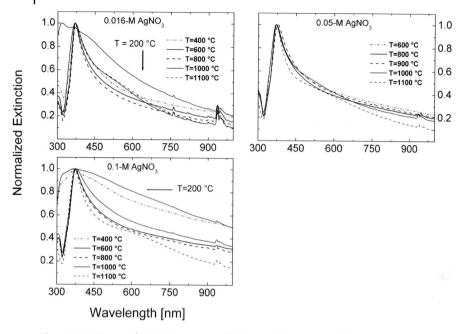

Figure 6.10 Measured extinction spectra of silver particles generated by spray pyrolysis.

are displayed in Figure 6.10. In all graphs the extinction is normalized to the maximum, because different particle concentrations would render the interpretation of the spectra more difficult.

In all spectra with $T > 400\,°C$, the absorption is dominated by a distinct maximum which can clearly be assigned to the SPP in small silver particles in air. The peak position is between 365 and 375 nm. At 200 °C, the spectrum is due to $AgNO_3$ salt particles which are formed due to the drying of the spray in the diffusion dryer. At 400 °C, first deviations from the salt spectrum can be recognized. The reason is that $AgNO_3$ dissociates at 444 °C with separation of silver. For temperatures above the melting point of silver (962 °C), extinction is increased in the wavelength region between 500 and 650 nm, exhibiting a broad extinction maximum.

Two unexpected features are to be seen at $\lambda = 761$ nm and in the wavelength range between 930 and 950 nm. They belong to molecular absorption bands of oxygen ($\lambda = 761$ nm) and water. Because of the long pathlength in the aerosol chamber, molecular absorption bands can be detected when measuring extinction. Water comes from the aqueous precursor solutions and is not completely removed when the aerosol enters the White cell. Oxygen is partly contained in the carrier gas and comes from the ambient air.

Comparing the spectra with calculated spectra provided in Figure 6.2, it is evident that the plasmon peak belongs to particles with diameters $2R \leq 60$ nm. However, in the experimental data obviously the plasmon peak is broadened at longer wavelengths, particularly for higher temperatures. The reason is larger particles in the aerosols. Aggregates of particles could be excluded from TEM.

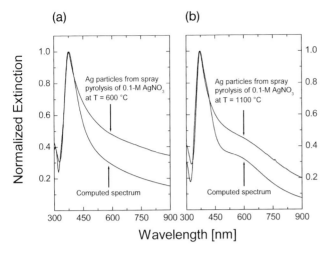

Figure 6.11 Computed extinction spectra of silver particles with a bimodal size distribution in comparison with measured spectra. The parameters are given in the text.

Assuming larger particles in the aerosol, the increase in the extinction at longer wavelengths in the measured high-temperature spectra can be qualitatively explained using a bimodal size distribution. This has been done in computations for particles with median size $2R_1 = 30$ nm and $2R_2 = 190$ nm and a log-normal distribution with standard deviation $\sigma = 2$ for both median sizes. The corresponding spectra for each size distribution were added and weighted with the factors 0.7 and 0.3, respectively. The obtained spectrum is plotted in Figure 6.11a in comparison with the measured spectrum of silver particles generated from a 0.1 M AgNO$_3$-solution at 600 °C. It exhibits the same features as the measured spectrum. The magnitude of the extinction at long wavelengths is less than that in the experiment. This is believed to come from salt particles (AgNO$_3$) which remain unchanged in the spray pyrolysis and contribute to the measured spectrum. A second spectrum is obtained with the weighting factors 0.6 and 0.4. It is plotted in Figure 6.11b in comparison with the measured spectrum of the same silver aerosol system at 1100 °C. It becomes obvious that now also the second maximum at wavelengths between about 550 and 650 nm clearly appears in the calculation.

6.2.2.6 Further Experiments

Experiments with Silver Nanoparticles Russell et al. [133] produced silver nanoparticles by microlithography and measured optical absorbance spectra in the wavelength range 300–600 nm. The particle size was varied between 40 and 70 nm. They calculated spectra using optical constants from [39] and compared them with the measured spectra. They were in a fairly good agreement with respect to peak positions and the number of contributing multipole orders. Improvements could have been achieved by taking into account a size distribution in the calculations.

Banerjee and Chakravorty [134] prepared silver nanoparticles by electrodeposition in a silica gel matrix. The mean particle sizes were varied between 10 and 26 nm. TEM analysis showed that the particles are not ideal spheres and are size distributed. Taking the size distribution into account in Mie calculations, they found very good agreement between the measured and calculated spectra.

The optical properties of Ag and Au nanoparticles dispersed within the pores of monolithic mesoporous silica were studied by Cai et al. [135]. They are governed in a peculiar way by particle–ambient interactions. Depending on the annealing-induced variation of particle size, a characteristic shift of the SPP resonance frequency is observed that differs for both metals and from the corresponding cases of fully embedded particles. A two-layer core–shell model was introduced to account for the effects of local porosity at the particle–matrix interface and of the free particle surface being in contact with the ambient air. The good agreement of model calculations with the experimental findings confirms that a combination of both effects dominates the observed TM dipole resonance evolution with particle size.

Experiments with Gold Nanoparticles Wilk and Schreiber [136] examined gold nanoparticles in acetate glasses. In the case of lead–lithium acetate glass the absorption spectra obtained at high molar ratio showed additional absorption at longer wavelengths, which was explained by the authors by larger particles (50 nm in diameter) leading to the purple color of the glass. However, particle size analysis was not given. We believe that aggregation of nanoparticles could also have led to the change in the color (see also Chapter 11).

Sönnichsen et al. [137] investigated the optical properties of spherical gold and silver nanoparticles with diameters of 20 nm and larger. The light scattering spectra of individual nanoparticles were measured using dark-field microscopy, thus avoiding inhomogeneous broadening effects. The dipolar SPP resonances of the particles were found to have nearly Lorentzian line shapes. With increasing size the resonance shifted to longer wavelengths (red shift). Agreement with Mie theory was reasonably good for the gold particles and less satisfactory for the silver particles, possibly due to particle faceting or chemical effects.

Experiments with Copper Nanoparticles Papavassilion and Kokkinakis [138] studied the optical absorption spectra of colloidal solutions of Cu with gelatin as a stabilizer. One characteristic absorption peak was found at 580 nm (2.14 eV) that could be attributed to the SPP of Cu nanoparticles.

Doremus et al. [139] examined small copper particles grown in a soda-lime silicate glass that was heat treated at 600 °C after cooling to room temperature. The optical absorption of the glass has a local maximum at a wavelength of 565 nm, as expected for copper particles much smaller than the wavelength of light. Electron micrographs show small particles with a mean diameter of $2R = 4.5$ nm.

Peña et al. [140] obtained Au and Cu nanoparticles by ion implantation in silica and subsequent thermal annealing in air. They determined the size distribution of the particles by a newly developed method and compared the results with those given by TEM and grazing-incidence small-angle X-ray scattering (GISAXS) geometry. The average radius obtained by all the three techniques was almost the same for the two metals studied, but the radius distribution was considerably underestimated by TEM.

Ila *et al.* [141] studied both linear and nonlinear optical properties of Suprasil-1 by implanting 2.0 MeV copper, 350 keV tin, 1.5 MeV silver, and 3.0 MeV gold. Changes in the optical properties occurred by spontaneous particle formation and by subsequent thermal annealing. The recorded absorption spectra showed the characteristic surface plasmon polariton peak for Cu, Ag, and Au, but no specific spectral feature for Sn.

Umeda *et al.* [142] studied the thermal stability of nanoparticles, fabricated by high-current negative ions, focusing on optical applications. Negative Cu ions of 60 keV were irradiated to silica glasses at high dose rates up to $50 \mu A\, cm^{-2}$. The high-current implantation caused a bimodal distribution of Cu nanoparticles. Thermal annealing at 773–1273 K for 1 h was applied to the specimens. For each step, optical absorption was measured in the wavelength range 200–2500 nm and nanoparticle morphology was evaluated by cross-sectional TEM. Thermal annealing below 873 K gave no discernible changes in either nanoparticle morphology or absorption spectra. Above 1073 K, pronounced coarsening of Cu particles occurred, with enhancement of the bimodal distribution.

Plaksin *et al.* [143] implanted 60 keV copper ions with fluxes from 5 to $50 \mu A\, cm^{-2}$ into aluminum oxide. The formation of copper nanoparticles was efficient up to fluences of $2 \times 10^{17}\, ions\, cm^{-2}$ or more and most efficient for a flux of $50 \mu A\, cm^{-2}$. Optical absorption spectra were recorded to monitor the particle formation.

Experiments with Other Nanoparticles with SPP Amekura *et al.* [144] prepared metallic zinc nanoparticles 5–15 nm in diameter in silica glass by Zn ion implantation at 60 keV. The spectra showed a strong ultraviolet absorption peak at around 260 nm (4.8 eV), which could be assigned to the SPP of Zn nanoparticles, and another small peak at 1033 nm (1.2 eV), which had never been reported before. To identify the origin of the 1033 nm peak, the correlations of thermal stability between the two peaks and Zn nanoparticles were evaluated under annealing both in a vacuum (pure thermal stability) and in oxygen gas (thermal oxidation stability). The well-correlated stability between the 1033 nm peak, the 260 nm peak, and Zn nanoparticles indicates that the 1033 nm peak should not be ascribed to radiation-induced defects but to the Zn nanoparticles. The 1033 nm peak could be described as a second SPP in Zn nanoparticles.

6.3
Catalyst Metal Particles (Pt, Pd, Rh)

The next class of metals are of practical importance as catalysts for many chemical reactions. They develop their catalytic properties in the form of nanoparticles mostly brought up on substrates. Optical properties of single isolated

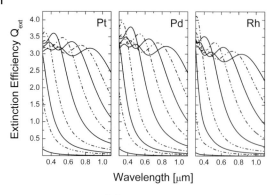

Figure 6.12 Spectra of platinum, palladium, and rhodium nanoparticles with diameter $2R = 10$, 50, 70, 100, 140, 180, 220, 260, and 300 nm.

nanoparticles of these metals were of less interest since the smaller nanoparticles do not exhibit a resonance which could be assigned to an SPP resonance. This can also be seen from the spectra of the smaller nanoparticles in Figure 6.12. The extinction efficiency of these particles only decreases from short to long wavelengths proportional to $1/\lambda$ (Rayleigh approximation of the extinction cross-section) without exhibiting any resonance-like feature. However, if the particle size increases, a broader resonance-like extinction band grows which shifts to longer wavelengths with increasing size, and can be assigned to the TM dipole mode. With increasing size, higher TM modes contribute to the spectrum with broad extinction bands which also shift to longer wavelengths. These resonances are determined more by the ratio of size to wavelength ratio than by the electronic structure of the metal, that is, its dielectric function. Therefore, the spectra of the larger particles are fairly similar to those of the larger particles of the Drude and noble metals where the size also becomes dominant for the width of the TM modes. The contributions of the TE modes are again negligible for particles smaller than 300 nm. The high-order TM modes contribute to the extinction mainly by absorption, whereas the low-order TM modes contribute mainly by scattering. Optical constants were taken from [36] for the computations.

An experimental example is given in the following. Aqueous suspensions of nanometer-sized Pd particles were prepared by chemical reduction from $PdCl_3$ using the method described by Turkevich and Kim [145]. In this way, colloidal palladium particle systems were prepared with different mean particle sizes of $2R = 27.2 \pm 5.0$ nm (Pd1), 27.2 ± 6.0 nm (Pd2), 28.8 ± 7.2 nm (Pd3), and 33.2 ± 5.8 nm (Pd4) [146]. Typical filling factors are in the order of $f \approx 10^{-6}$. Figure 6.13 shows the measured and corresponding calculated spectra of these samples.

The calculated and measured spectra agree very well and show the characteristic decrease from short to long wavelengths proportional to $1/\lambda$ which could be expected for such small particles. As mainly the violet and blue visible light is absorbed, the color of the colloidal solution is reddish.

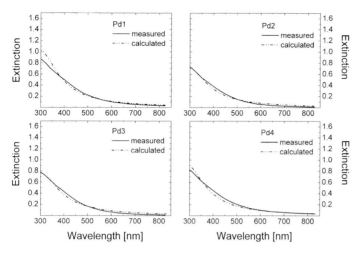

Figure 6.13 Measured and calculated extinctions of the Pd colloids Pd1, Pd2, Pd3, and Pd4.

6.4
Magnetic Metal Particles (Fe, Ni, Co)

Nanoparticles of magnetic metals are of practical interest for some technical applications (magnetic memory media, lubricants) and also in medical applications for cancer therapy. In all these applications the magnetic moments of these particles are of crucial importance. In contrast, the magnetic properties of the particles are unimportant when looking at their optical properties. The reason is simple: the magnetic moments are too inert as to follow the high-frequency external magnetic field of the incident electromagnetic wave. This holds true down to frequencies in the FIR spectral region. Therefore, we can consider the relative permeability to be $\mu = 1$.

The optical properties are fairly similar to those of the nanoparticles of catalyst metals. This can be recognized from the spectra in Figure 6.14, where extinction efficiency spectra are plotted for Fe, Ni, and Co particles (optical constants from [147]),

The extinction efficiency of these particles only decrease from short to long wavelengths proportional to $1/\lambda$ (Rayleigh approximation of the extinction cross-section) without exhibiting any resonance-like feature for the smaller particles. If the particle size increases, a broader resonance-like extinction band grows, which shifts to longer wavelengths with increasing size. This can be assigned to the TM dipole mode. With increasing size, higher TM modes contribute to the spectrum with broad extinction bands which also shift to longer wavelengths. These resonances are determined more by the ratio of size to wavelength than by the electronic structure of the metal, that is, its dielectric function. Therefore, the spectra of the larger particles are fairly similar to those of the larger particles of the Drude and noble metals where the size also becomes dominant for the width of the TM

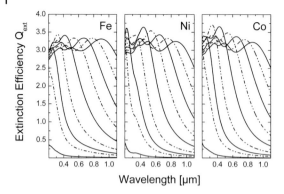

Figure 6.14 Spectra of iron, nickel, and cobalt nanoparticles with diameter 2R = 10, 50, 70, 100, 140, 180, 220, 260, and 300 nm.

modes. The contributions of the TE modes are again negligible for particles smaller than 300 nm. The high-order TM modes contribute to the extinction mainly by absorption, whereas the low-order TM modes contribute mainly by scattering.

Amekura *et al.* [148] fabricated magnetic nickel nanoparticles in silica glass (SiO_2) using high-flux implantation of 60 keV negative nickel ions. Photoabsorption measurements and cross-sectional transmission electron microscopy (XTEM) observations confirmed the formation of metallic Ni nanoparticles in SiO_2, and excluded the possible formation of Ni silicides (Ni_3Si, Ni_2Si, NiSi) and oxides (NiO) as major products. The mean diameter of the nanoparticles was 2.9 nm.

6.5
Rare Earth Metal Particles (Sc, Y, Er)

In this section we show spectra of Sc, Y, and Er nanoparticles as examples of rare earth metal particles.

Scandium, yttrium, and erbium have similar dielectric functions (optical constants from [149] for Sc and Er and from [103] for Y) and are good candidates for two resonances in nanoparticle extinction spectra: an electronic resonance at wavelengths below 450 nm due to an interband transition and an SPP resonance at wavelengths above 900 nm caused by the free carriers. As an example we refer again to Figures 5.17 and 5.18 for the dielectric function of Y and the possible resonances of yttrium nanoparticles.

The spectra of nanoparticles of Sc, Y, and Er presented in Figure 6.15 are fairly similar. Scandium nanoparticles exhibit two peaks in the spectral range between 270 and 350 nm, which can be assigned to the in principle two possible electronic resonances. In yttrium and erbium nanoparticles, only one the two possible electronic resonances is resolved (340–410 nm for Y, 300–380 nm for Er). For the Y

6.5 Rare Earth Metal Particles (Sc, Y, Er)

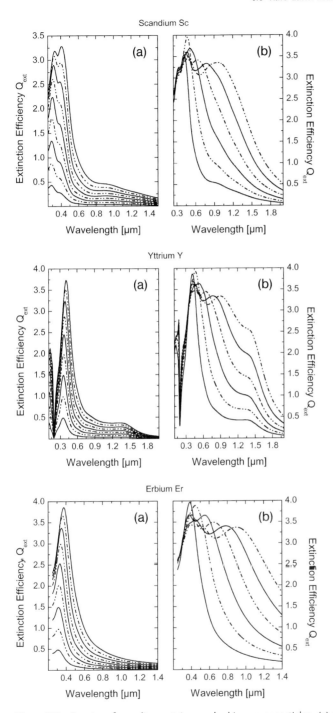

Figure 6.15 Spectra of scandium, yttrium and erbium nanoparticles. (a) $2R = 10$–$90\,\text{nm}$, stepwidth $10\,\text{nm}$; (b) $2R = 100$–$300\,\text{nm}$, stepwidth $40\,\text{nm}$.

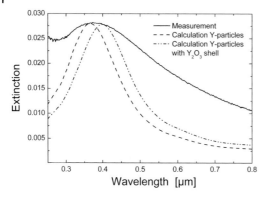

Figure 6.16 Measured and calculated Y nanoparticle spectra: measured (solid line), calculated with $2R = 23.5$ nm, $\sigma = 1.62$, log-normal distribution (dashed line) and with $2R = 19.5$ nm, $\sigma = 1.62$, log-normal distribution and oxide shell with $d = 2$ nm (dashed-dotted line).

nanoparticles, additionally the homogeneous band of interband transitions was observed below 195 nm with a very sharp edge, continuing to lower wavelengths (higher photon energies). This high-energy interband transition region remains substantially unchanged with increasing particle size except that the magnitude of the extinction efficiency increases.

Only for Sc and Y can a weak SPP band be recognized in the spectra at long wavelengths (around 950 nm for Sc and around 1350 nm for Y). For Er nanoparticles, this band cannot be recognized, although the condition for this resonance is fulfilled around 1500 nm. This is due to the high imaginary part of the dielectric function of Er in this spectral range, causing the SPP to be damped away.

Common features for all three metals are that with increasing particle size the position of the electronic TM dipole mode shifts to longer wavelengths and new resonances of higher order TM modes appear. For large particles, the weak SPP resonance becomes covered by the other resonances.

In the following experimental example, small yttrium nanoparticles were prepared by laser ablation followed by adiabatic expansion of the metal vapor into ultra-high vacuum through a small nozzle [150]. In this way, the metal atoms form nanoparticles by nucleation and condensation from the vapor phase. The optical extinction by the nanoparticles was measured using a multi-pass cell adjusted to use under vacuum conditions for measurement on the free particle jet. For size determination, particles were collected on grids for TEM. Figure 6.16 shows the optical extinction spectra of Y particles with median size $2R = 23.5$ nm and a standard deviation of $\sigma = 1.62$ (log-normal distribution). The measured spectrum (solid line) shows a broad extinction band which can be assigned to the interband transition at 325 nm (3.818 eV). The calculated spectrum (dashed line) was normalized to allow comparison with the measured spectrum. The peak position is slightly shifted to lower wavelengths compared with the peak position in the measurement. However, the most prominent difference between calculation and measurement is in the halfwidth of the peak, which is much smaller in the calculation. Obviously, neither the peak position nor the halfwidth of the extinction band can be reproduced in the calculations without further assumptions. One assumption is that the Y particles formed an oxide shell on their surface. The oxygen comes from the carrier gas argon in which oxygen appears as an impurity. The dashed-dotted line in Figure 6.16 shows the calculated spectrum of Y nanoparticles with a core of $2R = 19.5$ nm and an oxide shell of 2 nm thickness. It was calculated using the extension of Mie's theory on layered spheres, which will be introduced later in Chapter 7. In that case the peak position shifts to longer wavelengths but also cannot explain the halfwidth of the measured spectrum.

6.6
Transition Metal Particles (V, Nb, Ta)

Except for Al, Na, and K, all of the metals considered so far are transition metals. However, they have been classified into other classes because of their specific properties, for example Fe, Co, and Ni as genuine magnetic metals and Cu, Ag, and Au as noble metals. Therefore, we present here spectra of nanoparticles of transition metals that have not yet been classified. We selected the metals vanadium, niobium, and tantalum from the vanadium group.

As before for scandium, yttrium, and erbium, also vanadium, niobium, and tantalum are good candidates for two resonances in nanoparticle extinction spectra: an electronic resonance due to an interband transition and an SPP resonance caused by the free carriers. This follows from their dielectric function not shown here. The optical constants used in the calculations were taken from [149].

The spectra of nanoparticles of vanadium presented in Figure 6.17 are similar to those of Sc and Y nanoparticles. The interband transition leads to a particle resonance in the spectral range 280–350 nm. The weak SPP resonance lies around 1000 nm, but is difficult to recognize. For Nb, a distinct electronic resonance at 145 nm contributes to the spectrum of the smallest Nb nanoparticles, but an SPP resonance cannot be resolved from the spectra. In contrast, tantalum nanoparticles with size $2R < 150$ nm exhibit a clearly resolved SPP resonance at 640 nm. The electronic resonance appears at 150–160 nm and is less resolved. For larger sizes, the SPP becomes more and more superposed by the other size-determined resonances.

Common features are again that with increasing particle size the position of the electronic TM dipole mode shifts to longer wavelengths and new resonances of higher order TM modes appear. For large particles, the SPP, if present, becomes covered by the other resonances.

Takeda and Kishimoto [151] investigated the optical absorption and nonlinear optical response for nanoparticle composites in amorphous SiO_2 fabricated by negative Ta ion implantation at 60 keV. X-ray photoelectron spectroscopy was used to identify Ta and the oxide formation in the matrix. Optical absorption clearly indicated a broad SPP peak at 575 nm (2.156 eV) for the sample as implanted,

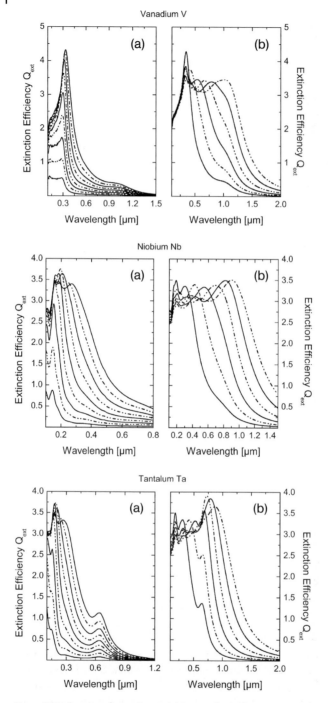

Figure 6.17 Spectra of vanadium, niobium and tantalum nanoparticles. (a) $2R = 10–90\,nm$, stepwidth 10 nm; (b) $2R = 100–300\,nm$, stepwidth 40 nm.

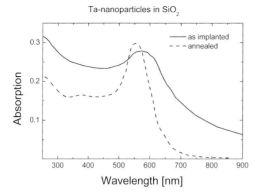

Figure 6.18 Optical absorption of Ta nanoparticles in amorphous SiO$_2$. After [151].

indicating the formation of nanoparticles embedded in the matrix (see Figure 6.18). After annealing, the peak shifts to 554 nm (2.24 eV) and becomes narrower. A size analysis is not available. The measured peak positions clearly differ from the peak position at 640 nm in the calculations. The reason is not yet known, but may arise from different optical constants. Takeda and Kishimoto measured the nonlinear absorption with a pump–probe method using a femtosecond laser system. The pumping laser transiently bleached the SPP band and led to the nonlinearity. The transient response recovered in a few picoseconds and behaved according to electron dynamics in metallic nanoparticles.

6.7
Summary of Metal Particles

Before turning to nonmetallic particles, we give a short summary for the metal particles.

1) For small metal particles of size less than 100 nm, both electronic resonances and SPP resonances can occur, which dominate the spectrum. These resonances can be explained on the one hand as a resonant excitation of the free electron plasma in the metals – a SPP – since these particles are in the order of or smaller than the penetration depth of light in metals. In other words, they only represent the *surface* for the light waves. On the other hand, interband transitions which can be described by a harmonic oscillator can lead to resonances in the optical extinction cross-section of small particles. In the framework of classical electrodynamics, the electronic resonances and the SPP are resonances of the TM dipole mode of a spherical particle with dielectric function ε in a surrounding medium with dielectric function ε_M. The resonances occur because for metals ε may become negative due to either the free electron contribution or to an interband transition. Then, the dielectric function of the metal approximately fits the condition $\varepsilon \approx -2\varepsilon_M$. This condition can

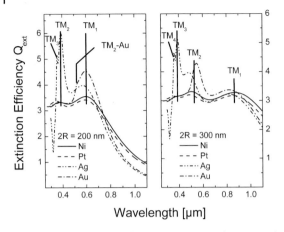

Figure 6.19 Comparison of extinction spectra of Ag, Ni, and Pt nanoparticles with 2R = 200 and 300 nm.

only be satisfied for small values of the imaginary part of ε. Therefore, only a few metals are well suited for such a TM dipole resonance. For the other metals, spheres smaller than 100 nm do not exhibit a resonance. The most prominent resonances in such small metal particles are the SPP resonances of aluminum, the Drude metals, and the noble metals.

2) The SPP leads to characteristic colors of finely dispersed particle systems of low concentration.

3) When the particle size exceeds $2R \approx 200$ nm, the differences among the spectra of metal particles increasingly vanish. The reason is that with increasing size more and more higher TM multipoles contribute to the spectrum. Their position and width are determined more by the ratio of the size to the wavelength than by the dielectric function of the metal. Nevertheless, if an electronic resonance or a SPP can be excited in the metal, also these resonances are sharper and higher than those of the other metals. This can clearly be recognized from Figure 6.19, where spectra of 200 and 300 nm sized particles of Ag, Au, Ni, and Pt are compared.

6.8
Semimetal Particles (TiN, ZrN)

TiN and ZrN belong to a larger group of compounds with extreme stability (hardness, chemical inertness, high melting point, high Young's modulus). The reason for the high stability is their bonding, which is an intricate mixture of covalent bonds and ionic contributions. They result in a *ceramic* behavior with respect to hardness and inertness, and a *metallic* behavior with high electrical and thermal

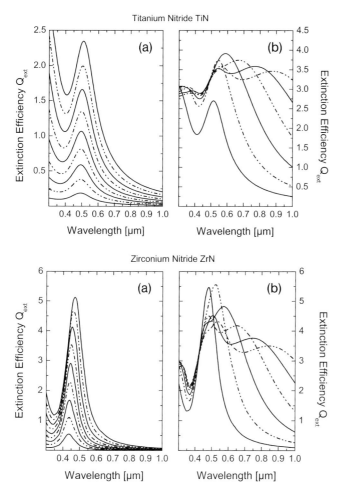

Figure 6.20 Spectra of titanium nitride and zirconium nitride nanoparticles. (a) 2R = 10–90 nm, stepwidth 10 nm; (b) 2R = 100–300 nm, stepwidth 40 nm.

conductivity and free electron-like optical behavior. In these transition metal nitrides, the d electrons just below the Fermi surface contribute to the electron gas, which is a striking result since it is unusual to associate d band compounds with free electron behavior.

The spectra of nanoparticles of TiN and ZrN in Figure 6.20 can be divided into two parts. At high photon energies, interband transitions from occupied electronic levels into unoccupied electronic levels contribute to the absorption of light by the particles. The onset wavelengths of these interband transitions reach into the visible spectral range, leading to the characteristic golden color of bulk TiN. The interband transitions are followed at lower photon energies by an extinction maximum at $\lambda = 490$ nm (2.52 eV) for TiN and at $\lambda = 435$ nm (2.85 eV) for ZrN.

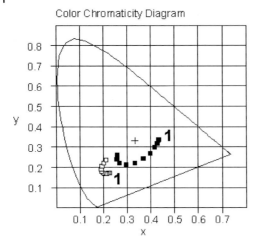

Figure 6.21 Chromaticity diagram showing the color change with increasing particle size for TiN nanoparticles (□) and ZrN nanoparticles (■). The "1" indicates the starting point of the curves.

This extinction band can be interpreted as SPP [101]. So far, the development of this peak and further structures in the spectra with increasing particle size is similar to that already discussed for Drude and noble metals. The optical constants for TiN and ZrN were taken from [101] and [102].

The color of assemblies of TiN and ZrN nanoparticles also changes with increasing size. This can be seen from Figure 6.21, where the color coordinates of TiN and ZrN nanoparticles with $2R \leq 100$ nm suspended in glass are plotted in a chromaticity diagram. The norm light source used was D65. The "1" in the chromaticity diagram indicates the start of the curves.

It is obvious that the apparent color of assemblies containing ZrN nanoparticles changes from red to violet with increasing particle size. The chromaticity coordinates from the calculations lie in a loop in the chromaticity diagram. For assemblies containing TiN nanoparticles the dependence on the particle size is still weaker. The color of assemblies containing small particles is already blue and remains blue with increasing size.

Reinholdt et al. [152, 153] examined the structural, compositional, and optical properties of TiN nanoparticles prepared by laser ablation under ultra high-vacuum conditions. The recorded optical spectra could be successfully compared with spectra of TiN nanospheres, although the structure of the particles is cubic.

Cortie et al. [154] also recently examined the optical properties of nanostructures comprised of TiN. The dielectric properties of TiN_{1-x} depend upon stoichiometry and are favorable for plasmon resonance phenomena in the mid-visible to near-infrared (NIR) range of the spectrum and for $x \approx 0$. They analyzed the optical phenomena operating in such structures using a combination of experiment and simulation and showed that semi-shells of TiN exhibit a tunable localized plasmon resonance with light. The material is, however, unsuitable for applications in which a long-distance SPP is desired.

6.9
Semiconductor Particles (Si, SiC, CdTe, ZnSe)

6.9.1
Calculations

The research on semiconductor nanoparticles is younger than that on metal nanoparticles. Much of the knowledge and understanding obtained with metal nanoparticles could be directly applied to semiconducting materials. The most prominent effect in semiconductor nanoparticles is the *quantum size effect*, which describes the effect that the wavelength for photoluminescence shifts to shorter wavelengths with decreasing particle size. A simple practical description of this effect is the following. UV illumination generates an electron–hole pair, which is called an *exciton*. When the exciton recombines, light with the corresponding wavelength is emitted. The energy of the exciton can be described within Bohr's atomic model. Here, a larger Bohr radius means low energy or a long emission wavelength. If the nanocrystal becomes smaller than this Bohr radius, it must shrink because it must lie within the crystal. A smaller radius means higher energy and therefore a shorter wavelength for the emitted light. This allows for the first time control of the wavelength for photoluminescence by preparing particles with different size. The great advantage of the semiconductor nanoparticles is that they do not lose their luminance but increase it up to its maximum.

Improvements in synthesis in the last 20 years have led to the production of quantum dots of different materials – controlling their size, shape, and surface properties. Examples are CdS, CdTe, InP, GaAs, PbS, PbSe, and CdSe. CdSe is the most studied system. An example of the size dependence of the fluorescence of quantum dots is given in Figure 6.22.

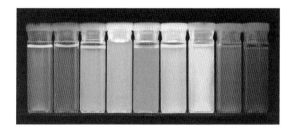

Figure 6.22 Fluorescence of CdSe–ZnS nanoparticles. The size of the quantum dots increases from about 2 nm to about 6 nm (left to right). The emission spans across the visible part of the electromagnetic spectrum from the blue (2 nm nanocrystals) to the red (6 nm nanocrystals). A color version of this figure can be found in the color plates at the end of the book. Reprinted from [155] with permission of Elsevier, copyright 2009.

Table 6.1 Complex refractive index values of crystalline and polycrystalline silicon and of silicon carbide.

Wavelength (nm)	Photon energy (eV)	n(Si) crystalline	n(Si) polycrystalline	n(SiC)
400	3.100	5.581 + i0.366	5.381 + i0.515	2.765 + i0.0002
500	2.480	4.299 + i0.070	4.243 + i0.135	2.686
600	2.066	3.948 + i0.026	3.905 + i0.052	2.645
700	1.771	3.784 + i0.012	3.741 + i0.013	2.622
800	1.550	3.693 + i0.006	3.651 + i0.006	2.603

The control of the fluorescence is often complicated. Due to the high surface-to-volume ratio, impurities often grow at the surface that lead to recombination of the exciton without emission or the emission lies below the bandgap. This is prevented by surface modification either by adsorption of special molecules or by forming core–shell structures such as CdS–CdOH$_2$, CdSe–CdS, CdSe–ZnSe, InP–ZnS, and InAs–CdSe.

In the following, we do not consider photoluminescence further but concentrate on absorption and scattering by semiconducting nanoparticles for the semiconductors silicon (Si), silicon carbide (SiC), cadmium telluride (CdTe), and zinc selenide (ZnSe).

Both Si and SiC exhibit strong interband transitions in the near-UV range close to the visible spectral region. The result is that both materials have high refractive indices in the visible spectral region while the absorption index is small. For example, in Table 6.1 some refractive indices are summarized for Si and SiC at various wavelengths (photon energies). Optical constants were taken from [36]. Hence both materials are good candidates for electronic resonances in the interband transition region and for morphology-dependent resonances in the visible and NIR region, even for nanoparticles. This can also be followed from the spectra in Figure 6.23. With increasing particle size, sharp MDRs develop very rapidly in the spectra for both TM and TE partial waves. This is in contrast to the metal particles considered earlier, where the main contribution comes almost solely from the TM modes. The shift of the resonance positions to longer wavelengths is similar to that already observed for the metals.

The spectral region is extended for Si down to wavelength 0.1 μm to demonstrate the influence of the strong interband transition at 285 nm (4.35 eV). At this position the dielectric function of Si becomes negative according to the harmonic oscillator model and an electronic resonance appears in the spectrum of nanometer-sized Si particles. With this resonance and the development of higher multipoles which shift to longer wavelengths with increasing size, in the wavelength range 0.1–0.3 μm Si nanoparticles behave similarly to aluminum nanoparticles (compare Figure 6.1).

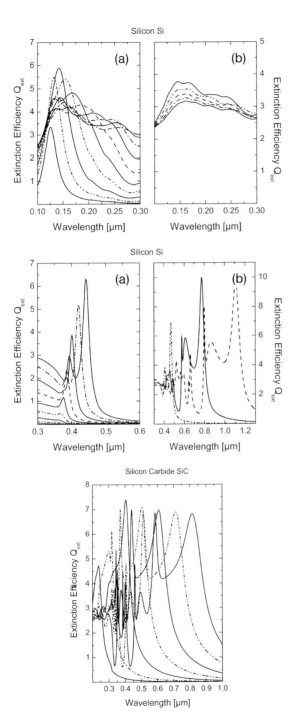

Figure 6.23 Spectra of silicon and silicon carbide nanoparticles. Silicon: (a) $2R = 10$–90 nm, stepwidth 10 nm; (b) $2R = 100, 200,$ and 300 nm. The traces at the top show the wavelength range between 0.1 and 0.3 μm in particular. Silicon carbide: $2R = 70$ nm, $2R = 100$–300 nm, stepwidth 40 nm.

6 Application of Mie's Theory

Table 6.2 Complex refractive index values of cadmium telluride and zinc selenide.

Wavelength (nm)	Photon energy (eV)	n(CdTe)	n(ZnSe)
400	3.100	$3.644 + i0.455$	$2.886 + i0.478$
500	2.480	$3.275 + i0.258$	$2.741 + i0.086$
600	2.066	$3.066 + i0.070$	$2.577 + i0.061$
700	1.771	$2.939 + i0.053$	$2.507 + i0.050$
800	1.550	$2.871 + i0.044$	$2.466 + i0.044$

In the following, two further semiconductor materials, CdTe and ZnSe, are considered as examples. Like Si and SiC, also CdTe and ZnSe, and practically all other semiconducting materials, exhibit strong interband transitions in the near-UV range close to the visible spectral region. The result is that these materials have high refractive indices in the visible spectral region while the absorption index is small. In Table 6.2 some refractive indices are summarized for CdTe and ZnSe at various wavelengths (photon energies). Optical constants were taken from [156] for CdTe and from [157] for ZnSe. They are high as in nanoparticles of these materials MDRs develop in the spectral region beyond the interband transitions where electronic resonances dominate the spectra (see Figure 6.24).

6.9.2
Experimental Examples

6.9.2.1 Si Nanoparticles in Polyacrylene

Crystalline silicon nanoparticles were prepared by W. Hoheisel (personal communications) using plasma-enhanced chemical vapor deposition (PECVD) and were embedded in polyacrylene at different concentrations. Optical density spectra were recorded and are compared with calculated spectra in Figure 6.25. Investigation of the particles using electron microscopy showed the crystalline structure and almost spherical particles. The calculated spectra coincide fairly well with the experimental spectra.

6.9.2.2 Quantum Confinement in CdSe Nanoparticles

Alivisatos and his group examined CdSe nanoparticles in detail [158–160]. They arrived at very small particles with $2R$ between 0.9 and 3.5 nm. Optical extinction spectra were recorded for 12 colloidal solutions with varying size [159]. These spectra shown in Figure 6.26 clearly exhibit a resonance caused by the exciton. The peak position depends on the particle size and clearly shifts from 492m to 615 nm with increasing size.

Robel et al. [161] proposed and built solar cells of CdSe nanoparticles (quantum dots) with diameter of 3 nm assembled on films of titanium dioxide nanoparticles of diameter 40–50 nm. The quantum dots are well suited for solar energy conver-

6.9 Semiconductor Particles (Si, SiC, CdTe, ZnSe)

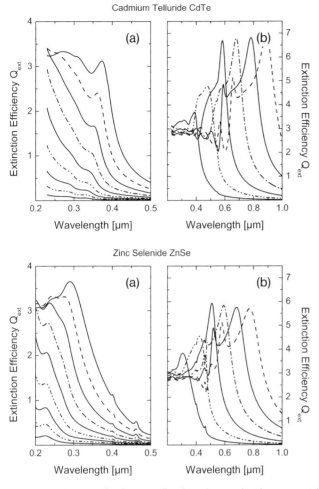

Figure 6.24 Spectra of cadmium telluride and zinc selenide nanoparticles. (a) $2R = 10$–90 nm, stepwidth 10 nm; (b) $2R = 100$, 140, 180, 220, 260, and 300 nm.

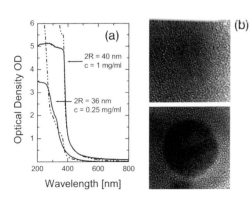

Figure 6.25 (a) Measured and calculated optical density spectra of Si nanoparticles with $2R = 36$ and 40 nm in polyacrylene. (b) Electron micrograph of an Si nanoparticle. Data and electron micrograph courtesy of W. Hoheisel.

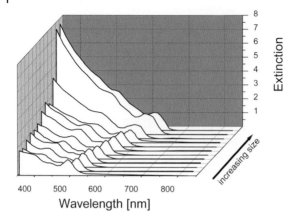

Figure 6.26 Optical extinction spectra of small CdSe nanoparticles (quantum dots) with varying particle size 2R. The size amounts to 2R = 0.93, 1.0, 1.04, 1.07, 1.14, 1.23, 1.32, 1.41, 1.59, 1.8, 2.4, and 3.35 nm from bottom to top. The spectra are shifted along the ordinate for better presentation. After [158].

sion because they have a high power conversion efficiency and the absorption maximum can be spectrally tuned.

6.10
Carbonaceous Particles

The optical properties of particles that consist of carbonaceous material are of interest in many fields of research such as aerosol science and astrophysics. In both, astrophysical systems (e.g., interstellar clouds, circumstellar shells and disks) and terrestrial atmospheric aerosols they directly influence the energy balance of the system by absorption and re-emission of radiation. From the astrophysical point of view, the ultraviolet region is also extremely important since the π–π^* electronic transitions in such particles are observed in the interstellar extinction curve [162, 163].

Carbon can have covalent bindings with different sp-hybridization – interaction of the s and p electrons of carbon atoms – for which a large variety of organic and inorganic carbonaceous species exist. The sp hybridization dominantly determines the crystalline structure and also electronic and optical properties. This is manifested in the different properties of graphite (sp^2, aromatic or olefinic bondings) on the one hand and diamond (sp^3, aliphatic bondings) on the other. In Figures 6.27–6.29 we summarize the spectra of nanoparticles for the three classes of carbonaceous materials: amorphous carbon, graphite, and diamond. For graphite, spectra for both crystalline directions (parallel and perpendicular to the crystalline c-axis) are presented. The optical constants are taken from Draine and Lee [164, 165], where also an overview of optical constants of graphite is given. For the

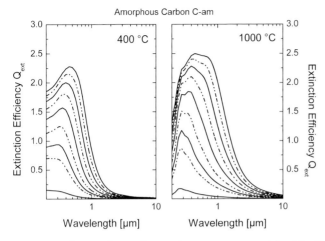

Figure 6.27 Spectra of amorphous carbonaceous nanoparticles. 2R = 10, 50, and 70 nm, and 2R = 100–300 nm, stepwidth 40 nm. From pyrolysis of cellulose at 400 and 1000 °C.

amorphous carbon we used optical constants from Jäger et al. [166], which were obtained from measurements on amorphous carbons produced by pyrolysis of cellulose. These data have the advantage that they cover a range of different microstructures (sp^2:sp^3 hybridization ratios) and, therefore, electrical conductivity and optical properties.

Amorphous carbon can be considered as a mixture of sp^2 and sp^3 hybridization with additional sp^1 hybridization (acetylenic bondings). As a result, distinct interband transitions do not exist, nor are the materials transparent in any spectral region. This can be seen from the curves in Figure 6.27. The spectra of the nanoparticles exhibit only a broad extinction structure which appears slightly size dependent.

Graphite (sp^2 hybridization) can be considered as a semiconducting material with a relatively low density of free electrons. Therefore, a SPP resonance occurs in the FIR region, fairly well separated from the interband transitions in the UV region. Since graphite is further not uniaxial, the optical properties parallel and perpendicular to the crystalline c-axis differ. Graphite is strongly absorbing at wavelengths longer than 80 nm (photon energies less than 20 eV). The corresponding spectra of graphite nanoparticles are shown in Figure 6.28.

According to the optical constants tabulated by Draine and Lee [164, 165], for graphite perpendicular to the crystalline c-axis, there are two electronic transitions in the UV region at 280 nm (4.43 eV) and 90 nm (13.6 eV), and furthermore an interband transition in the IR region at 62 μm (0.02 eV). The transitions in the UV region correspond to transitions between the valence band and the conduction band and can be assigned to π electrons (~4 eV, one electron per C atom) and σ electrons (~14 eV, three electrons per C atom). In the FIR region, the dielectric function is determined by free electrons with $\omega_p = 4.08 \times 10^{14}\,\text{s}^{-1}$ and $\gamma_{fc} = 4.7 \times 10^{13}\,\text{s}^{-1}$. For nanoparticles with these optical constants, the interband

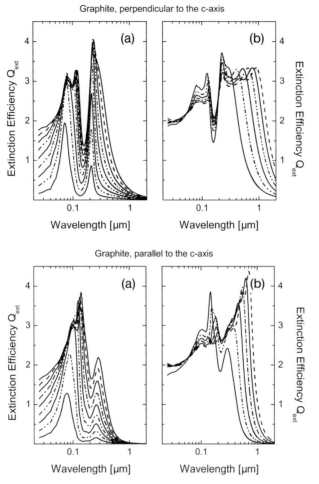

Figure 6.28 Spectra of graphitic carbonaceous nanoparticles, perpendicular and parallel to the c-axis. (a) $2R = 10–90$ nm, stepwidth 10 nm; (b) $2R = 100–300$ nm, stepwidth 40 nm.

transitions result in resonances of the TM dipole mode at 212 nm (5.85 eV) and 72 nm (17.22 eV) for the smallest particles, which shift to longer wavelengths with increasing particle size. Moreover, the evolution of higher TM modes and their shift to lower energies with increasing size can be recognized.

For graphite parallel to the c-axis, the parameters of the free electrons are $\omega_p = 1.53 \times 10^{14}\,\text{s}^{-1}$ and $\gamma_{fc} = 7.1 \times 10^{13}\,\text{s}^{-1}$. Also for graphite parallel to the c-axis there are two electronic transitions in the near-UV region, but with maxima now at approximately 300 nm (4.13 eV) and 110 nm (11.25 eV). They can also be assigned to the π and σ electrons as for graphite perpendicular to the c-axis. For nanoparticles with these optical constants, the interband transitions also result in resonances of the TM dipole mode, now at $\lambda = 260$ nm (4.77 eV) and 79 nm (15.69 eV)

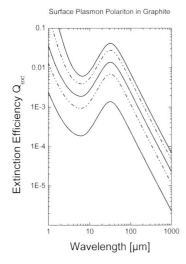

Figure 6.29 SPP in graphite parallel to the c-axis. $2R = 10, 50, 100, 200,$ and $300\,nm$.

for the smallest particles, which shift to longer wavelengths with increasing particle size. Moreover, the evolution of higher TM modes and their shift to lower energies with increasing size can be recognized.

The resonance of the TM dipole mode at $\lambda = 212\,nm$ (5.85 eV) for graphite perpendicular to the c-axis is about five times larger in magnitude than the resonance at $\lambda = 300\,nm$ (4.13 eV) for graphite parallel to the c-axis. It plays an important role in the interpretation of the interstellar extinction curve [162, 163] since it corresponds to the π–π* electronic transitions in such nanoparticles.

The amount of free electrons should lead to a SPP peak in the corresponding wavelength range. Indeed, for graphite parallel to the c-axis a weak, but recognizable, SPP peak appears at $\lambda = 31\,\mu m$ (0.04 eV), as shown in Figure 6.29. The peak position is independent of the size because the particle size of less than 300 nm is very small compared with the wavelength of about 10 μm. For graphite perpendicular to the c-axis the SPP cannot be resolved.

Diamond (sp^3 hybridization) exhibits extreme properties, including ultra-high strength, unmatched thermal conductivity, ultra-high melting temperature, radiation hardness, high magnetic field compatibility, low friction and adhesion, biocompatibility, and chemical inertness. Diamond is a wide-gap insulator, but once electrons have been excited above the gap, they exhibit the highest mobility of any known material.

Optically, diamond exhibits high absorption in the UV region due to interband transitions. It is nonabsorbing in the visible and NIR regions, with a high refractive index of 2.2–2.5 in the visible spectral region. Hence diamond particles are good candidates for morphology-dependent resonances in nanoparticles. The features that can be recognized in the calculated spectra in Figure 6.30 are actually morphology-dependent resonances.

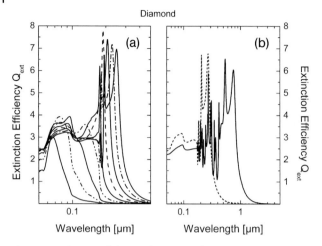

Figure 6.30 Spectra of diamond nanoparticles. (a) $2R$ = 10–90 nm, stepwidth 10 nm; (b) $2R$ = 100, 200, and 300 nm.

In the following experimental example, carbonaceous particles were prepared by an experimental technique based on the extraction of carbon grains from different sources (flames, bulk evaporation) by free-jet expansion and the subsequent isolation of these particles in noble gas solid matrices for *in situ* spectroscopy [167]. The particles were collected at different distances from the nozzle (15, 20, and 25 mm). At short distances the particles remained separated in the argon matrix, whereas at larger distances they formed agglomerates. This can be seen from the TEM images in Figure 6.31a and b. The corresponding measured absorption spectra are shown in Figure 6.31c. They exhibit a peak at λ = 236 nm (5.254 eV). This extinction peak can also be recognized in the spectra of graphitic nanoparticles (see Figure 6.28) with sizes less than $2R$ = 100 nm. The sample AR15 (isolated particles) can be fitted fairly well with a computed spectrum with particles of $2R$ = 10 nm and optical constants of amorphous carbon. The deviations of the other samples AR20 and AR25 are caused by particle aggregates.

A shift of the UV bump of amorphous carbon nanoparticles towards lower wavelengths can be obtained by hydrogen incorporation. This was investigated in a second series of matrix-isolated carbonaceous particles [167] in which hydrogen was added to the argon gas. Here, we compare the resulting spectrum with the interstellar extinction curve which exhibits a hump at 217.5 nm (5.7 eV). This hump is the dominant and most controversial feature of the interstellar extinction curve, discovered by Stecher in 1965 [162]. It has been attributed to π–π^* band transitions in small graphite particles or amorphous carbon grains.

In Figure 6.32 the interstellar extinction curve is shown (dots) together with the spectrum of other interstellar dust particles (diamonds) as calculated from Li and Greenberg [168]. The difference curve (dashed-dotted-dotted line) can fitted very well with a spectrum of spherical nanoparticles ($2R$ = 10 nm) of hydrogenated (33%) amorphous carbon (solid line).

6.10 Carbonaceous Particles

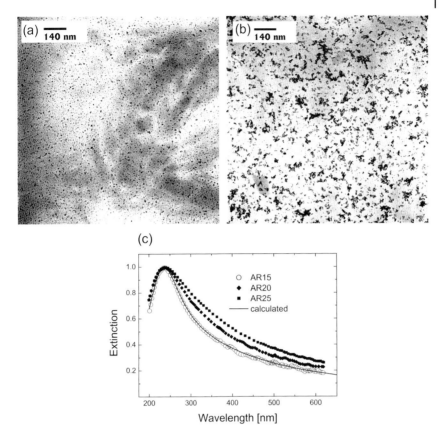

Figure 6.31 Transmission electron micrographs of matrix-isolated carbonaceous particles collected at distance $d =$ (a) 15 and (b) 25 mm from the nozzle. Reproduced from [167] with permission of the American Astronomical Society. (c) Measured absorbance spectra of the samples AR15, AR20, and AR25 with comparison with a calculated spectrum ($2R = 10$ nm). Data courtesy of H. Mutschke.

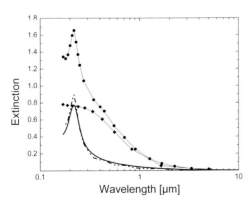

Figure 6.32 Interstellar extinction curve (circles), interstellar dust particle extinction (diamonds) and comparison of the difference spectrum (dashed-dotted-dotted line) with the spectrum of hydrogenated amorphous carbon nanoparticles (solid line).

6.11
Absorbing Oxide Particles (Fe$_2$O$_3$, Cr$_2$O$_3$, Cu$_2$O, CuO)

6.11.1
Calculations

In Figure 6.33 we summarize the calculated spectra of iron oxide, chromium(III) oxide, copper(I) oxide, and copper(II) oxide nanoparticles for various particle sizes. Optical constants were taken from [60] for Fe$_2$O$_3$ and Cr$_2$O$_3$ and from [36] for Cu$_2$O and CuO.

Nanoparticles of α-Fe$_2$O$_3$ (hematite) are well known as red pigment particles. For small nanoparticles the decrease in the extinction efficiency from wavelengths

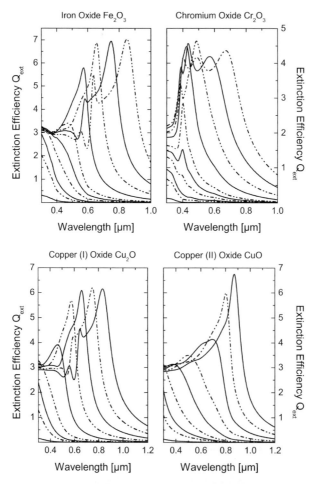

Figure 6.33 Spectra of Fe$_2$O$_3$, Cr$_2$O$_3$, Cu$_2$O, and CuO nanoparticles with diameter $2R$ = 10, 50, and 70 nm, and $2R$ = 100–300 nm, stepwidth 40 nm. For Fe$_2$O$_3$ and Cr$_2$O$_3$ also $2R$ = 120 nm.

in the near-UV to longer wavelengths results in a brownish red color. However, for application as pigments, the resonance of the TM dipole mode in the red visible spectral region for particles with $2R \approx 200\text{--}300\,\text{nm}$ is much more important. This resonance shifts to longer wavelengths with increasing particle size, whereby additional resonances at higher energies occur. Hematite is also found in cosmic dust and in natural atmospheric and polluted aerosols. Other prominent pigments based on Fe are α-FeOOH (goethite) as a yellow pigment and $M^I Fe^{II} Fe^{III}(CN)_6 \cdot H_2O$ (M = metal) as blue pigments (55–70 nm).

Analogously, chromium(III) oxide (Cr_2O_3) nanoparticles are used for green pigments. For particles with sizes $2R \approx 200\text{--}280\,\text{nm}$, the TM dipole resonance of the particles contributes optimally to the absorption and scattering, so that a green color could result in such nanoparticle systems. Another prominent pigment based on chromium is $(Ti,Cr,Sb)O_2$ as a yellow pigment ($2R \approx 200\,\text{nm}$).

The color of Fe_2O_3 and Cr_2O_3 nanoparticles, however, develops best for opaque particle systems such as lacquers. Then, the Kubelka–Munk theory must be applied to obtain the reflectance of the nanoparticle system. This is demonstrated later in Chapter 12.

Similarly to Fe_2O_3, Cu_2O nanoparticles could also serve as a red pigment, for particles with $2R > 200\,\text{nm}$. Again, the color develops best for opaque particle systems. Less spectacular are the spectra of CuO nanoparticles. They are peaked just for nanoparticles with $2R > 260\,\text{nm}$. Hence systems with smaller CuO nanoparticles are poorly colored and appear gray or black.

6.11.2
Experimental Examples

6.11.2.1 Aerosols of Fe_2O_3

Colloidal nanoparticles of α-Fe_2O_3 (hematite) can simply be prepared from thermolysis of $FeCl_3$ in an aqueous solution [169]. $FeCl_3$ is favorable because of its low melting point ($T_m = 37\,°C$) and its low boiling point ($T_b = 282\,°C$). The particles form readily when the temperature approaches the boiling point of water. Similarly to colloidal Fe_2O_3 nanoparticles, such nanoparticles were generated as aerosol particles by spray pyrolysis of aqueous $FeCl_3$ solutions of various concentrations [170]. The procedure for Fe_2O_3 aerosols was as follows:

- Preparation of an aqueous solution of $FeCl_3$.

- Spraying the solution.

- Drying the droplets in a diffusion drier.

- Heating the resulting salt particles in a furnace at different temperatures. The best results were obtained for temperatures between 600 and 800 °C. The solid salt particles which leave the drier are still wet, hence thermolysis leads to oxides.

- Measurement of the extinction of light by the aerosol.

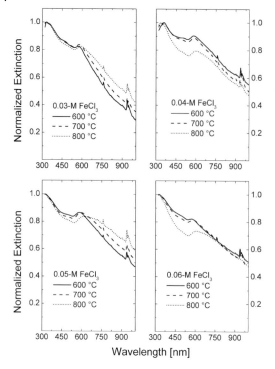

Figure 6.34 Extinction spectra of Fe_2O_3 aerosol particles from spray pyrolysis of $FeCl_3$. Concentrations in the precursor and temperatures in the furnace are indicated.

Fe_2O_3 nanoparticles were generated by spray pyrolysis of 0.03, 0.04, 0.05, and 0.06 M aqueous $FeCl_3$ solution. The extinction spectra were recorded with a multi-pass cell and a multi-channel spectrometer and are displayed in Figure 6.34. In all graphs the extinction is normalized to the maximum, because different particle concentrations would render the interpretation of the spectra more difficult.

At wavelengths between 300 and about 550 nm, the extinction decreases monotonically with increasing wavelength in all spectra. The decrease is the largest for the highest temperature. It is followed by a distinct extinction maximum at $\lambda = 570$–580 nm, which shifts towards longer wavelengths with increasing temperature. Simultaneously with the shift, the maximum is broadened with increasing temperature. At longer wavelengths the extinction again decreases monotonically.

Two unexpected features are to be seen at $\lambda = 761$ nm and in the wavelength range between 930 and 950 nm. These belong to molecular absorption bands of oxygen ($\lambda = 761$ nm) and water. Because of the long pathlength in the aerosol chamber, molecular absorption bands can be detected when measuring extinction. Water comes from the aqueous precursor solutions and is not completely removed when the aerosol enters the multi-pass cell. Oxygen is partly contained in the carrier gas and comes from the ambient air.

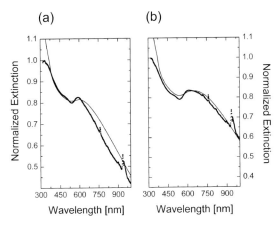

Figure 6.35 Comparison of two typical extinction spectra of Fe$_2$O$_3$ particles from spray pyrolysis of FeCl$_3$ (squares) with spectra computed assuming a bimodal log-normal size distribution of Fe$_2$O$_3$ particles (solid lines).

Comparing the spectra with those of Fe$_2$O$_3$ particles with $2R = 190\text{--}260$ nm (see also Figure 6.33), it is clear that these particles exhibit a sufficiently small extinction maximum in this wavelength range and can preferably be used for interpretation of the measured spectra. However, for explanation of the decrease in the spectra at short wavelengths, smaller particles must also be considered which mainly contribute in this spectral region, showing Rayleigh scattering and absorption. Therefore, it seems appropriate to use a bimodal log-normal distribution of smaller and larger particles to interpret the measured spectra. For interpretation of the two measured spectra in Figure 6.35, particles with $2R_1 = 20$ nm and $2R_2 = 160$ nm must be assumed for the spectrum in Figure 6.35a, whereas for the spectrum in Figure 6.35b the size of the larger particles must have been increased to $2R_2 = 180$ nm. The corresponding amount of smaller particles was 94% in both computed spectra and 6% for the larger particles. This result shows that the particle size of the larger particles increases with increasing temperature.

6.11.2.2 Aerosols of Cu$_2$O and CuO

Copper oxide nanoparticles were generated as aerosol particles by spray pyrolysis of aqueous CuSO$_4$ solutions of various concentrations [170]. The procedure for copper oxide aerosols was as follows:

- preparation of an aqueous solution of copper sulfate (CuSO$_4$)
- spraying the solution
- drying the droplets in a diffusion drier
- heating the resulting salt particles in a furnace at different temperatures
- measurement of the extinction of light by the aerosol.

Copper oxide nanoparticles were generated as aerosol particles by spray pyrolysis of 0.05, 0.1, 0.188, and 0.2 M aqueous CuSO$_4$ solutions. Extinction spectra were

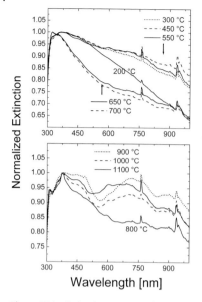

Figure 6.36 Extinction spectra of Cu_xO_y particles generated from spray pyrolysis of a 0.188 M $CuSO_4$ solution.

recorded with a multi-pass cell and a multi-channel spectrometer. They are displayed in Figure 6.36 for only the 0.188 M solution. The results for the other concentrations were similar.

The spectra change distinctly with increasing temperature. Starting at 200 °C, the spectrum is due to $CuSO_4$ salt particles. With increasing temperature, mainly the extinction at longer wavelengths increases. On approaching 650 °C, the spectrum changes significantly. Now the extinction decreases monotonically from 300 to 1000 nm. This behavior is maintained at higher temperatures (800 °C). At 900 °C, the spectrum again changes significantly. The spectra at temperatures above 900 °C can be divided approximately into three parts: part 1 from 300 to 500 nm, part 2 from 600 to 800 nm, and part 3 from 850 to 1000 nm. They are separated by minima in the extinction, of which the deepest is between parts 1 and 2. The decrease of all spectra at wavelengths below about 320 nm is caused by the halogen lamp used, which does not provide sufficient intensity in this spectral region.

The features at $\lambda = 761$ nm and in the wavelength range between 930 and 950 nm can again be assigned to molecular absorption bands of oxygen and water. $CuSO_4$ decomposes at 650 °C with separation of oxygen. This oxygen also contributes to the oxygen content in the aerosol, by which CuO particles are formed. For that reason, the spectrum changes when the temperature approaches 650 °C. Formation of Cu particles or Cu particles with a CuO shell can be excluded because the oxidation of copper is complete before nucleation from the gas phase can lead to condensed particles. When the temperature exceeds 900 °C, the CuO particles undergo a reduction process to Cu_2O, which is well known in chemistry. This is

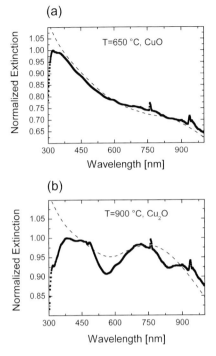

Figure 6.37 Comparison of two measured spectra (T = 650 and 900 °C) with bimodal log-normal distributions with $2R_1$ = 60 nm and $2R_2$ = 220 nm with σ = 1.5.

the reason for the second change in the spectra. As the optical constants of CuO and Cu_2O are different, the spectra of particles of same size must differ at different temperatures. However, now the formation of shells cannot be excluded, and the particles may consist of a CuO core and a Cu_2O shell.

Two spectra at 650 and 900 °C are compared in Figure 6.37 with computed spectra of CuO and Cu_2O particles with a bimodal distribution with $2R_1$ = 60 nm and $2R_2$ = 220 nm (dashed lines in Figure 6.37). For the spectrum of CuO particles at 650 °C, the contribution of the smaller particles amounts 94%, and in the spectrum at 900 °C (Cu_2O) their contribution is 85%. The corresponding contributions for the larger particles are 6% and 15%. In both cases the agreement is good. This means that the differences between the two measured spectra are actually not caused by the particle size but by the transition from Cu_2O to CuO.

6.11.2.3 Colloidal Fe_2O_3 nanoparticles

Detailed experimental and theoretical investigations on hematite particles with modal diameters of $2R$ = 100, 120, 130, 150, 160, and 510 nm were carried out by Hsu and Matijévic [171]. Experimental spectra are shown in Figure 6.38. Comparison was made by the authors with calculations according to Mie's theory, with very good agreement.

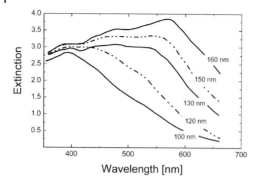

Figure 6.38 Experimental absorption spectra of hematite nanoparticles with modal diameters $2R = 100$, 120, 130, 150, and 160 nm. After [171].

6.12
Transparent Oxide Particles (SiO_2, Al_2O_3, CeO_2, TiO_2)

In this section, we turn to nonabsorbing oxidic particles, defining *nonabsorbing* here with missing absorption in the visible and NIR spectral region. Optical constants for SiO_2 and Al_2O_3 are from [36], for CeO_2 from [172] and for TiO_2 from [173].

Silica (SiO_2) has long been established as the workhorse nanoparticle. However, the seemingly endless functionalization combinations attract ever-increasing attention to SiO_2 nanoparticles. From enhancing the mechanical properties (scratch resistivity) to changing the viscosity and thixotropy of a formulation, to the lotus effect and easy-to-clean functionality, SiO_2 nanoparticles are important contributors.

Alumina (Al_2O_3) is a common ceramic material known to have many crystalline phases. Corundum (α-Al_2O_3) is the only phase that is thermodynamically stable at all temperatures. Being very hard, corundum is used as an abrasive in many applications, such as abrasive papers and chemical–mechanical polishing (CMP).

TiO_2 pigments are used in large amounts in almost all composites, such as lacquers, colors, plastics, and synthetic fibers. Nanoparticles of TiO_2 have been available for only a couple of decades. Intended to be used as UV absorbers, today they are also used for coloring effects based on selective scattering of blue light. Typical mean sizes of TiO_2 particles in sun protection creams (UV absorption) are 10–25 and 300 nm. As white pigments the size ranges between 180 and 350 nm.

Ceria nanoparticles with primary particle sizes of less than 10 nm and aggregates smaller than 200 nm are used as a polish in CMP or in the catalysis of diesel exhaust.

Another prominent oxide with technical applications as nanoparticles is zinc oxide (ZnO). It is used in ointments for wound healing, for UV protection of plastics, leather, wood, and paintings, and also in sun creams. Typical primary

6.12 Transparent Oxide Particles (SiO$_2$, Al$_2$O$_3$, CeO$_2$, TiO$_2$)

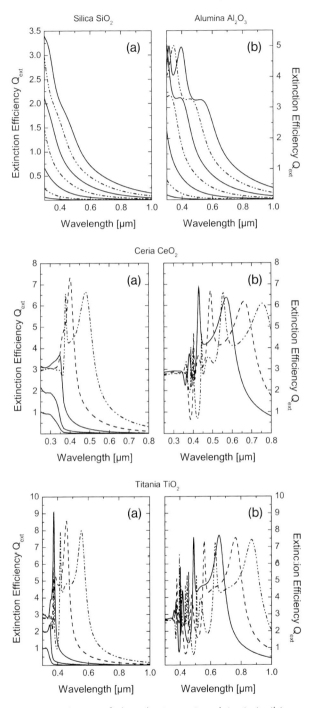

Figure 6.39 Spectra of silica, alumina, ceria and titania (rutile) nanoparticles. Particle sizes for SiO$_2$ and Al$_2$O$_3$: $2R = 70$ nm and $2R = 100$–300 nm, stepwidth 40 nm. Particle sizes for CeO$_2$ and TiO$_2$: $2R = 50, 70$ nm and $2R = 100$–300 nm, stepwidth 40 nm.

Table 6.3 Complex refractive index values of TiO_2 and CeO_2.

Wavelength (nm)	Photon energy (eV)	$n(TiO_2)$	$n(CeO_2)$
400	3.1	3.40	2.749 + i0.0004
500	2.48	3.03	2.541
600	2.066	2.90	2.464
700	1.771	2.83	2.423
800	1.55	2.79	2.404

particle sizes are 25 nm with aggregates of 100–250 nm in size. ZnO is not further considered here, but is similar to CeO_2.

For nonabsorbing materials, only morphology–dependent resonances can occur. However, for silica (SiO_2) and alumina (Al_2O_3), the refractive index is still too low for MDRs in nanoparticles with sizes below 300 nm. Therefore, only broad resonances of low-order multipoles can be expected in the spectra, as can actually be seen from Figure 6.39.

Titania (TiO_2) and ceria (CeO_2) absorb strongly in the near-UV region due to the onset of direct interband transitions. These close-lying interband transitions cause a fairly high refractive index in the visible spectral region for these materials, similarly to the semiconducting particles considered earlier. Exemplary values are given in Table 6.3. For these highly refractive materials, MDRs develop for nanoparticles with increasing particle size. Similarly to the semiconductors, both TM and TE modes exhibit MDRs which shift to longer wavelengths with increasing particle size.

Crystalline and amorphous ceria nanoparticles prepared by Nyacol (Ashland, MA, USA) by chemical reduction in poly(vinyl alcohol) (PVA) at different concentrations were examined by W. Hoheisel (personal communication). Optical density spectra were recorded and compared with calculated spectra. They are presented in Figure 6.40 together with a characteristic TEM image. The investigation of the particles with electron microscopy showed almost spherical particles. The calculated spectra coincide fairly well with the experimental spectra.

6.13
Particles with Phonon Polaritons (MgO, NaCl, CaF_2)

Interaction of light with matter is dominantly determined by the interaction with the electronic system. In all the examples considered so far, this interaction also determined the optical properties of nanoparticles. We turn now to examples where the coupling of light with TO phonons leads to a significant change in the dielectric function of the material. In almost all ionic crystals this coupling can be interpreted in the framework of harmonic oscillators. As already seen and discussed, for example, for yttrium, the real part of the dielectric constant may become negative close to the resonance frequency of the harmonic oscillator.

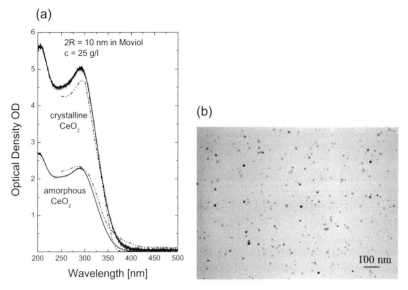

Figure 6.40 (a) Measured and calculated optical density spectra of ceria nanoparticles with $2R = 10$ nm in PVA. (b) Transmission electron micrograph of ceria nanoparticles. Data and electron micrograph courtesy of W. Hoheisel.

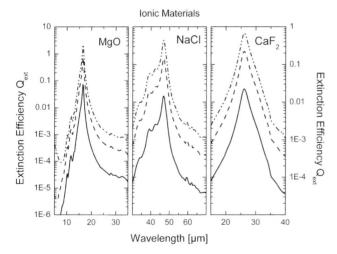

Figure 6.41 Spectra of MgO, NaCl, and CaF_2 nanoparticles with $2R = 10$, 100, and 300 nm. Logarithmic scale.

Indeed, the resonance condition $\varepsilon = -2\varepsilon_M$ for the TM dipole mode may be approached twice, depending on the steepness of the dielectric function in the corresponding spectral region. In the spectra of NaCl and MgO nanoparticles in Figure 6.41, one large peak and several smaller peaks can be clearly seen. For CaF_2, only one resonance structure can be recognized. Optical constants were taken from [36].

In contrast to the various electronic resonances considered earlier for the metal and semiconductor nanoparticles, the resonances here show a weak size dependence with respect to the peak position. Only the magnitude of Q_{ext} increases with increasing size. The reason is simply that even the largest particle size considered of $2R = 300$ nm is very small compared with the wavelength (15–60 µm) in these wavelength ranges. Therefore, these particles can be treated in the Rayleigh limit, in which the particle size does not influence the resonance position.

6.14
Miscellaneous Nanoparticles (ITO, LaB$_6$, EuS)

In this last section we give with ITO, lanthanum hexaboride (LaB$_6$) and europium sulfide (EuS) three examples of nanoparticle materials that could not be clearly assigned to any of the particle material classes considered earlier.

Particularly ITO is of distinct interest because in this material free electrons contribute to a certain electrical conductivity of the material, but the material remains transparent at wavelengths in the visible spectral range. Hence ITO films are often used in applications where an electrical current is used for switching a process, and where in addition the film must be transparent, for example, in flat panel displays or organic light emitting diodes (OLEDs) or as transparent conductive oxides (TCOs) in thin-film photovoltaics. Owing to the increased price of indium, today alternatives to ITO have been developed such as aluminum-doped zinc oxide (AZO), fluorine-doped tin oxide (FTO) and antimony-doped tin oxide (ATO). The highest optical transparency and the highest electrical conductivity, however, are still obtained for ITO.

The dielectric function ε of an ITO film is not unique, since it depends on the quality of the film and the preparation conditions. However, in all cases, it can be approximated in the photon energy range 0.5–6 eV by a sum of a harmonic oscillator in the UV region and a Drude susceptibility with plasma frequency ω_p lying in the NIR region. In between, in the visible spectral range, the contributions of both to the imaginary part ε_2 of the dielectric constant are rather small, so that absorption in the visible region becomes negligible for thin films with thickness <1 µm. On the other hand, the close-lying harmonic oscillator in the UV region leads to a high refractive index in the visible spectral range of $n_{ITO} \approx 1.8$–1.9. For nanoparticles of ITO this means that (i) morphology-dependent resonances may evolve with increasing particle size and (ii) a SPP resonance can be expected in the NIR region.

Indeed, it can be seen from the calculated spectra of ITO nanospheres in Figure 6.42 that many MDRs develop in the spectral range from 0.2 to 0.6 µm, extending over the visible spectral range. This part is similar to the spectra for the high refractive index materials TiO$_2$ and CeO$_2$. For still larger particles, more MDRs can be expected, their peaks shifting to longer wavelengths. In the NIR spectral region, the SPP can clearly be recognized at wavelength $\lambda = 1.766$ µm (0.702 eV) for the smallest particle. With increasing particle size, the plasmon peak

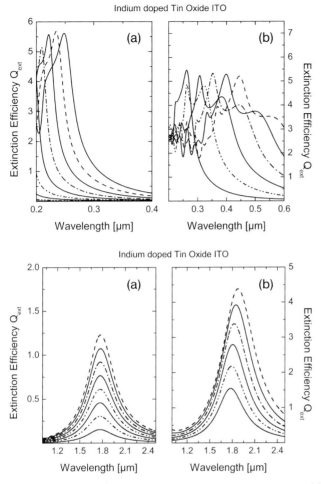

Figure 6.42 Spectra of ITO nanoparticles. (a) $2R = 10-90$ nm, stepwidth 10 nm; (b) $2R = 100-300$ nm, stepwidth 40 nm. The graphs at the bottom show the SPP resonance.

undergoes a slight red shift to $\lambda = 1.875\,\mu m$ (0.6613 eV) for 300 nm large particles. Optical constants for ITO were taken from [174].

The first materials used for NIR absorbers based on nanoparticles were also the conductive oxides ATO and ITO. One especially interesting group of materials for infrared absorbers consists of the rare earth hexaborides and lanthanum hexaboride (LaB_6). They belong to the quasi-metals and have been of interest for some time. LaB_6 has attracted the most attention in this group, because it is of interest also for other applications. For example, its low work function prompts its use for electron sources in electron microscopes and possibly as an electron source for a free electron laser. The NIR absorption, caused by the excitation of SPPs, prompts its use as an infrared absorber.

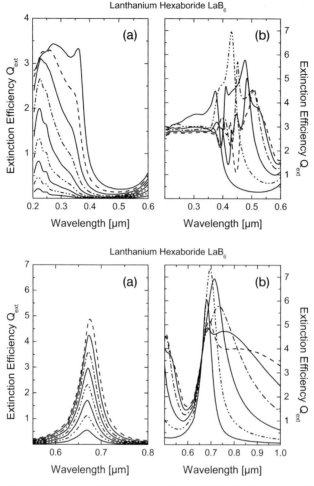

Figure 6.43 Spectra of LaB$_6$ nanoparticles. (a) $2R = 10–90$ nm, stepwidth 10 nm, (b) $2R = 100–300$ nm, stepwidth 40 nm. The graphs at the bottom show the SPP resonance.

Comparing the calculated spectra of LaB$_6$ nanospheres in Figure 6.43 with the spectra of ITO, the spectral region below 0.6 μm is comparable to that of ITO. The SPP can also clearly be recognized but now at the wavelength $\lambda = 0.67$ μm for the smallest particle. With increasing particle size, the plasmon peak undergoes a red shift and higher TM multipole modes contribute at longer wavelengths. This is in contrast to ITO, where more or less only the TM dipole mode exhibits a SPP resonance. In addition, the MDRs shift to the wavelength where the SPP is peaked. It can be expected that for still larger particles MDRs and SPPs will mix in the wavelength range above 0.6 μm. Optical constants for LaB$_6$ were taken from [175].

Europium chalcogenides have been studied extensively since the 1960s because of their unique magnetic and semiconducting properties. Nanoparticles of EuS

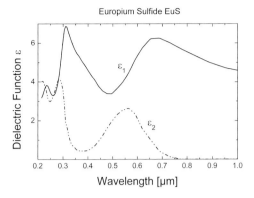

Figure 6.44 Dielectric function of europium sulfide (EuS).

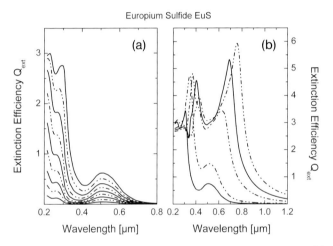

Figure 6.45 Spectra of EuS nanoparticles. (a) $2R = 10\text{--}90$ nm, stepwidth 10 nm; (b) $2R = 100\text{--}300$ nm, stepwidth 40 nm.

exhibit increased luminescence efficiency compared with the bulk material resulting from quantum size confinement.

Looking at the dielectric function of EuS in Figure 6.44 (data from P. Fumagalli, personal communications), it is composed of at least two well-separated harmonic oscillators at $\lambda = 286$ and 565 nm. In between, the real part ε_1 has a minimum at $\lambda = 495$ nm.

Although ε_1 has only positive values in the spectral region shown, the curvatures of ε_1 and ε_2 lead to a resonance-like behavior in the extinction efficiency of small EuS particles, as can be seen in Figure 6.45. The spectra of EuS nanoparticles exhibit a broad extinction band at a wavelength of about $\lambda = 500$ nm (2.5 eV), leading to an orange–red color impression for such nanoparticle systems. In the

UV region, further extinction bands can be recognized which hardly influence the color of the system provided that the particles are less than about 100 nm in diameter. For larger particles, these TM resonances shift to longer wavelengths and more TM resonances appear. Their contributions finally dominate the extinction efficiency also around $\lambda = 500$ nm.

To understand the resonance-like structure at $\lambda = 500$ nm, it seems appropriate to repeat the equation for the extinction efficiency in the Rayleigh approximation:

$$Q_{ext} \propto \frac{R}{\lambda} \frac{\varepsilon_2}{(\varepsilon_1 + 2\varepsilon_M)^2 + \varepsilon_2^2} \tag{6.2}$$

Inserting the values of the dielectric function from Figure 6.44 in this equation, it can be calculated that Q_{ext} is maximum at $\lambda = 500$ nm, and decreases continuously to the right and left of this position. Therefore, Q_{ext} shows a feature similar to a resonance but without being a resonance caused by the electronic structure of EuS, neither an electronic resonance nor a SPP resonance.

7
Extensions of Mie's Theory

7.1
Coated Spheres

The most interesting extension of Mie's theory for a single compact sphere is the extension on a coated sphere, since they can often be found in realistic nanoparticle systems due to condensation of gaseous components from the ambient, or as natural oxides of the particle material. The first extensions on a sphere with a single layer trace back on Aden and Kerker [176] in 1951 and Güttler [177] in 1952, who were interested in aerosols and interstellar dust. Their results were later extended to multiple layers by several groups [178–185].

7.1.1
Calculations

Consider a multilayered spherical particle consisting of $r-1$ concentric shells of arbitrary materials and of different thickness d_s, $s = 2, \ldots, r$, and a spherical core with radius R_{core} ($s = 1$) (Figure 7.1). The total diameter of the sphere is $2R$. The refractive indices n_1 of the core and n_s of each shell may be complex numbers, which include absorption in the core and in each shell. The particle is assumed to be embedded in a nonabsorbing medium with refractive index n_M and to be illuminated by an incident plane wave with wavenumber $k_M = (2\pi n_M)/\lambda$. The parameters $m_s = n_s/n_{s+1}$ represent the relative refractive indices from shell s to shell $s+1$. For $s = r$, the refractive index n_{r+1} corresponds to the refractive index n_M of the embedding medium. In each medium, the wavenumber is $k_s = (2\pi n_s)/\lambda$. The corresponding size parameters x_s are defined as

$$x_s = k_{s+1}\left(R_{core} + \sum_{j=2}^{s} d_j\right) \tag{7.1}$$

The electromagnetic fields of the incident wave, the scattered wave, and the waves in the core and in each shell are expressed in terms of vector spherical harmonics. The expansions of the incident and the scattered wave and also the expansion of the wave inside the core are already known from Equations 5.8–5.10 in Section 5.1. The expansion of the wave inside a shell is

Optical Properties of Nanoparticle Systems: Mie and Beyond. Michael Quinten
Copyright © 2011 WILEY-VCH Verlag GmbH & Co. KGaA, Weinheim
ISBN: 978-3-527-41043-9

7 Extensions of Mie's Theory

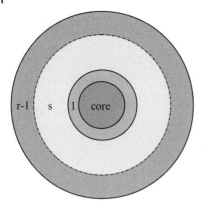

Figure 7.1 Spherical particle with $r-1$ concentric shells.

$$E_s = E_0 \sum_{n=1}^{\infty} \frac{i^n(2n+1)}{n(n+1)} \{[\beta_n^s \mathbf{M}_{o n 1}^{(1)}(k_s) + \delta_n^s \mathbf{M}_{o n 1}^{(2)}(k_s)]$$
$$- i[\alpha_n^s \mathbf{N}_{e n 1}^{(1)}(k_s) + \gamma_n^s \mathbf{N}_{e n 1}^{(2)}(k_s)]\} \quad (7.2\text{a})$$

$$H_s = \frac{-k_s E_0}{\omega \mu_0} \sum_{n=1}^{\infty} \frac{i^n(2n+1)}{n(n+1)} \{[\alpha_n^s \mathbf{M}_{e n 1}^{(1)}(k_s) + \gamma_n^s \mathbf{M}_{e n 1}^{(2)}(k_s)]$$
$$+ i[\beta_n^s \mathbf{N}_{o n 1}^{(1)}(k_s) + \delta_n^s \mathbf{N}_{o n 1}^{(2)}(k_s)]\} \quad (7.2\text{b})$$

where α_n^s and γ_n^s represent the expansion coefficients of the TM modes and β_n^s and δ_n^s represent the expansion coefficients of the TE modes in the shells ($s > 1$).

The use of spherical Neumann functions y_n (superscript 2 of the vector harmonics) for the fields inside each shell accounts for the continuity of these functions in these limited domains. Note that in the literature [180, 183] alternatively the fields inside the shells are expanded using spherical Hankel functions of the first kind. Then, the corresponding expansion coefficients formally differ from the above, but the numerical results for the scattering coefficients and the cross-sections are identical.

At each boundary between two distinct media, the electromagnetic fields have to satisfy Maxwell's boundary conditions. For a multilayered sphere, in principle, there are three different boundaries which must be discussed in detail:

- the boundary between the core and the innermost shell
- the boundary between two shells
- the boundary between the particle and the embedding medium.

For the latter, they connect the electromagnetic fields of the incident wave and the scattered wave with the wave in the outermost shell.

Starting with the innermost boundary, the stepwise resolution of the boundary conditions yields the following equations for the scattering coefficients a_n and b_n of a multilayered spherical particle with $r-1$ shells:

$$a_n = \frac{\psi_n(x_r)[\psi'_n(m_r x_r) + S_n^{r-1}\chi'_n(m_r x_r)] - m_r\psi'_n(x_r)[\psi_n(m_r x_r) + S_n^{r-1}\chi_n(m_r x_r)]}{\xi_n(x_r)[\psi'_n(m_r x_r) + S_n^{r-1}\chi'_n(m_r x_r)] - m_r\xi'_n(x_r)[\psi_n(m_r x_r) + S_n^{r-1}\chi_n(m_r x_r)]} \quad (7.3)$$

$$b_n = \frac{m_r\psi_n(x_r)[\psi'_n(m_r x_r) + T_n^{r-1}\chi'_n(m_r x_r)] - \psi'_n(x_r)[\psi_n(m_r x_r) + T_n^{r-1}\chi_n(m_r x_r)]}{m_r\xi_n(x_r)[\psi'_n(m_r x_r) + T_n^{r-1}\chi'_n(m_r x_r)] - \xi'_n(x_r)[\psi_n(m_r x_r) + T_n^{r-1}\chi_n(m_r x_r)]} \quad (7.4)$$

with

$$S_n^s = -\frac{\psi_n(x_s)[\psi'_n(m_s x_s) + S_n^{s-1}\chi'_n(m_s x_s)] - m_s\psi'_n(x_s)[\psi_n(m_s x_s) + S_n^{s-1}\chi_n(m_s x_s)]}{\chi_n(x_s)[\psi'_n(m_s x_s) + S_n^{s-1}\chi'_n(m_s x_s)] - m_s\chi'_n(x_s)[\psi_n(m_s x_s) + S_n^{s-1}\chi_n(m_s x_s)]} \quad (7.5)$$

$$T_n^s = -\frac{m_s\psi_n(x_s)[\psi'_n(m_s x_s) + T_n^{s-1}\chi'_n(m_s x_s)] - \psi'_n(x_s)[\psi_n(m_s x_s) + T_n^{s-1}\chi_n(m_s x_s)]}{m_s\chi_n(x_s)[\psi'_n(m_s x_s) + T_n^{s-1}\chi'_n(m_s x_s)] - \chi'_n(x_s)[\psi_n(m_s x_s) + T_n^{s-1}\chi_n(m_s x_s)]} \quad (7.6)$$

and the starting values $S_n^0 = T_n^0 = 0$. $\psi_n(x) = xj_n(x)$, $\chi_n(x) = xy_n(x)$, and $\xi_n(x) = xh_n^{(1)}(x)$ are Riccati–Bessel, Riccati–Neumann and Riccati–Hankel functions, respectively. The prime denotes derivation with respect to the argument. Note that the solution for a homogeneous sphere is obtained for $r = 1$.

Extinction and scattering cross-sections, σ_{ext} and σ_{sca}, the efficiencies Q_{ext} and Q_{sca}, respectively, of a multilayered sphere are formally the same as for a compact sphere (Equations 5.26 and 5.27), except for the scattering coefficients a_n and b_n.

Also for multilayered spheres, the corresponding TM and TE modes and, hence, the cross-sections may be resonant. The nth TM mode with expansion coefficient a_n is resonant if

$$m_r \frac{\xi'_n(x_r)}{\xi_n(x_r)} = \frac{\psi'_n(m_r x_r) + S_n^{r-1}\chi'_n(m_r x_r)}{\psi_n(m_r x_r) + S_n^{r-1}\chi_n(m_r x_r)} \quad (7.7)$$

and the nth TE mode with expansion coefficient b_n is resonant if

$$\frac{1}{m_r} \frac{\xi'_n(x_r)}{\xi_n(x_r)} = \frac{\psi'_n(m_r x_r) + T_n^{r-1}\chi'_n(m_r x_r)}{\psi_n(m_r x_r) + T_n^{r-1}\chi_n(m_r x_r)} \quad (7.8)$$

For discussion of the electronic and plasmonic resonances of coated spheres, we consider a sphere (core size R) coated with one shell (thickness d) in the Rayleigh approximation, so that only the TM dipole mode ($n = 1$) contributes to the cross-section:

$$\sigma_{ext}(\lambda,R,d) = \frac{8\pi^2(R+d)^3\sqrt{\varepsilon_M}}{3\lambda} \cdot \text{Im}\begin{bmatrix} (\varepsilon_{sh}-\varepsilon_M)(\varepsilon_c+2\varepsilon_{sh}) \\ +\left(\dfrac{R}{R+d}\right)^3(\varepsilon_c-\varepsilon_{sh})(\varepsilon_M+2\varepsilon_{sh}) \\ \hline (\varepsilon_c+2\varepsilon_{sh})(\varepsilon_{sh}+2\varepsilon_M) \\ +2\left(\dfrac{R}{R+d}\right)^3(\varepsilon_c-\varepsilon_{sh})(\varepsilon_{sh}-\varepsilon_M) \end{bmatrix} \quad (7.9)$$

where ε_c, ε_{sh}, and ε_M are the dielectric functions of the core, the shell, and the surrounding medium, respectively, and Im means the imaginary part of the expression in brackets. It is easy to prove that for vanishing shell thickness $d \to 0$ a homogeneous sphere with $\varepsilon_c = \varepsilon_{sh}$ and size $2R$ is obtained. For nearly identical dielectric functions of the core and the shell, a homogeneous sphere but with size $2(R+d)$ is obtained.

We can distinguish two cases in which resonances appear:

1) The dielectric function ε_{sh} of the shell becomes negative in a certain wavelength range and the denominator of Equation 7.9 becomes minimum. This is the case for dielectric particles coated with a metallic shell.

2) The dielectric function ε_c of the core material becomes negative in a certain wavelength range and the denominator of Equation 7.9 becomes minimum. This is the case for metal particles coated with a dielectric shell.

Both will be discussed in the following on the basis of Equation 7.9.

7.1.1.1 Metallic Shells on a Transparent Core

Assuming ε_c to be real valued, that is, the core particle is nonabsorbing, and ε_{sh} to be the dielectric function of a metal, for example, silver, then the denominator in Equation 7.9 becomes minimum at two wavelengths, because the denominator is quadratic in ε_{sh}. This is similar to a thin metal film, where two surface plasmons can be induced due to the coupling among the surfaces of the film [97, 98].

For a small shell thickness, that is, for $R/(R+d) \approx 1 - d/R$, the resonance conditions can be approximately derived as

$$\varepsilon_{sh} \approx -\frac{R}{2d}(\varepsilon_c + 2\varepsilon_M) = -2\varepsilon_M \frac{R}{2d}\left(1 + \frac{\varepsilon_c}{2\varepsilon_M}\right) \quad (7.10)$$

$$\varepsilon_{sh} \approx -\frac{1}{2}\varepsilon_c \quad (7.11)$$

The first condition (Equation 7.10) means that the real part of dielectric function ε_{sh} must becomes more negative than for a compact metal sphere, mainly because of the factor R/d. A more negative value is reached at longer wavelengths, meaning that the resonance is red shifted. The red shift is the larger the smaller is the shell thickness d. The second condition (Equation 7.11) is caused by the inner boundary between the core and shell.

In Figure 7.2 we present extinction spectra of silver-coated silica particles with different core size $2R = 20$, 40, 60, 80, and 100 nm *in vacuo* calculated with the full extended Mie theory. The silver shell thickness is $d = 5$ nm.

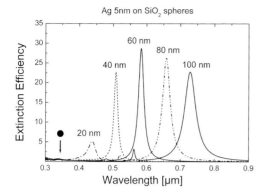

Figure 7.2 Computed extinction efficiency spectra of silver-coated SiO$_2$ spheres with core sizes $2R$ = 20, 40, 60, 80, and 100 nm and a silver shell thickness d = 5 nm.

There are at least two resonances to observe, which correspond to the excitation of surface plasmon polaritons (SPPs). The small peak at λ = 341 nm (marked with the black circle) corresponds to the SPP according to Equation 7.11. The most prominent peak belongs to the SPP according to Equation 7.10. Its wavelength position depends on the size of the core, and it is possible to assign the wavelength to the core size. Its width is dominated by the gradient of the real part $\varepsilon_{1,sh}$ of the dielectric function of the shell. With increasing core size, additional extinction maxima appear between these two SPPs. They belong to higher multipolar orders $n \geq 2$, which now become clearly resolved in the spectrum. They cannot be followed from Equation 7.9, since we restricted consideration there to only the dipole contribution. However, they are included in the full theory for layered spheres. In the Rayleigh approximation, their resonance condition (of the TM multipole of order n) is given from the solution of the following equation:

$$\left(\varepsilon_{sh} + \frac{n+1}{n}\varepsilon_M\right)\left(\varepsilon_c + \frac{n+1}{n}\varepsilon_{sh}\right) + \frac{n+1}{n}\left(\frac{R}{R+d}\right)^3 (\varepsilon_{sh} - \varepsilon_M)(\varepsilon_c - \varepsilon_{sh}) = 0 \quad (7.12)$$

In Figure 7.3, the resonance positions of the first three TM multipoles of a silver sphere with $2R$ = 22 nm are compared with the resonance position of a silica sphere with $2R$ = 20 nm and a silver film with thickness d = 1 nm.

It is easy to recognize that the resonance positions of the multipoles are clearly separated for the silver-coated particle (open symbols) whereas the peak positions of the compact silver particle (full symbols) are close together. The discussion so far makes it clear that resonance positions of a silver-coated sphere may be tuned within a wide range of wavelengths by adjusting the thickness of the shell.

In some further calculations on silver-coated silica particles, the dependence of the observable resonance positions on core size was examined for thickness d = 1, 2, 5, and 10 nm [186]. The results are depicted in Figure 7.4. The lines through the calculated positions are only guidelines to the eye, but to a good approximation a linear relationship is observed. For d = 10 nm, the evolution of spectra with particle core size from $2R$ = 40 to 300 nm is shown in Figure 7.5. For clarity, the spectra are

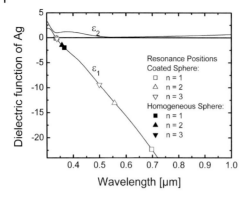

Figure 7.3 Resonance positions of Mie coefficients a_n with $n \leq 3$ for silver spheres of diameter $2R = 22$ nm and for silver-coated silica spheres with core diameter $2R = 20$ nm and shell thickness $d = 1$ nm. The surrounding medium is vacuum ($n_M = 1$).

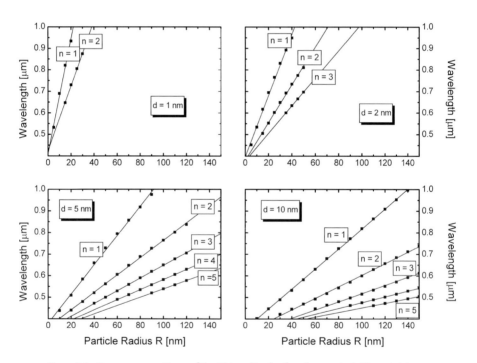

Figure 7.4 Resonance positions of the TM multipoles for silver-coated silica particles as a function of core radius and thickness of the shell. The straight lines are guidelines to the eye.

shifted vertically by a constant amount. The individual multipolar contributions are identified. The line marked with a star arises from all multipoles appearing in the spectrum and corresponds to the resonance given by Equation 7.11.

For small silica particles the spectrum is dominated by the TM dipole mode indicated as TM1, but already from $2R = 100$ nm the TM-quadrupole (TM2)

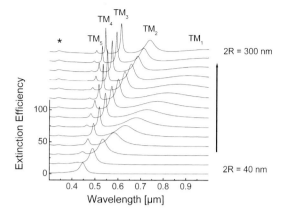

Figure 7.5 Evolution of spectra with core size for silica particles coated with $d = 10$ nm Ag.

contributes to the spectra. With still increasing size more TM multipoles up to TM5 can be recognized. They all shift to longer wavelengths with increasing size.

For large shell thickness, that is, for $R/(R + d) \ll 1$, the resonance conditions can be approximately derived as

$$\varepsilon_{sh} \approx -2\varepsilon_M \tag{7.13}$$

$$\varepsilon_{sh} \approx -\frac{1}{2}\varepsilon_c \tag{7.14}$$

The second condition is the same as Equation 7.11 for the case of thin shells. The first condition (Equation 7.13) coincides with the condition for a homogeneous sphere. It is the limiting value for large shell thickness. Then, the metal-coated sphere with dielectric core behaves like a compact metal sphere. However, as for large thickness d the total size of the particle also increases and therefore the particle do not further fulfill the Rayleigh approximation, this limit will not be completely approached, but the extinction efficiency of the coated sphere will deviate from the extinction efficiency of the compact sphere.

In the Figure 7.6 we present extinction spectra of silver-coated silica particles with core size 100 nm and increasing thickness $d = 5, 10, 20$, and 30 nm *in vacuo*. For comparison, the extinction efficiency of a compact silver sphere with $2R = 160$ nm is also plotted. It is to be compared with the coated sphere with $d = 30$ nm. For computation, full Mie-theory is applied.

Actually, the spectrum of the coated particle with $d = 30$ nm is similar to the spectrum of the compact sphere for wavelengths $\lambda > 500$ nm. At smaller wavelengths, the differences are larger except for the interband transitions.

Another example of metal-coated dielectric particles is given in Figure 7.7, which depicts the spectra of gold-coated silica nanoparticles with increasing core size and a shell thickness of $d = 5$ nm. The spectra are similar to that in Figure 7.2 for silver-coated silica spheres.

However, the short-wavelength resonance cannot be resolved and the contribution of the quadrupole resonances is weaker than for silver-coated silica spheres.

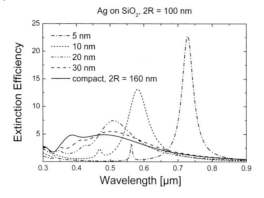

Figure 7.6 Computed extinction efficiency spectra of silver-coated SiO$_2$ spheres with core size $2R = 100$ nm and a silver shell thickness $d = 5$, 10, 20, and 30 nm.

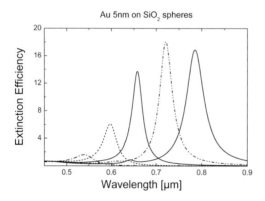

Figure 7.7 Computed extinction efficiency spectra of gold-coated SiO$_2$ spheres with core sizes $2R = 20$, 40, 60, 80, and 100 nm and a gold shell thickness $d = 5$ nm.

Further calculation examples for copper-coated dielectric particles can be found in [187].

7.1.1.2 Oxide Shells on Metal and Semiconducting Core Particles

For the inverse system – dielectric shell on a metal core particle – the denominator in Equation 7.9 must be resolved for the dielectric function ε_c of the core. This is a simpler task since the denominator is linear in ε_c. The resonance condition is

$$\varepsilon_c = -2\varepsilon_{sh} \frac{(\varepsilon_{sh} + 2\varepsilon_M) - \left(\dfrac{R}{R+d}\right)^3 (\varepsilon_{sh} - \varepsilon_M)}{(\varepsilon_{sh} + 2\varepsilon_M) + 2\left(\dfrac{R}{R+d}\right)^3 (\varepsilon_{sh} - \varepsilon_M)} \qquad (7.15)$$

The limiting case for small d is

$$\varepsilon_c \approx -2\varepsilon_M \qquad (7.16)$$

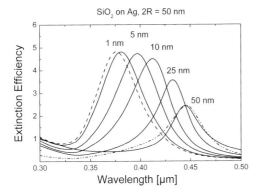

Figure 7.8 Computed extinction efficiency spectra of silica-coated Ag spheres with core size $2R = 50$ nm and shell thickness $d = 1, 5, 10, 25,$ and 50 nm. The dashed line corresponds to the uncoated silver sphere and the dashed-dotted-dotted line to an uncoated silver sphere embedded in silica. This spectrum is divided by 5.1.

which corresponds to the uncoated ($d = 0$) sphere. Vice versa, for large d, the resonance position is

$$\varepsilon_c \approx -2\varepsilon_{sh} \qquad (7.17)$$

This corresponds to a compact sphere with dielectric function ε_c embedded in a surrounding medium with dielectric function ε_{sh}. Note that unlike the surrounding medium with real-valued ε_M the dielectric function of the shell may be complex valued, that is, absorption in the shell is possible. Moreover, this limiting case cannot actually be approached, since for large d the total size of the sphere exceeds the limits of the Rayleigh approximation and full Mie theory must be applied.

Figure 7.8 comprises the extinction efficiency spectra for an SiO$_2$-coated silver sphere with $2R = 50$ nm core radius. The thickness of the silica shell is 1, 5, 10, 25, and 50 nm. For comparison, the spectrum of an uncoated silver sphere (dashed line) and the spectrum of an uncoated silver sphere in silica as surrounding medium (dashed-dotted-dotted line) are plotted. The latter is divided by 5.1. It can clearly be recognized that even a very thin film of only 1 nm leads to a red shift of the SPP resonance. The increasing thickness of the shell leads to a further red shift of the SPP resonance of the core particle. If the thickness is large, the spectrum is comparable to that of a silver particle embedded in silica, except for the magnitude. To allow comparison with the spectrum of a silver sphere in silica, this spectrum must be divided by 5.1.

Unlike the noble metals platinum, gold, and silver, for most of the metals oxidation leads to natural oxide shells, even for the noble metal copper. Semiconducting materials are also affected by this natural oxidation. For nanoparticles, this oxidation process is even accelerated and one can expect that nanoparticles of most metals and semiconductors have a natural oxide shell. In the following, we give examples of the optical properties of chromium, copper, and silicon nanoparticles with an oxide shell of $d = 5$ nm.

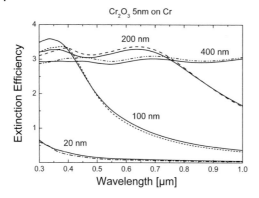

Figure 7.9 Computed extinction efficiency spectra of Cr_2O_3-coated chromium spheres with core size $2R_c = 20, 100, 200,$ and $400\,nm$ with a shell thickness $d = 5\,nm$. The solid lines correspond to compact chromium spheres with $2R = 2(R_c + d)$.

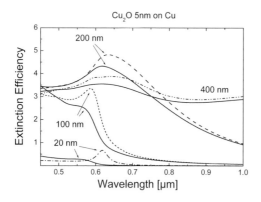

Figure 7.10 Computed extinction efficiency spectra of Cu_2O-coated copper spheres with core size $2R_c = 20, 100, 200,$ and $400\,nm$ with a shell thickness $d = 5\,nm$. The solid lines correspond to compact copper spheres with $2R = 2(R_c + d)$.

We start with the spectra of chromium nanoparticles with core diameters $2R_c = 20, 100, 200,$ and $400\,nm$ with a chromium oxide Cr_2O_3 shell with $d = 5\,nm$. For comparison, the spectra of compact chromium nanoparticles with $2R = 2(R_c + d)$ are plotted as solid lines. The spectra are summarized in Figure 7.9.

Independently of the core size, the effect of the 5 nm thick oxide shell on the spectral response of the particles is small.

For copper nanoparticles, we assumed a copper(I) oxide (Cu_2O) shell on the nanoparticles. This is the natural oxide that forms first when copper is oxidized in the ambient atmosphere. Later, copper(II) oxide (CuO) is formed by further oxidation of Cu_2O. The color of a Cu_2O film is similar to that of pure copper, whereas CuO is black. Core size and shell thickness are the same as for the chromium nanoparticles. The results of the computations are shown in Figure 7.10.

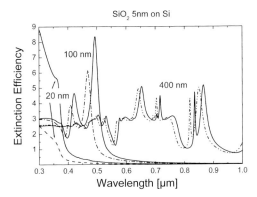

Figure 7.11 Computed extinction efficiency spectra of SiO_2-coated silicon spheres with core size $2R_c = 20$, 100, and 400 nm with a shell thickness $d = 5$ nm. The solid lines correspond to compact silicon spheres with $2R = 2(R_c + d)$.

Obviously, the influence of the copper oxide shell is remarkable for sizes up to 400 nm. The differences between uncoated and coated nanoparticles are not negligible as before for the chromium nanoparticles. As earlier in Figure 7.8 for the coated silver nanoparticles, the oxide shell leads to a clear red shift of the SPP resonance in the small copper spheres. Moreover, the SPP resonance becomes better resolved due to the presence of this oxide shell.

Bulk silicon exhibits a high reflectivity caused by the high refractive index of silicon. If a transparent coating such as silica is on top of a bulk silicon surface, this reflectivity is reduced because the transparent coating acts like an antireflective coating.

Turning to silicon nanoparticles, we have already shown in Chapter 6 that silicon nanoparticles exhibit morphology-dependent resonances in the visible and NIR spectral region, due to the high refractive index. When coating silicon nanoparticles with a thin natural oxide film of SiO_2, it can be expected that the morphology-dependent resonances (MDRs) are affected. Similarly to the bulk silicon and the antireflective SiO_2 coating, we can expect that the silica shell reduces the optical contrast between the silicon nanoparticles and the surrounding. Then, the MDRs are blue shifted. This can actually be recognized from the spectra shown in Figure 7.11. The spectra for the small particles with $2R = 20$ nm are multiplied by a factor of 30 for better presentation. They do not exhibit MDRs. Nevertheless, the influence of the silica shell is also obvious. It strongly reduces the efficiency of the particle.

7.1.2
Experimental Examples

7.1.2.1 Ag–Au and Au–Ag Core–Shell Particles
Silver-coated gold nanoparticles were investigated in 1964 by Morriss and Collins [188] and later in more detail by Quinten [114] and Sinzig et al. [181, 182, 189]. The inverse system – gold-coated silver nanoparticles – was studied by Sinzig [181, 182, 189] and Mulvaney et al. [190]. As silver and gold have many almost identical physi-

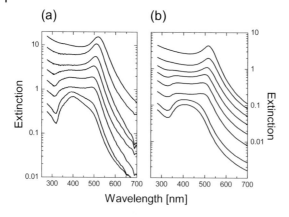

Figure 7.12 (a) Measured and (b) computed extinction spectra of silver-coated gold particles, starting with the gold particles at the top of the graphs. Sizes and shell thicknesses as described in the text.

cal properties (e.g., density of free electrons, lattice parameters) but different work functions, the question arises of whether in heterogeneous particles of silver and gold an inner boundary exists which is similar to a grain boundary. This boundary may be absent because gold and silver are completely alloyable, so that an intermediate region between shell and core could be established consisting of an alloy. For an alloy of gold and silver, the optical extinction spectra are dominated by only one SPP excitation (see Section 5.3.3). On the other hand, the work functions of Ag and Au are different, so that the electronic systems may remain well separated and two clearly separated SPPs from the core particle and the shell may be obtained.

Silver-coated gold particles or gold-coated silver particles were prepared in two steps:

1) preparation of a colloidal gold/silver solution with particles of mean diameter $2R$
2) precipitation of silver/gold from chemical reduction of a silver salt/gold salt.

The amount of chemically reduced silver/gold determines the resulting thickness d of the shell.

With this method, a series of coated nanoparticle systems was prepared and the extinction spectra were recorded. The size of the particles was determined by electron microscopy, and so also the thickness of the shell. For comparison, extinction spectra were calculated according to the extended Mie theory, using optical constants from [40]. Measured and computed spectra are compiled in Figure 7.12 for silver-coated gold particles and in Figure 7.13 for gold-coated silver particles. For better presentation, the spectra are shifted along the ordinate by arbitrary factors.

The spectra in Figure 7.12 start with the bare gold core particle with mean diameter $2R = 17.2$ nm at the top of the graphs. The extinction of the gold particle is governed by the absorption due to interband transitions at short wavelengths and the corresponding SPP at longer wavelengths. The SPP is peaked at $\lambda = 517$ nm. With increasing shell thickness, the SPP of the core particle shifts to shorter wavelengths and its magnitude decreases. Simultaneously, extinction is increased

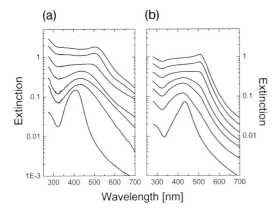

Figure 7.13 (a) Measured and (b) computed extinction spectra of gold-coated silver particles, starting with the bare silver particles at the bottom of the graphs. Sizes and shell thicknesses as described in the text.

at shorter wavelengths due to excitation of SPPs in the silver shell. In the bottom spectrum, there is an extinction maximum at $\lambda = 400$ nm which can be assigned to the most prominent SPP in the silver shell, and a shoulder at about $\lambda = 496$ nm caused by the gold core. At short wavelengths, the interband transitions in silver now dominate the spectrum. In these spectra the silver shell thickness amounts to $d = 0, 0.2, 0.5, 0.9, 1.2, 1.3, 3.0$, and 3.6 nm from top to bottom.

In Figure 7.13, the spectrum of the bare silver core particles with mean diameter $2R = 18$ nm at the bottom of the graphs is dominated by interband transitions in silver and the SPP of small silver particles. The SPP is peaked at $\lambda = 411$ nm. It shifts to longer wavelengths and its magnitude decreases with increasing overlayer thickness. Finally, in the top spectrum it has vanished. Instead, a new extinction maximum occurs at $\lambda = 500$ nm which can be assigned to the most prominent SPP in the gold shell. Additionally, the interband transitions become superposed by the interband transitions in gold. In these spectra, the gold shell thickness amounts to $d = 0, 1.3, 1.9, 2.4, 3.8, 4.6$, and 5.0 nm from bottom to top.

The comparison of the measured spectra with the computed spectra of coated particles with the same size and shell thickness as in the experimental data shows excellent agreement. For that reason, we can argue that the core and the shell must be well phase separated in the coated particles by a boundary similar to a grain boundary.

In 2002, Moskovitz et al. [191] studied bimetallic Ag–Au nanoparticles consisting of an Ag core and an Ag–Au alloy shell, prepared by chemical reduction of Ag and Au on Ag nanoparticles. They recorded optical extinction spectra and analyzed them for the amount of Au in the Ag–Au alloy and for the optical constants of Ag–Au alloys.

7.1.2.2 Multishell Nanoparticles of Ag and Au

Liz-Marzan and co-workers [192, 193] synthesized multishell nanoparticles of Ag and Au with impressive coloring effects by citrate reduction of gold from

190 | *7 Extensions of Mie's Theory*

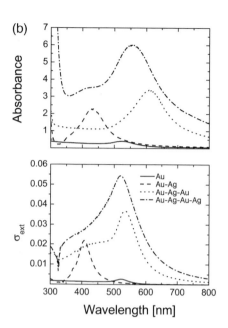

Figure 7.14 (a) Series of pictures for the multishell nanoparticle colloid Ag–Au–Ag–Au. Reproduced from [193] by permission of The Royal Society of Chemistry. (b) Measured absorbance spectra and calculated extinction efficiency spectra. A color version of this figure can be found in the color plates at the end of the book. Data courtesy of Luis M. Liz-Marzan.

chlorogold acid and silver from silver nitrate. They applied TEM for structural characterization and measured the absorbance (extinction) spectra of the colloidal solutions. The spectra were compared with computations according to the multilayer algorithm introduced by Sinzig and Quinten [182]. Figure 7.14 shows a series of pictures of the system Ag–Au–Ag–Au from the starting gold colloid to the final multishell nanoparticle colloid and the measured and calculated absorbance spectra.

The electron micrographs adjacent to the pictures of the colloidal suspensions show that the preparation of multilayer gold–silver nanoparticles leads to particles of polygonal shape with sharp edges and tips, and not only to spherically shaped particles. This is the main reason for the differences between the measured and computed extinction spectra, which, however, agree fairly well. In Chapter 9 we demonstrate that the nonspherical shape is a reason for broadened plasmon polariton peaks as already observed here.

7.1.2.3 Optical Bistability in Silver-Coated CdS Nanoparticles

Cadmium sulfide (CdS) is a material in which the nonlinear susceptibilities $\chi^{(2)}$, $\chi^{(3)}$, ... , are enormously increased at visible wavelengths due to resonances. Therefore, CdS is a potential candidate for easy observation of high-order nonlinearities. Nanoparticles are promising objects for studying optical nonlinearities since the inner electric fields can be increased enormously with respect to the external field (e.g., [194, 195]). Hence nanoparticles of CdS should allow easy observation of optical nonlinearities. Additional enhancement is possible by covering the particles with thin shells of silver, exhibiting strong SPP excitation in the visible spectral region. Then, large inner electric fields at low input laser power can be achieved by adjusting the thickness of the silver film properly, so that the maximum excitation of SPPs occurs at the wavelength of the incident laser beam. This enhancement of optical nonlinearities by combination of nonlinear materials with materials exhibiting surface resonances was predicted by Haus et al. [196], Kalyaniwalla et al. [197], Bergman et al. [198], and Levy-Nathanson and Bergman [199].

In the following experiments, the optical bistability, a third-order optical nonlinearity, is considered. Optical bistability means that at high incident intensities a certain optical property (reflection, absorption, etc.) can take two different values. Which one of them will occur depends on the history of illumination. For silver-coated CdS clusters it appears as two different absorption cross-sections of the whole particle, depending on the history of illumination.

For a silver-coated particle, optical bistability enters the formalism via the nonlinear dielectric constant ε_c of the core particle, which is assumed to be homogeneous and isotropic. Then ε_c can be written as

$$\varepsilon_c = 1 + \chi_c^{(1)} + \chi_c^{(3)} |E_c|^2 = \varepsilon_{c0} + \chi_c^{(3)} |E_c|^2 \tag{7.18}$$

where $\chi^{(1)}$ is the linear susceptibility of the core particle material and $\chi^{(3)}$ is the third-order nonlinear susceptibility. Second-order susceptibility is omitted due to the spherical symmetry of the nanoparticle. E_c is the electromagnetic field amplitude inside the core. It is connected to the incident field amplitude E_{inc} by the enhancement factor γ, which follows from the boundary conditions. In the Rayleigh limit it is [197]

$$\gamma = \frac{E_c}{E_{inc}} = \frac{9\varepsilon_M \varepsilon_{sh}}{(\varepsilon_c + 2\varepsilon_{sh})(\varepsilon_{sh} + 2\varepsilon_M) + 2\left(\frac{R}{R+d}\right)^3 (\varepsilon_c - \varepsilon_{sh})(\varepsilon_{sh} - \varepsilon_M)} \tag{7.19}$$

where ε_{sh}, the complex dielectric constant of the shell, and ε_M, the real dielectric constant of the surrounding matrix, are assumed to remain linear; d is the shell thickness and $2R$ is the particle core diameter.

The full calculation, which is omitted here, shows that the intrinsic field intensity X follows the incident field intensity Y in a hysteresis loop. An example is given in Figure 7.15a for $\zeta = [R/(R+d)]^3 = 0.3$, computed at wavelength $\lambda = 514.5$ nm, assuming a silver-coated CdS particle. The optical constants were $\varepsilon_{CdS} = 7.4$ and $\varepsilon_{Ag} = -11.2 + i0.4$. As the local field intensity X inside the particle cannot decrease when the applied field intensity Y is increased over Y_1, X spontaneously takes the value X_1. With further increase in Y, X follows the upper part of the curve. On the other hand, when Y decreases again below Y_2, the intensity X cannot increase, but spontaneously takes the value X_2. Therefore, variation of Y results in a hysteresis loop.

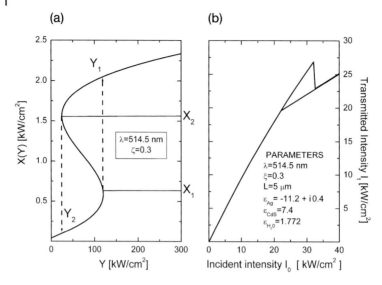

Figure 7.15 (a) Hysteresis loop of the intrinsic field intensity in kW cm^{-2} as a function of the incident field intensity for a silver-coated CdS particle. (b) Computed transmitted intensity I_t versus incident intensity I_0 for a collection of silver-coated CdS particles. Parameters for the computation are given in the inset of the graph.

CdS particles spontaneously form from Cd^{2+} ions in aqueous solution at pH 9–10 when S$^-$ ions are added. The growth of the particles and their size distribution can be controlled to some extent. Coated particles are prepared using a colloidal suspension of CdS particles with size $2R$, on the surface of which silver atoms from added silver nitrate solution are precipitated. Optical bistability in Ag-coated CdS clusters was measured with the set-up described in [200, 201].

Experiments were carried out on dilute aqueous suspensions (volume fraction $f \leq 10^{-4}$) on which the transmitted intensity was recorded versus the incident intensity. For this quantity the corresponding hysteresis loop is slightly different from that shown in Figure 7.15a. Figure 7.15b shows an example of the computed transmitted intensity I_t as ca function of the incident intensity I_0 for a sample with a thickness of $L = 5\,\mu m$ containing silver-coated CdS particles [200, 201]. Other parameters are given in the plot. In the computation the slowly varying envelope approximation (SVEA) [202] was used. The most important result is that power densities of less than about 50 kW cm^{-2} are required for experimental proof of intrinsic optical bistability.

Figure 7.16 shows the hysteresis loops of the measured and computed transmitted intensity of one sample versus the incident intensity for the sample under consideration. The optical bistability is clearly recognized. The recorded switching height amounts to 23% in the maximum absorption. Computation was performed according to the SVEA model with $\varsigma = 0.3$. The third-order susceptibility $\chi^{(3)}$ was varied. Best agreement with the experimental result was obtained for $\chi^{(3)} = 3.1 \times 10^{-11}\,m^2\,V^{-2}$. The mean value $\chi^{(3)} = (5.1 \pm 2.3) \times 10^{-11}\,m^2\,V^{-2}$ of several

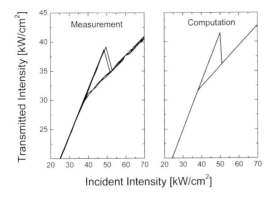

Figure 7.16 Measured and computed dependences of the transmitted intensity on the incident intensity for the system of silver-coated CdS particles.

samples examined agrees excellently with the value $\chi^{(3)}_{\text{theor}} = 4.4 \times 10^{-11}\,\text{m}^2\,\text{V}^{-2}$ predicted in the more comprehensive theoretical paper of Kalyaniwalla et al. [197].

It is emphasized that strong contributions of thermally induced bistability of the samples can be ruled out, despite the long laser pulses of minimum 0.5 ms. This may be due to the high specific heat capacity of the water surrounding the particles in the colloidal suspensions.

In addition to the intensity-dependent extinction at the laser wavelength, the wavelength spectrum of the extinction at common light intensities was recorded for the pure CdS particles and the Ag-coated CdS particles in aqueous suspension in the wavelength range 300–850 nm. The results were compared with corresponding computations to estimate the size and shell thickness of the various particles. Electron microscopy was applied to obtain the total particle size.

Optical extinction spectra of bare CdS nanoparticles and silver-coated CdS particles are plotted in Figure 7.17. In contrast to the structureless spectrum of the pure CdS particles, the coated particle spectrum is dominated by a broad resonance-like structure, due to the SPP in the silver shell. The peak position is $\lambda = 514$ nm, which is very close to the laser wavelength. The dotted line represents the computed spectrum for a silver-coated CdS particle. Roughly, the features of the measured spectrum can be found in the computed spectrum with core size $2R = 10.7$ nm and a silver shell of thickness $d = 2.6$ nm. However, the halfwidth is smaller and the extinction is increased at shorter wavelengths. These discrepancies are believed to result from the optical constants of particularly CdS. The separate formation of pure silver particles can be excluded from the spectra, since then the SPP of compact silver nanoparticles of $2R = 14$ nm would lie at wavelength $\lambda = 418$ nm, as shown with the dashed-dotted spectrum in Figure 7.17.

7.1.2.4 Ag and Au Aerosols with Salt Shells

Coated aerosol particles with metallic core were generated in two steps. In step 1, aqueous colloidal suspensions of gold or silver particles were prepared. Their size

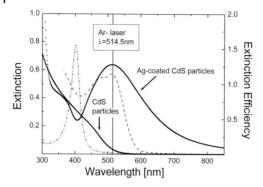

Figure 7.17 Measured extinction spectra of CdS clusters and silver-coated CdS clusters (full line), and computed extinction efficiency spectra of silver-coated CdS clusters (dotted line, $2R = 10.7$ nm, $d = 2.55$ nm) and a mixture of CdS particles ($2R = 10.7$ nm) and silver particles ($2R = 14$ nm) (dashed-dotted line).

distribution was obtained from electron micrographs. In step 2, the aqueous suspensions were sprayed and dried, resulting in gold or silver particles coated with salt from the electrolyte and in empty salt particles [129–132].

When spraying or drying colloidal suspensions, aggregation of particles may occur, which is avoided in the suspension by an ionic double layer formed during the preparation of the colloidal particles. However, as can be seen in the experimental results, in the main the authors were aware of the problem of aggregation of particles.

For determination of the salt shell thickness, broadband optical extinction spectra of the aqueous suspension were measured with a commonly used spectrometer, and for the corresponding aerosol with a multi-pass cell. As the metal particles show an SPP resonance in the extinction, the shell thickness could easily be determined from comparison of the spectra obtained and evaluation with computations according to the extended Mie theory for layered spheres. The contribution of the empty salt particles to the extinction spectrum of the coated particles was resolved by separate preparation of such salt particles and measuring their extinction spectrum. This spectrum was subtracted from the aerosol spectra and also taken as a reference spectrum when measuring the aerosol spectra.

Figure 7.18 shows spectra of nanoparticles with a mean silver core of $2R = 42$ nm. The spectrum in Figure 7.18a belongs to uncoated silver particles in the colloidal suspension and the spectrum in Figure 7.18b to the aerosol, containing the coated silver particles. This spectrum is already corrected for the contribution of the salt particles to the extinction of the aerosol. The spectra are dominated by the SPP. In the aerosol, the scattering of salt particles modifies the spectra at longer wavelengths. Comparing the two spectra, the peak position in the aerosol ($\lambda = 418$ nm) is almost the same as that in the aqueous suspension ($\lambda = 422$ nm), indicating that

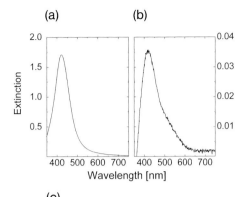

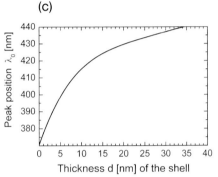

Figure 7.18 (a) Extinction spectrum of colloidal silver particles in suspension ($2R = 42$ nm). (b) Extinction spectrum of aerosol particles with silver core of the same size, corrected for the contribution of salt particles. (c) Dependence of the plasmon peak position of coated silver particles on the NaCl shell thickness.

the silver particles are coated with a dielectric shell. As the refractive index of the surrounding medium has changed from water to air, the peak position should shift to a wavelength of $\lambda = 370$ nm for the uncoated particle.

From computations for salt-coated Ag particles, the plot in Figure 7.18c was derived showing the resonance peak position versus the thickness of the salt shell. The shell thickness is determined from this plot as 12 nm with an accuracy of 1 nm. In this case, the SPP peak position is shifted from $\lambda = 370$ nm (uncoated) to 418 nm (coated).

Similar experimental results on aerosol particles with a gold core of mean size $2R = 56$ nm are shown in Figure 7.19: the extinction spectrum of the colloidal gold suspension in Figure 7.19a and the corrected gold aerosol spectrum in Figure 7.19b. The peak position of the SPP is $\lambda = 530$ nm for the colloidal suspension and 542 nm for the aerosol, indicating again the formation of a dielectric layer around the gold particles. From Mie theory, the predicted peak position in air is $\lambda = 502$ nm for uncoated gold particles. From computations for gold particles coated with a salt shell (Figure 7.19c), the shell thickness is obtained as $d = 30 \pm 1$ nm. This is

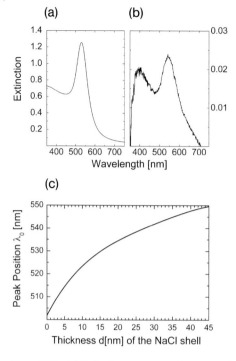

Figure 7.19 (a) Extinction spectrum of colloidal gold particles in suspension (2R = 56 nm). (b) Extinction spectrum of aerosol particles with gold core of the same size, corrected for the contribution of salt particles. (c) Dependence of the plasmon peak position of coated gold particles on the NaCl shell thickness.

larger than in the case of the silver; however, the gold particles are also larger than the silver particles. The influence of the pure salt particles can be effectively suppressed when using the salt spectrum as reference spectrum. Aggregation of particles could not be established from the spectrum.

7.1.2.5 Further Experiments

The group of Halas at the Department of Chemistry at Rice University (Houston, TX, USA) is very active in the preparation and application of noble metal-coated nanoparticles. Here, we report on three experiments from this group.

Oldenburg et al. [203] measured infrared extinction spectra of silica nanoparticles coated with a gold shell. The particles were prepared in two steps. First, small Au particles with $2R = 2$–3 nm were prepared and attached to silica particles with $2R_c = 420$ nm. Then, further gold was reduced and precipitated on these particles to form closed shells of thickness d. Optical extinction spectra were recorded in the wavelength range 400–2600 nm. The measured spectra were compared with calculated spectra for gold-coated silica particles. The authors found quantitative agreement between measurement and calculation, which is demonstrated in Figure 7.20.

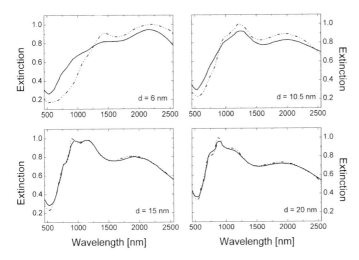

Figure 7.20 Comparison of measured and calculated spectra of gold-coated silica particles. The dash-dot-dotted lines are calculated according to Mie's theory. After [203].

Similarly, Sershen et al. [204] prepared gold-coated silica nanoparticles and recorded spectra in the wavelength range 400–1100 nm. Again, the measured and calculated spectra coincided fairly well.

To establish silver shells on a dielectric core, Jackson and Halas [205] used three different chemical methods. In particular, one reduction method based on a variation of the method reported by Zsigmondy [206] produced silver nanoshells with smooth and highly uniform morphologies. The main idea of Zsigmondy was to use very small gold nanoparticles as nuclei for further precipitation of silver by chemical reduction from silver nitrate solution. Here, first gold-decorated dielectric silica particles, that is, silica nanoparticles with a thin gold overlayer, are used as nuclei (see again the experiments described in [203] and [204] for the gold-coated silica particles). The optical extinction spectra were recorded in the wavelength range 400–1000 nm and corresponded quantitatively to spectra according to the Mie scattering theory for coated particles, as can be seen from Figure 7.21. The experimentally achieved mean total sizes and core sizes are given in the graph.

In early experiments, Barnickel and Wokaun [207] tried to coat polystyrene latex spheres with Ag to prepare particles with high efficiency for surface-enhanced Raman scattering.

Mayer et al. [208] evaluated several *in situ* chemical reduction methods with a view to the formation of silver–polystyrene latex composites, and in particular with respect to the task of coating colloidal latex spheres with uniform thin layers of silver. The samples were investigated by TEM and UV–VIS spectroscopy. Most of the chemical reduction methods resulted in free, nonimmobilized Ag particles separately from the polystyrene spheres. Partly the Ag -nanoparticles were seeded on the latex spheres forming inhomogeneous coatings and even aggregates. An

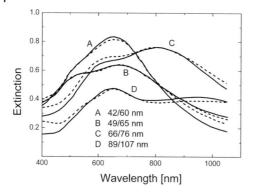

Figure 7.21 Comparison of measured and calculated spectra of silver-coated silica nanoparticles with gold shell. The dashed lines are calculated according to Mie's theory. After [205].

especially uniform silver particle deposition was obtained using a two-step reduction procedure, involving NaH_2PO_2 as initial reducing agent. The TEM images showed a homogeneous distribution of deposited silver particles exhibiting a fairly narrow size distribution. The coated latex particles appeared orange–brown and showed a pronounced hydrophobic character.

Kan et al. [209] synthesized Au core–Pd shell nanoparticles by successive and simultaneous sonochemical irradiation of the metal precursors in ethylene glycol. First, Pd nanoparticles were obtained from $Pd(NO_3)_2$. Then, $HAuCl_4$ solution was added and gold shells were formed. Calculated spectra for gold-coated Pd nanoparticles according to Mie's model were in excellent agreement with experimental results.

Several approaches have been used to make silica- [210, 211], titania- [212], and zirconia-covered [213] noble metal nanoparticles.

An interesting method to prepare dielectric–dielectric nanoparticles was described by Nair et al. [214]. First, core–shell Ag–ZrO_2 (ZrO_2-coated Ag) nanoparticles were prepared. Then, the metal core was removed by a reaction in which halocarbons, generally chlorides, oxidized the metal core and leached out the metal ions. The reduction in the surface plasmon excitation intensity and decrease in the voltammetric current during metal core removal were used to study the process. The reaction with CCl_4 led to ZrO_2-coated AgCl nanobubbles.

7.2
Supported Nanoparticles

A classical method to generate nanoparticles is by deposition of atoms on planar supports with subsequent surface diffusion, nucleation, and growth. The size and topology of the particles obtained are governed by statistics, and the total coverage must remain well below one nanoparticle monolayer to avoid coalescence processes.

Another method to produce well-defined nanoparticles and nanostructures on supports, even with a regular arrangement of the particles, is nanolithography. Alternatively, preformed nanoparticles from a liquid colloidal system or from a particle jet can be deposited on a substrate.

In all cases, the nanoparticles on or near the surface of a substrate scatter and absorb the incident light in a different way to the free particles. Several attempts have been made to account for this situation and here we summarize some of the solutions.

A first attempt to take into account the vicinity of an interface with ε_{sub} is to replace the surrounding medium with ε_M by a medium with some averaged effective dielectric function $\varepsilon_M^{(a)}$. A common approach is

$$\varepsilon_M^{(a)} = p\varepsilon_{sub} + (1-p)\varepsilon_M \tag{7.20}$$

and

$$\frac{1}{\varepsilon_M^{(a)}} = \frac{p}{\varepsilon_{sub}} + \frac{1-p}{\varepsilon_M} \tag{7.21}$$

where p is a parameter between 0 and 1, often chosen as $p = 0.5$. Equation 7.20 is similar to the effective dielectric constant in a capacitor which is built of two parallel capacitors, and matches best the situation of s-polarized incident fields. Equation 7.21 is similar to the effective dielectric constant of a capacitor that consists of two serial capacitors, and matches best the situation of p-polarized incident fields. We can expect this approximation to yield only qualitative results, as the interaction of the particle with the support is treated rather poorly.

A more elaborate approach was taken by Wind et al. [215] in the Rayleigh approximation. The main conclusion of their work is summarized in the dipole polarizability of the supported sphere:

$$\alpha = 4\pi R^3 \frac{\varepsilon - \varepsilon_M}{\varepsilon_M + G(\varepsilon_M, \varepsilon_{sub})(\varepsilon - \varepsilon_M)} \tag{7.22}$$

with a factor

$$G(\varepsilon_M, \varepsilon_{sub}) = \frac{1}{3}\left(1 - \frac{1}{8}\frac{\varepsilon_{sub} - \varepsilon_M}{\varepsilon_{sub} + \varepsilon_M}\right) \tag{7.23}$$

This means that the influence of the substrate–embedding medium interface is similar to depolarization of the sphere response. Since $G(\varepsilon_M, \varepsilon_{sub}) \leq 1/3$, the resonance condition for SPPs or electronic resonances of nanoparticles is red shifted compared with the unsupported sphere with $G = 1/3$. The red shift increases with increasing optical contrast $\varepsilon_{sub} - \varepsilon_M$ between substrate and embedding medium. For illustration, we show in Figure 7.22 the calculated extinction cross-section spectra of Ag nanoparticles with $2R = 10\,nm$ supported on various substrates with increasing refractive index n_{sub}. The red shift of the peak position with respect to the unsupported particle is up to 8 nm for $n_{sub} = 5$.

Yamaguchi et al. [216] introduced a mirror-image dipole within the quasi-static approximation which causes van der Waals-like interactions with the particle to

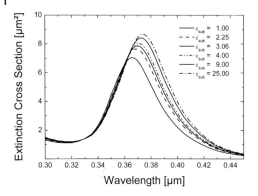

Figure 7.22 Extinction cross-section of a silver nanoparticle with $2R = 10\,\text{nm}$ supported on a dielectric substrate with increasing refractive index, varying from $n_{sub} = 1$ ($\varepsilon_{sub} = 1$) to $n_{sub} = 5$ ($\varepsilon_{sub} = 25$).

investigate the optical absorption of Ag particles deposited on poly(vinyl alcohol) (PVA) substrates. As a result, the absorption peak also shifts to longer wavelengths.

Ruppin [217] calculated the absorption spectrum of an NaCl sphere above metallic and nonmetallic substrates in the quasi-static regime using bispherical coordinates. Again, the interaction with the substrate leads to a shift of the resonance peak to longer wavelengths. This method is, however, restricted to finite distances between sphere and surface.

Lindell et al. established in [218] an exact-image theory formulation for the problem of a small scattering object close to an interface. In the continuation of that work, they concentrated on the special case of a scatterer that is small compared with the wavelength, permitting the electric-dipole approximation, and to have a scalar polarizability [219]. After the derivation of the dipole moment, investigations concentrated on far-field scattering. Backscattering enhancement and reversal of linear polarization were confirmed through statistical averaging over scatterer height and system orientation.

All the above solutions are restricted to nanoparticles that are small compared with the wavelength of the incident light. The full scattering pattern from a sphere in the vicinity of a plane surface can be obtained taking into account the electromagnetic interaction between the sphere and the substrate. In a number of studies many authors have addressed this problem in various approximations [220–236]; this is not a complete list of references but it gives a good overview. Among them Bobbert and co-workers [220, 221] developed a rigorous solution, which was extended by Fucile et al. [237, 238] in 1997 to a nonvanishing field propagating along the surface. Another similar rigorous solution was developed by Johnson [235], which, however, is an approximation with respect to the reflection of scattered light on the substrate surface. In the following, we briefly review the methods and point to the approximations and extensions made by the above-mentioned authors.

Figure 7.23 shows the geometry of the sphere–substrate system with representative light rays from the incident, reflected, and scattered electromagnetic fields.

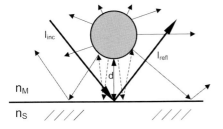

Figure 7.23 Sketch of the geometry of the sphere–substrate system and representative light rays from the incident, reflected, and scattered fields.

At first glance the scattering problem seems similar to the classical Mie theory: the interior fields must match the incident, reflected, and scattered fields. However, for a spherical particle in the vicinity of a plane surface, there are two important differences.

The first important difference is that the presence of an interface requires the establishment of a coordinate system in which the direction of incidence is fixed. In consequence, the scattering becomes polarization dependent, as the incident fields are composed of p- and s-polarized components:

$$E_{inc} = \sum_{m=-\infty}^{\infty} \sum_{n=\mu}^{\infty} [p_{nm}^{inc} M_{nm}^{(1)}(k_M) + q_{nm}^{inc} N_{nm}^{(1)}(k_M)] \quad (7.24a)$$

$$H_{inc} = \frac{-k_M}{\omega\mu_0} \sum_{m=-\infty}^{\infty} \sum_{n=\mu}^{\infty} [p_{nm}^{inc} N_{nm}^{(1)}(k_M) + q_{nm}^{inc} M_{nm}^{(1)}(k_M)] \quad (7.24b)$$

with

$$p_{nm}^{inc} = -i^{n+1}(-1)^{n+m}[E_{po}\pi_{nm}(\alpha) + iE_{s0}\tau_{nm}(\alpha)]\exp(-ik_M d\cos\alpha) \quad (7.25a)$$

$$q_{nm}^{inc} = i^{n+1}(-1)^{n+m}[E_{po}\tau_{nm}(\alpha) + iE_{s0}\pi_{nm}(\alpha)]\exp(-ik_M d\cos\alpha) \quad (7.25b)$$

where α is the angle of incidence. The angular dependent functions $\tau_{nm}(\theta)$ and $\pi_{nm}(\theta)$ are defined as $\tau_{nm}(\theta) = \partial P_{nm}/\partial\theta$ and $\pi_{nm}(\theta) = P_{nm}m/\sin\theta$. $P_{nm}(\cos\theta)$ are associated Legendre polynomials. The phase factor $\exp(-ik_M d\cos\alpha)$ takes into account the distance d of the particle center from the plane surface.

As a consequence, now contributions with $m \neq \pm 1$ must also be taken into account in the field expansions with spherical harmonics. Moreover, the total fields incident on the particle are composed by the incident fields (E_{inc}, H_{inc}) and the fields reflected on the surface (E_{ref}, H_{ref}). The fields (E_{ref}, H_{ref}) are defined via

$$E_{ref} = \sum_{m=-\infty}^{\infty} \sum_{n=\mu}^{\infty} [p_{nm}^{ref} M_{nm}^{(1)}(k_M) + q_{nm}^{ref} N_{nm}^{(1)}(k_M)] \quad (7.26a)$$

$$H_{ref} = \frac{-k_M}{\omega\mu_0} \sum_{m=-\infty}^{\infty} \sum_{n=\mu}^{\infty} [p_{nm}^{ref} N_{nm}^{(1)}(k_M) + q_{nm}^{ref} M_{nm}^{(1)}(k_M)] \quad (7.26b)$$

with

$$p_{nm}^{\text{ref}} = -i^{n+1}\left[r_p(\alpha)E_{po}\pi_{nm}(\alpha) - ir_s(\alpha)E_{so}\tau_{nm}(\alpha)\right]\exp(ik_M d\cos\alpha) \quad (7.27\text{a})$$

$$q_{nm}^{\text{ref}} = -i^{n+1}\left[r_p(\alpha)E_{po}\tau_{nm}(\alpha) - ir_s(\alpha)E_{so}\pi_{nm}(\alpha)\right]\exp(ik_M d\cos\alpha) \quad (7.27\text{b})$$

The superposition of the incident and the reflected fields leads to standing waves in which the particle is located.

The second important difference with respect to classical Mie theory is that part of the scattered light is reflected by the plane surface and can hit the particle again. This part defines the particle–surface interaction, contained in the interaction fields (E_{int}, H_{int}):

$$E_{\text{int}} = \sum_{m=-\infty}^{\infty}\sum_{n=\mu}^{\infty}\left[r_s(\beta)(-1)^{n+m}b_{nm}M_{nm}^{(3)}(k_M r_2)\right.$$
$$\left. + r_p(\beta)(-1)^{n+m}a_{nm}N_{nm}^{(3)}(k_M r_2)\right] \quad (7.28\text{a})$$

$$H_{\text{int}} = -\frac{k_M}{\omega\mu_0}\sum_{m=-\infty}^{\infty}\sum_{n=\mu}^{\infty}\left[r_s(\beta)(-1)^{n+m}a_{nm}M_{nm}^{(3)}(k_M r_2)\right.$$
$$\left. + r_p(\beta)(-1)^{n+m}b_{nm}N_{nm}^{(3)}(k_M r_2)\right] \quad (7.28\text{b})$$

where a_{nm} and b_{nm} are the scattering coefficients, also used in the expansion of the scattered wave, and β is the angle of incidence of the scattered light at the interface.

We point to the fact that the reference point of the vector spherical harmonics in these fields is not the center of the sphere, as for the scattered wave, but is the mirror image of the particle center (indicated in the above equations by r_2). Hence these functions must first be transformed into waves with the particle center as reference point, before they can be inserted into Maxwell's boundary conditions at the surface of the particle. As the spherical vector harmonics M_{nm} and N_{nm} represent a complete set of orthogonal functions, it is possible to expand these vector functions into series of vector functions describing an incident wave at the site of the particle, using vector addition theorems [239, 240]:

$$M_{nm}^{(3)}(kr_2) = \sum_{q=\mu}^{\infty} A_{qn}^m M_{qm}^{(1)}(kr_1) + B_{qn}^m N_{qm}^{(1)}(kr_1) \quad (7.29)$$

$$N_{nm}^{(3)}(kr_2) = \sum_{q=\mu}^{\infty} A_{qn}^m N_{qm}^{(1)}(kr_1) + B_{qn}^m M_{qm}^{(1)}(kr_1) \quad (7.30)$$

These transformations introduce a matrix formalism in which the geometry (distance between the particle center and its mirror image) is considered in the matrix elements A_{qn}^m and B_{qn}^m.

Maxwell's boundary conditions can now be resolved for the unknown scattering coefficients, resulting in two linear sets of equations:

$$a_{nm} = a_n\left[p_{nm}^{\text{inc}} + p_{nm}^{\text{ref}} + \sum_{q=\mu}^{\infty} r_p(\beta)A_{qn}^m a_{qm} + r_s(\beta)B_{qn}^m b_{qm}\right] \quad (7.31)$$

$$b_{nm} = b_n\left[q_{nm}^{\text{inc}} + q_{nm}^{\text{ref}} + \sum_{q=\mu}^{\infty} r_p(\beta)B_{qn}^m a_{qm} + r_s(\beta)A_{qn}^m b_{qm}\right] \quad (7.32)$$

where a_n and b_n are the scattering coefficients as already known from classical Mie theory. The cross-sections finally follow from Poynting's law as

$$\sigma_{ext} = \frac{4\pi}{k_M^2} \sum_{n=1}^{\infty} \sum_{m=-n}^{n} \left[\left|\frac{a_{nm}}{a_n}\right|^2 \mathrm{Re}(a_n) + \left|\frac{b_{nm}}{b_n}\right|^2 \mathrm{Re}(b_n) \right] \qquad (7.33)$$

$$\sigma_{sca} = \frac{4\pi}{k_M^2} \sum_{n=1}^{\infty} \sum_{m=-n}^{n} \left(\left|\frac{a_{nm}}{a_n}\right|^2 |a_n|^2 + \left|\frac{b_{nm}}{b_n}\right|^2 |b_n|^2 \right) \qquad (7.34)$$

Most critical for the numerical evaluation of Equations 7.31–7.34 is the summation over the multipole orders n and q, since they run to infinity. However, analogous to the classical Mie theory, it can be restricted to a maximum order n_{MAX} for numerical reasons. Then, for each order m a set of $(2n_{MAX} + 1 - m)$ linear equations has to be solved. The maximum number n_{MAX}, however, must be larger than the largest order in the calculation for a sphere in classical Mie theory, since the transformations of vector spherical harmonics in Equations 7.29 and 7.30 introduce higher multipoles.

A further critical point is the reflection coefficients $r_s(\beta)$ and $r_p(\beta)$, since the angle β varies from $\beta = 0°$ to a maximum of 30°. The latter is obtained for a particle in contact with the plane surface. Bobbert and co-workers [220, 221] and Fucile et al. [237] considered all angles β by $r_s(\beta)$ and $r_p(\beta)$. In contrast, Johnson [235] argued that for many substrate materials of practical importance $r_s(\beta)$ and $r_p(\beta)$ differ from each other only in the range $0° \leq \beta \leq 30°$, and even can be replaced by the value $r = -r_p(0) = r_s(0)$. This assumption reduces enormously the calculation effort without introducing large errors in the results.

In Figure 7.24 we give an example of a silver nanoparticle with $2R = 200$ nm on the surface of a glass prism. The angle of incidence for p-polarized incident light was varied as 0°, 45° and 89°. The extinction and scattering cross-section spectra are compared with those for the Ag nanoparticle in free space.

For the larger silver particles, surface plasmon resonances with higher multipolar orders become important and are obviously enhanced and shifted to longer wavelengths compared with the particle in free space. The excitation with p-polarized incident light leads to a larger cross-section compared with the particle in free space. However, we point to the limited meaning of any definition of cross-sections for an excitation that varies significantly over the particle surface, as in the case of particles in the vicinity of a surface. The red shift of the SPP resonances (which are TM mode resonances) was confirmed by calculations by Liu et al. [241]. In contrast, the TE mode resonances (which do not exist for such small Ag nanoparticles) exhibit a blue shift.

We compared the results of Johnson's model with the approximation of Wind et al. [215] (see Equations 7.22 and 7.23 and Figure 7.22). For that purpose, we calculated the spectra of silver spheres with $2R = 10$ nm supported on substrates with n_{sub} varying between 1 and 5. The surrounding matrix was vacuum with $n_M = 1$. From the spectra we took the resonance positions of the SPP and they are summarized in Figure 7.25.

For $\varepsilon_{sub} = 1$, the particle is in free space and the peak position is the same for both models. With increasing ε_{sub} the peak position is red shifted, as expected

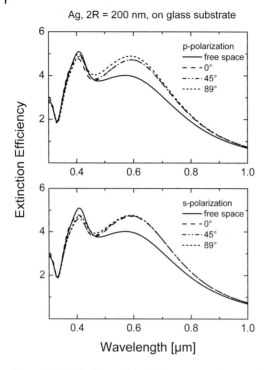

Figure 7.24 Extinction and scattering cross-section spectra for a spherical Ag nanoparticle with $2R = 200$ nm. The angle of incidence of the p-polarized plane wave is varied as 0°, 45°, and 89°. The spectra are compared with the spectra for the Ag particle in free space.

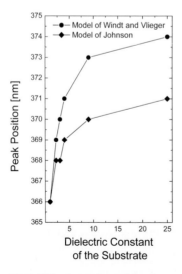

Figure 7.25 Comparison of the rigorous model according to Johnson [235] and the approximation of Wind et al. [215] for silver particles with $2R = 10$ nm on a substrate with ε_{sub} varying between $\varepsilon_{sub} = 1$ and $\varepsilon_{sub} = 25$.

from both models. However, the red shift is less in the rigorous model than in the approximation. The difference amounts to a 3 nm wavelength difference ($\Delta E = 0.027$ eV) for $\varepsilon_{sub} = 25$ ($n_{sub} = 5$).

Kim et al. measured the polarization and intensity of light scattered by polystyrene latex and copper spheres with various diameters deposited on silicon substrates with various thicknesses of oxide films with monochromatic light [242, 243]. The results are compared with the theory of Bobbert and co-workers [220, 221] and extended to include coatings on the substrate.

Bosi [244, 245] calculated the optical response of thin films of spherical particles on a dielectric substrate including higher multipolar interactions among the spheres and their mirror images and multiple reflections. The particles of $2R = 16$ nm are arranged on a square lattice with lattice parameter $d = 50$ nm. Then, the distance among neighboring particles is about $6R$, which is high as interactions among the particles are negligible. This is the main reason for the failure to explain the measured transmittance spectra of the films.

An interesting experiment was reported by Giessen and co-workers [246, 247]. The samples for the experiments were prepared by electron-beam lithography on quartz glass or ZnSe substrates covered with a 140 nm thick ITO film. Two sets of samples were manufactured, arranging gold nanoparticles in rectangular arrays with periods a_x and a_y along the x- and y-axis, respectively. Although the nanoparticles are not spherical, but form a rectangular parallelepiped, they exhibit a (broad) SPP resonance. The lattice distances a_x and a_y are still large as electromagnetic interactions among the particles can be neglected. An ITO film of certain thickness on a quartz substrate acts like a waveguide. Then, coupling of the particle SPP resonance to the waveguide modes can occur, resulting in almost complete suppression of light extinction in narrow spectral bands within the SPP band. This is demonstrated in Figure 7.26a. The ITO film with thickness $d = 140$ nm acts as waveguide

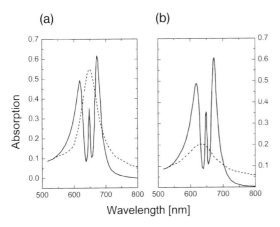

Figure 7.26 Extinction of gold nanoparticles arranged with periods of $a_x = 400$ nm along the x-axis and $a_y = 300$ nm along the y-axis on an ITO film. (a) ITO film on quartz substrate. Solid line, 140 nm ITO; dashed line, 30 nm ITO. (b) Solid line, 140 nm ITO on quartz substrate; dashed line, 140 nm ITO on ZnSe substrate.

whereas the 30 nm thick film does not affect the SPP. Figure 7.26b demonstrates again that only if the ITO film acts as a waveguide does this suppression occur. In this case, the substrate has changed from quartz, having a lower refractive index than ITO, to zinc selenide (ZnSe), having a higher refractive index than ITO. Then, the ITO film does not further act as a waveguide and no suppression of the SPP occurs.

Girard [248] studied electromagnetic optical interactions between a small metal sphere and a metallic surface by using a self-consistent approach in the presence of an external field. The intensity scattered by the metal particle was given for different polarizations of the incident field. This quantity, determined from a local treatment of the response function of the two interacting systems, exhibits a spatial dependence with respect to the approach distance close to that obtained from recent experimental studies.

7.3
Charged Nanoparticles

Charged nanoparticles can often be found in hydrosols, aerosols, and vacuosols[1]. However, also when static charge transfer occurs for embedded particles, the particles appear to be charged. In this section we discuss the influence of a static charge on the surface of a sphere on the scattering by the spherical particle.

In presence of a surface charge, Maxwell's boundary conditions change to [249]

$$(E_{inc} + E_{sca} - E_p) \times n\big|_{surface} = 0 \tag{7.35}$$

$$(H_{inc} + H_{sca} - H_p) \times n\big|_{surface} = j_s \tag{7.36}$$

where n is the vector normal to the surface and j_s is a surface current density parallel to the surface, caused by the surface charge; j_s solely affects the magnetic fields because of Maxwell's equation:

$$\text{rot} H = j_s \tag{7.37}$$

As the surface current density is assumed to be limited to an infinitesimally thin layer at the particle surface, it is

$$j_s = \sigma_s E_t \tag{7.38}$$

where σ_s is the corresponding surface conductivity and E_t is the tangential component of the electric field. Using this relation in Maxwell's boundary conditions, the scattering coefficients a_n and b_n for a homogeneous particle follow as

$$a_n = \frac{\psi_n(x)\psi'_n(mx) - m\psi'_n(x)\psi_n(mx) - i\sigma_s\sqrt{\frac{\mu_0}{\varepsilon_0\varepsilon_M}}\psi'_n(x)\psi'_n(mx)}{\xi_n(x)\psi'_n(mx) - m\xi'_n(x)\psi_n(mx) - i\sigma_s\sqrt{\frac{\mu_0}{\varepsilon_0\varepsilon_M}}\xi'_n(x)\psi'_n(mx)} \tag{7.39}$$

---

1) A vacuosol is an ensemble of particles in vacuum, as prepared for example by supersonic nozzle beam expansion in ultra-high-vacuum chambers.

$$b_n = \frac{m\psi_n(x)\psi'_n(mx) - \psi'_n(x)\psi_n(mx) - i\sigma_s\sqrt{\frac{\mu_0}{\varepsilon_0\varepsilon_M}}\psi_n(x)\psi_n(mx)}{m\xi_n(x)\psi'_n(mx) - \xi'_n(x)\psi_n(mx) - i\sigma_s\sqrt{\frac{\mu_0}{\varepsilon_0\varepsilon_M}}\xi_n(x)\psi_n(mx)} \quad (7.40)$$

The uncharged particle is obtained for $\sigma_s = 0$. However, for a discussion of the changes in extinction and scattering due to a surface charge, an appropriate model for σ_s is needed. Within the Drude model for N free excess surface charges q_s with damping constant γ_s and plasma frequency ω_s, the Drude dielectric constant ε_s of the excess charge is

$$\varepsilon_s(\omega) = \frac{Nq_s^2}{2\pi R^3 m\varepsilon_0 \omega} \frac{-\omega + i\gamma_s}{\omega^2 + \gamma_s^2} = \frac{-\omega_s^2}{\omega^2 + \gamma_s^2} + i\frac{\gamma_s}{\omega}\frac{\omega_s^2}{\omega^2 + \gamma_s^2} \quad (7.41)$$

from which the surface conductivity σ_s is obtained.

The damping constant γ_s is a free parameter which is assumed to be much lower than the frequencies in the visible spectral region. The sign of q_s is unimportant because q_s^2 enters the calculation.

Free surface charges can often be found in

- colloidal suspensions, where ions in the solution form a cloud around the particles and lead to an effective surface charge
- aerosols and vacuosols, where the particles become charged by friction, collisions, and other processes.

In colloidal suspensions, the number N of excess surface charges can be estimated as a function of the particle size as

$$N(R) = AR + BR^2 \quad (7.42)$$

Typical values of A and B are 3–5 nm^{-1} and 0.5 nm^{-2}, respectively. We used this dependence to estimate the maximum effect of the excess charge on the optical spectra of Ag nanoparticles in aqueous suspension. As an example, Figure 7.27 shows computed extinction efficiency spectra for four particle sizes [250]. The number N of maximum excess charges according to Equation 7.42 is given in the plots. For computation, the optical constants from [37] were used for silver and ε_s was calculated according to Equation 7.41. The relaxation frequency γ_s was chosen as $\gamma_s = 10^{13}$ s^{-1}, which is in the same order as the relaxation frequency of free electrons in bulk. Therefore, this value is an upper limit, yielding the strongest effect in ε_s. For free charges on the particle surface much lower frequencies are possible without affecting the results presented. In each plot in Figure 7.27, the dashed line is the extinction efficiency spectrum of the corresponding uncharged nanoparticle.

It can be recognized that the plasmon polariton peak position shifts towards shorter wavelengths (blue shift) for the charged particles. The maximum shift occurs for the smallest particles, amounting to 3 nm. Although the number N increases with the surface area $4\pi R^2$ of the nanoparticle, this shift rapidly vanishes with increasing size and already for $2R = 10$ nm it has completely vanished. The peak height has hardly decreased. It is difficult to recognize from the plots that

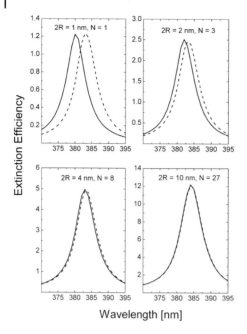

Figure 7.27 Optical extinction efficiency spectra of charged silver nanoparticles (solid lines) and uncharged silver clusters (dashed lines) with diameters $2R = 1$, 2, 4, and 10 nm. The maximum number N of elementary charges is given in the plots.

the peak halfwidth has changed. Actually, from evaluation of the curves, it follows that the halfwidth is slightly increased for the smallest nanoparticles. For the larger particles the halfwidth has not changed.

In terms of the dielectric constants ε of the particle and ε_s of the surface charge, the resonance condition for the surface plasmon of small silver particles reads

$$\omega^2 = \frac{\omega_p^2}{1 + \chi_\infty + 2\varepsilon_M} \tag{7.43}$$

for uncharged nanoparticles and

$$\omega^2 = \frac{\omega_p^2 + \omega_s^2}{1 + \chi_\infty + 2\varepsilon_M} \tag{7.44}$$

for charged nanoparticles. That means that for a charged particle the resonance occurs at shorter wavelength (higher frequency) compared with the uncharged particle. This holds true also for other metal nanoparticles, such as Na, Au, and Cu.

We point out that the model of free surface charges can also be applied to dielectric particles. However, since nanometer-sized dielectric particles do not show any resonances in the visible spectral region, the effect of a surface charge is smaller than for particles which exhibit a resonance like the SPP.

In contrast to dielectric particles, an excess charge on the surface of a metal particle due to adsorption of foreign atoms may also contribute to the density of free electrons in the nanoparticle volume. In consequence, the plasma frequency ω_p must be affected, rather than an additional dielectric constant ε_s being created. This is discussed in the following.

For that purpose, consider a spherical silver particle with diameter $2R$. For an estimation of the effect, it is assumed, that the half next neighbor distance a_0 in metals and the electron density n in the cluster are the same as in the bulk. In the bulk, the number of free electrons per atom is approximately 1 for silver. Then, a spherical particle having $N_0 = 0.7405(R/a_0)^3$ atoms has also N_0 free electrons. An excess charge $Q = \pm Ne_0$, corresponding to N excess electrons or N missing electrons, then modifies the plasma frequency according to

$$\omega_p(N) = \omega_p(0)\sqrt{\frac{N_0 \pm N}{N_0}} \qquad (7.45)$$

This means that the plasma frequency is increased by a negative surface charge and is decreased by a positive surface charge with respect to the uncharged nanoparticle. In consequence, the corresponding SPP and also the bulk plasmon must be blue shifted. Vice versa, a positive excess charge $Q = +Ne_0$ lessens ω_p, leading to a red shift of the plasmons.

For a quantitative discussion, numerical computations were carried out for monodisperse Ag nanoparticles. The number N of excess charges was assumed to be the same as for a static surface charge above to obtain the maximum effect. The optical constants of the silver particle were modified according to

$$\varepsilon_1(\omega) = \varepsilon_{1,\text{JC}}(\omega) - \varepsilon_{1,\text{Drude}}[\omega_p(0)] + \varepsilon_{1,\text{Drude}}[\omega_p(N)] \qquad (7.46a)$$

$$\varepsilon_2(\omega) = \varepsilon_{2,\text{JC}}(\omega) - \varepsilon_{2,\text{Drude}}[\omega_p(0)] + \varepsilon_{2,\text{Drude}}[\omega_p(N)] \qquad (7.46b)$$

for each frequency ω. The subscript JC indicates the use of optical constants of Johnson and Christy [37].

In Figure 7.28 the wavelength λ_0 where the SPP is peaked is plotted for four different sizes (squares) in comparison with the uncharged particle (triangles) and the

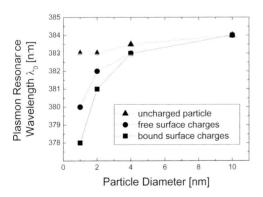

Figure 7.28 Effect of static charge transfer and a static surface charge on the plasmon polariton peak position of Ag nanoparticles.

model of free surface charges (circles). The blue shift is stronger than in the model of free surface charges. In fact, however, the magnitude of the shift is also fairly small, amounting to a maximum of 5 nm for the smallest particle. Further computations for other particle materials, for example, Na, Au, and Cu, yielded similar results.

7.4
Anisotropic Materials

7.4.1
Dichroism

Up to this point only isotropic materials were considered for the nanoparticles for which the dielectric function is a scalar. However, in many crystals the dielectric function is different along the principal crystal axis and perpendicular to this axis, given by the two dielectric functions $\varepsilon_\parallel$ and $\varepsilon_\perp$. For a plane interface it is well known that the refracted beam is then split into an *ordinary* beam ($\varepsilon_\perp$) and an *extraordinary* beam ($\varepsilon_\parallel$), with the ordinary beam obeying Snell's law. In light scattering, the coexistence of two independent propagating electromagnetic waves in the particle must lead to two separate solutions in Mie scattering. Therefore, a well-established (and so far, to our knowledge, the only) solution is to calculate the optical response by the sphere (or another particle) in two steps: first calculate the optical cross-sections σ for the particle using $\varepsilon_\parallel$ and $\varepsilon_\perp$ separately, and then add up the results according to the *1/3–2/3 rule*:

$$\sigma_k = \frac{1}{3}\sigma_k(\varepsilon_\parallel) + \frac{2}{3}\sigma_k(\varepsilon_\perp), \quad k = \text{ext}, \text{sca} \tag{7.47}$$

An example is given in Figure 7.29, where the extinction efficiencies of hematite spherical particles with $2R = 400$ nm in vacuum are plotted versus

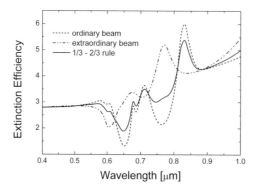

Figure 7.29 Extinction efficiency spectra of hematite (Fe_2O_3) nanoparticles with $2R = 400$ nm, computed with optical constants of hematite for the ordinary and the extraordinary beam [36]. The solid line is the result of the applied 1/3–2/3 rule.

the wavelength. The dashed line corresponds to the efficiency for hematite particles with dielectric function $\varepsilon_\perp$ (ordinary beam) and the dashed-dotted line corresponds to the efficiency of particles with dielectric function $\varepsilon_\parallel$ (extraordinary beam). There are clear differences between the two efficiencies, which result only from the dielectric constants. The solid line, finally, gives the result of the 1/3–2/3 rule. Particularly at wavelengths around 750 nm the effect of this average can clearly be recognized: the minimum in the ordinary beam efficiency is smeared out by the maximum in the extraordinary beam efficiency.

Monzon [251] described a field expansion for the solution of three-dimensional electromagnetic problems in material regions characterized by the most general rotationally symmetric anisotropy. The analysis (in spherical coordinates) revealed that the angular dependence is dictated by spherical harmonics, whereas the radial functions satisfy a set of coupled second-order ordinary differential equations. Due to birefringence, four sets of radial functions (each set representing mode coupling) were obtained. The expansion was applied to the problem of scattering by an anisotropic sphere.

7.4.2
Field-Induced Anisotropy

Anisotropy may also caused by external fields, for example, a constant magnetic field and a sphere of magneto-active (Faraday-active) material. In this case, the dielectric function of the particle material becomes a tensor of second rank in the external magnetic field $\varepsilon = \varepsilon(B_{ext})$

$$\varepsilon(B) = I + [(\varepsilon(B=0) - 1)I + \varepsilon_F T] \qquad (7.48)$$

with the antisymmetric tensor

$$T = i\delta_{ijk} B_k \qquad (7.49)$$

and the coupling parameter

$$\varepsilon_F = \frac{2V_0}{\Omega}\sqrt{\varepsilon} \qquad (7.50)$$

where V_0 is the Verdet constant and Ω is the frequency of the applied magnetic field. Lacoste et al. [252] presented in 1998 an exact calculation for the scattering of light from a single sphere made of a Faraday-active material, based on the above relations. Here, we just refer the reader to this paper.

7.4.3
Gradient-Index Materials

A further important anisotropy is given for gradient-index materials, that is, materials in which the refractive index varies along a certain direction. These materials are well known from applications in microoptics (GRIN lenses) and telecommunication (GRIN fibers).

We give here a brief introduction to the problems involved when solving Maxwell's equations for local complex dielectric functions $\hat{\varepsilon}(\omega,r)$. For the sake of simplification, we restrict considerations to only time-harmonic waves. The complex dielectric function is again composed by two real, measurable quantities, the permittivity ε and the conductivity σ:

$$\hat{\varepsilon}(\omega,r) = \varepsilon(\omega,r) + i\frac{\sigma(\omega,r)}{\omega\varepsilon_0} \tag{7.51}$$

which are now also local quantities. Maxwell's equations now read

$$\mathrm{div}[\varepsilon(r)E] = \varepsilon(r)\mathrm{div}E + E\mathrm{grad}[\varepsilon(r)] = 0 \tag{7.52}$$

$$\mathrm{div}B = 0 \tag{7.53}$$

$$\mathrm{curl}E = i\omega\mu_0 H \tag{7.54}$$

$$\mathrm{curl}H = [-i\varepsilon(r)\varepsilon_0\omega + \sigma(r)]E \tag{7.55}$$

omitting here and in the following the time dependence $\exp(-i\omega t)$ and the frequency dependence of the permittivity and conductivity in the equations.

As a first conclusion, the electric field need not be further divergence free, as in the case of a nonlocal dielectric function. Moreover, the vector wave equations for the electric and the magnetic field are affected in a different way. From the curl of Equation 7.54 and using Equation 7.55, we obtain

$$\mathrm{curl}(\mathrm{curl}E) = \frac{\omega^2}{c^2}\hat{\varepsilon}(\omega,r)E \tag{7.56}$$

This equation is similar to the Helmholtz equation for the electric field in the case of nonlocal dielectric functions. However, the resulting electric field must exhibit a gradient in direction of r, since the dielectric function varies in direction of r.

From the curl of Equation 7.55 and inserting Equation 7.54, we obtain

$$\mathrm{curl}(\mathrm{curl}H) = \frac{\omega^2}{c^2}\hat{\varepsilon}(\omega,r)H - i\omega\varepsilon_0 E\mathrm{grad}[\hat{\varepsilon}(\omega,r)] \tag{7.57}$$

The Helmholtz equation for the magnetic field is obviously more affected than that for the electric field. In particular, the solution for H is not independent of the solution for E. Hence TM and TE solutions are now coupled. A general solution is not yet available, to our knowledge, except for quadratic profiles $k(r) = k_0 - 0.5k_2r^2$ with $k^2 = \omega^2/c^2\varepsilon(\omega,r)$ [253].

Perelman [254] avoided a general solution of this problem by dividing the sphere with radially variable refractive index into many shells, that is, he treated this sphere as a multilayer sphere with each layer having a slightly different refractive index according to the refractive index profile of the whole particle.

7.4.4
Optically Active Materials

Optically active media are those in which plane harmonic waves can propagate without a change in polarization. In this sense, they are isotropic. However, for circularly polarized light, the complex refractive indices for left-circularly (L) and right-circularly (R) polarized waves are different. Although now being an anisotropic material, a solution for light scattering and absorption can be derived in this special case of optically active media.

A sufficient macroscopic description of optical activity is given by extending the constitutive relations between $\mathbf{D}$ and $\mathbf{E}$ and between $\mathbf{B}$ and $\mathbf{H}$ to

$$\mathbf{D} = \varepsilon_0 (\varepsilon \mathbf{E} + \beta \varepsilon \cdot \mathrm{rot} \mathbf{E}) \tag{7.58}$$

$$\mathbf{B} = \mu_0 (\mathbf{H} + \beta \cdot \mathrm{rot} \mathbf{H}) \tag{7.59}$$

with

$$\beta = \frac{1}{2}\left(\frac{1}{k_R} - \frac{1}{k_L}\right) \tag{7.60}$$

and

$$\omega\sqrt{\varepsilon} = \frac{1}{\frac{1}{2}\left(\frac{1}{k_R} + \frac{1}{k_L}\right)} \tag{7.61}$$

where k_R and k_L are the corresponding wavenumbers for L- and R-polarized light.

We can now introduce these relations into Maxwell's equations to find solutions for $\mathbf{E}$ and $\mathbf{H}$ that belong to L- and R-polarized incident, scattered, and interior waves of an optically active sphere. The solutions obtained are inserted in Maxwell's boundary conditions and cross-sections for extinction and scattering are calculated. We omit the details of the lengthy calculation and give only the results. For a more detailed discussion, see Bohren [255] and the book by Bohren and Huffman [8].

The cross-sections for L- and R-polarization differ and read

$$\sigma_{\mathrm{ext,L}} = \frac{2\pi}{k_M^2} \sum_{n=1}^{\infty} (2n+1)\mathrm{Re}(a_n + b_n - 2ic_n) \tag{7.62}$$

$$\sigma_{\mathrm{sca,L}} = \frac{2\pi}{k_M^2} \sum_{n=1}^{\infty} (2n+1)\{|a_n|^2 + |b_n|^2 + 2|c_n|^2 - 2\mathrm{Im}[(a_n + b_n)c_n^*]\} \tag{7.63}$$

$$\sigma_{\mathrm{ext,R}} = \frac{2\pi}{k_M^2} \sum_{n=1}^{\infty} (2n+1)\mathrm{Re}(a_n + b_n + 2ic_n) \tag{7.64}$$

$$\sigma_{\mathrm{sca,R}} = \frac{2\pi}{k_M^2} \sum_{n=1}^{\infty} (2n+1)\{|a_n|^2 + |b_n|^2 + 2|c_n|^2 + 2\mathrm{Im}[(a_n + b_n)c_n^*]\} \tag{7.65}$$

The meaning of the coefficients a_n and b_n is the same as in classical Mie theory. In addition, due to the R–L dependence, a coefficient c_n must be considered which is the same for R- and L-polarization, except for a minus sign. The values of a_n and b_n, however, clearly differ from classical Mie theory and are polarization dependent:

$$a_n = \frac{V_n(R)A_n(L)+V_n(L)A_n(R)}{W_n(L)V_n(R)+V_n(L)W_n(R)} \tag{7.66}$$

$$b_n = \frac{W_n(L)B_n(R)+W_n(R)B_n(L)}{W_n(L)V_n(R)+V_n(L)W_n(R)} \tag{7.67}$$

$$c_n = i\frac{W_n(R)A_n(L)-W_n(L)A_n(R)}{W_n(L)V_n(R)+V_n(L)W_n(R)} \tag{7.68}$$

with

$$W_n(J) = m\xi_n'(x)\psi_n(m_J x) - \xi_n(x)\psi_n'(m_J x) \tag{7.69}$$

$$V_n(J) = \xi_n'(x)\psi_n(m_J x) - m\xi_n(x)\psi_n'(m_J x) \tag{7.70}$$

$$A_n(J) = m\psi_n'(x)\psi_n(m_J x) - \psi_n(x)\psi_n'(m_J x) \tag{7.71}$$

$$B_n(J) = \psi_n'(x)\psi_n(m_J x) - m\psi_n(x)\psi_n'(m_J x) \tag{7.72}$$

where J is L or R. The relative refractive indices m_L, m_R and the mean refractive index m are defined as follows:

$$m_L = \frac{n_L}{n_M};\ m_R = \frac{n_R}{n_M};\ \frac{1}{m} = \frac{1}{2}\left(\frac{1}{m_R}+\frac{1}{m_L}\right) \tag{7.73}$$

7.5
Absorbing Embedding Media

7.5.1
Calculations

In some cases of practical importance in various fields of application and engineering, the surrounding medium is absorbing, that is, ε_M or n_M is a complex number. Examples are the scattering and absorption by particles in polymers, semiconducting materials, carbon films, solid dye films, and many others. In these cases, a new treatment of the Mie theory is required. In some previous papers [256–258], solutions were obtained in the *far-field* of the scattering sphere. However, the absorption in the host restricts the solution of the problem either to very weakly absorbing media or to finite distances from the encapsulated particle. Here, we give a solution that is valid for all distances from the surface of the encapsulated spherical particle [146, 170, 259].

Consider a spherical volume with diameter $2R_{cs}$ containing an absorbing medium and a particle of diameter $2R$ with a reference frame in the center of both spheres (Figure 7.30).

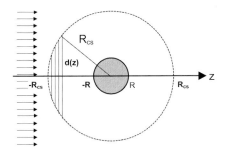

Figure 7.30 Definition of the coordinates when calculating W_0 in a spherical volume with diameter $2R_{cs}$. The hatched region corresponds to the encapsulated particle.

The intensity of a plane wave propagating in this volume along the z-axis decreases according to Lambert's law as

$$I(z) = I_0 \exp[-2\mathrm{Im}(k_M)(z + R_{cs})] \qquad (7.74)$$

starting at $z = -R_{cs}$. The energy absorption rate for the whole volume is given by the integral

$$W_0 = \int_{-R_{cs}}^{R_{cs}} A(z)\,dI(z) \qquad (7.75)$$

of the intensity $I(z)$ on a surface $A(z) = \pi(R_{cs}^2 - z^2)$, with the result

$$W_0 = 4\pi I_0 \exp[-2\mathrm{Im}(k_M)R_{cs}]\left\{R_{cs}^2\left(\frac{\cosh[2\mathrm{Im}(k_M)R_{cs}]}{2\mathrm{Im}(k_M)R_{cs}} - \frac{\sinh[2\mathrm{Im}(k_M)R_{cs}]}{[2\mathrm{Im}(k_M)R_{cs}]^2}\right)\right.$$
$$\left. - R^2\left(\frac{\cosh[2\mathrm{Im}(k_M)R]}{2\mathrm{Im}(k_M)R} - \frac{\sinh[2\mathrm{Im}(k_M)R]}{[2\mathrm{Im}(k_M)R]^2}\right)\right\} \qquad (7.76)$$

The second term with argument $\mathrm{Im}(k_M)R$ accounts for the absorption in a volume of the size of the particle filled with absorbing host medium. This part has to be subtracted from the first term to obtain the absorption solely in the embedding medium.

W_0 can be calculated also from the integral Equation 5.18 in Section 5.2. The result of the integration yields the same as Equation 7.76 only if the coefficients of the incident wave are modified to $i^n\{(2n + 1)/[n(n + 1)]\}\exp[-\mathrm{Im}(k_M)R_{cs}]$. Using these coefficients, the following equation for W_0 is calculated from Equation 5.18:

$$W_0 = \frac{4\pi}{|k_M|^2} I_0 \exp[-2\mathrm{Im}(k_M)R_{cs}]$$
$$\sum_{n=1}^{\infty}(2n+1)\mathrm{Im}[\psi_n(k_M R_{cs})\psi_n^{*\prime}(k_M R_{cs}) - \psi_n(k_M R)\psi_n^{*\prime}(k_M R)] \qquad (7.77)$$

The next step is the calculation of W_{ext} and W_{sca} from the corresponding Equations 5.19 and 5.20 in Section 5.2:

$$W_{\text{ext}} = -\frac{2\pi}{|k_M|^2} I_0 \exp[-2\text{Im}(k_M)(R_{cs}-R)]$$

$$\times \sum_{n=1}^{\infty} (2n+1) \Big\{ \text{Re}(a_n+b_n) \text{Im}[\xi_n(k_M R_{cs}) \psi_n'^*(k_M R_{cs}) - \xi_n'(k_M R_{cs}) \psi_n^*(k_M R_{cs})]$$

$$+ \text{Im}(a_n+b_n) \text{Re}[\xi_n(k_M R_{cs}) \psi_n'^*(k_M R_{cs}) - \xi_n'(k_M R_{cs}) \psi_n^*(k_M R_{cs})]$$

$$+ \frac{\text{Im}(k_M)}{\text{Re}(k_M)} \text{Re}(a_n-b_n) \text{Re}[\xi_n(k_M R_{cs}) \psi_n'^*(k_M R_{cs}) + \xi_n'(k_M R_{cs}) \psi_n^*(k_M R_{cs})]$$

$$- \frac{\text{Im}(k_M)}{\text{Re}(k_M)} \text{Im}(a_n-b_n) \text{Im}[\xi_n(k_M R_{cs}) \psi_n'^*(k_M R_{cs}) + \xi_n'(k_M R_{cs}) \psi_n^*(k_M R_{cs})] \Big\}$$

(7.78)

$$W_{\text{sca}} = \frac{2\pi}{|k_M|^2} I_0 \exp[-2\text{Im}(k_M)(R_{cs}-R)]$$

$$\sum_{n=1}^{\infty} (2n+1) \Big\{ (|a_n|^2 + |b_n|^2) \text{Im}\big[\xi_n(k_M R_{cs}) \xi_n^{*\prime}(k_M R_{cs})\big]$$

$$+ \frac{\text{Im}(k_M)}{\text{Re}(k_M)} \big[|a_n|^2 - |b_n|^2\big] \text{Re}[\xi_n(k_M R_{cs}) \xi_n^{*\prime}(k_M R_{cs})] \Big\}$$

(7.79)

The far-field solutions in [256–258] are obtained from (7.76–7.79) for large arguments kR_{cs}. For large arguments $\text{Im}(k_M)R_{cs}$, both hyperbolic functions in Equation 7.76 can be approximated by the same exponential $\exp[2\text{Im}(k_M)R_{cs}]$. Then, W_0 increases proportionally to R_{cs} and dominates the total absorption rate W_{abs}. In this case, the intensity of the incident wave may completely vanish at the site of the particle and scattering cannot occur. To prevent such a case, the intensity I_0 must also be increased arbitrarily, causing serious problems with nonlinearities.

Extinction and scattering rates in the far-field are obtained using the approximations of the Riccati–Bessel and Riccati–Hankel functions for large arguments. Then

$$W_{\text{ext}} = -\frac{2\pi}{|k_M|^2} I_0 \exp[-2\text{Im}(k_M)(R_{cs}-R)]$$

$$\times \sum_{n=1}^{\infty} (2n+1) \Big\{ \text{Re}(a_n+b_n) \exp[-2\text{Im}(k_M) R_{cs}]$$

$$+ \frac{\text{Im}(k_M)}{\text{Re}(k_M)} \text{Im}\big[(-1)^n (a_n-b_n) \exp(i2\text{Re}(k_M) R_{cs})\big] \Big\}$$

(7.80)

and

$$W_{\text{sca}} = \frac{2\pi}{|k_M|^2} I_0 \exp[-2\text{Im}(k_M)(R_{cs}-R)]$$

$$\sum_{n=1}^{\infty} (2n+1) \big(|a_n|^2 + |b_n|^2 \big) \exp[-2\text{Im}(k_M) R_{cs}]$$

(7.81)

These results coincide with the results of Chýlek [257] except for the factor $\exp[-2\text{Im}(k_M) R_{cs}]$ which appears in the above equation due to the modified expansion coefficients of the incident wave.

Obviously, for absorbing host media, the intensity of the incoming light decreases on its way through the host medium to the encapsulated particle and the extinction rate W_{ext} and the scattering rate W_{sca} depend explicitly on the size of the spherical volume of the absorbing medium around the particle. Then, definitions of cross-sections as quantities which are specific only for the particle must be carefully discussed. We point out here the limited meaning of any definition of cross-sections for an excitation that varies significantly over the particle surface, as in all applications the scattered or absorbed power is the relevant quantity. Hence suggestions for the cross-sections are

$$\sigma_{ext} = \frac{W_{ext}}{I_0 \exp[-2\text{Im}(k_M)(R_{cs} - R)]} \qquad (7.82)$$

$$\sigma_{sca} = \frac{W_{sca}}{I_0 \exp[-2\text{Im}(k_M)(R_{cs} - R)]} \qquad (7.83)$$

They take into account that the intensity incident on the particle is already attenuated at the particle surface. For the total extinction caused by the absorption in the host and the extinction by the particle, the following definition of a cross-section is suggested:

$$\sigma_{ext,total} = \frac{W_{ext} + W_0}{I_0 \exp[-2\text{Im}(k_M) R_{cs}]} \qquad (7.84)$$

Lebedev [260] and Sudiarta and Chýlek [261–263] discussed the problem of absorbing surrounding media again. Their solutions are contained in the above equations if R_{cs} is set to the particle radius R, that is, $R_{cs} = R$. Then, the above equations are drastically reduced. In a similar way Fu and Sun [264] obtained cross-sections for scattering and extinction of a sphere in an absorbing medium. Their results differ from those presented here and obtained by Lebedev or Sudiarta and Chýlek. Videen and Sun [265] examined the scattering properties of particles contained in absorbing media by examination of the extinction from its fundamental definition: the energy removed from the plane wave, or incident beam. The resulting energy received by a detector contains two terms: one the result of the incident beam traversing through the medium that would have occurred if the particle were not present, and a correction term due to the presence of the particle. They argued that the method of Lebedev or Sudiarta and Chýlek to set the integration sphere to the particle sphere ($R_{cs} = R$) actually removes some of the ambiguity of the integrating sphere, but it is not at all satisfactory for yet another reason: Additional absorption can be realized by the medium, because the absorbing medium itself plays an active role in the scattering process, and this additional absorption contributes to the total extinction. So far, their ansatz is closer to the presented solution with $R_{cs} > R$ as in all other solutions.

Figure 7.31 depicts an example for a silver particle with $2R = 20$ nm in an embedding medium with complex refractive index $n_M = 1.33 + i\kappa$, with varying absorption index κ [259]. The size of the conceptual sphere around the silver particle is $2R_{cs} = 200$ nm. In the computations the optical constants of silver from [37]

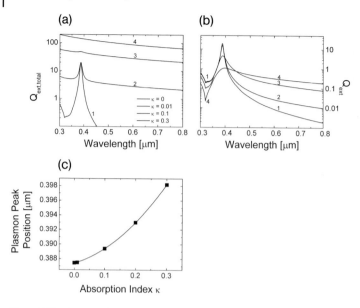

Figure 7.31 Computed extinction spectra $\sigma_{\text{ext,total}}(\lambda)$ (a) and $\sigma_{\text{ext}}(\lambda)$ (b) of a nanometer-sized silver sphere ($2R = 20\,\text{nm}$) in a spherical shell with refractive index $n_M = 1.33 + i\kappa$ ($2R_{cs} = 200\,\text{nm}$). The curves are consequently numbered according to the increase of the absorption index κ. (c) Dependence of the spectral position of the SPP resonance of the small silver sphere.

were used. Curve 1 corresponds to the extinction spectrum of the silver particle in nonabsorbing medium ($\kappa = 0$). Then, $\sigma_{\text{ext,total}}$ and σ_{ext} are identical quantities. The spectrum shows the SPP with its maximum at the wavelength $\lambda = 0.388\,\mu\text{m}$. With increasing κ the total extinction is dominated by the absorption in the embedding medium, so that the SPP vanishes (Figure 7.31a).

Looking at the extinction by the particle σ_{ext}, the increase in κ results in an increase of the extinction at longer wavelengths. Simultaneously, the characteristic resonance of the TM dipole mode decreases and becomes broader (Figure 7.31b). A shift of the surface plasmon resonance can be recognized from this plot, but on plotting the same data in a narrower wavelength range, a shift of the plasmon polariton peak with increasing κ can be clearly seen. This is shown in Figure 7.31c, where the dependence of the peak position on the absorption index κ is summarized. The peak position shifts from $\lambda = 0.388$ to $0.398\,\mu\text{m}$ when the absorption index is increased to $\kappa = 0.3$.

Further examples can be found in Ruppin [266] for (i) MgO inclusions in an NaCl matrix in spectral regions where NaCl exhibits absorption due to phonon polaritons and (ii) pores in aluminum. He used the models of Sudiarta and Chýlek [262, 263] and Fu and Sun [264].

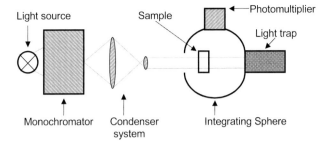

Figure 7.32 Schematic diagram of the integrating-sphere spectrometer.

7.5.2
Experimental Examples

7.5.2.1 Absorption of Scattered Light in Ag and Au Colloids

In 1991, broadband optical scattering spectra of mixtures of colloidal gold particles and colloidal polystyrene particles were measured and interpreted by Salzmann [267] and Hoffmann [268]. Total scattering of samples [$\sigma_{sca}(\lambda)$] was measured at wavelengths between 350 and 700 nm using a modified integrating-sphere spectrometer, depicted in Figure 7.32. A polychromatic parallel light beam enters an integrating sphere and is passed through the sample (rectangular cell for colloidal suspensions, plane plate for solid samples). The transmitted light beam is caught in a light trap. Only the scattered light homogeneously brightens the interior of the integrating sphere and is measured with a photon counter.

The integrating sphere spectrometer can also be used for measuring the extinction of light by the same samples by removing the light trap and closing the integrating sphere with a cover. Extinction is measured with the sample in front of the integrating sphere. Then, the transmitted light brightens the interior of the sphere.

The observed peak shift and the observed spectral behavior were confirmed later by Stier [269] for silver nanoparticle colloids. Figure 7.33 shows exemplarily the measured extinction and scattering spectra of a sample containing silver spheres with $2R = 36$ nm. The measured scattering exhibits a peak shift of $\Delta\lambda = 82$ nm to longer wavelengths compared with the extinction spectrum. This shift is unexpected from calculations according to Mie's theory. However, as it is beyond any error in measurement, it needs a careful explanation.

First, multiple scattering can be neglected, since the volume fraction of dispersed particles is in the order of $f \approx 10^{-6}$. Multiple scattering begins to contribute if the volume fraction exceeds $f > 0.01$. Hence the differences between extinction and scattering must be caused by partial absorption of the light scattered by a sphere by all other spheres in the measuring volume on its way to the detector. To prove this assumption, the Mie theory extended to absorbing surrounding media is used. The dielectric constant ε_M of the surrounding medium is calculated, assuming that the water with the suspended silver particles forms an effective medium. The main

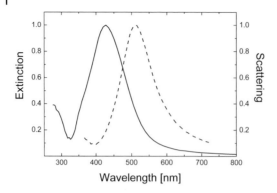

Figure 7.33 Extinction and scattering spectrum of silver particles with diameter $2R = 36$ nm.

idea of an effective medium theory is to describe the linear optical response of inhomogeneous matter, that is, particles suspended in a matrix material, by a global dielectric constant ε_{eff}. Effective medium models are summarized in Chapter 14. The simplest approach, known as the Maxwell-Garnett model, was given in 1904 by Garnett [270] for diluted systems and was applied here.

With these approximated, now complex-valued optical constants for the surrounding medium, the total absorption (Equation 7.84) and the scattering (Equation 7.83) of a single spherical particle with $2R = 36$ nm embedded in this absorbing medium were computed. The size $2R_{\text{cs}}$ of the conceptual sphere was $2R = 20\,\mu\text{m}$, $200\,\mu\text{m}$, 2 mm, and 10 mm. The two last values correspond to typical sizes of the cells used for measuring the extinction and scattering from colloidal suspensions. The results are shown in Figure 7.34.

The absorption in the host, W_0, dominates the total absorption, resulting in a shift of the peak towards the wavelength where the absorption index of the effective medium is maximum ($\lambda = 400$ nm). The spectra are similar for all sizes of the conceptual sphere volume. In consequence, the measured extinction will also be determined from W_0. For better presentation, the spectra are shifted along the ordinate of the semilogarithmic plot by arbitrary factors.

The scattering in Figure 7.34b is modified by the absorption in the host. With increasing volume of surrounding absorbing material the absorption of the scattered light is increased and finally leads to a strong reduction at those wavelengths where the absorption is maximum. Then, the peak in the scattering appears to be red shifted, similarly to the measured spectra. The shift amounts to $\Delta = 50$ nm, which is close to the experimental shift of $\Delta = 82$ nm. For better agreement with the measured spectrum, an average must be taken over spectra with different sizes of the conceptual sphere. Then, the sharp contours in the spectrum at the bottom in Figure 7.34b will be smeared out.

7.5.2.2 Ag and Fe Nanoparticles in Fullerene Film

Carbon can have covalent bindings with different sp hybridization, for which a large variety of organic and inorganic carbonaceous species exist. The most

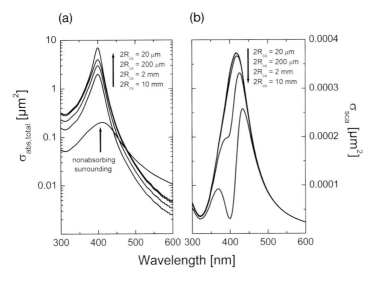

Figure 7.34 (a) Total absorption for a spherical silver particle with $2R = 36\,nm$ embedded in an effective medium with optical constant according to the Maxwell-Garnett model [270]. (b) Corresponding scattering cross-sections. The sizes of the conceptual sphere are varied from $2R_{cs} = 20\,\mu m$ to $10\,mm$.

prominent are graphite (sp^2) and diamond (sp^3). They are supplemented by a new kind of carbon molecule with a specific mixture of sp^2 and sp^3 hybridization: the fullerenes. Two of the most prominent representatives are C_{60} and C_{70} molecules. Fullerenes belong to a class of carbon nanoparticles with exciting structural and electronic properties, and fullerene films exhibit many distinct properties. One of them is to bind metal nanoparticles strongly, by which they are well suited as support materials in heterogeneous catalysis.

The strong binding between C_{60} and metal atoms at the surface of metal nanoparticles causes changes in the electronic system of the nanoparticles, mainly a static charge transfer from the metal to the carbon species. This may cause changes in the optical extinction by the metal nanoparticles (see Section 7.3). As fullerene materials are also absorbing, the optical properties of the embedded metal nanoparticles are additionally influenced by the absorption of the embedding fullerene film. Hence Mie's theory extended to absorbing embedding media and to charged particles must be applied to explain the measured spectra.

In the following two examples, nanoparticles of Ag with $2R = 2.6\,nm$ [150, 271] and of Fe with $2R = 20\,nm$ [150, 272] were embedded in C_{60} films. The Ag particles were prepared in the thermal cluster beam apparatus THECLA [273] whereas the Fe nanoparticles were produced by laser ablation in the ablation beam apparatus LUCAS [150]. In both cases the fullerene was co-evaporated to embed the particles in the fullerene film. The optical extinction spectra were recorded with a common spectrometer. Figure 7.35 shows the results for the Ag nanoparticles in comparison with calculations according to the Mie theory extended to absorbing

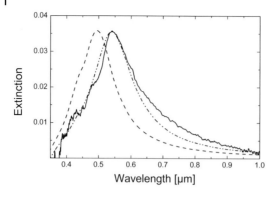

Figure 7.35 Optical extinction of Ag nanoparticles ($2R = 2.6$ nm) in fullerene film in comparison with calculated spectra according to Mie's theory extended to absorbing embedding media with and without static charge transfer. Data courtesy of M. Gartz.

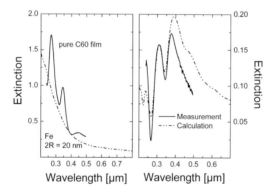

Figure 7.36 (a) Optical extinction of a pure C_{60} film and Fe nanoparticles in air. (b) Measured optical extinction of Fe nanoparticles ($2R = 20$ nm) in C_{60} film in comparison with a spectrum calculated according to Mie's theory extended to absorbing embedding media. Data courtesy of M. Gartz.

embedding media. As can be recognized from the dashed line, the theoretical spectrum already shows the features of the measured spectrum (solid line). However, the peak position is shifted by about 45 nm towards lower wavelengths. This indicates that the fullerene film strongly influences the SPP of the Ag particles. Therefore, in addition, the static charge transfer was considered. This reduces the number of free electrons contributing to the surface plasmon. $N = -0.2 N_0$, and rather good agreement of the computed and the measured spectrum is obtained, as can be seen from the dashed-dotted-dotted line.

In the case of Fe nanoparticles, the situation is more complex. First, the spectrum of nanoparticles of Fe smaller than $2R < 100$ nm do not exhibit any significant feature but only decrease monotonically from short to long wavelengths, and a static charge transfer from Fe to C_{60} will not be recognizable in the spectrum. The second effect is that the spectrum of the composite is dominated by the features of the pure C_{60} film spectrum shown in Figure 7.36a. In Figure 7.36b, we show a comparison of the measured spectrum with the calculated spectrum. A static charge transfer was not taken into account. The agreement is fairly good, as before for the Ag nanoparticles in C_{60} film.

7.6
Inhomogeneous Incident Waves

The incident electromagnetic wave has become increasingly important in light scattering by small particles because in common experimental set-ups and in applications nowadays laser light is used. The field distribution of a laser, however, is inhomogeneous even in the ground mode of the laser, the *Gaussian mode*. In another field of applications dealing with optical sensors, inhomogeneous *evanescent* waves, as obtained for example by total internal reflection, are used. For these two examples we briefly repeat the basics of the extension of Mie's theory to inhomogeneous incident waves.

The first treatment of inhomogeneous incident waves as converging beams stems, however, from Möglich [79] in 1933. He derived integral expressions for the expansion coefficients of a converging beam and applied them to particles in the Rayleigh approximation. The main result is an additional factor in the series expansion of the electromagnetic fields of the incident wave for each multipole order. These factors obey a recurrence relation and can be calculated from two initial values. Möglich gave explicit expressions for these coefficients as a function of the size, wavelength, and beam shape. This represents the first work on the calculation of scattering by particles in an arbitrarily shaped beam, long before the Mie theory was extended to Gaussian laser beams.

Before we start with the extension to Gaussian beams, we want to point to an interesting extension with plane waves. If one uses two counterpropagating plane waves for particle illumination, then the sphere is in the fields of a standing electromagnetic wave. This standing wave causes an inhomogeneous field distribution, but with predictable expansion coefficients as described in [274].

7.6.1
Gaussian Beam Illumination

Provided that the scattering particle is small compared with the dimensions of a Gaussian beam of a laser, it can be expected that the scattering by a particle will not exhibit substantial deviations from the scattering of a plane wave. In a focused beam, however, the particle size and the beam waist may become comparable in size. Then, the extinction and scattering are modified in a distinct way. This was first described by Yeh and co-workers [275–277] by expanding the incident focused beam field in terms of its plane wave spectrum.

A comprehensive theory for the scattering of Gaussian beams was developed by Gouesbet, Gréhan, and co-workers [278–286] starting in 1982. Their theory is best known as the *generalized Lorenz–Mie theory* (GLMT) and has found many applications, which is reflected in the huge number of publications on this subject. Rigorous justification of the GLMT was done by Lock and Gouesbet [287, 288]. Further developments of the Mie theory for Gaussian beams stem from Barton *et al.* [289] and Khaled *et al.* [290, 291].

The calculation of the scattering of an arbitrarily shaped beam of electromagnetic radiation by spherical particles is very similar to the classical Mie theory. All fields are expanded into spherical harmonics. As, however, the incident wave is not a plane wave, we have to admit also associated Legendre polynomials with $m \neq \pm 1$ in the solution of the scalar wave equation. Moreover, the expansion coefficients of the incident wave must be changed. The introduction of so-called g-factors that contain information on the type of the incoming wave and the position of particles in the wave is the main difference between classical Mie theory and the GLMT. For an arbitrarily shaped beam, the calculation of the g-factors is a complicated and time-consuming task. A simple method is to use the *localization principle* explained by van de Hulst [6]. It attributes to each term of order n in the multipole expansion a ray passing the origin at a distance of $(n + \frac{1}{2})\lambda/2\pi$. In this approach, the amplitudes $g_{n,m}^{TE}$, $g_{n,m}^{TM}$ of a particular ray can now be found from the known intensity profile of the incident beam, as shown by Gouesbet and coworkers [278–286]. When the scattering particle is centered on the axis of a Gaussian beam, the GLMT simplifies dramatically. All coefficients $g_{n,m}^{TE}$, $g_{n,m}^{TM}$ with $m \neq \pm 1$ become equal to zero. The coefficients with $m = \pm 1$ may be expressed as a single set of beam shape coefficients g_n:

$$g_n = \exp\left[-\frac{s^2}{4}(2n+1)^2\right] \tag{7.85}$$

where s is the beam confinement parameter or beam shape factor, given by the wavenumber k of the incident wave and the beam waist radius w_0:

$$s = \frac{1}{kw_0} \tag{7.86}$$

$s = 0$ means a plane wave, which is obtained for $w_0 \to \infty$. On the other hand, the largest s is approximately 0.26, which is obtained for a Gaussian beam focused to its physical diffraction limit.

This so-called *localized beam approximation* is valid from $w_0/R = 0.1$ to 100 to a high degree of accuracy, with $2R$ being the particle diameter.

In the localized beam approximation, the cross-sections for extinction and scattering become

$$\sigma_{\text{ext}} = \frac{2\pi}{k^2}\frac{1}{N}\sum_{n=1}^{\infty}(2n+1)|g_n|^2 \operatorname{Re}(a_n + b_n) \tag{7.87}$$

$$\sigma_{\text{sca}} = \frac{2\pi}{k^2}\frac{1}{N}\sum_{n=1}^{\infty}(2n+1)|g_n|^2 \left(|a_n|^2 + |b_n|^2\right) \tag{7.88}$$

and the scattering intensities are

$$i_{\text{per}}(\theta) = \left| \sum_{n=1}^{\infty} \frac{2n+1}{n(n+1)} g_n \left[a_n \pi_n(\theta) + b_n \tau_n(\theta) \right] \right|^2 \quad (7.89)$$

$$i_{\text{par}}(\theta) = \left| \sum_{n=1}^{\infty} \frac{2n+1}{n(n+1)} g_n \left[a_n \tau_n(\theta) + b_n \pi_n(\theta) \right] \right|^2 \quad (7.90)$$

As the intensity of the incident beam is not constant over the cross-section of the particle, the extinguished and scattered power have to be normalized on the corresponding incident intensity to obtain cross-sections as particle-specific quantities which can be compared with the plane wave case. This has already been discussed in [292–294] and is taken into account in Equations 7.87 and 7.88 by the factor N. It becomes for a Gaussian beam [292]

$$N = 1 + \sum_{k=1}^{\infty} \frac{(-1)^k}{(k+1)!} \left(\frac{2R^2}{w_0^2} \right)^k \quad (7.91)$$

As a general rule of thumb, it can be stated that GLMT and classical Mie theory give the same results when the diameter of the laser beam is 10 times larger than the particle diameter ($w_0/R \geq 5$) and the particle is placed in the center of the laser beam.

In Figure 7.37 we show an example of the deviations in extinction and scattering for a silver sphere with $2R = 200\,\text{nm}$ placed in the center of a focused laser beam with $w_0/R = 0.61$. This is the extreme case of physical diffraction limit for the laser beam focus. Obviously, the efficiencies become smaller than for illumination with a plane wave. However, we point to the limited meaning of this difference in the efficiencies. In all applications the scattered or absorbed power is the relevant quantity and not the efficiency. This power is much larger for the focused laser beam than for a plane wave, and compensates for the effect of reduced efficiency.

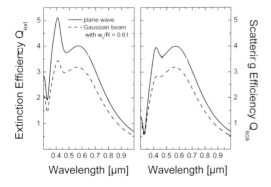

Figure 7.37 Extinction and scattering efficiencies of a silver nanoparticle with $2R = 200\,\text{nm}$. The solid lines are for plane wave excitation and the dashed lines for Gaussian beam illumination with a (beam waist):radius ($w_0:R$) ratio of 0.61 (physical diffraction limit).

Lock [295] derived an alternative form for the localized beam coefficients that is more convenient for computer computations and that provides physical insight into the details of the scattering process. Doicu and Wriedt [296] developed a method to evaluate the beam-shape coefficients in the general case of an off-axis location of the scatterer by starting from the set of beam-shape coefficients for an on-axis location and using the addition theorem for the spherical vector wave functions of the first kind under a translation of the coordinate origin.

Extensions of the GLMT on coated spheres were made by Khaled et al. [297] and Gouesbet, Gréhan, and co-workers [298, 299]. Finally, Barton [300, 301] extended the GLMT to higher order Gaussian beams, that is, to Hermite–Gaussian beams and donut mode beams of four different polarizations.

Shifrin and Zolotov [302] generalized the Mie theory for the case of a spherical particle irradiated by a pulse with a finite length L that is transferred by a carrier wavelength λ_0. Two cases were physically distinguished, depending on radiation-receiver properties: quasi-stationary scattering (a receiver integrates the entire signal over time) and nonstationary scattering, when a receiver is capable of recording scattered signal changes with time. General equations that allow one to calculate optical characteristics for both scattering cases and for an arbitrary ratio L/λ_0 were derived.

7.6.2
Evanescent Waves from Total Internal Reflection

Chew et al. [303] outlined the theory for scattering of evanescent waves by a spherical particle, which essentially consists in the analytical continuation of the case of plane-wave excitation to complex angles of incidence. An independent approach to the scattering of evanescent waves within the framework of Mie's theory was made in 1993 by Barchiesi and Labeke [304]. The theory of Chew et al. [303] was slightly improved by Liu et al. [305, 306] and comprehensively discussed for cross-section spectra by Quinten et al. [186, 307], and for morphology-dependent resonances by Liu et al. [305, 306]. In the following, we briefly repeat the derivation of cross-sections and scattering intensities. We assume a spherical particle in front of a glass prism of refractive index n_P within a medium of real refractive index n_M at distance d from the prism surface (see Figure 7.38).

Multiple scattering at the particle and prism surfaces is neglected at this point. A plane wave propagating in the prism and incident at a subcritical angle θ_i is partly reflected and partly refracted at the glass–air boundary, the refraction being described by Snell's law. In the reference frame of the particle, the refracted wave is incident at an angle θ_k to the z-axis, as given by Snell's law.

Similarly to the Gaussian beam, we have to change the expansion coefficients of the incident wave. For the general case $\theta_k \neq 0$ contributions of Legendre polynomials P_{nm} with $m \neq \pm 1$ are obtained which are not required in standard Mie theory where $\theta_k = 0$. Then, the expansion coefficients of the incident wave are as follows [305]:

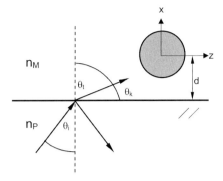

Figure 7.38 Geometry in the case of evanescent wave excitation.

$$\alpha_{TM}^s(n,m) = 2i^n n_M \left[\frac{\pi(2n+1)(n-m)!}{n(n+1)(n+m)!} \right]^{\frac{1}{2}} \frac{mP_{nm}(\cos\theta_k)}{\sin\theta_k} E_t^s \tag{7.92}$$

$$\alpha_{TE}^s(n,m) = 2i^{n+1} \left[\frac{\pi(2n+1)(n-m)!}{n(n+1)(n+m)!} \right]^{\frac{1}{2}} \frac{dP_{nm}(\cos\theta_k)}{d\theta_k} E_t^s \tag{7.93}$$

for an s-polarized incident beam and

$$\alpha_{TM}^p(n,m) = 2i^{n+1} n_M \left[\frac{\pi(2n+1)(n-m)!}{n(n+1)(n+m)!} \right]^{\frac{1}{2}} \frac{dP_{nm}(\cos\theta_k)}{d\theta_k} E_t^p \tag{7.94}$$

$$\alpha_{TE}^p(n,m) = -2i^n \left[\frac{\pi(2n+1)(n-m)!}{n(n+1)(n+m)!} \right]^{\frac{1}{2}} \frac{mP_{nm}(\cos\theta_k)}{\sin\theta_k} E_t^p \tag{7.95}$$

for a p-polarized incident beam. E_t is the complex magnitude of the electric field vector of the refracted wave, which is determined using Fresnel's equations. For the scattered fields outside the sphere a similar expansion may be set up using expansion coefficients $\beta_{TM}^{s,p}(n,m)$ and $\beta_{TE}^{s,p}(n,m)$, and also a similar expansion may be set up for the fields in the interior of the sphere. Applying Maxwell's boundary conditions at the surface of the spherical particle results in the scattering coefficients of the sphere, given by

$$\beta_{TM}^{s,p}(n,m) = a_n \alpha_{TM}^{s,p}(n,m) \tag{7.96}$$

$$\beta_{TE}^{s,p}(n,m) = b_n \alpha_{TE}^{s,p}(n,m) \tag{7.97}$$

where a_n and b_n are the scattering coefficients from classical Mie theory. We can now calculate the cross-sections for extinction and scattering of the refracted plane wave from Poynting's law and obtain

$$\sigma_{ext}^s = \frac{2\pi}{k_M^2} \frac{1}{N} \sum_{n=1}^{\infty} (2n+1) \operatorname{Re}(a_n \Pi_n + b_n T_n) \tag{7.98}$$

$$\sigma_{\text{sca}}^{\text{S}} = \frac{2\pi}{k_M^2} \frac{1}{N} \sum_{n=1}^{\infty} (2n+1) \left(|a_n|^2 \Pi_n + |b_n|^2 T_n \right) \tag{7.99}$$

for s-polarized light and

$$\sigma_{\text{ext}}^{\text{P}} = \frac{2\pi}{k_M^2} \frac{1}{N} \sum_{n=1}^{\infty} (2n+1) \text{Re}(a_n T_n + b_n \Pi_n) \tag{7.100}$$

$$\sigma_{\text{sca}}^{\text{P}} = \frac{2\pi}{k_M^2} \frac{1}{N} \sum_{n=1}^{\infty} (2n+1) \left(|a_n|^2 T_n + |b_n|^2 \Pi_n \right) \tag{7.101}$$

for p-polarized light, with the definitions

$$\Pi_n(\theta_k) = \frac{2}{n(n+1)} \sum_{m=-n}^{n} \frac{(n-m)!}{(n+m)!} \left| m \frac{P_{nm}(\cos\theta_k)}{\sin\theta_k} \right|^2 \tag{7.102}$$

$$T_n(\theta_k) = \frac{2}{n(n+1)} \sum_{m=-n}^{n} \frac{(n-m)!}{(n+m)!} \left| \frac{dP_{nm}(\cos\theta_k)}{d\theta_k} \right|^2 \tag{7.103}$$

The normalization factor N is equal to one for plane waves, but assumes a different value for evanescent waves.

The cross-sections are now dependent on the polarization of the incoming light, and are determined by the behavior of the functions T_n and Π_n. For an incident plane wave (angles θ_k with $\cos\theta_k \leq 1$) it is easy to prove that $T_n = \Pi_n = 1$ for all multipolar orders n and angles θ_k, using the addition theorem of the associated Legendre polynomials. In this case, the cross-sections do not differ for s- and p-polarization and Equations 7.98–7.101 are the well-known results from standard Mie-theory.

For $\theta_i > \theta_c$, it follows from Snell's law that $\cos\theta_k > 1$, hence $\sin\theta_k$ becomes purely imaginary. Then, $T_n, \Pi_n > 1$, except for $\Pi_1 = 1$. Furthermore, $T_n > \Pi_n$ for all n [307]. The polarization dependence of the cross-sections is now due to the fact that p- and s-polarized evanescent waves are not related to each other by a simple rotation, that is, a symmetry operation of the sphere, in contrast to the corresponding plane waves: for p-polarized waves the electric field is rotating in the plane of incidence, due to the complex phase shift associated with total internal reflection, whereas for s-polarized waves it is oscillating perpendicular to it. A similar polarization dependence of the cross-sections is already observed when the particle resides inside the prism, that is, in the standing wave resulting from total internal reflection. The same holds for a particle in the standing wave in front of a metallic mirror where also a complex phase shift occurs in reflection. In both cases, however, the polarization effects are much smaller than for evanescent waves.

As the component $k_M \sin\theta_k$ of the wavevector becomes imaginary, the magnitudes of the electromagnetic fields at distance d from the prism surface have decreased exponentially by a factor $\exp(-\kappa d)$ from their values at the prism surface where the attenuation constant is given by

$$\kappa = k_M \sin\theta_k = \frac{2\pi}{\lambda} (n_P^2 \sin^2\theta_i - n_M^2)^{\frac{1}{2}} \tag{7.104}$$

The energy flux given by the time-averaged Poynting vector is then not constant over the geometric cross-section of the sphere due to the exponential decrease of the fields. Therefore, in the calculation of optical cross-sections for evanescent wave excitation the incident intensity I_0 has to be redefined. The corresponding quantity $\tilde{I}_0$ cannot be uniquely determined. For evanescent wave excitation it is preferable to normalize to the total power incident on the particle, in order to avoid diverging efficiencies for large particles where the intensity of the evanescent wave at the center of the spherical particle is close to zero. As $\tilde{I}_0$ we choose the incident intensity, averaged over the cross-sectional area of the sample perpendicular to the Poynting vector of the evanescent wave:

$$\tilde{I}_0 = \frac{1}{\pi R^2} \iint \mathbf{S}_{\text{inc}} \cdot \mathbf{n} \, dA = I_0 \exp(-2\kappa d) \frac{n_P}{n_M} \sin(\theta_i) \frac{I_1(2\kappa R)}{\kappa R} \qquad (7.105)$$

where $I_1(2\kappa R)$ is the modified Bessel function of order 1 with argument $2\kappa R$, for which the series expansion

$$\frac{I_1(2\kappa R)}{\kappa R} = 1 + \sum_{m=1}^{\infty} \frac{(\kappa R)^{2m}}{m!(m+1)!} \qquad (7.106)$$

may be used. For a plane wave it is $\tilde{I}_0 = I_0$.

As the multipole expansions for the incident and scattered fields are made in a coordinate system, the origin of which coincides with the center of the sphere, the factor E_t in the multipole coefficients in Equations 7.92–7.95 has to be replaced by $E_t \exp(-\kappa d)$ for an evanescent wave. Therefore, for evanescent wave excitation the factor $I_0 \exp(-2\kappa d)$ in Equation 7.105 cancels out in the calculation of optical cross-sections. The remaining factor

$$N = \frac{n_P}{n_M} \sin\theta_i \frac{I_1(2\kappa R)}{\kappa R} \qquad (7.107)$$

is the normalization factor which must be taken into account in the cross-sections.

In the following we give examples of the extinction cross-section of silver spheres (with SPP resonances) and diamond spheres (with MDRs).

For the larger silver particles, SPP resonances with higher multipolar orders are important and are enhanced for evanescent waves as compared with plane waves. This leads to the structured spectra in Figure 7.39, which are not simply in a constant ratio to each other any longer. The excitation with p-polarized evanescent waves leads to a larger cross-section and the s-polarized evanescent wave to a smaller cross-section compared with the plane wave. However, we point to the limited meaning of any definition of cross-sections for an excitation that varies significantly over the particle surface, as in all applications the scattered or absorbed power is the relevant quantity. This power is the largest for excitation with a plane wave.

Figure 7.40 displays the decomposition of the extinction spectrum into the different multipolar contributions for p-polarized evanescent waves (Figure 7.40a)

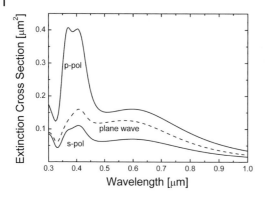

Figure 7.39 Extinction cross-section of a silver sphere with diameter $2R = 200\,\text{nm}$ for plane wave excitation and evanescent wave excitation.

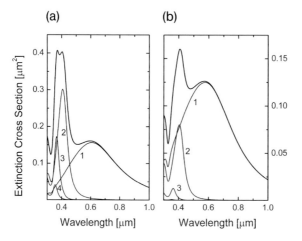

Figure 7.40 Decomposition of the extinction cross-section of a silver sphere with diameter $2R = 200\,\text{nm}$. (a) Evanescent wave excitation, p-polarization; (b) plane wave excitation.

and for plane waves (Figure 7.40b), the numbers indicating the multipolar order. The strong enhancement of higher multipoles for evanescent wave compared with plane wave excitation is evident. For example, the octupolar contribution to the spectrum of the large silver sphere in Figure 7.40a can now clearly be resolved and leads to the double peak structure at short wavelengths. For s-polarization the higher multipoles are also enhanced with respect to plane wave excitation but less than for p-polarization (compare Figure 7.39).

A second example is shown in Figure 7.41, which displays the scattering cross-sections of a spherical diamond particle of diameter $2R = 600\,\text{nm}$. In this case the extinction and scattering cross-sections are, of course, identical, as the absorption of pure crystalline diamond vanishes in this wavelength range. Sharp MDRs of

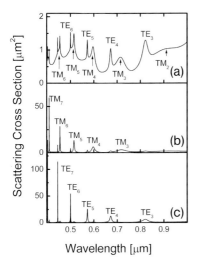

Figure 7.41 Extinction cross-section of a diamond sphere with diameter $2R = 600$ nm. (a) Plane wave excitation; (b) evanescent wave excitation, p-polarization; (c) evanescent wave excitation, s-polarization.

order $n \leq 7$ in the visible spectrum exhibit strong enhancements in the evanescent field and, as in the case of the silver particles, TM modes are favored for p-polarized excitation and TE modes for s-polarized excitation, the difference increasing with order n of the multipoles. The spectral dependence of the scattering cross-sections in Figure 7.41 demonstrates clearly the large enhancements of contributions from specific higher multipoles due to the large field gradients of evanescent waves. When comparing the very large absolute values of the cross-sections for evanescent wave excitation (Figure 7.41b and c) with those for plane wave excitation (Figure 7.41a), it should be kept in mind that the cross-sections have been normalized.

8
Limitations of Mie's Theory – Size and Quantum Size Effects in Very Small Nanoparticles

The classical Mie theory presumes a single uncharged spherical particle which is placed in a surrounding nonabsorbing medium so that electromagnetic interaction with other particles or a surface of a macroscopic body, for example, a substrate, does not occur at all. In Chapter 7 we gave some extensions of Mie's theory to coated particles, supported particles, charged particles, anisotropic particle materials, absorbing surrounding media, and incident waves which deviate from a plane wave. Interaction among particles is considered in Chapters 10–12.

Mie's theory and its extensions are exact within the frame of classical electrodynamics, where the electromagnetic fields inside and outside the particle undergo a smooth transition at a sharp, infinitely thin boundary between particle and surrounding. However, do these step-like Maxwell boundary conditions still hold for sufficiently small nanoparticles? Furthermore, can the dielectric functions of the bulk material still be used in calculations for such small particles?

From Table 3.1 in Section 3.4, we know that the skin depth in metals lies in the range 20–30 nm. This means that a metal particle with size less than this skin depth is mainly *surface*, not *bulk* for an electromagnetic wave. Already from this classical electrodynamic view it becomes obvious that at least the optical properties of such small nanoparticles must undergo size effects, particularly if the electronic system of the particle forms an easily polarizable medium as in the case of metals. In fact, size and quantum size effects are most prominent for metal particles, for which the following considerations are concerned with metal nanoparticles.

8.1
Boundary Conditions – the Spill-Out Effect

In electrochemistry and nanoscience the *jellium model* [28] is a well-established simple model for separating the conduction electrons of a metal body from the ions. In this model the ions appear as a constant positive charge background with the task of compensating for the integrated negative charge of the electrons. Whereas the *ionic* body has a an atomically sharp surface, the *electronic* body exhibits a smooth transition of the electron density due to the finite length of the electron waves, resulting in negative charge density *outside* the body. This is called

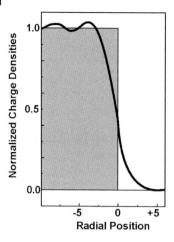

Figure 8.1 Normalized charge densities in the jellium model.

the *spill-out effect*. The electron density varies in an oscillating manner, the so-called *Friedel* oscillations. This is illustrated in Figure 8.1, where the gray region is the constant positive charge background. For calculation of the electron density, see for example [308].

As a consequence of the jellium model, the surface of a metal nanoparticle is *soft*. The polarizability and, hence, the corresponding dielectric function vary locally with the radial coordinate, both inside and outside the *ionic* particle, and an electric charge double layer is created with the negative sign outside. Therefore, the surface region can no longer be approximated by a sharp two-dimensional plane, but forms a three-dimensional extended area. In electrochemistry, this surface region can be assigned a capacitance in the order of 30–50 μF cm^{-2}.

8.2
Free Path Effect in Nanoparticles

The bulk optical material functions are determined by quantities such as ω_{Pj}^2, ω_j, and γ_j for harmonic oscillators and ω_P^2 and γ_{fe} for free electrons. All these quantities will develop size dependences on going to nanoparticles of sufficiently small size. Most strongly, the parameters ω_P^2 and γ_{fe} of free electrons change due to *size effects*, for which mainly metal and semiconductor nanoparticles are affected.

In a classical approach, the relaxation frequency τ^{-1}:

$$\tau^{-1} = \sum_i \tau_i^{-1} = \tau^{-1}_{\text{point defects}} + \tau^{-1}_{\text{dislocations}} + \tau^{-1}_{\text{grain boundaries}}$$
$$+ \tau^{-1}_{\text{surface/interface}} + \tau^{-1}_{\text{e-phonon}} + \tau^{-1}_{\text{e-e}} \quad (8.1)$$

can be expressed by introducing the bulk *mean free path* l_∞ between memory-canceling collisions of the single free electrons. Provided that these collisions are only slightly inelastic–dissipating only the excess energy or momentum (drift momentum, drift energy) due to an external field–solely electrons close to the Fermi surface contribute and, hence,

$$l_\infty = v_F \tau \tag{8.2}$$

Using the known Fermi velocities v_F and relaxation times, one finds for example for aluminum $l_\infty(\text{Al}) = 16\,\text{nm}$, for sodium $l_\infty(\text{Na}) = 34\,\text{nm}$, and for silver $l_\infty(\text{Ag}) = 52\,\text{nm}$ at 273 K. If the particle size becomes comparable to or smaller than l_∞, the collisions of the conduction electrons with the particle surface become important as an additional scattering process. In the frame of Matthiessen's rule (Equation 8.1) they are usually added to the former collision processes [115, 309]. The effective mean free path is then smaller than l_∞ in all nanoparticles with $d \leq l_\infty$. This is called the *free path effect*. It was predicted theoretically in 1939 [310] and 1954 [311], but first identified experimentally in 1958 [312–315]. Although more experiments followed [316–320], the first quantitative comparison with theory was not reported until 1969 [115].

The common feature of all predictions is that the additional particle surface scattering contribution to γ_{fe}, which we call $\Delta\gamma_{fe}(R)$, is proportional to $1/R$. Therefore, γ_{fe} in the Drude susceptibility in Equation 3.19 is replaced by

$$\gamma_{fe}(R) = \gamma_{fe} + \Delta\gamma_{fe}(R) \quad \text{with} \quad \Delta\gamma_{fe}(R) = A\frac{v_F}{R} \tag{8.3}$$

The $1/R$ dependence reflects the ratio of surface scattering probability being proportional to the surface area $4\pi R^2$ and the number of electrons being proportional to the volume $4\pi R^3/3$.

All cited approaches assume that surface scattering is the only size-dependent contribution to Matthiessen's rule (Equation 8.1). However, there are further contributions coming from

- the modified phonon spectrum of very small nanoparticles [321]
- strong (Coulomb) electron–phonon coupling by external field-induced surface charges
- point and lattice defect concentrations
- structural phase transitions (e.g., the transition from icosahedral to fcc structure in Au nanoparticles [322])
- scattering of electrons on grain boundaries [323]
- stress in the embedding matrix [324]
- lattice contraction [325].

If the particles become small as details of the electronic and geometric particle structure come into play, quantum mechanical corrections to the classical description of the particle material have to be applied. For an early review, see [326].

Early quantum mechanical concepts for nanoparticles started from the solid-state physics point of view, in particular from the tight binding approximation. Soon afterwards Fröhlich pointed out [327] that the continuous electronic conduction band of a solid metal should break up into discrete states if the dimensions of the metal become small enough. This effect is called the *quantum size effect*

(QSE). So far, however, these discrete eigenstates could not be observed directly in optical experiments on metal nanoparticles, in contrast to semiconductor nanoparticles where they are present in quantum dots.

Kawabata and Kubo [328] pointed out that in the quantum mechanical treatment the surface no longer acts as an additional scatterer but–via the boundary conditions–determines the electron energies as a whole. So, the observed $1/R$ law in Equation 8.3 can be considered as a QSE, and this was shown in several theoretical studies [329–350]. Indeed, the $1/R$ dependence of quantum size effects appears to be a fundamental law as far as extrapolations of solid-state models hold: it reflects the surface to volume ratio. However, numerical details are nonfundamental but instead characteristic for particular geometries and materials.

Improvements of the calculations of Kawabata and Kubo were made by Zaremba and Persson [351]. In their paper, the influence of a surface on the polarizability of a small metal particle in vacuum was considered in the random phase approximation (RPA). In summary, the surface induces an additional contribution to the Drude dielectric constant of the particle, separately from the bulk dielectric constant, making it nonlocal. Using the jellium model for the metal and introducing a surface polarization potential, which is missing in the work of Kawabata and Kubo, the authors found a damping rate proportional to $1/R$. Assuming a step potential for concrete calculations, the surface-induced relaxation time is obtained as $\tau = R/(3v_F)$, corresponding to an A parameter of 3.

In the following, we give some results obtained for the dielectric function of gold and silver nanoparticles by applying the free path effect in the calculations.

The size-dependent dielectric functions of small Ag and Au particles were investigated by Kreibig [111, 116, 309, 352] and Quinten [40] using the following procedure. This procedure is based on the assumption that the optical extinction $E(\omega)$:

$$E(\omega) = \sigma_{ext}(\omega, R)\frac{N}{V}\log(e)d \tag{8.4}$$

of light with photon energy $\hbar\omega$ by an ensemble of noninteracting spherical particles of diameter $2R$ corresponds to the real part of a complex function $F(\omega) = E(\omega) + iP(\omega)$, of which the real and imaginary parts satisfy Kramers–Kronig relations. Then, $P(\omega)$ can be deduced from $E(\omega)$ at $\omega = \omega_0$ via

$$P(\omega_0) = -\frac{1}{\pi}\wp\int_{-\infty}^{\infty}\frac{E(\omega)}{\omega-\omega_0}d\omega + P(\infty)$$

$$= \frac{2\pi c^2}{\omega_0^2 \varepsilon_M(\omega_0)}\sum_{n=1}^{\infty}(2n+1)\mathrm{Im}[a_n(\omega_0) + b_n(\omega_0)] \tag{8.5}$$

where Im means the imaginary part and $\wp$ means the principal value of the integral. Recall that ω takes only real, positive values. Then, $P(\omega_0)$ simply becomes

$$P(\omega_0) = -\frac{2\omega_0}{\pi}\wp\int_0^{\infty}\frac{E(\omega)}{\omega^2-\omega_0^2}d\omega + P(\infty) \tag{8.6}$$

since $E(-\omega) = E(\omega)$. $F(\omega)$ is regular for $\omega \to 0$ and $\omega \to \infty$.

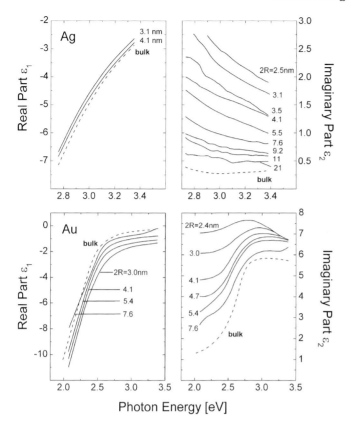

Figure 8.2 Dielectric constants of various gold and silver nanoparticles, obtained in the quasi-static approximation by Kramers–Kronig analysis of measured extinction data. For comparison, the dielectric constants of the data from Otter [100] are plotted as the dashed line.

Applying Equation 8.5 to measurements on systems containing Ag or Au nanoparticles in suspension, one must take into account that

- The spectral region in which the measured data are available is limited; consequently, the Kramers–Kronig integral must be divided into parts
- Outside the measured spectral range, the Kramers–Kronig integral can only be approximated by computations according to the Mie theory, using bulk optical constants.
- The missing optical constants at high and low photon energies hinder this computation. The corresponding integrals must be approximated.

After determination of $P(\omega)$ from $E(\omega)$, the complex dielectric constant $\varepsilon(\omega) = \varepsilon_1(\omega) + i\varepsilon_2(\omega)$ can be determined. This method was applied to several Au and Ag nanoparticle colloids. Figure 8.2 shows the size-dependent functions $\varepsilon_1(\omega, R)$ and $\varepsilon_2(\omega, R)$ of silver and gold nanoparticles calculated in the quasi-static approximation on

Table 8.1 Mean diameters 2R and mean free path parameters A for several gold and silver clusters.

Sample	Mean diameter 2R (nm)	Mean free path parameter A
Au 1	6.9	0.6
Au 2	12.6	0.7
Au 3	16	0.6
Au 4	30.6	1.3
Au 5	38	2
Ag 1	16.6	2.5
Ag 2	17.8	2.3
Ag 3	20.4	2.5
Ag 4	27.8	2
Ag 5	32	3

the basis of the bulk optical functions $\varepsilon_{\text{bulk}}(\omega)$ from [100] obtained by Kreibig [116, 352].

These first investigations on the optical properties of small silver and gold particles were extended later by Quinten [40] to the range 1.5–4.5 eV (275–825 nm) and to larger particles. Furthermore, full Mie theory was applied. For determination of the complex dielectric constant $\varepsilon(\omega) = \varepsilon_1(\omega) + i\varepsilon_2(\omega)$ from $P(\omega)$ and $E(\omega)$ the generalized Newton–Raphson iteration method was used. As an initial guess for the iteration, the dielectric constants from [37] were used.

The mean particle sizes of the examined Au and Ag colloids are given in Table 8.1. The samples are numbered consecutively with respect to increasing size. In addition, a parameter A is given that accounts for the mean free path effect.

In Figure 8.3, the results for ε_1 and ε_2 of the various gold and silver particles are presented. For comparison, the data from Johnson and Christy [37] are given as the dashed line.

Apparently, the main deviation with respect to the bulk values occurs for the imaginary part ε_2. A clear dependence on particle size is recognized. With increasing particle size, ε_2 decreases, but does not approach the bulk value.

Although we found dielectric constants for several silver and gold clusters of various size, it is of more practical interest to have only one unique set of optical constants for all gold or silver nanoparticles, which can be easily changed according to the model of the mean free path effect. These *bulk* dielectric constants of gold and silver nanoparticles are plotted in Figure 8.4 in comparison with the data from Johnson and Christy. Again, it can be recognized that the real part ε_1 coincides fairly well with the real part of the data from Johnson and Christy. The main differences are in the imaginary parts. The band edges of gold at 2.5 eV and of silver at 4.0 eV are similar to those obtained from evaluation of the data of Johnson and Christy. However, the interband transitions begin to contribute to the dielectric constant already at lower photon energies, yielding an increased imaginary part ε_2 at photon energies of the corresponding cluster plasmon. For silver there

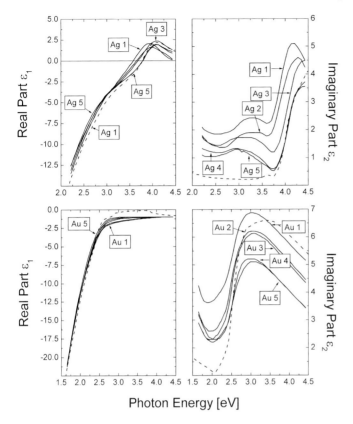

Figure 8.3 Dielectric constants of various gold and silver clusters, obtained by Kramers–Kronig analysis of measured extinction data. For comparison, the dielectric constants of the data from Johnson and Christy [37] are plotted as the dashed line.

is a small peak in ε_2 between 3 and 3.3 eV. This feature is also apparent in the data of Johnson and Christy, but at 3.65 eV photon energy, and its magnitude is fairly small. The origin of this feature is still unknown. At photon energies lower than 1.8 eV for gold and 2.4 eV for silver, the imaginary part again increases with decreasing photon energy. This increase is caused by the Drude dielectric constant, which only contributes to ε_2 in this region. Therefore, it was possible to fit the data on pure Drude dielectric constants, resulting in $\chi_{\text{interband},1}(\omega = 0) = 7.2 \pm 0.1$, $\omega_p = 1.365 \times 10^{16} \text{s}^{-1}$, $\gamma_{\text{fe}} = 7.0 \times 10^{13} \text{s}^{-1}$ for gold and $\chi_{\text{interband},1}(\omega = 0) = 2.25 \pm 0.1$, $\omega_p = 1.365 \times 10^{16} \text{s}^{-1}$, $\gamma_{\text{fe}} = 2.3 \times 10^{13} \text{s}^{-1}$ for silver. The advantage of these unique sets of optical constants is that the position of the SPP is now fitted very well for each particle size, in contrast to the data of Johnson and Christy.

Dalacu and Martinu [353] also examined optical constants of Au nanoparticles with varying size in the wavelength range 300–850 nm (1.46–4.1 eV). They produced the gold particles by plasma deposition in an SiO_2 nanocomposite film. Analysis was performed using the effective medium approach of Garnett

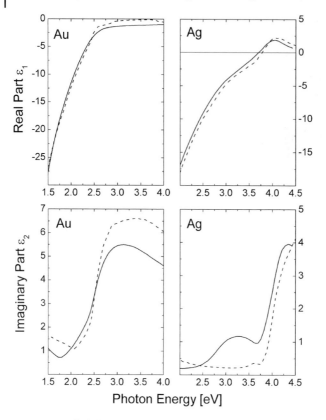

Figure 8.4 Bulk dielectric constants of gold and silver clusters, derived from the dielectric constants in Figure 8.3. For comparison, the dielectric constants of the data from Johnson and Christy [37] are plotted as the dashed line.

(Maxwell-Garnett model; see Chapter 14). Although it was claimed by the authors that the results are in disagreement with the results of Kreibig or Quinten, actually their results reflect the same behavior as shown in Figures 8.2 and 8.3. A parameters were not determined but additional evaluation of the interband edge was given.

8.3
Chemical Interface Damping – Dynamic Charge Transfer

As it turns out from the work of Apell and co-workers [342, 343], the additional damping due to the scattering of the free electrons at the surface and thus the corresponding A parameter are material dependent, because the scattering depends on the electron density in the particle. Hence the damping should be influenced by physisorption, chemisorption, or even chemical reactions at the particle surface, which induce changes in the density profile normal to the surface. There is strong evidence for such an influence from pioneering experiments on

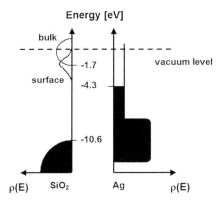

Figure 8.5 Density of states $\rho(E)$ of the occupied valence band and the lowest conduction band of bulk SiO_2 (left side). Due to the interaction with a silver nanoparticle, the local density of states in the vicinity of the nanoparticle is shifted as indicated by the dashed line. The right side depicts a schematic density of states for bulk silver. After [356].

Ag nanoparticles embedded in solid CO_2, O_2, and various solid rare gas matrices by Charlé et al. [354, 355], and on Ag particles created in photosensitive glass by Kreibig [309]: adsorbates, substrates, or embedding matrices are supposed to change the energy level distributions and, in particular, they may be sources of additional damping mechanisms.

Persson [356] discussed the origin of this effect and extended the work of Zaremba and Persson [351] to include the influence of a layer of atoms or molecules associated with the matrix environment of the particle. Persson's idea is as follows. The influence of adsorbate molecules is taken into account by diffusive electron scattering cross-sections σ_{diff}. In the inelastic scattering process, optically excited nanoparticle electrons are transferred into affinity levels of the adsorbate (surrounding) and – after residence times of typically 10^{-14} s – back to the nanoparticle. The dominant contribution to σ_{diff} occurs if the adsorbate has a resonance state or virtual level at a distance $\sim\hbar\omega$ from the Fermi level, due to adsorbate–substrate coupling.

For example, Figure 8.5 depicts a schematic energy level diagram showing the density of states for Ag nanoparticles in an SiO_2 matrix. The valence band maximum of SiO_2 lies at -10.6 eV whereas the lowest conduction band of bulk SiO_2 has its minimum at -1.7 eV. In comparison, the Fermi level in Ag lies at -4.3 eV.

The general form of the mean free path effect in Equation 8.3 remains valid in this case; however, the details of the interaction processes lead to a broadening which is increased by a contribution $A_{surface}$ of the surface:

$$A_{total} = A_{size} + A_{surface} \tag{8.7}$$

Persson treated this *chemical interface damping* (CID) or *dynamic charge transfer* in the jellium approximation by introducing a coefficient ΔA, defined by

$$\gamma_{fe}[eV] = \gamma_{fe,\infty}[eV] + \frac{A_v + \Delta A}{R[nm]} \tag{8.8}$$

with $A_v = A_{vacuum} \approx 0.25$ eV nm [356]. This equation is related by $A_v + \Delta A = \hbar v_F A$ to Equation 8.3, the $\hbar$ being due to the choice of dimensions (A has no dimensions); v_F is the Fermi velocity of the conduction electrons. Persson introduced a complex self-energy correction into the spherical oscillator polarizability which encloses both peak shifts and broadening effects. Two different mechanisms for power absorption due to the field-induced motion of the particle electrons normal and parallel to the surface are distinguished. Both contribute additively to the size parameter A: $A = A_\| + A_\perp$, analogous to the model of Apell et al. [342]. The $1/R$ dependence is motivated by the R^3 dependence of the electrons involved in the collective excitation and the R^2 dependence of the number of adsorbate locations at the surface.

In Table 8.2, experimental values from Charlé et al. [354, 355] and Kreibig [309] are compared with theoretical values obtained by Persson [356] according to the

Table 8.2 Theoretical and experimental values of the A parameter for dynamic charge transfer.

Matrix material	Theory	Experiment
Free nanoparticles	0.29	0.25
Solid Ne	0.29	0.25
Solid Ar	0.35	0.30
Solid Kr	0.325	
Solid Xe	0.42	
Solid N_2	0.395	
Solid O_2	0.557	0.50
Solid C_2H_4	0.70	
Solid CO	1.01	0.90
Soda-lime glass	0.97	1.00
Ice		0.50
LiF_2		0.72
CaF_2		0.76
MgF_2		0.80
Fullerite (C_{60})		1.00
SiO_2		1.30
ITO		1.50
$SrTiO_3$		1.50
Al_2O_3		1.70
TiO_2		1.80
$P(C_6H_5)_3$		2.00
Sb_2O_3		2.00
Si		≈3.00
Cr_2O_3		≈3.00
MgF_2 substrate		0.55
ITO substrate		0.57
SiO_2 substrate (at 110 K)		0.55
SiO_2 substrate (at 300 K)		0.59
Cr_2O_3 substrate		0.59

model described above. Note that Persson originally obtained A parameters having the dimensions eV nm. The values given in Table 8.2 are calculated from these values using the Fermi velocity $v_F = 1.38 \times 10^{14}$ nm s^{-1} for Ag.

Kreibig's group have investigated the effects of dynamic charge transfer in a long-term research project in Saarbrücken and Aachen for Ag and Au particles since 1987. They used a wide range of different liquid and solid surroundings and substrates. Experimental verification of Persson's model was first published in 1993 [357] and experimental details can be found in [11, 273, 358–369]. The most recent experimental results were obtained with the thermal cluster source apparatus THECLA, described in detail in [363, 364]. Very small Ag particles of mean diameter 2–4 nm were produced by thermal evaporation and supersonic nozzle beam expansion into vacuum. Different seeding gases were used (Ar, Kr, and Xe), resulting in mean sizes $2R = 2.1$, 3.7, and 4.0 nm, respectively. Three kinds of optical absorption spectra of the created Ag particles were recorded *in situ*:

- in the free beam, that is, with a clean, uncontaminated free surface
- after deposition on to arbitrarily chosen solid substrates
- after embedding in solid surrounding media which were co-deposited from two electron gun sources.

Interface effects after deposition or embedding were *calibrated* with the results on the free particles. Influences of the size distributions on the optical spectra were controlled numerically by Mie calculations introducing TEM size histograms.

Figure 8.6 shows exemplary spectra of Ag nanoparticles with $2R = 2.1$ nm embedded in various co-sputtered solid surroundings.

Evaluation of peak position and halfwidth and direct comparison with calculations according to Mie's theory using the A parameter as a free parameter, Kreibig's group achieved a quantitative determination of the dynamic charge transfer. The results are given in Figure 8.7 and Table 8.2.

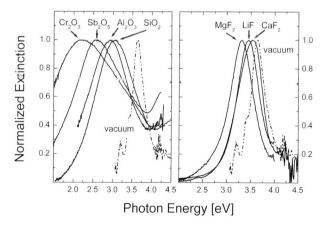

Figure 8.6 Spectra of Ag-nanoparticles produced with THECLA and embedded in various solid oxide and fluoride matrices. Data courtesy of U. Kreibig.

8 Limitations of Mie's Theory – Size and Quantum Size Effects in Very Small Nanoparticles

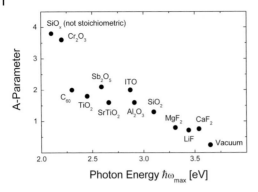

Figure 8.7 Peak position and A parameter from many experiments on silver nanoparticles deposited on and embedded in various solid matrices (see also [365]). Sample preparation and optical spectrum recording were performed *in situ* with THECLA.

Figure 8.7 shows the peak position and A parameter obtained from evaluation of many spectra of Ag nanoparticles deposited on and embedded in various solid matrices. Sample preparation and optical spectrum recording were performed *in situ* with THECLA. Table 8.2 summarizes numerical values of the A parameter from these experiments. Data for deposited particles are added.

9
Beyond Mie's Theory I – Nonspherical Particles

The most serious limitation of Mie's theory is its restriction to spherical particles. However, in nature particularly solid particles are mostly nonspherical. These particles scatter light in a way different from spheres. An example is shown in Figure 9.1 for gold nanoparticles grown in nonspherical geometry by chemical reduction of gold from chlorogold acid in aqueous solution. The obvious deviations from the spherical shape also affect the optical extinction of the colloidal solution: instead of a narrow surface plasmon polariton (SPP) extinction band, the spectrum in Figure 9.1 is dominated by a broad extinction band. The color of this colloidal solution turned to a clear blue without scattering.

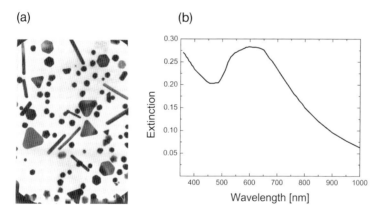

Figure 9.1 (a) Transmission electron micrograph of nonspherical colloidal gold nanoparticles; (b) optical extinction spectrum of the colloidal gold solution.

A first rough approximation for describing light scattering and absorption by nonspherical particles is to replace the particle by either the surface area-equivalent sphere or the volume-equivalent sphere. Chýlek and Ramaswamy [42] showed that for nonspherical particles with particle size parameters $x = 2\pi R/\lambda < 0.6$ the sphere with the equivalent surface area yields a better approach to the extinction cross-section of the nonspherical particle than the volume-equivalent sphere. However, the difference is small. The extinction cross-section of the sphere with the same

Optical Properties of Nanoparticle Systems: Mie and Beyond. Michael Quinten
Copyright © 2011 WILEY-VCH Verlag GmbH & Co. KGaA, Weinheim
ISBN: 978-3-527-41043-9

surface area deviates by less than 2% from the extinction cross-section of the nonspherical particle, while the volume-equivalent sphere deviates to less than 6%. For larger size parameters they stated that the volume-equivalent sphere is a rather good approximation. This approximation works fairly well for transparent dielectric particles provided that resonances can be excluded in these particles. These resonances occur as morphology-dependent resonances (MDR). For absorbing particles the approximation often fails, particularly if resonances such as the SPP resonance occur.

Another attempt to describe light scattering by nonspherical particles is by the application of pertubation theory. This was first derived by Yeh [370, 371] and Erma [372–374]. The first-order pertubation ansatz is to replace the radius R of the particle by

$$R = R_s[1 + \varepsilon f(\theta, \varphi)] \tag{9.1}$$

where R_s is the radius of the unpertubed sphere and $f(\theta,\varphi)$ describes the shape of the irregularity; ε is a small number, $\varepsilon \ll 1$, and $|\varepsilon f(\theta,\varphi)| < 1$. Typically, the shape of the irregularity is assumed to be a nth-order Chebyshev polynomial T_n. Then, the first-order pertubation yields for the scattering coefficients a_n and b_n of the pertubed sphere

$$a_n = a_n^{(0)} + \varepsilon a_n^{(1)} \qquad b_n = b_n^{(0)} + \varepsilon b_n^{(1)} \tag{9.2}$$

The superscript (0) belongs to the coefficients of the unpertubed sphere. Results and applications can be found in, for example, [9]. In 1993 and 1995 Martin [375] revisited the pertubation theory.

The exact starting point for calculating all electrodynamic problems is Maxwell's equations. In the case of scattering by particles, we have to solve a boundary condition value problem. Appropriate light scattering models that take into account explicitly the corresponding particle shape are restricted to a few nonspherical bodies, namely the spheroidal particle and the infinitely long cylinder. For these bodies, closed solutions exist which we will discuss in the following. Approximate closed solutions on the basis of the Rayleigh approximation, that is, for particle dimensions small compared with the wavelength, also exist for triaxial ellipsoids and cubes.

For all other geometries, either approximations for large particles and certain distinct geometries (e.g., platonic bodies) are available, or numerical methods have been established to calculate the optical response of an irregularly shaped body with certain algorithms. The best evolved and often applied methods are the *discrete dipole approximation* (DDA) and the *T-matrix method* or *extended boundary condition method* (EBCM). They and some others will also be introduced and discussed in the following sections. As for the numerical methods a huge number of papers exist where they are applied to various small and large bodies, we restrict considerations here to those papers that are relevant for the understanding of the theoretical ansatz and for nanoparticles. A systematic and unified discussion of light scattering by nonspherical particles and its practical applications can be found in the books by Mishchenko and co-workers [12, 15] and Borghese et al.

[14]. They represent the state-of-the-art of this important research field and are very practical for scientists and engineers in geophysics, remote sensing, and planetary and space physics. However, they deal predominantly with dielectric particles larger than 1000 nm and mainly consider the spatial (angular) distribution of the scattered light. Spectral behavior is not discussed.

The solutions for spheroids, ellipsoids, cylinders, and cubes are supplemented by calculations for various particle materials and by experimental results. The numerical methods are supplemented mainly by experimental results where the method was applied and compared with the experimental data.

9.1
Spheroids and Ellipsoids

The particle most related to the sphere is the ellipsoidal particle. Hence it seems appropriate to study first the ellipsoidal particle in the pot-pourri of nonspherical particles. However, to our knowledge, exact analytical solutions for the light scattering are not available for triaxial ellipsoidal particles with half-axes A, B, and C. Only if two of the three axes are identical, that is, for ellipsoids of revolution or spheroids, Asano and Yamamoto [376] developed an exact solution which will be introduced first. Then, triaxial ellipsoidal particles are treated in the Rayleigh approximation.

9.1.1
Spheroids (Ellipsoids of Revolution)

Asano and Yamamoto [376] and Onaka [377] solved independently the problem of absorption and scattering by an isotropic, homogeneous spheroid of arbitrary size by expanding the incident, scattered, and internal fields in *vector spheroidal harmonics*. So, at first glance, their solution appears to be very similar to the Mie theory for a sphere. However, the spheroidal harmonics do not represent a complete set of orthogonal functions as the spherical harmonics do. Therefore, the solution is even more complicates. Sinha and MacPhie also gave a solution for the scattering by spheroidal particles [378, 379], but only for conducting prolate spheroids. Later, Kurtz and Salib [380] revisited the work of Asano and Yamamoto in 1993 to improve it using a slightly different method for solving the boundary conditions. A new exact solution was developed by Voshchinnikov and Farafonov [381] in 1993 which the authors claimed to be more efficient than that of Asano and Yamamoto from the computational point of view [382].

A spheroidal particle may be either *prolate* or *cigar-like shaped* or is *oblate* or *pancake-like shaped*. For both kinds of spheroid, let A be the major axis and B the minor axis. Compared with spherical coordinates r, θ, φ, the spheroidal coordinates ξ, η, φ correspond to the reciprocal of the eccentricity e, $\xi = 1/e$, η corresponding to the cosine of the polar angle θ, and φ is the azimuthal angle as in spherical coordinates (Figure 9.2).

9 Beyond Mie's Theory I – Nonspherical Particles

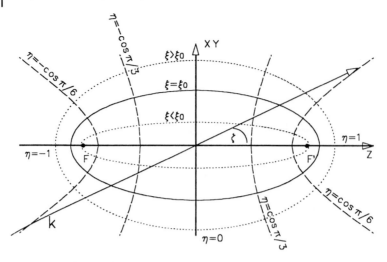

Figure 9.2 Definition of spheroidal coordinates.

In terms of the coordinate ξ, the major axis is $A = F\xi$ and the minor axis is $B = F\sqrt{\xi^2 - 1}$. The eccentricity e is defined via

$$e^2 = 1 - \left(\frac{B}{A}\right)^2 \tag{9.3}$$

If the half-axes become identical, that is, $A = B = R$, the eccentricity e and the half focal distance F approach zero and both prolate and oblate spheroids become spheres.

9.1.1.1 Electromagnetic Fields
In spheroidal coordinates ξ, η, φ, the scalar wave equation

$$\left(\nabla^2 + k^2\right)\Phi = 0 \tag{9.4}$$

has an infinite number of independent solutions Φ_{nm}:

$$\Phi_{\substack{e\\o}nm}(c, \xi, \eta, \varphi) = R_{nm}^{(j)}(c, \xi) S_{nm}(c, \eta) \begin{matrix}\cos(m\varphi)\\ \sin(m\varphi)\end{matrix} \tag{9.5}$$

with $n = 1, 2, \ldots, \infty$ and $m = 0, \pm 1, \pm 2, \ldots, \pm n$, obtained with a separation of variables method. The additional variable c is defined as

$$c = kF \tag{9.6}$$

The subscripts "o" for odd and "e" for even account for the symmetry of the functions when changing the azimuthal angle φ to $-\varphi$. The functions S_{nm} and R_{nm} are defined for a prolate spheroid as

$$S_{nm}(c, \eta) = \sum_{r=0,1}^{\infty} d_r^{nm}(c) P_{m+r,m}(\eta) \tag{9.7}$$

$$R_{nm}^{(j)}(c,\xi) = \frac{\left(\frac{\xi^2-1}{\xi^2}\right)^{\frac{m}{2}}}{\sum_{r=0,1}^{\infty} \frac{(r+2m)!}{r!} d_r^{nm}(c)} \sum_{r=0,1}^{\infty} i^{r+n-m} \frac{(r+2m)!}{r!} d_r^{nm}(c) z_{m+r}(c\xi) \quad (9.8)$$

The summation over r must be carried out either over even values r if $|n - m|$ is even, or over odd values if $|n - m|$ is odd. The functions z_{m+r} are any of the four spherical Bessel functions, either spherical Bessel functions j_n ($j = 1$), or spherical Neumann functions y_n ($j = 2$), or spherical Hankel functions of the first kind $h_n^{(1)}$ ($j = 3$) or of the second kind $h_n^{(2)}$ ($j = 4$). $P_{m+r,m}(\eta)$ are associated Legendre polynomials. For more detailed information about spherical Bessel functions and associated Legendre polynomials, we refer to the *Handbook of Mathematical Functions* by Abramovitz and Stegun [83].

The corresponding functions for an oblate spheroid are obtained by replacing c by $-ic$ and ξ by $i\xi$ in Equations 9.7 and 9.8.

The coefficients $d_r^{nm}(c)$ satisfy the recurrence relation

$$A_r^m(c) d_{r+2}^{nm}(c) + \left[B_r^m(c) - \lambda_{nm}(c)\right] d_r^{nm}(c) + C_r^m(c) d_{r-2}^{nm}(c) \quad (9.9)$$

with the coefficients

$$A_r^m(c) = \frac{(2m+r+2)(2m+r+1)}{(2m+2r+3)(2m+2r+5)} c^2 \quad (9.10)$$

$$B_r^m(c) = \frac{2(m+r)(m+r+1) - 2m^2 - 1}{(2m+2r-1)(2m+2r+3)} c^2 + (m+r)(m+r+1) \quad (9.11)$$

and

$$C_r^m(c) = \frac{r(r-1)}{(2m+2r-3)(2m+2r-1)} c^2 \quad (9.12)$$

The eigenvalues $\lambda_{nm}(c)$ can be calculated according to the method of Hodge [383]. In practice, this is the longest lasting term in the calculation of spheroidal wavefunctions.

Spheroidal vector wavefunctions can now be defined in different ways. For an overview we refer to Flammer [384]. The most common definitions are given in the following.

With $a = e_x$, e_y, or e_z, a set of linear independent vector harmonics is created from

$$M_{\substack{e \\ o}nm}^x = \text{curl}\left(\Phi_{\substack{e \\ o}nm} e_x\right) \quad (9.13)$$

$$N_{\substack{e \\ o}nm}^x = k^{-1} \text{curl} M_{\substack{e \\ o}nm}^x \quad (9.14)$$

Another common definition is

$$M_{\substack{e \\ o}nm}^r = \text{curl}\left(\Phi_{\substack{e \\ o}nm} r\right) \quad (9.15)$$

$$N_{\substack{e \\ o}nm}^r = k^{-1} \text{curl}\left(M_{\substack{e \\ o}nm}^r\right) \quad (9.16)$$

The vector $\boldsymbol{r}$ is the position vector, which becomes in spheroidal coordinates

$$r = \mp \frac{1}{2} d\eta \left(\frac{1-\eta^2}{\xi^2 \pm \eta^2} \right)^{\frac{1}{2}} \boldsymbol{e}_\eta + \frac{1}{2} d\xi \left(\frac{\xi^2 \pm 1}{\xi^2 \pm \eta^2} \right)^{\frac{1}{2}} \boldsymbol{e}_\xi \qquad (9.17)$$

The greatest disadvantage of any set of vector spheroidal wavefunctions is that the corresponding vector functions are neither orthogonal among themselves nor, in general, orthogonal to those of the other sets. Therefore, the computation of light scattering and absorption by spheroids becomes a serious challenge. In the following we give a brief derivation of underlying theory which is restricted to only a few equations. For more details we refer to Asano and Yamamoto [376].

For a compact spheroidal particle illuminated by a plane wave, the fields of the incident, scattered, and interior waves are expanded into corresponding vector spheroidal harmonics defined in Equations 9.15 and 9.16. For oblique incidence two cases are distinguished, transverse magnetic (TM) fields and transverse electric (TE) fields:

incident wave, TE:

$$\boldsymbol{E}_{inc}^{TE} = E_0 \sum_{n=0}^{\infty} \sum_{m=n}^{\infty} i^n \left[g_{nm}(\xi) \boldsymbol{M}_{enm}^{r(1)}(k_M) + if_{nm}(\xi) \boldsymbol{N}_{onm}^{r(1)}(k_M) \right] \qquad (9.18a)$$

$$\boldsymbol{H}_{inc}^{TE} = \frac{-k_M E_0}{\omega \mu_0} \sum_{n=0}^{\infty} \sum_{m=n}^{\infty} i^n \left[f_{nm}(\xi) \boldsymbol{M}_{onm}^{r(1)}(k_M) - ig_{nm}(\xi) \boldsymbol{N}_{enm}^{r(1)}(k_M) \right] \qquad (9.18b)$$

incident wave, TM:

$$\boldsymbol{E}_{inc}^{TM} = E_0 \sum_{n=0}^{\infty} \sum_{m=n}^{\infty} i^n \left[f_{nm}(\xi) \boldsymbol{M}_{onm}^{r(1)}(k_M) - ig_{nm}(\xi) \boldsymbol{N}_{enm}^{r(1)}(k_M) \right] \qquad (9.19a)$$

$$\boldsymbol{H}_{inc}^{TM} = \frac{k_M E_0}{\omega \mu_0} \sum_{n=0}^{\infty} \sum_{m=n}^{\infty} i^n \left[g_{nm}(\xi) \boldsymbol{M}_{enm}^{r(1)}(k_M) + if_{nm}(\xi) \boldsymbol{N}_{onm}^{r(1)}(k_M) \right] \qquad (9.19b)$$

scattered wave, TE:

$$\boldsymbol{E}_{sca}^{TE} = E_0 \sum_{n=0}^{\infty} \sum_{m=n}^{\infty} i^n \left[b_{nm}^{TE} \boldsymbol{M}_{enm}^{r(3)}(k_M) + ia_{nm}^{TE} \boldsymbol{N}_{onm}^{r(3)}(k_M) \right] \qquad (9.20a)$$

$$\boldsymbol{H}_{sca}^{TE} = \frac{-k_M E_0}{\omega \mu_0} \sum_{n=0}^{\infty} \sum_{m=n}^{\infty} i^n \left[a_{nm}^{TE} \boldsymbol{M}_{onm}^{r(3)}(k_M) - ib_{nm}^{TE} \boldsymbol{N}_{enm}^{r(3)}(k_M) \right] \qquad (9.20b)$$

scattered wave, TM:

$$\boldsymbol{E}_{sca}^{TM} = E_0 \sum_{n=0}^{\infty} \sum_{m=n}^{\infty} i^n \left[a_{nm}^{TM} \boldsymbol{M}_{onm}^{r(3)}(k_M) - ib_{nm}^{TM} \boldsymbol{N}_{enm}^{r(3)}(k_M) \right] \qquad (9.21a)$$

$$\boldsymbol{H}_{sca}^{TM} = \frac{k_M E_0}{\omega \mu_0} \sum_{n=0}^{\infty} \sum_{m=n}^{\infty} i^n \left[b_{nm}^{TM} \boldsymbol{M}_{enm}^{r(3)}(k_M) + ia_{nm}^{TM} \boldsymbol{N}_{onm}^{r(3)}(k_M) \right] \qquad (9.21b)$$

interior wave, TE:

$$\mathbf{E}_{\text{int}}^{\text{TE}} = E_0 \sum_{n=0}^{\infty}\sum_{m=n}^{\infty} i^n \left[\alpha_{nm}^{\text{TE}} \mathbf{M}_{enm}^{r(1)}(k) + i\beta_{nm}^{\text{TE}} \mathbf{N}_{onm}^{r(1)}(k) \right] \qquad (9.22\text{a})$$

$$\mathbf{H}_{\text{int}}^{\text{TE}} = \frac{-kE_0}{\omega\mu_0} \sum_{n=0}^{\infty}\sum_{m=n}^{\infty} i^n \left[\beta_{nm}^{\text{TE}} \mathbf{M}_{onm}^{r(1)}(k) - i\alpha_{nm}^{\text{TE}} \mathbf{N}_{enm}^{r(1)}(k) \right] \qquad (9.22\text{b})$$

interior wave, TM:

$$\mathbf{E}_{\text{int}}^{\text{TM}} = E_0 \sum_{n=0}^{\infty}\sum_{m=n}^{\infty} i^n \left[\alpha_{nm}^{\text{TM}} \mathbf{M}_{onm}^{r(1)}(k) - i\beta_{nm}^{\text{TM}} \mathbf{N}_{enm}^{r(1)}(k) \right] \qquad (9.23\text{a})$$

$$\mathbf{H}_{\text{int}}^{\text{TM}} = \frac{kE_0}{\omega\mu_0} \sum_{n=0}^{\infty}\sum_{m=n}^{\infty} i^n \left[\beta_{nm}^{\text{TM}} \mathbf{M}_{enm}^{r(1)}(k) + i\alpha_{nm}^{\text{TM}} \mathbf{N}_{onm}^{r(1)}(k) \right] \qquad (9.23\text{b})$$

The use of Bessel functions j_n (superscript 1) in the expansions of the electromagnetic fields in the interior of the particle and of the incident wave guarantees the continuity of the fields at the origin of the reference frame which is centered in the spheroid. The use of Hankel functions $h_n^{(1)}$ (superscript 3) in the expansion of the scattered wave takes into consideration outgoing spheroidal waves. The abbrevations $f_{nm}(\xi)$ and $g_{nm}(\xi)$ stand for the expansion coefficients of the incident wave which are given, for example, in Equations 35 and 36 in [376].

9.1.1.2 Scattering Coefficients

Usually, the expansion coefficients of the scattered wave $a_{nm}^{\text{TE,TM}}$ and $b_{nm}^{\text{TE,TM}}$ are resolved from Maxwell's boundary conditions. This can be done also for spheroidal particles. However, as already stated above, the spheroidal functions $S_{nm}(\eta)$ are not orthogonal. Hence, unlike for spheres, the individual terms in the sums over n cannot be matched term by term. A solution is to expand the terms with $S_{nm}(\eta)$ into series of associated Legendre polynomials. These expansions are very lengthy and are not given here, but can be read in detail in [376] (Equations 47–74). This additional expansion with additional index t leads to an infinite system of coupled linear equations which must be resolved for the unknown coefficients $a_{nm}^{\text{TE,TM}}$ and $b_{nm}^{\text{TE,TM}}$. In practice, this system of equations can be truncated to a finite number n_{MAX} of equations, because the infinite system converges, as shown by Siegel et al. [385] and Wait [386]. For each $m = 0, \pm 1, \pm 2, \ldots, \pm n_{\text{MAX}}$ one obtains $n_{\text{MAX}} - m + 1$ linear equations:

for $(n - m) + t$ even:

$$a_m^{\text{TE}} = -\frac{\left[\hat{X}_m^{(1)}(c)\right]^{-1}\hat{X}_m^{(1)}(c_M) - \frac{c}{c_M}\left[\hat{V}_m^{(1)}(c)\right]^{-1}\hat{V}_m^{(1)}(c_M)}{\left[\hat{X}_m^{(1)}(c)\right]^{-1}\hat{X}_m^{(3)}(c_M) - \frac{c}{c_M}\left[\hat{V}_m^{(1)}(c)\right]^{-1}\hat{V}_m^{(3)}(c_M)} f_m \qquad (9.24)$$

$$b_m^{\text{TE}} = -\frac{\frac{c}{c_M}\left[\hat{X}_m^{(1)}(c)\right]^{-1}\hat{X}_m^{(1)}(c_M) - \left[\hat{V}_m^{(1)}(c)\right]^{-1}\hat{V}_m^{(1)}(c_M)}{\frac{c}{c_M}\left[\hat{X}_m^{(1)}(c)\right]^{-1}\hat{X}_m^{(3)}(c_M) - \left[\hat{V}_m^{(1)}(c)\right]^{-1}\hat{V}_m^{(3)}(c_M)} g_m \qquad (9.25)$$

for $(n - m) + t$ odd:

$$a_m^{TE} = -\frac{\left[\hat{U}_m^{(1)}(c)\right]^{-1}\hat{U}_m^{(1)}(c_M) - \frac{c}{c_M}\left[\hat{Y}_m^{(1)}(c)\right]^{-1}\hat{Y}_m^{(1)}(c_M)}{\left[\hat{U}_m^{(1)}(c)\right]^{-1}\hat{U}_m^{(3)}(c_M) - \frac{c}{c_M}\left[\hat{Y}_m^{(1)}(c)\right]^{-1}\hat{Y}_m^{(3)}(c_M)} f_m \quad (9.26)$$

$$b_m^{TE} = -\frac{\frac{c}{c_M}\left[\hat{U}_m^{(1)}(c)\right]^{-1}\hat{U}_m^{(1)}(c_M) - \left[\hat{Y}_m^{(1)}(c)\right]^{-1}\hat{Y}_m^{(1)}(c_M)}{\frac{c}{c_M}\left[\hat{U}_m^{(1)}(c)\right]^{-1}\hat{U}_m^{(3)}(c_M) - \left[\hat{Y}_m^{(1)}(c)\right]^{-1}\hat{Y}_m^{(3)}(c_M)} g_m \quad (9.27)$$

In both cases we have

$$a_m^{TM} = b_m^{TE} \quad \text{and} \quad b_m^{TM} = a_m^{TE} \quad (9.28)$$

The elements of the matrices $\hat{X}_m$, $\hat{Y}_m$, $\hat{U}_m$, and $\hat{V}_m$ are functions of c and c_M with different coefficients similar to those in Equations 9.10–9.12 and are determined by the behavior of the radial functions $R_{nm}^{(j)}(c,\xi)$ with indices $j = 1$ and 3. They are not given here in detail, but we refer the reader to the original contribution of Asano and Yamamoto [376].

9.1.1.3 Cross-sections

Analogously to the sphere, the rates at which energy is absorbed and scattered are calculated from Poynting's law (energy conservation law). Dividing these rates by the intensity I_0 of the incident light, the optical cross-sections σ for scattering and absorption are determined for TE and TM polarization. Again, the calculations are very extensive and we only give the results.

$$\sigma_{ext}^{TE,TM} = -\frac{4\pi}{k_M^2} \text{Re} \sum_{m=0}^{\infty} \sum_{n=m}^{\infty} \left[a_{nm}^{TE,TM} \frac{m}{\sin\xi} S_{nm}(\cos\xi) + b_{nm}^{TE,TM} \frac{\partial S_{nm}(\cos\xi)}{\partial \xi} \right] \quad (9.29)$$

$$\sigma_{sca}^{TE,TM} = \frac{\pi}{k_M^2} \sum_{m=0}^{\infty} \sum_{n=m}^{\infty} \sum_{q=m}^{\infty} \Pi_{nq}^m \text{Re}\left[a_{nm}^{TE,TM}\left(a_{qm}^{TE,TM}\right)* + b_{nm}^{TE,TM}\left(b_{nm}^{TE,TM}\right)* \right] \quad (9.30)$$

with

$$\Pi_{nq}^m = \begin{cases} 0 & |n-q| = \text{odd} \\ \sum_{r=0,1} \frac{2(r+m)(r+m+1)(r+2m)!}{(2r+2m+1)r!} d_r^{nm} d_r^{qm} & |n-q| = \text{even} \end{cases} \quad (9.31)$$

9.1.1.4 Resonances

Also for spheroidal particles, MDRs and also material resonances (SPPs, electronic resonances) occur if the denominators of the scattering coefficients $a_{nm}^{TE,TM}$ and $b_{nm}^{TE,TM}$ become small. Unfortunately, due to the complexity of Equations 9.24 and 9.25, we cannot give here an equation for the resonance conditions but will discuss SPP and electronic resonances in the following section on triaxial ellipsoids in the Rayleigh approximation and here with numerical results obtained for prolate spheroidal silver nanoparticles with varying ratio $A{:}B$ of long half-axis A to short half-axis

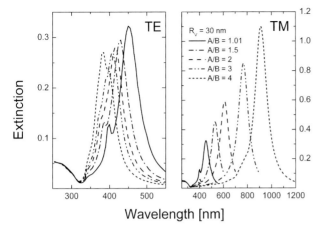

Figure 9.3 TM and TE modes of a prolate spheroidal Ag nanoparticle with $2R_V = 60$ nm and varying ratio $A:B = 1.01, 1.5, 2, 3,$ and 4, embedded in glass with $n_M = 1.517$. Data courtesy of J. Porstendorfer.

B from Porstendorfer [387]. Analogous numerical results for gold spheroids can be found in [388]. Remember that according to Equation 9.3 the eccentricity is calculated from the ratio $B:A$. In the computations for Figure 9.3, the particle volume corresponding to the equivalent volume of a spherical particle of $2R_V = 60$ nm. The spectra are separated into TM (polarization along the long axis of the prolate spheroid) and TE (polarization along the short axis) contributions. The ratio $A:B = 1.01$ almost corresponding to a sphere for which the TM and TE contributions are identical. The particle is assumed to be embedded in glass with $n_M = 1.517$.

The spectra in Figure 9.3 exhibit two resonances, one in TE and one in TM polarization. They can be assigned to a SPP resonance excited along the short axis B (TE polarization) and along the long axis A (TM polarization). The nonsphericity of the spheroid obviously leads to more than one SPP in the nanoparticle. This holds true also for electronic resonances, as will be seen for triaxial ellipsoids in the following section. The TE plasmon polariton shifts from 450 to 385 nm with increasing ratio $A:B$. The TM plasmon polariton shifts from 450 to 915 nm with increasing ratio $A:B$. As the TE plasmon polariton is blue shifted, only the absorption of light by the TM plasmon polariton leads to a characteristic color, if for example the spheroids are embedded in a glass. This can be seen in Section 9.1.4, where experimental results are presented.

In addition to the calculations in Figure 9.3, we show in Figure 9.4 the comparison of the calculations according to the theory of Asano and Yamamoto [376] with the EBCM (described in Section 9.4). For that purpose, again a prolate spheroidal silver particle with $2R_V = 60$ nm and $A:B = 1.5$ and 2 was considered. As the EBCM is more time consuming than the direct calculation, only for several striking spectral features were the values calculated [387]. Both calculations were in excellent agreement.

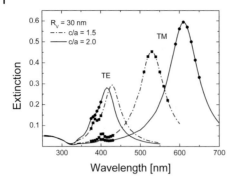

Figure 9.4 Comparison of the TM and TE spectra of a prolate spheroidal Ag nanoparticle with $2R_V = 60$ nm and ratio $A:B = 1.5$ and 2, embedded in glass with $n_M = 1.517$, calculated with the theory according to Asano and Yamamoto [376] and the EBCM. Data courtesy of J. Porstendorfer.

9.1.1.5 Numerical Examples

An extensive numerical discussion of spheroidal particles was presented by Asano [389]. He discussed the scattering cross-section of spheroids with various $A:B$ ratios as a function of the size parameter $x_V = (2\pi R_V)/\lambda$, where R_V is the volume-equivalent sphere radius. He also examined the influence of the angle of incidence and considered the angular distribution of the scattered light in the scattering plane and in the perpendicular plane, and in particular the forward and backward scattering. These investigations were supplemented by numerical studies on randomly oriented spheroidal particles by Asano and Sato [390].

Some benchmark results including also absorbing spheroids were published by Voshchinnikov et al. [391].

All these authors, however, did not treat particles with electronic resonances at all.

9.1.1.6 Extensions

At the end of this section, we give some references to extensions of the scattering theory for spheroidal particles.

Cooray and Ciric [392] in 1992 treated the scattering of electromagnetic waves by a coated dielectric spheroid and studied the angular distribution of the scattered light in the far-field.

Farafonov et al. [393] extended their formalism (see [381]) for a spheroidal particle on light scattering by a core-mantle spheroidal particle.

Although a homogeneous spheroid and core–mantle spheroid are already difficult, Gurwich et al. [394] made an extension to multilayer spheroids with a recursive calculation from one layer to the next.

Han and Wu [395, 396] developed an approach to expand a Gaussian beam in terms of the spheroidal wavefunctions in spheroidal coordinates. The beam-shape coefficients of the Gaussian beam in spheroidal coordinates can be computed conveniently by use of the known expression for beam-shape coefficients in spherical coordinates. The unknown expansion coefficients of scattered and internal electromagnetic fields are determined by a system of equations derived from the boundary conditions for continuity of the tangential components of the electric and magnetic vectors across the surface of the spheroid.

Barton [397, 398] developed a spheroidal coordinate separation-of-variables solution for the determination of internal, near-surface, and scattered electromagnetic fields of a layered spheroid with arbitrary monochromatic illumination.

Han et al. [399] evaluated the beam-shape coefficients of arbitrary off-axis Gaussian beams in spheroidal coordinates with a generalized Lorenz–Mie theory (GLMT). The light-scattering properties of absorbing and nonabsorbing homogeneous spheroidal particles, such as the angular distribution of scattered intensity for a wide range of particles sizes and different complex refractive indices versus the magnitude and location of the beam waist, were investigated.

Bobbert and Vlieger in 1987 considered a spheroidal particle on a surface [400].

In 2000, Roman-Velazquez et al. [401] developed a spectral formalism to study the effective polarizability of a spheroidal particle lying over a substrate, including multipolar effects. With the help of the spectral representation, they discussed the optical response in terms of the excitation of the multipolar modes of the system.

Wave scattering by a chiral spheroid was studied in 1993 by Cooray and Ciric [402].

9.1.2
Ellipsoids (Rayleigh Approximation)

Treating the ellipsoidal particle in the Rayleigh approximation as done by Gans [403], we can restrict considerations to excitation of a dipole moment in the particle. For an arbitrary triaxial ellipsoidal particle with major axes A, B, and C, the polarizability tensor $\underline{\alpha}$ only has diagonal elements α_{jj}, with [8]

$$\alpha_{jj} = \frac{4\pi}{3} ABC \frac{\varepsilon - \varepsilon_M}{\varepsilon_M + G_j(\varepsilon - \varepsilon_M)} = V_P \frac{\varepsilon - \varepsilon_M}{\varepsilon_M + G_j(\varepsilon - \varepsilon_M)} \quad (9.32)$$

The geometry factors G_j take into account the geometry of the ellipsoid. They satisfy the conditions

$$\sum_j G_j = 1 \quad G_j \leq 1 \quad \forall j \quad (9.33)$$

and can be calculated from [8]

$$G_j = \frac{ABC}{2} \int_0^\infty \frac{dq}{(J^2 + q)\sqrt{(A^2 + q)(B^2 + q)(C^2 + q)}} \quad (9.34)$$

where J stands for the three axes A, B, or C, corresponding to $j = 1$, 2, or 3, respectively.

According to the optical theorem, each of the three diagonal elements α_{jj} contributes to the extinction cross-section of the ellipsoid by the same amount:

$$\sigma_{ext} = kV_P \, \text{Im} \sum_{j=1}^{3} \frac{1}{3} \frac{(\varepsilon - \varepsilon_M)}{\varepsilon_M + G_j(\varepsilon - \varepsilon_M)} + \frac{k^4 V_P^2}{6\pi} \sum_{j=1}^{3} \frac{1}{3} \left| \frac{(\varepsilon - \varepsilon_M)}{\varepsilon_M + G_j(\varepsilon - \varepsilon_M)} \right|^2 \quad (9.35)$$

where the second term on the right-hand side corresponding to the scattering cross-section:

$$\sigma_{sca} = \frac{k^4 V_P^2}{6\pi} \sum_{j=1}^{3} \frac{1}{3} \left| \frac{(\varepsilon - \varepsilon_M)}{\varepsilon_M + G_j(\varepsilon - \varepsilon_M)} \right|^2 \quad (9.36)$$

The cross-section exhibits material resonances (SPPs, electronic resonances) if the complex dielectric function $\varepsilon = \varepsilon_1 + i\varepsilon_2$ of the particle material fulfills the conditions

$$\varepsilon_1 = -\frac{1 - G_j}{G_j} \varepsilon_M, \quad \varepsilon_2 \approx 0 \quad (9.37)$$

Aluminum is well suited for sharp SPP resonances in the wavelength range between about 80 and 700 nm because of the undisturbed Drude-like behavior of its dielectric function in this spectral range. Therefore, we will demonstrate the effect of the shape exemplarily with calculations for Al nanoparticles.

Figure 9.5 shows the extinction cross-section spectra for triaxial aluminum nanoellipsoids with the three axes (A, B, C) amounting to (20, 10, 5) nm in Figure 9.5a, (20, 5, 10) nm in Figure 9.5b, and (10, 20, 5) nm in Figure 9.5c *in vacuo*. The spectra are divided into p- and s-polarized light contributions to show the differences that occur due to the permutation in the axis lengths.

For comparison, the spectrum of the volume-equivalent sphere, that is, with $2R = 20$ nm, is also shown as the dotted line. For better comparison the sphere cross-section is divided by 2.

Three extinction maxima can be clearly recognized, corresponding to the three resonances for G_1, G_2, and G_3. All three resonances depend on the polarization of the incident light and the values of A, B, and C. For the incoming plane wave traveling along the positive z-axis, the p-polarized light excites the ellipsoid along the half-axis B. This is also clearly recognized from Figure 9.5. The long wavelength peak can be assigned to the longest axis and the smallest geometry factor G. Vice versa, the short wavelength peak belongs to the shortest axis and the largest geometry factor G. It can be assumed that the combination (A, B, C) = (20, 10, 5) yields the same spectrum as (5, 10, 20) and (20, 5, 10) = (10, 5, 20) and (10, 20, 5) = (5, 20, 10), which was proved by further calculations. For unpolarized incident light the contributions of p- and s-polarization must be averaged.

A reduction is obtained if two of the three axes of the ellipsoid coincide, that is, for *spheroids* or *ellipsoids of revolution*. For prolate spheroids in the Rayleigh approximation, the geometry factor $G_1(e)$ is

$$G_1(e) = -g(e)^2 \left[1 - \frac{1}{2e} \ln\left(\frac{1+e}{1-e} \right) \right] \quad (9.38)$$

with the eccentricity *e* given in Equation 9.3. For oblate spheroids, $G_1(e)$ is

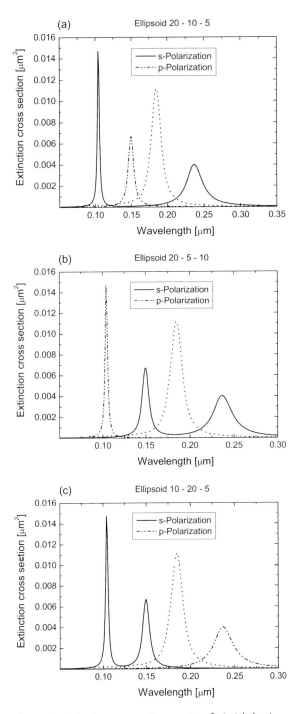

Figure 9.5 Extinction cross-section spectra of triaxial aluminum nanoellipsoids with half-axes 5, 10, and 20 nm. For comparison, the cross-section of the volume-equivalent sphere (2R = 20 nm, dotted line) is also plotted. It is divided by 2 for better comparison.

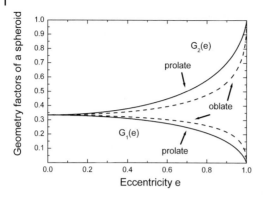

Figure 9.6 Geometry factors $G_1(e)$ and $G_2(e)$.

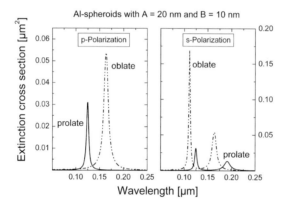

Figure 9.7 Spectra of prolate and oblate aluminum nanospheroids with half-axes $A = 20\,\text{nm}$ and $B = 10\,\text{nm}$.

$$G_1(e) = \frac{g(e)}{2e^2}\left\{\frac{\pi}{2} - \tan^{-1}[g(e)]\right\} - \frac{g(e)^2}{2} \quad (9.39)$$

In both Equations 9.38 and 9.39

$$g(e)^2 = \frac{1-e^2}{e^2} \quad (9.40)$$

$G_2(e)$ follows from Equation 9.33. In Figure 9.6, both geometry factors are depicted. In the limit $e^2 \to 0$ they start with $G_1(0) = G_2(0) = 1/3$, that is, with the geometry factor of the sphere. With increasing eccentricity $G_1(e)$ decreases and approaches zero for $e^2 = 1$. Vice versa, G_2 increases and ends with $G_2(1) = 1$ for $e^2 = 1$.

As we have only two different geometry factors G_1 and G_2, then only two resonances are possible. Figure 9.7 shows the spectra of a prolate and an oblate aluminum spheroid with $A = 20\,\text{nm}$ and $B = 10\,\text{nm}$. Compared with a triaxial ellipsoid we now have $(A, B, C) = (20, 10, 10)$ nm for the prolate spheroid and $(20, 20, 10)$ nm for the oblate spheroid. Correspondingly, for p-polarized light the prolate spheroid exhibits only the short-wavelength resonance and the oblate spheroid exhibits only the long-wavelength resonance. In s-polarization both spheroids exhibit both resonances. Comparing the prolate spheroid with the oblate spheroid, it has to be taken into account that the volume of the oblate spheroid is twice the volume of the prolate spheroid.

9.1.3
Numerical Examples for Ellipsoids

In this section we present numerical results for elongated ellipsoids of various materials. Similarly to the calculations presented in Chapter 6 for spherical particles, we use the following categories:

- metal particles
- semimetal and semiconductor particles
- carbonaceous particles
- oxide particles
- particles with phonon polaritons
- miscellaneous particles.

We concentrate on particles where resonances (electronic resonances and SPP resonances) play a role also in the Rayleigh approximation. The numerical examples are mostly for elongated ellipsoids with $B = 10$ nm and $C = 5$ nm, and the long axis A variable from 20 to 50 nm in most cases. We point out that the surrounding is always assumed to have a constant refractive index $n_M = 1$ (vacuum), whereas in experiments it may be larger. Then, the influence of the particle shape may be enhanced by the red shift of resonances for surroundings with $n_M > 1$. In the graphs the optical extinction cross-section is always plotted. Note that the extinction cross-section is proportional to the particle volume. Therefore, an increase of any of the three axes automatically increases the volume and the optical extinction cross-section. Comparison is always made between the largest ellipsoidal volume and the corresponding volume-equivalent sphere.

9.1.3.1 Metal Particles

We start the numerical examples with calculations for metal nanoellipsoids. In the first examples for aluminum in the following three figures we discuss the influence of the values of the three axes A, B, and C on the optical extinction cross-section. Aluminum is well suited because the three resonances keep well separated from each other on changing the values of the three axes.

Figure 9.8 shows the changes in the extinction cross-section when the long axis A is increased. The ellipsoidal particles exhibit three resonances which can be assigned to SPP resonances due to the excitation of the free carriers along the three half-axes. Two of them, the resonances along B and C, are peaked at wavelengths lower than that of the spherical particle. The resonance from the excitation

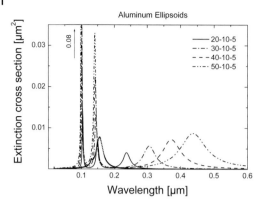

Figure 9.8 Ellipsoids of aluminum with the three half-axes (A, B, C) = (20, 10, 5), (30, 10, 5), (40, 10, 5), and (50, 10, 5) nm. For comparison, the spectrum of a volume equivalent sphere with radius R = 13.57 nm is also plotted [curve •, corresponding to the volume of the ellipsoid with (A, B, C) = (50, 10, 5) nm].

along the long axis A is shifted to longer wavelengths compared with the sphere with 2R = 27.14 nm. The spherical particle exhibits two resonances that can be assigned to the dipolar and the quadrupolar SPPs. On increasing the size of the long axis, the corresponding resonance undergoes a significant red shift to longer wavelengths. The increase in the magnitude is caused by the increase in the particle volume. The other two resonances undergo only a slight blue shift to shorter wavelengths. The red shift of the SPP along the axis A is so large that it is even peaked in the visible spectral region around 450 nm. Then, these nanoparticles would lead to a yellow or orange color.

For studying the changes on increasing the shorter axis B, we start with the ellipsoid with (A, B, C) = (50, 10, 5) nm and vary B from 10 to 40 nm. Then, the ellipsoid becomes more and more comparable to an oblate spheroid with A = B = 50 nm and C = 5 nm. In consequence, two of the three resonances – along the axis A and along the axis B – must converge, while the position of the resonance along the axis C should remain almost unaffected. This can be clearly recognized in Figure 9.9. As the volume of the ellipsoid with (A, B, C) = (50, 40, 5) nm is larger by a factor of 4 than that of the ellipsoid with (A, B, C) = (50, 10, 5) nm, we compare the ellipsoids with a volume-equivalent sphere with 2R = 43.1 nm.

The third case, keeping A and B constant and varying C, is presented in Figure 9.10. Starting with the ellipsoid with (A, B, C) = (50, 30, 5) nm, the axis C is varied as 10 and 20 nm. In that case, the ellipsoid becomes more and more of a prolate spheroid with long axis A = 50 nm and short axis B = C = 30 nm. Also in this case two of the three resonances – for the axes B and C – must converge. The resonance along the long axis A also shifts to lower wavelengths. This is caused by the reduction of the A:B ratio. For comparison, the spectrum of a volume-equivalent sphere with 2R = 62.14 nm is also plotted. It exhibits three resonances, corresponding

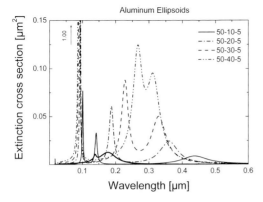

Figure 9.9 Ellipsoids of aluminum with the three half-axes $(A, B, C) = (50, 10, 5)$, $(50, 20, 5)$, $(50, 30, 5)$, and $(50, 40, 5)$ nm. For comparison, the spectrum of a volume equivalent sphere with radius $R = 21.55$ nm is also plotted [corresponding to the volume of the ellipsoid with $(A, B, C) = (50, 40, 5)$ nm].

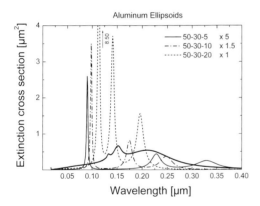

Figure 9.10 Ellipsoids of aluminum with the three half-axes $(A, B, C) = (50, 30, 5)$, $(50, 30, 10)$, and $(50, 30, 20)$ nm. For comparison, the spectrum of a volume equivalent sphere with radius $R = 31.07$ nm is also plotted [curve •, corresponding to the volume of the ellipsoid with $(A, B, C) = (50, 30, 20)$ nm].

to the dipolar, quadrupolar, and octupolar surface plasmon resonance in the aluminum sphere. For better comparison the spectra are multiplied by the given factors. Also the spectrum of the sphere is multiplied by a factor of 3 for better presentation.

In the following, we present further examples for metallic ellipsoidal nanoparticles with optical extinction cross-section spectra of sodium, potassium, silver, gold, iron, platinum, yttrium, and tantalum nanoellipsoids. The short axes B and C are fixed at $B = 10$ nm and $C = 5$ nm. The changes in the spectra result only from the variation in the length of the long axis A.

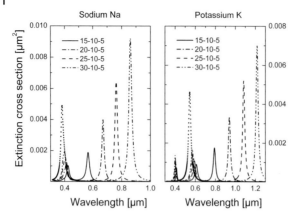

Figure 9.11 Ellipsoids of sodium and potassium with the three half-axes (A, B, C) = (15, 10, 5), (20, 10, 5), (25, 10, 5), and (30, 10, 5) nm. For comparison, the spectrum of a volume equivalent sphere with radius R = 11.45 nm is also plotted [curve •, corresponding to the volume of the ellipsoid with (A, B, C) = (30, 10, 5) nm].

The next example, in Figure 9.11, is for sodium and potassium ellipsoids where SPP resonances also contribute to the spectrum.

The resonance excited along the long axis A dominates the spectra of both alkali metal particles. Its position shifts rapidly to longer wavelengths with increasing axis length and passes through the complete visible spectral range. Therefore, it can be expected that ellipsoidal and spheroidal particles of Na or K will lead to a color. However, as these metals are very reactive, they can only prepared under vacuum conditions, which restricts their applicability as color pigments.

Unlike alkali metal particles, nanoparticles of gold and silver are well known as color pigments, as already seen in Chapter 6. The optical extinction spectra also exhibit SPP resonances. On going to ellipsoidal and spheroidal nanoparticles of gold and silver, at least two resonances can be observed, as shown in Figure 9.12. For better presentation, the ordinate is logarithmically scaled. The resonance excited along the long axis A shifts to longer wavelengths with increasing axis length. Its position passes through the complete visible range and, hence, must lead to varying color if these ellipsoidal particles are embedded in, for example, a glass. This is demonstrated in Figure 9.13, where the extinction cross-section was used to calculate first the optical density of an ensemble of noninteracting prolate spheroidal particles (B = C) embedded in the glass N-BK7. From the optical density the transmittance of the composite is calculated. The long axis A and the short axis B are chosen so that the particle volume remains constant and is equal to the volume of a sphere with $2R = 60$ nm. If $r = A/B$ is the ratio of long axis to the short axis, A is calculated from $A = r^{2/3} \times 30$ nm and the short axis B from $B = r^{1/3} \times 30$ nm. The transmittance spectra and the resulting color are depicted in Figure 9.13. In the color chromaticity diagram the black points give the color coordinates of the resulting colors.

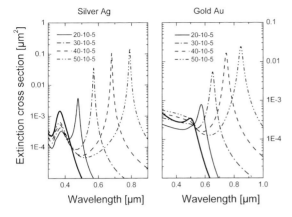

Figure 9.12 Ellipsoids of silver and gold with the three half-axes (A, B, C) = (20, 10, 5), (30, 10, 5), (40, 10, 5), and (50, 10, 5) nm. For comparison, the spectrum of a volume equivalent sphere with radius R = 13.57 nm is also plotted [curve •, corresponding to the volume of the ellipsoid with (A, B, C) = (50, 10, 5) nm].

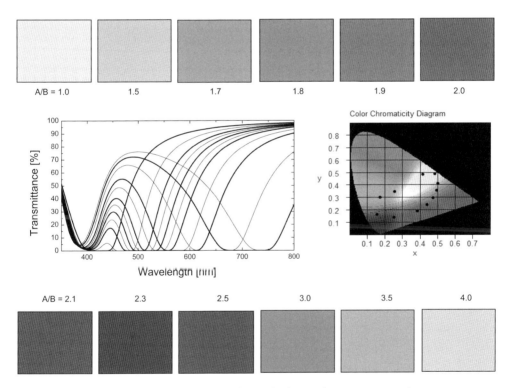

Figure 9.13 Transmittance spectra of ensembles of noninteracting prolate spheroidal particles in glass N-BK7 with varying ratio r = A/B. The corresponding colors and color coordinates are given in the rectangular boxes and the color chromaticity diagram. A color version of this figure can be found in the color plates at the end of the book.

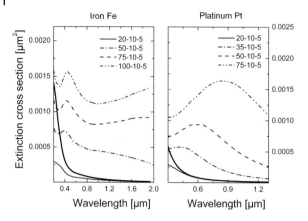

Figure 9.14 Ellipsoids of iron and platinum with the three half-axes (A, B, C) = (20, 10, 5), (50, 10, 5), (75, 10, 5), and (100, 10, 5) nm for Fe and (A, B, C) = (20, 10, 5), (35, 10, 5), (50, 10, 5), and (75, 10, 5) nm for Pt. For comparison, the spectrum of a volume equivalent sphere with radius $R = 17.1$ nm (Fe) and $R = 15.54$ nm (Pt) is also plotted [curve •, corresponding to the volume of the ellipsoid with (A, B, C) = (100, 10, 5) nm (Fe) and (A, B, C) = (75, 10, 5) nm (Pt)].

An interesting result is obtained for iron and platinum ellipsoidal nanoparticles. Unlike the spherical particles with sizes less than 70 nm, the ellipsoidal particles exhibit broad resonance-like extinction bands (Figure 9.14). Their position shifts to longer wavelengths with increasing long axis A, similarly to the SPP before for Na, K, Ag, and Au. Indeed, they can be assigned to dipolar surface plasmon resonances excited along the long axis A that are strongly damped and broadened by the high absorption in these nanoparticles. For iron nanoellipsoids even a coloring effect can be expected because the SPP resonance lies in the visible spectral region.

Yttrium and tantalum spherical nanoparticles proved to be good candidates for two resonances in nanoparticle extinction spectra: an electronic resonance at low wavelengths due to an interband transition and a SPP resonance at longer wavelengths caused by the free carriers. The SPP resonance in spherical nanoparticles, however, was very low for yttrium, caused by the high imaginary part of the dielectric function of yttrium. In contrast, tantalum nanoparticles with size $2R < 150$ nm exhibit an SPP resonance at 640 nm. On turning to ellipsoidal particles, the spectra in Figure 9.15 show that particularly for yttrium the SPP becomes stronger and is clearly resolved with increasing long axis A due to the strong red shift of the SPP resonance.

This red shift can also be observed for small tantalum ellipsoids, resulting in a clearly recognizable resonance compared with spherical particles of the same volume. Also at shorter wavelengths the spectra of the ellipsoids clearly differ from those of spherical nanoparticles. The ellipsoids exhibit two electronic resonances in the spectral region where the spherical particle exhibits only one electronic resonance. For yttrium they clearly affect the visible spectral region and lead to

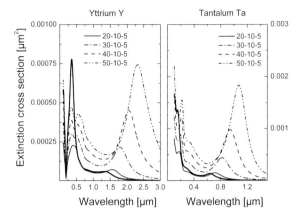

Figure 9.15 Ellipsoids of yttrium and tantalum with the three half-axes (A, B, C) = (20, 10, 5), (30, 10, 5), (40, 10, 5), and (50, 10, 5) nm. For comparison, the spectrum of a volume equivalent sphere with radius $R = 13.57$ nm is also plotted [curve •, corresponding to the volume of the ellipsoid with (A, B, C) = (50, 10, 5) nm].

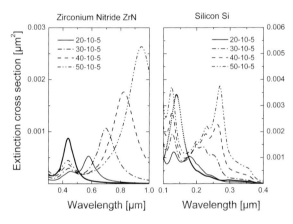

Figure 9.16 Ellipsoids of zirconium nitride and silicon with the three half-axes (A, B, C) = (20, 10, 5), (30, 10, 5), (40, 10, 5), and (50, 10, 5) nm. For comparison, the spectrum of a volume equivalent sphere with radius $R = 13.57$ nm is also plotted [curve •, corresponding to the volume of the ellipsoid with (A, B, C) = (50, 10, 5) nm].

coloring and also to reduced transparency. The effect is less for tantalum than for yttrium.

9.1.3.2 Semimetal and Semiconductor Particles

For the half-metal zirconium nitride (ZrN), the SPP of the spherical particle at around 430 nm is split into two resonances for the ellipsoidal particle (Figure 9.16). The resonance along the long axis clearly shifts to longer wavelengths with increasing axis length. The decrease in the absorption in the visible spectral region clearly reduces the color of ZrN nanoparticle systems. On the other hand, the strong resonance in the NIR region make them potential candidates for applications in surface-enhanced Raman scattering.

For the semiconducting material silicon, we only consider the wavelength range between 0.1 and 0.4 μm, where the electronic interband transitions lead to an electronic resonance for small Si spheres (Figure 9.16). For ellipsoidal nanoparticles this electronic resonance clearly splits into new resonances with its contributions shifting to longer wavelengths and also shorter wavelengths than the resonance position of the sphere. Moreover, the number of resonance-like spectral features increases with increasing long axis.

9.1.3.3 Carbonaceous Particles

For carbonaceous ellipsoidal nanoparticles it follows from the curves in Figure 9.17 that needles with high aspect ratio are necessary to have significant deviations from the spectrum of the spherical particle. For values of the long axis A below 50 nm the spectra of the ellipsoids and the sphere are almost identical. For $A \geq 100$ nm new resonances appear at longer wavelengths for graphite perpendicular to its crystalline c-axis and at short wavelengths for graphite parallel to its crystalline c-axis.

Roessler et al. [404] calculated the optical absorption by randomly oriented carbon spheroids and compared them with those of carbon spheres for all size regions. In general, they found that the absorption cross-section per unit volume is increased by axial elongation, particularly away from the resonance region.

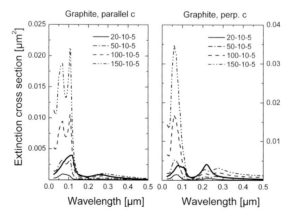

Figure 9.17 Ellipsoids of carbon parallel and perpendicular to the crystalline c-axis with the three half-axes $(A, B, C) = (20, 10, 5)$, $(50, 10, 5)$, $(100, 10, 5)$, and $(150, 10, 5)$ nm. For comparison, the spectrum of a volume equivalent sphere with radius $R = 19.57$ nm is also plotted [curve •, corresponding to the volume of the ellipsoid with $(A, B, C) = (150, 10, 5)$ nm].

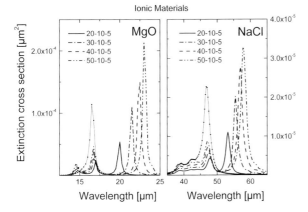

Figure 9.18 Ellipsoids of magnesia and sodium chloride with the three half-axes (A, B, C) = (20, 10, 5) (30, 10, 5), (40, 10, 5), and (50, 10, 5) nm. For comparison, the spectrum of a volume equivalent sphere with radius $R = 13.57$ nm is also plotted [curve •, corresponding to the volume of the ellipsoid with (A, B, C) = (50, 10, 5) nm].

9.1.3.4 Particles with Phonon Polaritons

In the spectra of MgO and NaCl nanoparticles in Figure 9.18, the spectrum of a spherical particle exhibits one peak and several smaller peaks. This is caused by the resonant coupling of the light to TO phonons in these ionic crystals for which their dielectric function can be described here by a harmonic oscillator.

In contrast to the spherical particle, the ellipsoidal MgO nanoparticle shows three clearly resolved resonances that can be assigned to the three different axes of the MgO ellipsoid. The long-wavelength resonance for the long axis A shifts to longer wavelengths with increasing axis length, whereas the positions of the other resonances remain almost unaffected. For sodium chloride only two resonances are resolved, with the resonance along the long axis again shifting to longer wavelengths with increasing axis length.

9.1.3.5 Miscellaneous Particles

In this subsection we look at the optical properties of ellipsoids of indium-doped tin oxide (ITO) and lanthanum hexaboride (LaB$_6$). Particularly ITO is of distinct interest because in this material free electrons contribute to a certain electrical conductivity of the material, but the material remains transparent at wavelengths in the visible spectral range. As the dielectric function ε can be approximated in the photon energy range 0.5–6 eV by the sum of a harmonic oscillator in the UV region and a Drude susceptibility with plasma frequency ω_p lying in the NIR region, we can expect a SPP for spherical particles which splits into at least one short-wavelength resonance and one long-wavelength resonance for ellipsoids. This can be recognized from the spectra of ellipsoidal nanoparticles in Figure 9.19.

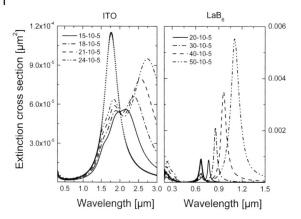

Figure 9.19 Ellipsoids of ITO and lanthanum hexaboride with the three half-axes (A, B, C) = (15, 10, 5)m, (18, 10, 5)m, (21, 10, 5), and (24, 10, 5) nm for ITO and (A, B, C) = (20, 10, 5)m, (30, 10, 5)m, (40, 10, 5), and (50, 10, 5) nm for LaB$_6$. For comparison, the spectrum of a volume equivalent sphere with radius $R = 10.63$ nm for ITO and $R = 13.57$ nm for LaB$_6$ is also plotted [curve •, corresponding to the volume of the ellipsoid with (A, B, C) = (24, 10, 5)m and (50, 10, 5) nm, respectively].

Lanthanum hexaboride (LaB$_6$) has attracted considerable attention for NIR absorption, caused by the excitation of SPP resonances. Indeed, the SPP of spherical particles lies in the red spectral range and shifts with increasing size to the NIR region. Another possibility for generating an NIR absorber is to use ellipsoidal nanoparticles of LaB$_6$, as can be seen in Figure 9.19. The SPP resonance of the spherical particle splits into a short-wavelength resonance that lies close to the sphere SPP, but is weaker than the sphere SPP, and a long-wavelength resonance in the NIR region. The position of the long-wavelength SPP shifts to longer wavelengths with increasing longer axis A. Then, the transparency of a nanoparticle system is increased and the NIR absorption is improved.

9.1.4
Experimental Results

9.1.4.1 Prolate Spheroidal Silver Particles in Fourcault Glass

In 1968, Stookey and Araujo considered applications of glasses containing uniformly oriented spheroidal Ag nanoparticles [405]. They stated that the two extinction bands from the two resonances of a spheroidal particle appear separately for light polarized along or perpendicular to the long axis. Moreover, the separation of both extinction bands increases with increasing eccentricity. This enables one to produce glasses with changing color and polarization of the transmitted light due to the absorption by the two extinction bands. They also described the preparation method. The first step is an ion exchange of Na ions in the soda-lime glass by Ag ions from silver nitrate. By subsequent tempering above 600 °C, spherical particles are formed in a thin layer of a few microns depth in the glass surface

(a)   (b)   (c)

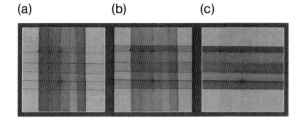

Figure 9.20 Five nanoparticle polarizers with different colors put perpendicular on five similar polarizers. (a) Incident light polarized vertically; (b) unpolarized incident light; (c) incident light polarized horizontally. A color version of this figure can be found in the color plates at the end of the book. Image courtesy K.-J. Berg.

[406]. In the next step, uniaxial stress is applied to the glass, which is heated to temperatures above the glass transition temperature T_g, forming uniformly elongated silver particles. This method is the only one so far to give glasses containing elongated silver nanoparticles for the production of polarizers. The experimental physics group at the University of Halle-Wittenberg has carried out research for a long period on such polarizing glasses [387, 388, 407–413] based on soda-lime glass. Figure 9.20 shows five parallel arranged polarizers with different colors on which five similar polarizers are placed perpendicular to the first polarizers. In (a) the incident light is polarized in the vertical direction, so that only the vertical polarizers can be recognized. In (c) the incident light is polarized horizontally, and in (b) the incident light is unpolarized.

In the following, we report exemplarily results for silver spheroidal nanoparticles in glass prepared and analyzed by Porstendorfer [387]. The radius of the preliminary spherical particles increased from the surface to the middle of the glass plate from 2 to 40–55 nm. Optical extinction spectra were recorded with a common spectrometer using polarized light. Figure 9.21 shows an exemplary set of measured spectra in comparison with corresponding calculated spectra. In the calculation it was assumed that the equivalent volume sphere had a diameter $R_V = 95$ nm and the ratio $A:B = 1.51$.

As can be seen, the peaks in the experimental curves are broader and the features especially of the TE mode are smeared out in comparison with the calculated spectra. These differences can be ascribed to the fact that in experiments the particles show size and shape distributions which were not considered in the calculations.

9.1.4.2 Plasma Polymer Films with Nonspherical Silver Particles

For interpretation of optical extinction spectra measured on plasma polymer films with embedded silver nanoparticles, the model of ellipsoids in the Rayleigh approximation was used. According to the assumption of elongated spheroidal particles, extensive analysis of TEM images was performed to determine the lengths of the major and the minor axes of the as-deposited and thermally annealed particles [414–416].

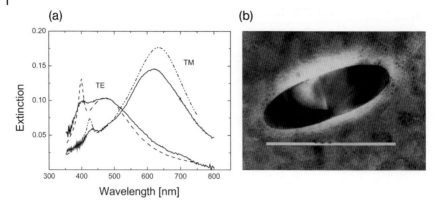

Figure 9.21 (a) Measured optical extinction spectra of spheroidal silver nanoparticles in Fourcault glass ($n_M = 1.517$) in comparison with calculated spectra for particles with $R_V = 95$ nm and $A/B = 1.51$. (b) Electron micrograph showing a typical spheroidal silver particle in the glass. Data courtesy J. Porstendorfer.

Plasma polymer thin films with embedded silver particles were deposited by simultaneous plasma polymerization and metal evaporation in a deposition reactor as described in [417]. The films used for this investigation were deposited from the monomer benzene (C_6H_6) at room temperature. The metal-containing plasma polymer films were deposited as multilayer systems with metal particles in only one plane between two plasma polymer layers. Electron micrographs of these multilayer systems showed well-separated single particles lying in one plane, which is necessary for determination of particle size and shape. TEM was also used for treating the samples with electron beam irradiation for thermal annealing. Extinction spectra were recorded with a commonly used spectrometer in the spectral range 200–2000 nm before and after thermal annealing. Annealing was performed under high-vacuum conditions (10^{-4} Pa) with a heating rate of about 3 K min^{-1} up to 480 K. To obtain quantitative results on the size and shape distribution of the embedded particles, TEM images were analyzed with an optical image processing system to determine the values for the half-axes A and B.

For a selected sample, 368 particles were analyzed both before and after annealing, and the results are given in three-dimensional histograms (Figure 9.22). The x-direction gives the minor axis B, the y-direction the major axis A and the z-direction the number of particles found in the A–B interval (count). Figure 9.22a shows the histogram for the sample as deposited and Figure 9.22b that of the annealed sample. Obviously, the shape of the particles approaches closer to spherical ($A = B$) when the sample is annealed.

The measured optical extinction spectra before and after thermal annealing are plotted in Figure 9.23a. The SPP resonance absorption is peaked at $\lambda = 520$ nm (2.384 eV) before annealing and at $\lambda = 484$ nm (2.562 eV) after annealing. This means that the extinction peak shifts by $\Delta\lambda = 36$ nm to shorter wavelengths. In

(a)

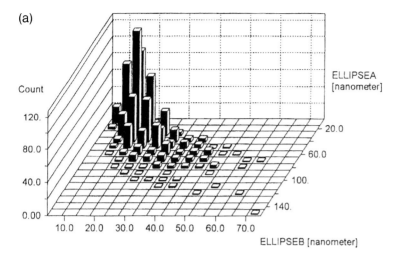

(b)

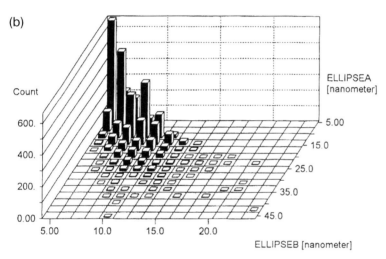

Figure 9.22 Three-dimensional particle histograms of the minor axis B, the major axis A, and the number of particles found in the A–B interval.

addition, the halfwidth of the plasma resonance absorption decreases from full width at half-maximum (FWHM) = 392 to 213 nm.

For comparison, optical extinction spectra for prolate spheroids were computed using optical constants of silver from [40], taking into account additional damping of the plasma resonance absorption according to the model of the mean free path effect. They are depicted in Figure 9.23b. For better comparison, the computed spectra are normalized. It is obvious that the spectrum of the sample after annealing is blue shifted with a shift of $\Delta\lambda = 57$ nm. Peak positions before and after annealing are at 520 nm (2.384 eV) and 463 nm (2.678 eV). Compared with the

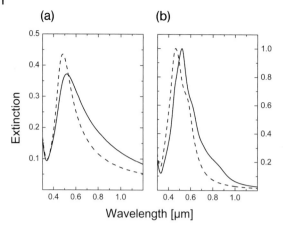

Figure 9.23 (a) Measured extinction spectra of embedded silver particles before (solid line) and after (dashed line) thermal annealing (480 K); (b) calculated extinction spectra according to the results of the optical image processing.

measured spectra, the shift is larger because the position of the peak after annealing is at higher energies than in the measurement. The halfwidths with FWHM = 235 and 211 nm are in the order of the halfwidth of the annealed sample.

9.1.4.3 Further Experiments

Götz et al. [418, 419] generated Na nanoparticles under ultra-high vacuum conditions in a particle jet. They were seeded on dielectric substrates (LiF$_2$, quartz, sapphire). Deformation of the particles when seeding on the substrate led to oblate spheroids. Polarization-dependent optical spectra were recorded to determine the aspect ratio. Laser desorption experiments were carried out on the Na spheroids.

Hövel et al. [420] generated Ag nanoparticles under ultra-high vacuum conditions by supersonic nozzle beam expansion and deposited them on quartz substrates. Polarization-dependent optical spectra were recorded and compared with the model of prolate and oblate spheroids. The axial ratio C:A was determined as 0.86 for the deposition-induced deformation when the particles hit the substrate with a velocity of 1500 m s^{-1}.

Hilger et al. [421] extended these investigations to determine also the contact area of the particle with the substrate. It amounted to less than 15% of the total particle surface in all experiments. The contact area proved to be strongly dependent upon the chemical nature of the substrate material and to influence the position of the SPP resonances of the spheroidal particle.

Hanarp et al. [422] examined the optical properties of gold nanodisk arrays prepared by colloidal lithography. The arrays exhibit short-range translational order and weak interparticle interactions. Tunable localized surface plasmon resonances were achieved by varying the diameter of the disks at constant disk height. The macroscopic optical properties were well described by modeling the gold disks as oblate spheroids in the electrostatic limit.

9.2
Cylinders

The first closed-form solutions of light scattering and absorption by an infinitely long cylinder – a wire – stem from von Ignatowski [67] in 1905 and Seitz [68] in 1906 and, hence, are as old as the solution for the sphere by Mie [17] in 1908. Scattering of plane waves at normal incidence for a homogeneous dielectric infinite cylinder was later independently solved by Lord Rayleigh in 1918 [423]. Wait [424] also published in 1955 an analytical solution for oblique incidence. Oblique incidence was reconsidered in 1966 by Lind and Greenberg [425], and tilted cylinders in 1982 by Cohen and Acquista [426].

9.2.1
Electromagnetic Fields and Scattering Coefficients

The symmetry of a cylindrical body is less than that of a spherical body. Hence the direction of the incident wave cannot always further be chosen along the positive z-axis. Instead, it includes in general an angle α between the x- and z-axes (Figure 9.24). Then, in cylindrical coordinates r, φ, z, the scalar wave equation has an infinite number of independent solutions Φ_n with

$$\Phi_n(r,\varphi,z) = Z_n(\rho)\exp(in\varphi)\exp(ihz) \qquad (9.41)$$

where $\rho = r\sqrt{k^2 - h^2}$ and $h = -k\cos\alpha$. The quantities Z_n are either Bessel functions J_n, Weber functions Y_n, or Hankel functions of the first kind $H_n^{(1)}$ or the second kind $H_n^{(2)}$. In contrast to the sphere, the order number n extends here from $-\infty$ to $+\infty$, and includes $n = 0$. The meaning of the order number n has obviously changed. For a sphere it corresponds directly to the multipole order, beginning with the dipole $n = 1$. For a cylinder it is more or less a number counting the partial waves.

Using the above solutions Φ_n and $\boldsymbol{a} = \boldsymbol{e}_z$ as an arbitrary constant vector, a set of linear independent cylindrical vector harmonics is created:

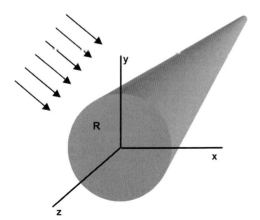

Figure 9.24 Sketch of the geometry in light scattering by an infinitely long cylinder.

$$\mathbf{M}_n = \mathrm{curl}(\Phi_n \mathbf{e}_z) = \sqrt{k^2 - h^2} \begin{bmatrix} in \dfrac{Z_n(\rho)}{\rho} \\ Z'_n(\rho) \\ 0 \end{bmatrix} \exp(in\varphi)\exp(ihz) \qquad (9.42)$$

$$\mathbf{N}_n = k^{-1}\mathrm{curl}\mathbf{M}_n = \dfrac{\sqrt{k^2 - h^2}}{k} \begin{bmatrix} ihZ'_n(\rho) \\ -hn\dfrac{Z_n(\rho)}{\rho} \\ \sqrt{k^2 - h^2}\, Z_n(\rho) \end{bmatrix} \exp(in\varphi)\exp(ihz) \qquad (9.43)$$

for $n \geq 0$. For negative n, the relation $Z_{-n}(\rho) = (-1)^n Z_n(\rho)$ leads to

$$\mathbf{M}_{-n} = (-1)^n \sqrt{k^2 - h^2} \begin{bmatrix} -in \dfrac{Z_n(\rho)}{\rho} \\ Z'_n(\rho) \\ 0 \end{bmatrix} \exp(-in\varphi)\exp(ihz) \qquad (9.44)$$

$$\mathbf{N}_{-n} = (-1)^n \dfrac{\sqrt{k^2 - h^2}}{k} \begin{bmatrix} ihZ'_n(\rho) \\ hn\dfrac{Z_n(\rho)}{\rho} \\ \sqrt{k^2 - h^2}\, Z_n(\rho) \end{bmatrix} \exp(-in\varphi)\exp(ihz) \qquad (9.45)$$

In the following we consider directly an infinitely long multilayered cylinder consisting of $r - 1$ concentric shells of arbitrary materials and different thickness d_s, $s = 2, \ldots, r$, and a circular core with radius R_{core} ($s = 1$) (see Figure 9.25). The total diameter of the cylinder is $2R$. Radially inhomogeneous infinite cylinders (coated

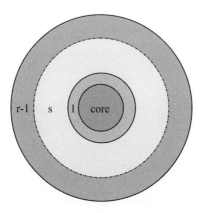

Figure 9.25 Cross-section of a cylindrical particle with $r - 1$ concentric shells.

cylinders) were treated earlier by Adey [427], Kerker and Matijevic [428], Evans et al. [429], Samaddar [430], Kai and d'Alessio [431], and Gurwich et al. [432].

The refractive indices n_1 of the core and n_s of each shell may be complex numbers, which include absorption in the core and in each shell. The cylinder is assumed to be embedded in an arbitrary nonabsorbing medium with refractive index n_M and to be illuminated by an incident plane wave with wavenumber $k_M = (2\pi n_M)/\lambda$. The parameters $m_s = n_s/n_{s+1}$ represent the relative refractive indices from shell s to shell $s+1$. For $s = r$, the refractive index n_{r+1} corresponding to the refractive index n_M of the surrounding medium. In each medium the wavenumber is $k_s = (2\pi n_s)/\lambda$. The size parameters are defined as

$$x_s = k_{s+1}\left(R_{core} + \sum_{j=2}^{s} d_j\right) \tag{9.46}$$

The expansions for the electromagnetic fields E and H of the incident wave, the scattered wave, and the wave inside the cylinder and any shell into transverse magnetic (TM) modes and transverse electric (TE) modes are as follows:

incident wave:

$$\mathbf{E}_{inc} = \frac{E_0}{k_M \sin\alpha} \sum_{n=-\infty}^{\infty} (-i)^n \left[\mathbf{N}_n^{(1)}(k_M) - i\mathbf{M}_n^{(1)}(k_M)\right] \tag{9.47a}$$

$$\mathbf{H}_{inc} = \frac{-ik_M}{\omega\mu_0} \frac{E_0}{k_M \sin\alpha} \sum_{n=-\infty}^{\infty} (-i)^n \left[\mathbf{N}_n^{(1)}(k_M) - i\mathbf{M}_n^{(1)}(k_M)\right] \tag{9.47b}$$

scattered wave:

$$\mathbf{E}_{sca} = \frac{E_0}{k_M \sin\alpha} \sum_{n=-\infty}^{\infty} (-i)^n \left[a_n \mathbf{N}_n^{(3)}(k_M) - ib_n \mathbf{M}_n^{(3)}(k_M)\right] \tag{9.48a}$$

$$\mathbf{H}_{sca} = \frac{-ik_M}{\omega\mu_0} \frac{E_0}{k_M \sin\alpha} \sum_{n=-\infty}^{\infty} (-i)^n \left[b_n \mathbf{N}_n^{(3)}(k_M) - ia_n \mathbf{M}_n^{(3)}(k_M)\right] \tag{9.48b}$$

wave inside the core:

$$\mathbf{E}_1 = \frac{E_0}{k_1 \sin\alpha} \sum_{n=-\infty}^{\infty} (-i)^n \left[\alpha_n^1 \mathbf{N}_n^{(1)}(k_1) - i\beta_n^1 \mathbf{M}_n^{(1)}(k_1)\right] \tag{9.49a}$$

$$\mathbf{H}_1 = \frac{-ik_1}{\omega\mu_0} \frac{E_0}{k_1 \sin\alpha} \sum_{n=-\infty}^{\infty} (-i)^n \left[\beta_n^1 \mathbf{N}_n^{(1)}(k_1) - i\alpha_n^1 \mathbf{M}_n^{(1)}(k_1)\right] \tag{9.49b}$$

wave inside any shells:

$$\mathbf{E}_s = \frac{E_0}{k_s \sin\alpha} \sum_{n=-\infty}^{\infty} (-i)^n \left\{\left[\alpha_n^s \mathbf{N}_n^{(1)}(k_s) + \gamma_n^s \mathbf{N}_n^{(2)}(k_s)\right] - i\left[\beta_n^s \mathbf{M}_n^{(1)}(k_s) + \delta_n^s \mathbf{M}_n^{(2)}(k_s)\right]\right\} \tag{9.50a}$$

$$\mathbf{H}_s = \frac{-ik_s}{\omega\mu_0} \frac{E_0}{k_s \sin\alpha} \sum_{n=-\infty}^{\infty} (-i)^n \left\{\left[\beta_n^s \mathbf{N}_n^{(1)}(k_s) + \delta_n^s \mathbf{N}_n^{(2)}(k_s)\right] - i\left[\alpha_n^s \mathbf{M}_n^{(1)}(k_s) + \gamma_n^s \mathbf{M}_n^{(2)}(k_s)\right]\right\} \tag{9.50b}$$

where α_n^s, γ_n^s are the expansion coefficients of the TM modes and β_n^s, δ_n^s are those of the TE modes in the shell s ($s = 1$ is the core), and a_n and b_n are the expansion coefficients of the TM and TE modes, respectively, of the scattered wave.

The use of Bessel functions J_n (superscript 1 of the vector harmonics) in the expansions of the electromagnetic fields inside the core and of the incident wave guarantees the continuity of the fields at the origin of the reference frame which is centered in the cylinder. The use of Hankel functions $H_n^{(1)}$ of first kind (superscript 3 of the vector harmonics) in the expansion of the scattered wave takes into consideration outgoing cylindrical waves. Further, the use of Weber functions Y_n (superscript 2 of the vector harmonics) for the fields inside each shell accounts for the continuity of these functions in these limited domains.

Applying Maxwell's boundary conditions at the interfaces and the cylinder surface, the expansion coefficients of all waves inside the multilayered cylinder and of the scattered wave can be resolved. The procedure is similar to that for a sphere. Here, we again restrict considerations to the coefficients a_n and b_n of the scattered wave. In general, they depend on the incidence angle α. The most important case is, however, perpendicular incidence, that is, $\alpha = 90°$. Then, the parameter $h = -k\cos\alpha$ becomes zero, $h = 0$, and Equations 9.42 and 9.45 simplify remarkably. In this case, the scattering coefficients are

$$a_n = \frac{\begin{array}{l} J_n(x_r)[J_n'(m_r x_r) + S_n^{r-1} N_n'(m_r x_r)] \\ - m_r J_n'(x_r)[J_n(m_r x_r) + S_n^{r-1} N_n(m_r x_r)] \end{array}}{\begin{array}{l} H_n(x_r)[J_n'(m_r x_r) + S_n^{r-1} N_n'(m_r x_r)] \\ - m_r H_n'(x_r)[J_n(m_r x_r) + S_n^{r-1} N_n(m_r x_r)] \end{array}} \quad (9.51)$$

$$b_n = \frac{\begin{array}{l} m_r J_n(x_r)[J_n'(m_r x_r) + T_n^{r-1} N_n'(m_r x_r)] \\ - J_n'(x_r)[J_n(m_r x_r) + T_n^{r-1} N_n(m_r x_r)] \end{array}}{\begin{array}{l} m_r H_n(x_r)[J_n'(m_r x_r) + T_n^{r-1} N_n'(m_r x_r)] \\ - H_n'(x_r)[J_n(m_r x_r) + T_n^{r-1} N_n(m_r x_r)] \end{array}} \quad (9.52)$$

with

$$S_n^s = -\frac{\begin{array}{l} J_n(x_s)[J_n'(m_s x_s) + S_n^{s-1} N_n'(m_s x_s)] \\ - m_s J_n'(x_s)[J_n(m_s x_s) + S_n^{s-1} J_n(m_s x_s)] \end{array}}{\begin{array}{l} J_n(x_s)[J_n'(m_s x_s) + S_n^{s-1} N_n'(m_s x_s)] \\ - m_s N_n'(x_s)[J_n(m_s x_s) + S_n^{s-1} J_n(m_s x_s)] \end{array}} \quad (9.53)$$

$$T_n^s = -\frac{\begin{array}{l} m_s J_n(x_s)[J_n'(m_s x_s) + T_n^{s-1} N_n'(m_s x_s)] \\ - J_n'(x_s)[J_n(m_s x_s) + T_n^{s-1} J_n(m_s x_s)] \end{array}}{\begin{array}{l} m_s J_n(x_s)[J_n'(m_s x_s) + T_n^{s-1} N_n'(m_s x_s)] \\ - N_n'(x_s)[J_n(m_s x_s) + T_n^{s-1} J_n(m_s x_s)] \end{array}} \quad (9.54)$$

and the starting values $S_n^0 = T_n^0 = 0$. The prime denotes derivatization with respect to the argument. The solution for a homogeneous cylinder is obtained for $r = 1$. Note that $a_{-n} = a_n$ and $b_{-n} = b_n$.

9.2.2
Efficiencies and Scattering Intensities

Analogous to the spherical particle, the rates at which energy is absorbed and scattered are calculated from Poynting's law (energy conservation law). Dividing these rates by the intensity I_0 of the incident light, the optical cross-sections σ for scattering and absorption are determined. However, these quantities are not appropriate for an infinitely long cylinder as they diverge with the infinite length L of the cylinder because they are proportional to the cylinder volume $\pi R^2 L$. On the other hand, the geometric cross-section $G = 2RL$ also tends to infinity with the length of the cylinder, but the ratio $\sigma{:}G$, which defines the dimensionless efficiency Q, remains finite. Hence it is only possible to calculate efficiencies. The explicit calculation yields:

$$Q_{ext} = \frac{2}{k_M R}\left[\text{Re}\left(a_0 + b_0 + 2\sum_{n=1}^{\infty} a_n + b_n\right)\right] \tag{9.55}$$

$$Q_{sca} = \frac{2}{k_M R}\left\{|a_0|^2 + |b_0|^2 + 2\sum_{n=1}^{\infty} |a_n|^2 + |b_n|^2\right\} \tag{9.56}$$

Note that in contrast to a spherical particle the index n runs from $n = -\infty$ to $n = \infty$, including $n = 0$. As for a given multipole n the functions $\mathbf{M}_{-n}$ and $\mathbf{N}_{-n}$ are related to the corresponding functions $\mathbf{M}_n$ and $\mathbf{N}_n$, the cross-sections and scattering intensities only contain $n \geq 0$. Whereas for a sphere n enumerates the multipoles that contribute to the fields and cross-sections, this interpretation fails for a cylinder. The order number $n = 0$ need not be interpreted as the contribution of an electrical or magnetic monopole! For example, in the Rayleigh approximation, the cross-sections and scattering intensities are dominated by the TM dipole mode a_1 for a spherical particle. The TE contribution b_1 is negligible. For a cylinder, the terms a_0, a_1, b_0 and b_1 contribute to almost the same extent, as can be seen in the following section.

The scattering intensities $i_{per}(\theta)$ and $i_{par}(\theta)$ perpendicular and parallel to the scattering plane, respectively, follow in the same way:

$$i_{per}(\theta) = \left|a_0 + 2\sum_{n=1}^{\infty} a_n \cos(n\theta)\right|^2 = |S_1(\theta)|^2 \tag{9.57}$$

$$i_{par}(\theta) = \left|b_0 + 2\sum_{n=1}^{\infty} b_n \cos(n\theta)\right|^2 = |S_2(\theta)|^2 \tag{9.58}$$

where $S_1(\theta)$ and $S_2(\theta)$ are known as (complex) *phase functions*.

Another quantity connected with $i_{per}(\theta)$ and $i_{par}(\theta)$ is the *asymmetry parameter* or weighted cosine of the scattering angle:

$$g = \langle\cos\theta\rangle = \frac{8}{\pi k_M R Q_{sca}} \int_0^\pi [i_{par}(\theta) + i_{per}(\theta)]\cos\theta\, d\theta \tag{9.59}$$

In Figure 9.26 we give examples of the scattering intensities i_{par} in the scattering plane and i_{per} in the plane perpendicular to the scattering plane for cylinders with

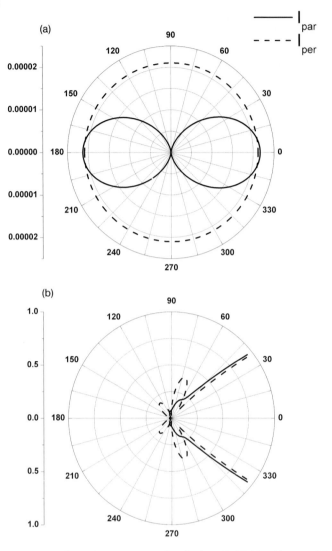

Figure 9.26 Scattering intensities of a cylinder with (a) $2R = 20$ nm and (b) $2R = 600$ nm.

$2R = 20$ and 600 nm. The thin cylinder acts as a dipole with i_{par} proportional to $(\cos\theta)^2$, and a constant scattering intensity i_{per}. If the diameter increases, the light is dominantly scattered in the forward direction. The scattering intensities become very similar close to the forward direction, but differ for larger angles due to the different contributions of a_n and b_n. This can be clearly recognized from Figure 9.26b. In this figure the scattering intensity in the forward direction is not completely plotted.

9.2.3
Resonances

Both MDRs and electronic resonances can occur if the denominators of the scattering coefficients a_n and b_n become small.

For a multilayered cylinder, the TM mode with expansion coefficient a_n is resonant if

$$m_r \frac{H'_n(x_r)}{H_n(x_r)} = \frac{J'_n(m_r x_r) + S_n^{r-1} N'_n(m_r x_r)}{J_n(m_r x_r) + S_n^{r-1} N_n(m_r x_r)} \tag{9.60}$$

and the TE mode with expansion coefficient b_n is resonant if

$$\frac{1}{m_r} \frac{H'_n(x_r)}{H_n(x_r)} = \frac{J'_n(m_r x_r) + T_n^{r-1} N'_n(m_r x_r)}{J_n(m_r x_r) + T_n^{r-1} N_n(m_r x_r)} \tag{9.61}$$

The resonances of the homogeneous cylinder are obtained for $r = 1$ (remember that $S_n^0 = T_n^0 = 0$):

$$m \frac{H'_n(x)}{H_n(x)} = \frac{J'_n(mx)}{J_n(mx)} \tag{9.62}$$

for the TM modes a_n and

$$\frac{1}{m} \frac{H'_n(x)}{H_n(x)} = \frac{J'_n(mx)}{J_n(mx)} \tag{9.63}$$

for the TE modes b_n.

To illustrate MDRs, Figure 9.27a shows the extinction (= scattering) efficiency Q_{ext} plotted versus size for cylinders with refractive index $n = 1.334$ in air. In this case, the wavelength is constant at $\lambda = 514.5$ nm and the cylinder diameter varies

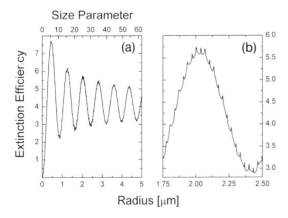

Figure 9.27 Extinction (= scattering) efficiency of cylinders in air showing MDRs. (a) Size range $2R = 2$–$10\,000$ nm, $n_{cylinder} = 1.334$, $\lambda = 514.5$ nm; (b) size range $2R = 3500$–5000 nm.

from $2R = 2$ to $10\,000$ nm in steps of 2 nm. The corresponding size parameter varies between $x = 0.01$ and 62.1. In Figure 9.27b, the part between $2R = 3500$ and 5000 nm is zoomed out to show the MDRs. Compared with the MDRs of a sphere, those of a cylinder are less pronounced.

Electronic resonances and SPP resonances are also possible. For discussion of these resonances, we assume very thin cylinders in the Rayleigh approximation, that is, $kR \ll 1$. In this case, only the contributions with $n = 0$ and 1 must be considered. The contribution for $n = 0$ is given by

$$\alpha_0 = a_0 + b_0 = \frac{\pi R}{2}\left\{\frac{1}{2}\left(\frac{\pi R}{\lambda}\right)^2(\varepsilon - \varepsilon_M)\right.$$

$$\left. + \frac{\varepsilon - \varepsilon_M}{\varepsilon_M\left[1 + 2\varepsilon\left(\frac{\pi R}{\lambda}\right)^2 \log\left(\frac{2\pi R}{\lambda}\sqrt{\varepsilon_M}\right) - i\pi\left(\frac{\pi R}{\lambda}\right)^2(\varepsilon - \varepsilon_M)\right]}\right\} \quad (9.64)$$

and that for $n = 1$ is given by

$$\alpha_1 = 2(a_1 + b_1) = \frac{\pi R}{2}\left[2\frac{\varepsilon - \varepsilon_M}{\varepsilon + \varepsilon_M} + \left(\frac{\pi R}{\lambda}\right)^2(\varepsilon - \varepsilon_M)\right] \quad (9.65)$$

both normalized to the geometric cross-section of the cylinder. The second term in braces in Equation 9.64 corresponding to b_0 is different to that in Equation 8.42 on p. 208 in the book by Bohren and Huffman [8]. It has been proven that for strongly absorbing materials, the term given by Bohren and Huffman is not a sufficient approximation.

In the Rayleigh limit, the efficiencies are

$$Q_{\text{ext}} = k_M \, \text{Im}(\alpha_0 + \alpha_1) + (k_M R) k_M^2 \left(|\alpha_0|^2 + |\alpha_1|^2\right) \quad (9.66)$$

$$Q_{\text{sca}} = (k_M R) k_M^2 \left(|\alpha_0|^2 + |\alpha_1|^2\right) \quad (9.67)$$

The terms a_0, b_0 and b_1 do not exhibit a resonance, since they are more or less proportional to the optical contrast $(\varepsilon - \varepsilon_M)$. Therefore, the corresponding spectrum is proportional to $\omega\varepsilon_2(\omega)/c = \varepsilon_2(\lambda)/\lambda$, which is similar to the absorption spectrum of a thin metal film. The term a_1 can exhibit a resonance if the dielectric function ε of the particle material fulfills the condition

$$\varepsilon_1 = -\varepsilon_M, \varepsilon_2 \approx 0 \quad (9.68)$$

This condition is different from that for a spherical particle ($\varepsilon_1 = -2\varepsilon_M$, $\varepsilon_2 \approx 0$) and leads to a blue-shifted resonance compared with the resonance of a sphere.

For illustration, Figure 9.28 shows the calculated extinction efficiency of an infinitely long cylinder of aluminum with $2R = 20$ nm with the contributions of a_1 and a_0 and of b_0 and b_1 resolved. For comparison, the efficiency spectrum of a sphere with $2R = 20$ nm is also plotted. The resonance of a_1 is clearly recognized

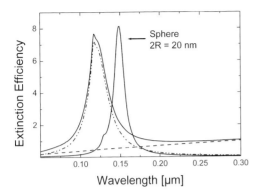

Figure 9.28 Extinction efficiency of an aluminum cylinder with $2R = 20$ nm in vacuum. The total spectrum (solid line) is divided into the contributions of the polarizations parallel (dashed line) and perpendicular (dashed-dotted line) to the cylinder axis. For comparison, the efficiency spectrum of a sphere with $2R = 20$ nm in vacuum is also plotted. It is divided by 3 to allow better comparison.

as a sharp extinction maximum which can be excited by s-polarized incident light (perpendicular to the cylinder axis). For p-polarized incident light (along the cylinder axis) the efficiency increases monotonically with increasing wavelength. The resonance of the cylinder is shifted to $\lambda = 117$ nm compared with the resonance of the sphere at $\lambda = 149$ nm.

9.2.4
Extensions

The theory of light scattering by infinite cylinders has been extended by several authors. Here, we consider some relevant publications.

Considering particularly metallic cylinders, the scattering theory was extended by Ruppin [433] and by Boustimi et al. [434] to include longitudinal plasmon resonances. The result is similar to that for a sphere, already shown in Figure 5.20 in Section 5.3.3, except that the resonance of the SPP is blue shifted for the cylinder.

Ruppin [435] extended the electromagnetic scattering formalism for a circular cylinder to the case in which the medium surrounding the cylinder is absorptive. The effects of the absorbing medium on the interference and ripple structures, which occur in the extinction, as a function of the size parameter, were demonstrated. The theory was applied to the geometry of a cylindrical cavity in a metal, in which peaks due to surface plasmons localized near the cavity appear in the spectra.

Sun et al. [436] also developed analytic solutions for the single-scattering properties of an infinite dielectric cylinder embedded in an absorbing medium with normal incidence, which include extinction, scattering, and absorption efficiencies, the scattering phase function, and the asymmetry factor.

A theoretical and numerical study of the optical near-field in metallic and semiconducting nanocylinders close to a dielectric surface was presented by Arias-Gonzalez and Nieto-Vesperinas [437, 438]. In the far-field zone, the scattering efficiency shows sharp peaks associated with the excitation of either plasmon resonances (in metallic particles) or MDRs (in semiconducting particles), which are further enhanced on incidence with an evanescent plane wave.

Videen and Ngo [439] derived a theory to calculate light scattering from a cylinder on or near a plane interface without restrictions on the values of the refractive indices of the system constituents. They used the approximation that the interference component of the electromagnetic fields strikes the plane interface at normal incidence.

Numerical simulation of the scattering of light and other electromagnetic waves from a cylinder in front of a plane was investigated by Madrazo and Nieto-Vesperinas [440] using the extinction theorem for multiply connected domains. Both media were assumed to be perfectly conducting. Far-fields and angular spectra were discussed.

The first extension of light scattering by cylinders on arbitrary incident beams and Gaussian beams was performed in 1982 by Kozaki [441, 442]. Later, Gouesbet and co-workers performed the extension within the framework of the GLMT [443–449]. Additionally, Lock [450, 451], Barton [452], and Mroczka and Wysoczanski [453] extended the theory to arbitrary incident beams.

Bohren [454] and Kluskens and Newman [455] treated cylinders of anisotropic optical materials.

At the end of this section, we come to the special case of a *finite cylinder*. The finite cylinder as a target for incident light is particularly difficult, and it is not even possible to construct an exact analytic solution. Instead, several numerical approaches have been worked out. An iterative approach to the scattering of light from a finite dielectric cylinder was applied by Cohen *et al.* [456] in 1982. The iteration converged with the first two orders of the iteration to within 1% when the aspect ratio (length/diameter) of the cylinder is as small as 20. More elaborate work on this subject appeared later in the 1990s, mainly developed by Waterman and Pedersen [457–459], Wang and van de Hulst [460], and Mishchenko *et al.* [461].

9.2.5
Numerical Examples

In this section, we concentrate on infinitely long cylinders with diameters between 10 and 300 nm. We present numerical results for a manifold of particle materials and wavelength regions. The restriction to particles with diameters less than 300 nm allows comparison with the results obtained in Chapter 6 for spherical particles of various diameters.

We point out that in the following calculated spectra the surrounding is always assumed to have a constant refractive index $n_M = 1$ (vacuum), whereas in experiments it may be larger. Nevertheless, provided that the refractive index n_M remains real valued, that is, the surrounding medium is not absorbing, comparison with the calculations is possible.

Similarly to the calculations presented earlier for spherical and ellipsoidal particles, we use the following categories:

- metal particles
- semimetal and semiconductor particles
- carbonaceous particles
- oxide particles
- particles with phonon polaritons
- miscellaneous particles.

We concentrate on particles where resonances (electronic resonances and SPP resonances) play a role also for an infinitely long cylinder.

In the examples, the spectral range comprises an extended spectral range including the visible spectral range, to present *genuine* optical properties. Exceptions are the materials with phonon polaritons, since these polaritons lie in the far-infrared region.

9.2.5.1 Metal Particles

As we already know from the calculations and experiments on metallic nanospheres in Chapter 6, for all metals TM multipole resonances develop with increasing size. They are caused by the ratio of particle size to the wavelength. In addition to these geometric resonances, electronic resonances and SPP resonances can also occur, even for such small spheres where the ratio of particle size to wavelength is too small to generate resonant spectral features. The reason is that the dielectric function of the metal fulfills certain conditions, namely

$$\varepsilon_1 = -2\varepsilon_M, \quad \varepsilon_2 \approx 0 \tag{9.69}$$

This condition can be fulfilled in the region of an interband transition described by a harmonic oscillator, or by free charge carriers, the free electron plasma. However, not all metals exhibit either an electronic resonance or a SPP resonance, because the imaginary part ε_2 of the dielectric function is large, so that the above condition is not fulfilled. For increasing particle size these resonances are further superposed by the geometric effects.

For metallic cylinders of infinite length but small diameter, the situation is similar. Here, we have the resonance condition for an electronic resonance or an SPP resonance given by

$$\varepsilon_1 = -\varepsilon_M, \quad \varepsilon_2 \approx 0 \tag{9.70}$$

In the following we concentrate on metals with electronic and SPP resonances and present the size dependence of the optical extinction efficiency for sodium (Na), potassium (K), gold (Au), silver (Ag), yttrium (Y) and tantalum (Ta) cylinders with radii $2R < 300\,\text{nm}$. In addition, we also present extinction efficiency spectra for platinum (Pt) and iron (Fe) cylinders. For these metals neither an electronic nor a SPP resonance could be observed for spherical particles.

The outstanding property of the alkali metals Na and K is that their dielectric function can be approximated fairly well by the Drude dielectric constant for free electrons:

$$\varepsilon(\omega) = 1 + \chi_1^{IB}(\omega) - \frac{\omega_p^2}{\omega^2} + i\left[\chi_2^{IB}(\omega) + \frac{\gamma_{fe}}{\omega}\frac{\omega_p^2}{\omega^2}\right] \tag{9.71}$$

The influence of the core electrons on the susceptibility of the interband transitions χ^{IB} is only perceptible for the real part χ_1^{IB} and is negligible for χ_2^{IB}. We can therefore expect SPP resonances which are hardly influenced by interband transitions.

Figure 9.29 summarizes the calculated spectra for Na and K nanoparticles. The optical constants used are from [94]. For thin nanocylinders, a single sharp resonance peak can be recognized that can be assigned to the resonance of the dipolar TM mode. It corresponds to the SPP in these small cylinders. Its position shifts to longer wavelengths (lower energies) with increasing diameter. Furthermore, the magnitude increases with increasing diameter, approaching its maximum for a certain size, which is material dependent. The maximum value is $Q_{ext} \approx 8$ for Na and $Q_{ext} \approx 9$ for K. This is less by a factor of 2 than those for the comparable spherical particles.

For larger sizes, the magnitude of the peak decreases again. In addition, new resonant-like structures appear at longer wavelengths. Their number increases with

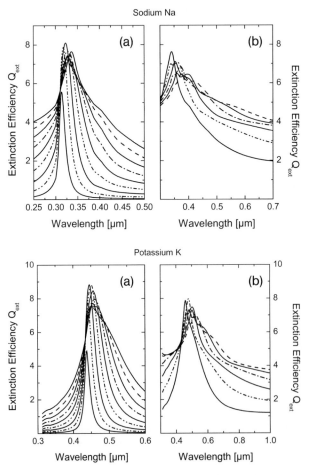

Figure 9.29 Spectra of sodium and potassium nanocylinders. (a) $2R = 10$–90 nm, stepwidth 10 nm; (b) $2R = 100$–300 nm, stepwidth 40 nm.

increasing particle size, and also their peak positions shift to longer wavelengths and their halfwidths increase. They can be assigned mainly to TM modes with higher multipole order. Their position and width are dominated more by the increasing size than by the electronic structure of the metal. The contribution of the nonresonant modes of the cylinder to the spectrum increases with increasing diameter and leads to a monotonic decrease of the efficiency at long wavelengths.

We should point out that the preparation of nanowires of Na or K is critical since these metals are chemically very reactive.

Similarly to the alkali metals, nanocylinders of noble metals also exhibit a SPP. In contrast to the alkali metals, the contribution χ^{IB} of interband transitions from the 4d (Ag) or 5d (Au) electrons to the hybridized 5sp (Ag) or 6sp (Au) band has an enormous influence on the positions of the SPPs. Figure 9.30 shows representative spectra for Ag and Au nanocylinders. Optical constants were taken from [37].

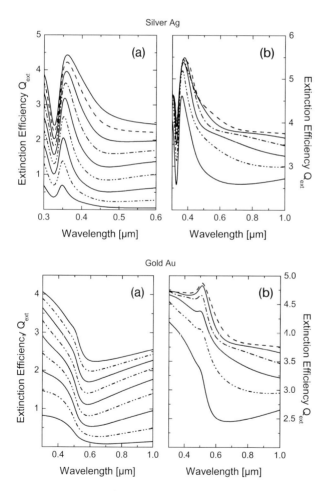

Figure 9.30 Spectra of silver and gold nanocylinders. (a) $2R = 10–90$ nm, stepwidth 10 nm; (b) $2R = 100–300$ nm, stepwidth 40 nm.

The largest influence of the interband transitions on the SPP resonance of nanocylinders can be recognized in the spectra of Au cylinders. The resonance at wavelengths around 500 nm is so strongly quenched by the close-lying interband transitions that it cannot be resolved in the spectra of cylinders with diameters $2R \leq 140$ nm. Only if the particle size is larger than $2R = 140$ nm the resonance can be clearly recognized. For the smaller nanocylinders the spectrum is dominated by the interband transitions and the nonresonant TM modes. The latter determine the spectral behavior at longer wavelengths. For Ag nanocylinders the resonance is clearly separated from the interband transitions and can be recognized in all spectra. Its position shifts slightly to longer wavelengths with increasing diameter. At wavelengths $\lambda > 450$ nm the nonresonant TM and TE modes determine the spectral behavior of the Ag nanocylinders.

As already demonstrated for spherical and ellipsoidal particles, yttrium and tantalum are good candidates for two resonances in nanoparticle extinction spectra: an electronic resonance at low wavelengths due to an interband transition and a SPP resonance at long wavelengths caused by the free carriers. The spectra presented in Figure 9.31 for nanocylinders of yttrium and tantalum, however, clearly deviate from the spectra of comparable spheres (see Chapter 6) and exhibit new features. For discussion of these spectra, we must also look at the dielectric function of yttrium and tantalum in Figure 9.32. They were taken from [103] for yttrium and [149] for tantalum.

In this graph, the horizontal dashed line and dashed-dotted-dotted line indicate the conditions $\varepsilon_1 = -2$ (for spheres) and $\varepsilon_1 = -1$ (for cylinders). It can be recognized that the wavelength distance of the two possible electronic resonances of a cylinder is larger than that for a spherical particle. Indeed, only one electronic resonance appears in Y spheres at wavelengths between 340 and 410 nm (as seen in Chapter 6). For Y cylinders, two electronic resonances can be recognized in the spectra at wavelengths 320–380 and 680–700 nm. On the other hand, the SPP which was already weak for spherical particles cannot be recognized in the spectra. It is shifted to shorter wavelengths where the imaginary part ε_2 of the dielectric function is even larger and therefore the SPP is damped away. The long-wavelength behavior of the extinction cross-section is finally determined by the nonresonant TM and TE contributions. A similar discussion can also made for tantalum cylinders. Whereas the electronic resonances becomes better separated and clearly resolved, the SPP is damped away and superposed by the contributions of the nonresonant TM and TE modes.

The next group of metals are of less interest for optical properties of nanoparticle systems since the smaller nanoparticles do not exhibit a resonance which could be assigned to a SPP or an electronic resonance. For spherical particles of Pt, Pd, Rh, Fe, Ni, Co, and many similar metals the extinction efficiency only decreases from short to long wavelengths proportionally to $1/\lambda$ (Rayleigh approximation of the extinction cross-section) without exhibiting any resonance-like feature. If the particle size increases, a broad resonance-like extinction band grows which shifts to longer wavelengths with increasing size. In addition, more relatively broad structures appear that also shift to longer wavelengths with increasing size. They can all be assigned to TM modes which develop resonance-like extinction bands due to the increasing ratio of size to wavelength.

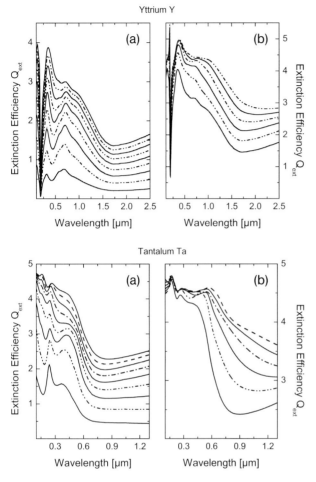

Figure 9.31 Spectra of yttrium and tantalum nanocylinders. (a) $2R = 10$–90 nm, stepwidth 10 nm; (b) $2R = 100$–300 nm, stepwidth 40 nm.

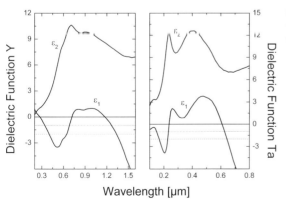

Figure 9.32 Dielectric function of yttrium and tantalum.

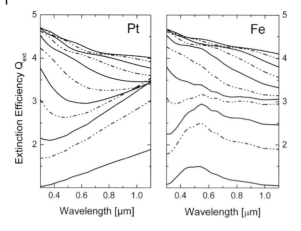

Figure 9.33 Spectra of platinum and iron nanocylinders. Diameters $2R$ = 10, 20, 30, 50, 70, 100, 140, 180, 220, 260, and 300 nm from bottom to top.

For nanocylinders, we have nonresonant TM and TE modes that contribute to the spectral behavior of the metal cylinders mainly in the long-wavelength range. They lead to a monotonic increase in the extinction efficiency with increasing wavelength, as could be recognized from the earlier examples. Therefore, we can expect for Pt and Fe nanocylinders a spectral dependence of the extinction efficiency that is determined by Rayleigh scattering at low wavelengths and a monotonic increase at long wavelengths. This is demonstrated with the exemplary spectra in Figure 9.33. Actually, the extinction efficiency of Pt cylinders shows the expected behavior. In contrast, Fe cylinders show a different behavior. Up to a diameter of approximately 60 nm, the cylinders exhibit a broad extinction band in the green visible spectral range. For thicker cylinders, the spectra become similar to those of the Pt cylinders. This is completely different to the behavior of spherical nanoparticles of Fe as shown in Chapter 6.

9.2.5.2 Semimetal and Semiconductor Particles

In TiN and ZrN the d electrons just below the Fermi surface contribute to the electron gas and lead to an extinction maximum for spherical nanoparticles at $\lambda \approx 490$ nm for TiN and at $\lambda \approx 440$ nm for ZrN. This extinction band can be interpreted as SPP [101]. At lower wavelengths, interband transitions from occupied electronic levels into unoccupied electronic levels contribute to the absorption of light by the particles.

On going to infinitely long nanocylinders, the shift of the SPP resonance to lower wavelengths (resonance condition $\varepsilon = -\varepsilon_M$) causes, for TiN nanocylinders of small diameters, the SPP resonance to be strongly damped because it now appears close to the interband transitions (Figure 9.34). Initially from approximately $2R = 100$ nm the plasmon resonance is clearly resolved. For wavelengths longer than 600 nm the nonresonant TM and TE modes of a nanocylinder determine the spectra. For ZrN nanocylinders the SPP is clearly resolved for all sizes because it

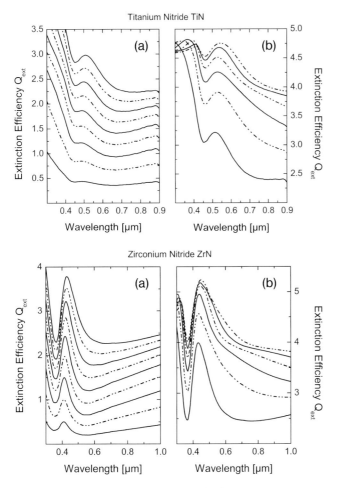

Figure 9.34 Spectra of TiN and ZrN nanocylinders. (a) $2R = 10–90$ nm, stepwidth 10 nm; (b) $2R = 100–300$ nm, stepwidth 40 nm.

lies far enough from the interband transitions. Also here, the long-wavelength extinction is determined by the nonresonant modes.

Silicon and cadmium telluride exhibit strong interband transitions in the near-UV region close to the visible spectral region that can be modeled by harmonic oscillators in the dielectric function. They cause both materials to have high refractive indices in the visible spectral region whereas the absorption index is small. Hence both materials are good candidates for electronic resonances in the interband transition region and for MDRs in the visible and NIR region, even in nanoparticles. Using optical constants from [36] for Si and [156] for CdTe, we calculated size-dependent spectra of nanocylinders, which are presented in Figure 9.35. The spectral region is extended for Si down to $0.1\,\mu m$ wavelength to demonstrate the influence of the strong interband transitions. With increasing particle size, sharp

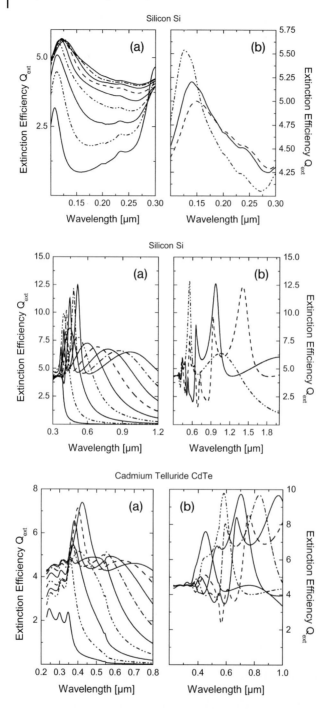

Figure 9.35 Spectra of silicon and cadmium telluride nanocylinders. Silicon: (a) $2R = 10–90$ nm, stepwidth 10 nm; (b) $2R = 100$, 200, and 300 nm. Cadmium telluride: (a) $2R = 10–90$ nm, stepwidth 10 nm; (b) $2R = 100–300$ nm, stepwidth 40 nm.

MDRs develop in the spectra for both TM and TE partial waves. The shift of the resonance positions to longer wavelengths is similar to that already observed for spherical nanoparticles of Si and CdTe.

9.2.5.3 Carbonaceous Particles

According to the optical constants of graphite [164, 165], for graphite parallel to the c-axis there are two electronic transitions in the near-UV region with maxima at approximately 300 nm (4.13 eV) and 110 nm (11.25 eV). They can be assigned to the π-electrons (≈4 eV, one electron per C atom) and σ-electrons (~14 eV, three electrons per C atom). For nanocylinders with these optical constants the interband transitions result in resonances of the TM dipole mode at λ = 289 nm (4.29 eV), 112 nm (11.07 eV) and 71 nm (17.46 eV) for the smallest cylinders, which shift to longer wavelengths with increasing particle size. This is one resonance more than in the case of spherical graphite particles. The corresponding spectra are depicted in Figure 9.36.

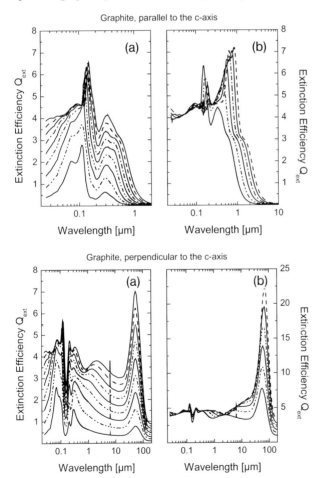

Figure 9.36 Spectra of graphitic nanocylinders. (a) 2R = 10–90 nm, stepwidth 10 nm; (b) 2R = 100–300 nm, stepwidth 40 nm.

For graphite perpendicular to the crystalline c-axis, there are two electronic transitions in the UV region at 280 nm (4.43 eV) and 90 nm (13.6 eV), and furthermore an interband transition in the IR region at 62 μm (0.02 eV). The transitions in the UV region correspond to transitions between the valence band and the conduction band and can also be assigned to π- and σ-electrons. For nanocylinders with these optical constants, the interband transitions result in resonances of the TM dipole mode at 294 nm (4.22 eV), 200 nm (6.2 eV) and 70 nm (17.71 eV) for the smallest cylinders, which shift to longer wavelengths with increasing particle size. Again, this is one more resonance than for spherical graphite particles with these optical constants. The corresponding spectra are plotted in Figure 9.36. Surprisingly, the cylinders exhibit a further strong resonance at 55 μm (0.0225 eV), which is absent for spherical nanoparticles. It belongs to an interband transition in graphite. The changed resonance condition for a cylinder allows this resonance to be clearly resolved whereas for spherical nanoparticles this resonance is not resolved.

9.2.5.4 Oxide Particles

Nanoparticles of α-Fe_2O_3 (hematite) are well known as red pigment particles. For application as pigments, however, spherical particles must have sizes in the range $2R \approx 200–300$ nm. Analogously, chromium(III) oxide (Cr_2O_3) nanoparticles are used for green pigments. For particles with sizes of $2R \approx 200–280$ nm, the TM dipole resonance of the particles contributes optimally to the absorption and scattering, so that a green color results in such nanoparticle systems.

On turning to nanocylinders, it can be recognized from Figure 9.37 that the extinction efficiency spectra are similar to those of spherical nanoparticles in Chapter 6. However, resonances of the TM and TE modes develop faster than

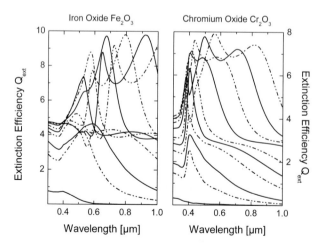

Figure 9.37 Spectra of Fe_2O_3 and Cr_2O_3 nanocylinders with diameter $2R = 10$, 50, and 70 nm, and $2R = 100–300$ nm, stepwidth 40 nm, respectively.

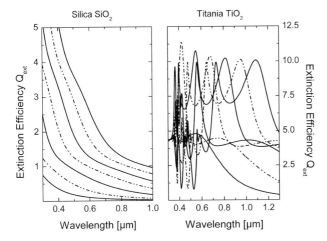

Figure 9.38 Spectra of silica (SiO$_2$) and titania (TiO$_2$) (rutile) nanocylinders. Particle sizes for SiO$_2$: 2R = 70 nm and 2R = 100–300 nm, stepwidth 40 nm. Particle diameters for TiO$_2$: 2R = 50, 70 nm and 2R = 100–300 nm, stepwidth 40 nm.

for spherical nanoparticles. Therefore, infinitely long nanocylinders of Fe$_2$O$_3$ or Cr$_2$O$_3$ are also well suited for color pigments, even with diameters smaller than for spheres.

For nonabsorbing materials only MDRs can occur at all. However, for SiO$_2$ the refractive index is still too low for MDRs in nanocylinders of sizes below 300 nm. Therefore, only broad resonances of low-order multipoles can be expected in the spectra. As actually can be seen from Figure 9.38, the extinction (which is identical with the scattering) only decreases monotonically with increasing wavelength for all diameters.

TiO$_2$ is strongly absorbing in the near-UV region due to the onset of direct interband transitions. These close-lying interband transitions cause a rather high refractive index in the visible spectral region for this material. In cylinders of such highly refractive materials already for nanocylinders MDRs develop with increasing particle diameter. Similarly to the semiconductors considered earlier, both TM and TE modes exhibit MDRs which shift to longer wavelengths with increasing particle diameter.

9.2.5.5 Particles with Phonon Polaritons

In ionic materials such as MgO or NaCl, the coupling of light with TO phonons leads to a significant change in the dielectric function of the material that can be interpreted in the framework of harmonic oscillators. Then, the real part of the dielectric constant may become negative close to the resonance frequency of the harmonic oscillator. Indeed, the resonance condition of a nanocylinder $\varepsilon = -\varepsilon_M$ for the TM dipole mode may be approached twice, depending on the steepness of the dielectric function in the corresponding spectral region. In the spectra of NaCl

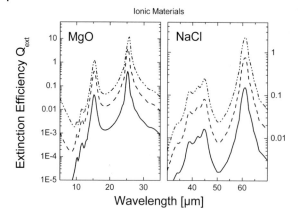

Figure 9.39 Spectra of MgO and NaCl nanocylinders with $2R = 10$, 100, and 300 nm. Logarithmic scale.

and MgO nanocylinders in Figure 9.39, actually two peaks and several smaller peaks can clearly be seen. This is in contrast to the spherical nanoparticles.

These resonances show a weak size dependence with respect to the peak position. Only the magnitude of Q_{ext} increases with increasing size. The reason is simply that even the largest diameter of $2R = 300$ nm is very small compared with the wavelength (15–60 μm) in these wavelength ranges. Therefore, these particles can be treated in the Rayleigh limit, in which the particle size does not influence the resonance position.

9.2.5.6 Miscellaneous Particles

In this last subsection we consider nanocylinders of indium-doped tin oxide (ITO) and lanthanum hexaboride (LaB_6).

In ITO, free electrons contribute to a certain electrical conductivity of the material, but the material remains transparent at wavelengths in the visible spectral range. This holds true also for spherical nanoparticles with sizes $2R < 100$ nm. For larger particles, higher multipole resonances develop and contribute to the extinction at wavelengths in the visible region. In contrast, as follows from Figure 9.40, nanocylinders with diameters $2R < 100$ nm exhibit a monotonically decreasing extinction already in the visible spectral range, which may lead to coloring of ensembles containing very long narrow cylinders. For larger diameters, higher multimode modes develop similarly to the spherical particles and determine the optical properties in the visible spectral range.

In the NIR spectral region, the SPP in ITO nanocylinders can be clearly recognized at wavelength $\lambda = 1.61$ μm (0.77 eV) for the smallest particle. With increasing particle size, the plasmon peak undergoes a slight red shift to $\lambda = 1.695$ μm (0.73 eV) for cylinders with diameter 300 nm. The SPP resonance is clearly blue shifted compared with the corresponding spherical particles, caused by the changed resonance condition. At wavelengths longer than about 2 μm finally

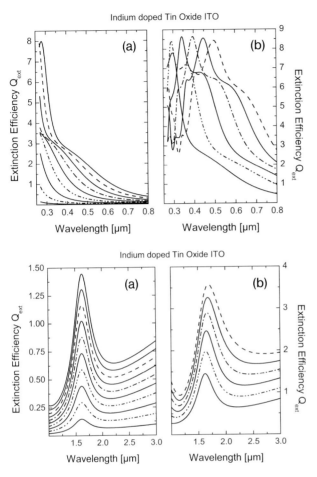

Figure 9.40 Spectra of ITO nanocylinders. (a) $2R = 10$–$90\,nm$, stepwidth $10\,nm$; (b) $2R = 100$–$300\,nm$, stepwidth $40\,nm$.

the nonresonant modes of a cylinder determine the spectral extinction of ITO nanocylinders.

The calculated spectra of LaB$_6$ nanocylinders in Figure 9.41 clearly differ from the corresponding spectra for nanospheres. Already for small diameters the extinction in the visible spectral region is determined by resonant-like extinction bands which become stronger with increasing diameter. The SPP of LaB$_6$ nanocylinders can also be clearly recognized but now at the wavelength $\lambda = 0.647\,\mu m$ for the smallest diameter. With increasing cylinder diameter, the plasmon peak undergoes a red shift to $\lambda = 0.671\,\mu m$ for the largest diameter. At longer wavelengths again nonresonant modes of the cylinder determine the spectrum of these nanocylinders.

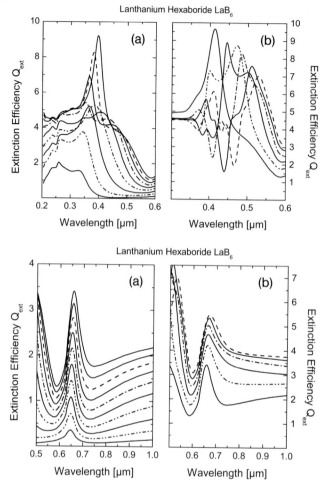

Figure 9.41 Spectra of LaB$_6$ nanocylinders. (a) $2R = 10–90$ nm, stepwidth 10 nm; (b) $2R = 100–300$ nm, stepwidth 40 nm.

9.3
Cubic Particles

9.3.1
Theoretical Considerations

Ionic materials such as MgO, NaCl, CaF$_2$, and many others prefer a cubic crystalline lattice and therefore preferably form also cubic nanoparticles. Analytical solutions for the scattering by cubes are not available, only a solution in the Rayleigh approximation was given by Napper [462] and Fuchs [463]. The reason is that these particles do not exhibit homogeneous polarization modes of the whole volume, in

Table 9.1 Oscillator strengths and geometry factors of the six most prominent contributions to the polarizability of a cube.

j	C_j	G_j
1	0.44	0.214
2	0.24	0.297
3	0.04	0.345
4	0.05	0.440
5	0.10	0.563
6	0.09	0.706

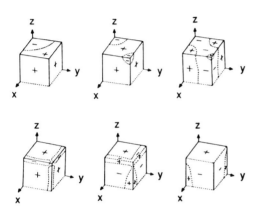

Figure 9.42 Illustration of the six most prominent dipole polarizabilities of a cube according to Fuchs [463].

contrast to the sphere, ellipsoid, or infinitely long cylinder. According to Fuchs, the dipolar polarizability of a cubic particle in the Rayleigh approximation is composed of N dipolar eigenmodes of the induced charge distribution in the cubic particle around the edges and corners which are always excited simultaneously:

$$\alpha = D^3 \sum_{j=1}^{N} \frac{C_j(\varepsilon - \varepsilon_M)}{\varepsilon_M + G_j(\varepsilon - \varepsilon_M)} \tag{9.72}$$

where D is the size of the cube and C_j are the oscillator strengths of the localized eigenmodes. The most prominent contributions are summarized in Table 9.1 and are shown in Figure 9.42.

These six eigenmodes have together an oscillator strength of $\sum_{j=1}^{6} C_j = 0.96$; only 4% of all oscillator strength belongs to eigenmodes not shown here. As we now have six different geometry factors G_j, we can expect also six electronic resonances or SPP resonances in the spectral range where the dielectric function $\varepsilon = \varepsilon_1 + i\varepsilon_2$ of the particle material fulfills the condition

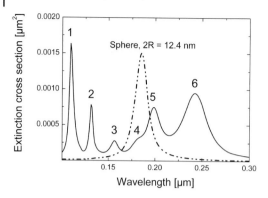

Figure 9.43 Extinction spectrum for an aluminum cube with size $D = 10\,\text{nm}$. For comparison, the cross-section of the volume equivalent sphere ($2R = 12.4\,\text{nm}$) divided by 2 is also plotted.

$$\varepsilon_1 = -\frac{1-G_j}{G_j}\varepsilon_M, \quad \varepsilon_2 \approx 0 \qquad (9.73)$$

Remember that for a sphere (one geometry factor $G = 1/3$) and an infinitely long cylinder ($G = 1/2$) only one resonance is apparent, and for a triaxial ellipsoid ($G_1 \neq G_2 \neq G_3 \neq 1/3$) three resonances appear for particles in the Rayleigh approximation. The resonance positions of a cubic particle are at those wavelengths where the real part ε_1 of the dielectric function has values in the interval $-3.68\varepsilon_M \leq \varepsilon_1 \leq -\varepsilon_M$.

The exemplary spectrum of a cubic aluminum nanoparticle with $D = 10\,\text{nm}$ *in vacuo* in Figure 9.43 exhibits five distinct maxima. The sixth resonance peak (No. 4 in the graph) appears as shoulder of a neighboring resonance peak. For comparison, the spectrum of the volume-equivalent sphere, that is, with $2R = 12.4\,\text{nm}$, is also plotted. It is divided by 2 for better comparison. Three of the resonances of the cube are blue shifted (shorter wavelength) compared with the resonance of a sphere. Two resonances contribute at wavelengths longer than the resonance wavelength of the sphere. The resonance No. 4 has almost the same position as the sphere.

In a direct application of Fuchs's theory to metallic cubic nanoparticles with a pure Drude-like dielectric function, Ruppin [464] also found six resonances in the extinction cross-section of the particles.

9.3.2
Numerical Examples

As we only consider here particles in the Rayleigh limit, only those materials are of interest where resonances can actually occur as in the above aluminum cubes. Hence, in the following examples we present spectra of cubic metal particles of Na, K, Ag, and Au, spectra of semimetals (ZrN) and semiconductors (Si), infrared spectra of cubic particles of MgO and NaCl, and finally spectra for the IR absorbers ITO and LaB$_6$. The edge length of the cube is $D = 10\,\text{nm}$. The cube spectrum in vacuum is compared with the volume-equivalent sphere with $2R = 12.4\,\text{nm}$ in vacuum.

9.3.2.1 Metal Particles

Small spheres of Na, K, Ag, and Au exhibit a SPP resonance in the near-UV and visible spectral region. Therefore, nanoparticle systems containing such particles show a characteristic color. On going to cubic nanoparticles of these metals, it can be expected that up to six resonances occur, similarly to the aluminum cubes considered earlier.

Actually, for Na four resonances and for K six resonances can be clearly recognized in Figure 9.44. The most prominent resonances contribute to the spectrum at longer wavelengths than the resonance of the volume-equivalent sphere. In consequence, the color of particle ensembles with such cubic particles must clearly change.

For Ag and Au nanocubes in vacuum, the cubic shape results mainly in a shift and a broadening of the extinction band. As can be seen in Figure 9.45, the resonances of the cubes cannot be clearly resolved. Only on embedding the cubes in a surrounding medium with higher refractive index can more resonances be resolved. This is, however, not shown here.

Y and Ta nanocubes exhibit a similar behavior to Ag and Au in the range of the electronic resonance and also in the spectral range of the SPP (also not shown here).

9.3.2.2 Semimetal and Semiconductor Particles

For ZrN as an example of semimetals and for Si as an example of semiconducting nanoparticles in vacuum, the cubic shape also mainly results in a red shift and a broadening of the extinction band (SPP and electronic resonance). The resonances of the cubes cannot be clearly resolved, as demonstrated in Figure 9.46.

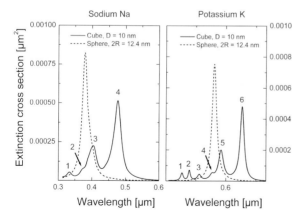

Figure 9.44 Extinction spectrum of cubic nanoparticles of sodium and potassium with edge length $D = 10\,\text{nm}$. The dashed line is the spectrum of the corresponding volume equivalent sphere with $2R = 12.4\,\text{nm}$.

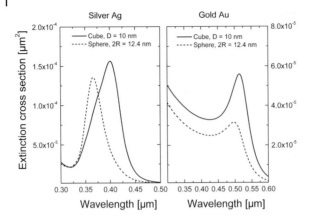

Figure 9.45 Extinction spectrum of cubic nanoparticles of silver and gold with edge length $D = 10$ nm. The dashed line is the spectrum of the corresponding volume equivalent sphere with $2R = 12.4$ nm.

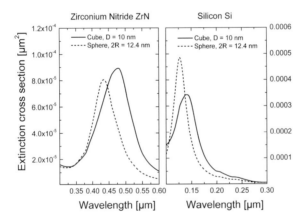

Figure 9.46 Extinction spectrum of cubic nanoparticles of ZrN and silicon with edge length $D = 10$ nm. The dashed line is the spectrum of the corresponding volume equivalent sphere with $2R = 12.4$ nm.

9.3.2.3 Particles with Phonon Polaritons

The most interesting materials for cubic nanoparticles are ionic materials such as magnesia (MgO) and sodium chloride (NaCl), because they initially form cubic particles. Interaction of light with these particles is dominantly determined by the coupling with TO phonons in the IR region. In the spectra of MgO and NaCl nanocubes in Figure 9.47, therefore, electronic resonances appears for both spherical and cubic nanoparticles. From the six possible resonances of a cube, only the long-wavelength resonances can be clearly resolved between 16 and 20 μm for

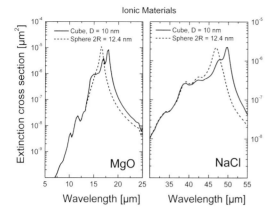

Figure 9.47 Extinction spectrum of cubic nanoparticles of MgO and NaCl with edge length $D = 10\,\text{nm}$. The dashed line is the spectrum of the corresponding volume equivalent sphere with $2R = 12.4\,\text{nm}$.

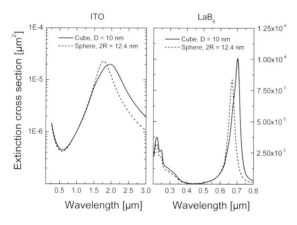

Figure 9.48 Extinction spectrum of cubic nanoparticles of ITO and LaB_6 with edge length $D = 10\,\text{nm}$. The dashed line is the spectrum of the corresponding volume equivalent sphere with $2R = 12.4\,\text{nm}$.

MgO and between 46 and 51 μm for NaCl. They are red shifted compared with the resonance of the corresponding volume-equivalent sphere.

9.3.2.4 Miscellaneous Particles

For the most important IR absorbers ITO and LaB_6, the shape of a cubic particle leads mainly to a red shift and broadening of the SPP. This can be followed from the spectra of cubic nanoparticles with $D = 10\,\text{nm}$ and the spectra of the corresponding volume-equivalent sphere in Figure 9.48.

9.4
Numerical Methods

For irregular nonspherical particles, either approximations for large particles and certain distinct geometries (e.g., platonic bodies) are available, or numerical methods have been established to calculate the optical response of an irregularly shaped body with certain algorithms. The best evolved and most often applied methods are the *discrete dipole approximation* (DDA) and the *T-matrix method* or *extended boundary condition method* (EBCM), and these are introduced and discussed in the next two sections. Some other methods will also be briefly introduced.

A systematic and unified discussion of light scattering by nonspherical particles and its practical applications can be found in the books by Mishchenko and coworkers [12, 15]. They represent the state-of-the-art of this important research field and are very practical for scientists and engineers in geophysics, remote sensing, and planetary and space physics. However, they deal predominantly with dielectric particles larger than 1000 nm and mainly consider the spatial (angular) distribution of the scattered light. Spectral behavior is not discussed.

9.4.1
Discrete Dipole Approximation

A very straightforward approach to overcome the problem of light scattering by nonspherical particles is to investigate the interaction of each individual molecule of the particle with the incident wave and with the re-radiation of all other excited molecules of the particle. The re-radiation of each molecule is not independent but is coupled to the radiation of all other molecules. This concept was used by Langbein [465] to calculate van der Waals attraction between molecules. Langbein's approach was not restricted to dipole–dipole coupling. Later, a method was developed that divides the scattering particle into N (spherical) subregions on a lattice which are small enough to be considered as radiating dipoles. This method is therefore designated the *discrete dipole approximation* (DDA) or *coupled dipole method* (CDM). This concept was first introduced and applied by Purcell and Pennypacker [466] and later by Draine [467]. In 1994, Draine and Flatau [468] reviewed the DDA method. A detailed description of the DDA can be found in [469] and is presented in the following.

In the DDA, the particle interacts with a plane incident wave. The total local field E_i at the position r_i of a dipole p_i is the sum of the incident field and the contribution of all dipole moments p_j of the other dipoles with $j \neq i$. According to Singham and Bohren [470], it can be written in the following form:

$$E_i = E_i^{\text{inc}} + \sum_{j \neq i}^{N} \left[A_{ij} \underline{\alpha}_j E_j + B_{ij} \left(\underline{\alpha}_j E_j r_{ji}^0 \right) r_{ji}^0 \right] \tag{9.74}$$

where r_{ji}^0 is a dimensionless unit vector pointing from the jth dipole to the location of the ith dipole. The electric field incident at particle i is

$$E_i^{\text{inc}} = E_0 \exp(ikr_{ij}) \tag{9.75}$$

The electric field $\mathbf{E}$ of a single dipole with dipole moment $\mathbf{p}$ in the direction of a dimensionless unit vector $\mathbf{r}_0 = \mathbf{r}/r$ is given by

$$\mathbf{E} = \frac{\exp(ikr)}{4\pi\varepsilon_M\varepsilon_0} \left\{ \underbrace{\frac{k^2}{r}[\mathbf{p}-(\mathbf{r}_0\cdot\mathbf{p})\mathbf{r}_0]}_{\text{far-field}} + \underbrace{[3\mathbf{r}_0(\mathbf{r}_0\cdot\mathbf{p})-\mathbf{p}]\left(\frac{1}{r^3}-\frac{ik}{r^2}\right)}_{\text{near-field}} \right\}$$

$$= \frac{\exp(ikr)}{4\pi\varepsilon_M\varepsilon_0}\left[\left(\frac{k^2}{r}-\frac{1}{r^3}+\frac{ik}{r^2}\right)\mathbf{p} + \left(-\frac{k^2}{r}+\frac{3}{r^3}-\frac{3ik}{r^2}\right)(\mathbf{r}_0\cdot\mathbf{p})\mathbf{r}_0\right] \quad (9.76)$$

Note that this is the electric field of a dipole including the *near-field zone* [the time dependence $\exp(-i\omega t)$ has been omitted]. The dipole moment $\mathbf{p}_i$ is related to the local field $\mathbf{E}_i$ by

$$\mathbf{p}_i = \varepsilon_0\varepsilon_M\underline{\underline{\alpha}}_i\mathbf{E}_i \quad (9.77)$$

where $\underline{\underline{\alpha}}_i$ is the polarizability of the ith dipole. Comparing Equation 9.74 with Equation 9.76 yields

$$A_{ij} = \left(k^2 - \frac{1}{r_{ij}^2} + \frac{ik}{r_{ij}}\right)\frac{\exp(ikr_{ij})}{r_{ij}}$$

$$B_{ij} = \left(\frac{3}{r_{ij}^2} - k^2 - \frac{3ik}{r_{ij}}\right)\frac{\exp(ikr_{ij})}{r_{ij}} \quad (9.78)$$

Once a solution for $\mathbf{E}_i$ has been found, the extinction and absorption cross-sections σ_{ext} and σ_{abs} may be evaluated from Poynting's law [470]:

$$\sigma_{\text{ext}} = \frac{4\pi k}{|E_0|^2}\sum_{j=1}^{N}\text{Im}\left[(E_j^{\text{inc}})^*\alpha_j E_j\right] \quad (9.79)$$

and

$$\sigma_{\text{abs}} = \frac{4\pi k}{|E_0|^2}\sum_{j=1}^{N}[\text{Im}(E_j\alpha_j E_j^*) - \frac{2}{3}k^3|\alpha_j E_j|^2] \quad (9.80)$$

The scattering cross-section is given by $\sigma_{\text{sca}} = \sigma_{\text{ext}} - \sigma_{\text{abs}}$.

The discrete dipole approximation has proven to be a powerful and flexible tool for computing the scattering of electromagnetic waves by arbitrary-shaped particles with size parameters $2\pi L/\lambda < 10$, that is, $L < 1000$ nm at wavelengths in the visible spectral range. This limitation is due to the available computer power since the set of linear equations for the $\mathbf{E}_i$ is of the rank $3N$, with N being the number of dipoles considered.

The most serious limitation of the DDA is its restriction to only dipole–dipole coupling, meaning that the spherical subregions are always small compared with the incident wavelength as only a dipole moment (the TM dipole) is induced. Furthermore, the outgoing dipole re-radiation that is incident at the site of a neighboring dipole is expressed only in terms of dipole radiation. An extension of the DDA including the TE dipole was given in 1994 by Mulholland *et al.* [471] to discuss the differential cross-section of soot particles.

Singham and Bohren [472] suggested a hybrid scheme where the particle is divided into domains larger than usual in the DDA and within which the fields were calculated exactly (or approximately) and then the iteration scheme of the DDA was applied or a lower order system of linear equations was solved. Generally, the same idea in a slightly modified form was realized in [473].

> **INFO – DDA and Clausius–Mossotti**
>
> A severe restriction of the DDA follows directly from its main idea. If one wants to take into account electromagnetic interactions among microscopic parts of the interior of a macroscopic body using a classical electrodynamics approach, it is necessary that the space between these subregions is filled with vacuum. In consequence, the wavenumber k in Equations 9.75–9.80 must be that of vacuum. Therefore, the DDA can only applied to particles in vacuum or air as surrounding medium.
>
> A second serious restriction originates from the determination of the dipole polarizability α of each of the N (spherical) subregions on a lattice. Purcell and Pennypacker [466] and later Draine and Flatau [468] used the Clausius–Mossotti relation [474, 475] between the (macroscopic) dielectric function ε of the homogeneous and isotropic target material and the microscopic dipole polarizability α_j of each atom/molecule[1]:
>
> $$\frac{\varepsilon-1}{\varepsilon+2} = \frac{1}{3\varepsilon_0}\sum_j N_j \alpha_j \qquad (9.81)$$
>
> This relation only holds for cubic crystalline structures for which the local electric field satisfies the Lorentz relation:
>
> $$E_{local} = E + \frac{1}{3\varepsilon_0}P \qquad (9.82)$$
>
> The Clausius–Mossotti relation can be used to calculate the macroscopic dielectric function from the microscopic polarizabilities of the point dipoles, but *not* vice versa. The Clausius–Mossotti relation in general does not yield the microscopic polarizabilities α_j! For example, for NaCl the dielectric constant ε_{NaCl} can be calculated from α_{Na} and α_{Cl}, but it is impossible to calculate α_{Na} and α_{Cl} from ε_{NaCl}. Only for the special case of identical polarizabilities of the point dipoles can Equation 9.81 be used to calculate the polarizability from the dielectric function. Therefore, the DDA is restricted to particles consisting of only one component such as carbon, and is ambiguous for materials such as H_2O, MgO, dust, and so on.
>
> In this respect, also the extension of the CDM by Dungey and Bohren [476] must be viewed critically. They used directly the TM dipole coefficient a_1 according to Mie's theory instead of the dipole polarizability in the calculations of the coupled dipole method.
>
> 1) Note that Purcell and Pennypacker and also Draine and Flatau used the cgs system in their original contributions.

A reformulation of the CDM with the result of increased numerical efficiency was made by Singham and Bohren [470, 477]. Their algorithm was based on an idea by Chiappetta [478, 479].

Mathematically equivalent to the DDA is the *digitized Green's function* (DGF) method [480] (originally proposed in [481]) or the *volume integral equation formulation* (VIEF) [482, 483]. The DGF/VIEF differs only in the use of another prescription for the dipole polarizability.

Piller [484, 485] improved the DDA of Draine and Flatau by introducing an appropriate discretization scheme. The convergence of the coupled-dipole approximation becomes much more regular and its accuracy is improved by as much as a factor of 24. This discretization scheme is simple, can be applied to arbitrarily shaped scatterers, and does not require additional computation time.

Unfortunately, the CDM becomes less accurate when scatterers with high permittivity are at hand. Refining the discretization mesh in that case does not necessarily increase the accuracy of the result, as was investigated in detail by Piller and Martin [486]. To overcome these limitations, Gay-Balmaz and Martin [487] introduced a modified CDM scheme based on sampling theory and using the so-called filtered Green's tensor [488]. This scheme was shown to handle scatterers accurately with a permittivity as large as 10, located in an infinite homogeneous background.

Lakhtakia [489] and Singham *et al.* [490–493] extended the DDA to anisotropic, chiral media.

The DDA was originally developed to model particles in free space. Taubenblatt and Tran [494] modified the method to model features on surfaces. The computation of scattering by features on a surface is accelerated using a two-dimensional fast Fourier transform technique as introduced by Schmehl *et al.* [495].

9.4.2
T-Matrix Method or Extended Boundary Condition Method

The T-matrix method or EBCM is ideally suited for calculating the electrodynamic interaction of particles of arbitrary shape with electromagnetic waves. It is based on a series of papers by Waterman [496–498]. The T-matrix method attracted the attention of optics scientists only after the publication of Barber and Yeh [499], who gave this method the name EBCM. In 1996, Mishchenko *et al.* [500] gave a first overview of T-matrix computations. More recently, Mishchenko *et al.* [501] gave a comprehensive database of T-matrix publications from the inception of the technique in 1965 through early 2004. We also refer here again to the books by Mishchenko and co-workers [12, 15] for further reading. The important advantage of the T-matrix method is its adaptation to almost all application fields of the light scattering theory.

The T-matrix solution begins with an expansion of the incident, scattered, and internal electromagnetic fields into corresponding vector spherical harmonics, just as with the Mie-theory for spheres. The internal field expansion coefficients c_n and d_n are given by the solution of a linear set of equations [499]:

$$\begin{pmatrix} K_{qn} + mJ_{qn} & L_{qn} + mI_{qn} \\ I_{qn} + mL_{qn} & J_{qn} + mK_{qn} \end{pmatrix} \begin{pmatrix} c_n \\ d_n \end{pmatrix} = (-i) \begin{pmatrix} e_n^{TM} \\ e_n^{TE} \end{pmatrix} \tag{9.83}$$

where e_n^{TM} and e_n^{TE} are the coefficients of the incident beam, summarized in a vector. The quantities I, J, K, and L are area integrals which must be evaluated numerically over the surface of the nonspherical particle, for example

$$K_{qn} = \frac{k^2}{\pi} \int_S [N_q^{(3)}(kr) \times M_n^{(1)}(mkr)] \cdot n_s \mathrm{d}S \tag{9.84}$$

Analogously, the scattered field expansion coefficients a_n and b_n are obtained from the solution of the linear set of equations

$$\begin{pmatrix} \tilde{K}_{qn} + m\tilde{J}_{qn} & \tilde{L}_{qn} + m\tilde{I}_{qn} \\ \tilde{I}_{qn} + m\tilde{L}_{qn} & \tilde{J}_{qn} + m\tilde{K}_{qn} \end{pmatrix} \begin{pmatrix} -ic_n \\ -id_n \end{pmatrix} = \begin{pmatrix} a_n \\ b_n \end{pmatrix} \tag{9.85}$$

with, for example,

$$\tilde{K}_{qn} = \frac{k_M^2}{\pi} \int_S [N_q^{(1)}(kr) \times M_n^{(1)}(mkr)] \cdot n_s \mathrm{d}S \tag{9.86}$$

Denoting the matrix in Equation 9.83 with $\underline{\underline{A}}$ and the matrix in Equation 9.85 with $\underline{\underline{B}}$, we have

$$\begin{pmatrix} a_n \\ b_n \end{pmatrix} = \underline{\underline{B}} \underline{\underline{A}}^{-1} \begin{pmatrix} e_n^{TM} \\ e_n^{TE} \end{pmatrix} = \underline{\underline{T}} \begin{pmatrix} e_n^{TM} \\ e_n^{TE} \end{pmatrix} \tag{9.87}$$

The matrix $\underline{\underline{T}}$ is the transition matrix that converts the incident field coefficients to the scattered field coefficients. Explicit formulations to calculate the matrix elements can be found in the book on remote sensing by Tsang et al. [502]. They are obtained by numerical integration as surface integrals, which is computationally expensive. Therefore, most implementations of the T-matrix method are restricted to axisymmetric particles, such as spheroids, finite circular cylinders, and even-order Chebyshev particles. Extensions to nonaxisymmetric particles were made by, for example, Scheider and Peden [503] and Wriedt and Doicu [504].

The standard, single spherical coordinate-based EBCM has numerical difficulties with scattering problems involving large aspect ratios. Therefore, a number of modifications to the standard EBCM have been suggested, including improved numerical methods [505–508] and formal modifications [504, 509]

We note that the number of applications of the T-matrix formalism is too large to be cited here. Instead, we refer again to the review paper by Mishchenko et al. [501], who gave a a comprehensive database of T-matrix publications from 1965 through early 2004.

9.4.3
Other Numerical Methods

9.4.3.1 Point Matching Method
Similarly to Mie's theory, the incident, scattered, and transmitted fields are expanded into spherical harmonics in a multipole expansion. The expansion coefficients of the scattered field are found by matching the electromagnetic fields point-by-point at the surface of the scattering body. This method was developed by Oguchi [510] and Morrison and Cross [511]. It is suitable for rotationally symmetric bodies.

9.4.3.2 Discretized Mie Formalism
Another differential equation approach was developed by Rother and Schmidt [512, 513], who named their method the *discretized Mie formalism*. It is based on the application of a difference operator from which a system of coupled ordinary differential equations results. Decoupling is achieved by a suitable transformation so that the equations can be solved analytically. In this way, it is a generalization of the separation of variables method used in Mie's theory. The discretized Mie formalism can be applied to nonspherical axisymmetric scatterers. An example can be found in [514].

9.4.3.3 Generalized Multipole Technique
Whereas in the Mie theory and the T-matrix method the spherical multipoles have their origin at the center of the particle, in the general multipole technique many origins are applied for multipole expansion. The coefficients of these expansions are the unknown values to be determined by applying the boundary conditions at the particle surface [515]. Al-Rizzo and Tranquilla [516] demonstrated the potential and versatility of this numerical procedure with calculations for a large category of problems. These problems involved spheroids of high axial ratio, sphere–cone–sphere geometries, peanut-shaped scatterers, and finite-length cylinders. Whenever possible, comparison with the T-matrix method was made, with excellent agreement.

As not only spherical multipoles but also each other set of *equivalent sources* can be used here, other names for similar concepts have been given, for example, *multiple multipole method* (MMP) [517], *discrete source method* (DSM) [518–520], and *fictitious sources method* [521]. The MMP method was developed for arbitrary three-dimensional electrodynamic problems.

9.4.3.4 Finite Difference Time Domain Technique
The finite difference time domain (FDTD) technique, pioneered by electrical engineers [522–526], is a direct implementation of Maxwell's time-dependent curl equations to solve the temporal variation of electromagnetic waves within a finite space that contains an object of arbitrary geometry and composition. In practice, the space is discretized by a grid mesh (Yee cells), and the existence of

the scattering particle is defined by properly assigning the electromagnetic constants, including permittivity, permeability, and conductivity, over the grid points. The Maxwell curl equations are subsequently discretized by using finite-difference approximations in both time and space. At the initial time a plane wave source, which does not require the harmonic condition, is turned on. The wave excited by the source will then propagate towards the scattering particle and eventually interact with it. The propagation and the scattering of the electromagnetic field are simulated by using the finite-difference analog of the Maxwell equations. Information on the convergent scattered field can be obtained when a steady-state field is established at each grid point if a continuous sinusoidal source is used, or when the field decreases to a significantly small value if a pulse source is used.

Sun and Fu [527] extended the FDTD method to light scattering by dielectric particles with large complex refractive indices to study silicon particles on silicon substrates. They checked various approximations and found that averaging the dielectric constant at grid points does not improve the accuracy of the FDTD scheme. It is recommended to use the local value of the permittivity for the FDTD simulation. Compared with Mie results for dielectric spheres with high permittivity, the errors in the extinction and absorption efficiencies are less than 4%. The errors in the scattering phase functions are less than 5%.

Sun et al. [528] used this method to simulate the scattering by an infinite dielectric column immersed in an absorbing medium.

9.5
Application of Numerical Methods to Nonspherical Nanoparticles

In this section, we give examples of the theoretical and practical application of the above and other numerical methods to nanoparticles and some experiments with nonspherical nanoparticles with shapes deviating from ellipsoids, cylinders, or cubes. As particularly nonspherical nanoparticles of noble metals have attracted considerable attention in the last couple of years, we distinguish in the following between metallic and nonmetallic nanoparticles. We start with nonmetallic nanoparticles.

9.5.1
Nonmetallic Nanoparticles

Although the maximum size of nonspherical particles slightly exceeds 1000 nm, the following two examples describe the interesting cases of *platonic bodies* and *helices*.

The single-scattering properties of the platonic shapes, namely the tetrahedron, hexahedron, octahedron, dodecahedron, and icosahedron, were investigated by use of the finite-difference time-domain method by Yang et al. [529]. These platonic shapes have different extents of asphericity in terms of the ratios of their volumes or surface areas to those of their circumscribed spheres. If R is the radius of the circumscribed sphere, R_a is the radius of the equivalent surface area sphere, R_v is the radius of the equivalent volume sphere, and $R_{eff} = 3V/A$ is the effective sphere in which V is the volume and A is the surface area of a given polyhedron, then the numerical results show that the derivations of the scattering properties of a nonspherical particle from its spherical counterpart depend on the definition of spherical equivalence. For instance, when the platonic and spherical particles have the same geometric dimensions, the phase function for a dodecahedron is more similar than that for an icosahedron to the spherical result even though an icosahedron has more faces than a dodecahedron. However, when the nonspherical and spherical particles have the same volume, the phase function of the icosahedral particle essentially converges to the phase function of the sphere, whereas the result for the dodecahedron is quite different from its spherical counterpart. Furthermore, the scattering calculations showed that the approximation of a platonic solid with a sphere based on V/A leads to larger errors than the spherical equivalence based on either volume or projected area. Note that the authors examined dielectric particles without resonances. It is not yet clear how the errors develop if the particles exhibit resonances, particularly in the case of electronic resonances.

Chiappetta and Torresani [530] applied a discrete model and a multiple scattering model to a dielectric helix whose dimensions are comparable to the incident radiation wavelength. They computed the differential scattering cross-sections of the helix. The numerical results were compared with experimental data on microwave scattering by a dielectric right-handed helix.

Quirantes and Delgado [531] used the T-matrix method to determine the size and shape in colloidal suspensions of hematite particles in water. The system was approximated by a suspension of randomly oriented, monodisperse, spheroidal particles with an equivalent volume diameter D_V and an axial ratio $e = a/b$ (with b = revolution axis). Both full and depolarized light scattering were measured and fitted to theoretical curves. Although both methods give good agreement on particle size, shape is not always characterized unequivocally with full light scattering. The problem is overcome by using depolarized light scattering, which, for spherical particles, is zero; fits between experimental depolarized light scattering and theoretical predictions give the correct value of eccentricity, and also a more accurate value of D_V.

Fournier and Evans [532, 533] made an analytic semi-empirical approximation to the extinction efficiency Q_{ext} for randomly oriented spheroids that is based on an extension of the anomalous diffraction equation. This approximation was verified for complex refractive indices, with $1.01 \leq n \leq 2.00$ and $0 \leq k \leq 1$ and aspect ratios from 0.5 to 4 in the first paper, and with $1.01 \leq n \leq \infty$ and $0 \leq k \leq \infty$ and aspect ratios from 0.2 to 5 in the second paper. The approximation is uniformly valid over all size parameters and aspect ratios and has the correct Rayleigh and large particle asymptotic behavior. The accuracy and limitations to this equation were extensively discussed by comparison with T-matrix calculations.

9.5.2
Metallic Nanoparticles

The properties of nonspherical nanoparticles of noble metals, especially those of gold and silver, have attracted considerable attention due to their wide range of applications, extending from surface-sensitive spectroscopic analysis to catalysis and microwave polarizers or biomedical applications, and their unique optical properties. The latter became obvious already from the calculations and experimental examples given in the previous chapters for spheroids, ellipsoids, infinitely long cylinders, and cubes.

In the case of spheroids and ellipsoids, at least one additional SPP resonance extinction band appeared at long wavelengths compared with the spherical particles with its single short-wavelength peak. From this we can expect that for all elongated particle geometries (finite cylinders, rods, disks) we obtain a similar optical response. The actual number of peaks, peak positions, and halfwidths will depend on the corresponding particle geometry.

In the case of cubes and for infinitely long cylinders also blue-shifted resonances with respect to the resonance of spherical particles were obtained. They can be expected for non-elongated particle geometries in addition to long-wavelength resonances.

In the following, we give some theoretical and experimental results on resonant nonspherical particles of gold and silver.

A nice theoretical study on the optical properties of silver cubes and decahedrons and some truncated cubic and decahedral morphologies was carried out by González and Noguez [534] using the DDA. For calculation, the size of the particles was assumed so that all particles have the same volume, namely the volume of a sphere with $2R = 20$ nm. This means for the cube that $D = 16$ nm. The calculated spectrum of the cube exhibits up to six more or less resolved resonances. All resonances are red shifted compared with the spherical particle. This is in contrast to the results obtained in Section 9.3.2 for cubes in the Rayleigh approximation, although the particles in this study can also be treated in the Rayleigh approximation. The red shift of *all* resonances is therefore not understandable. Truncation of the cube at the edges leads to a drastic reduction of the number of resonances and a blue shift of the remaining main peak with increasing truncation. In the second part of this study, regular decahedrons and their truncated morphologies were examined and compared with a sphere. The spectra of the decahedron and truncated decahedrons exhibit resonances at longer wavelengths than the sphere. The spectra are polarization dependent. With increasing truncation, again the number of resonances is drastically reduced and the remaining main peak shifts to shorter wavelength in the direction of the spherical particle.

Kottmann et al. [535, 536] investigated numerically the spectrum of SPP resonances for silver nanoparticles with a nonregular cross-section in the range 20–50 nm. The particles had geometric cross-sections corresponding to that of a circle, a hexagon, a pentagon, a square, and a triangle. The diameter of the circle was 20 or 50 nm, and all the particles had the same area, so that the optical cross-sections should be comparable. The spectra were calculated using a finite-element method described in [537] and are shown for 50 nm diameter in Figure 9.49.

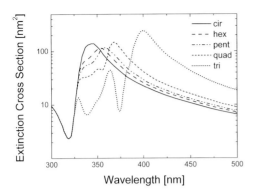

Figure 9.49 Extinction cross-sections of a circle, a hexagon, a pentagon, a square, and a triangle with 50 nm circle diameter. After [536].

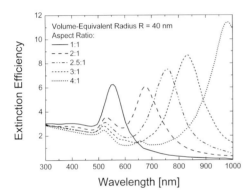

Figure 9.50 Extinction spectra of spheroidal Au nanospheroids with a volume equivalent radius of 40 nm. After [538].

The number of resonances increases strongly when the section symmetry decreases: A cylindrical wire exhibits one resonance, whereas we observe more than five distinct resonances for a triangular particle. The spectral range covered by these different resonances becomes very large, leading to distinct particle-specific colors.

Yin et al. [538] simulated optical properties (extinction efficiencies) of Au nanoparticles using the DDA. The influence of the nanoparticle size and shape on the SPP bands was investigated. For spherical particles the results are directly comparable to Mie's theory. For spheroidal and rod-like nanoparticles, there are two peaks in the extinction spectra corresponding to the transverse and longitudinal resonances. With increasing aspect ratio, the peak position of the transverse mode shifts to shorter wavelengths whereas that of the longitudinal mode shifts to longer wavelengths. The numerical results explain well the experimental data with nanorods and ellipsoidal particles. An example for spheroidal particles with a volume equivalent radius $R = 40$ nm is given in Figure 9.50, and an example for nanorods with the same volume equivalent radius is given in Figure 9.51.

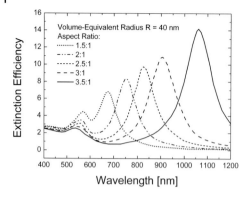

Figure 9.51 Extinction spectra of Au nanorods with a volume equivalent radius of 40 nm. After [538].

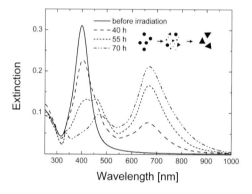

Figure 9.52 Time-dependent UV–VIS spectra showing the conversion of silver nanospheres to nanoprisms. The photoinduced conversion process is sketched in the inset. Data courtesy of G. C. Schatz and R. Jin.

Jin et al. [539] used a photo-process to convert large quantities of silver nanospheres into triangular nanoprisms. This was characterized by time-dependent UV–VIS spectroscopy and TEM, allowing the observation of several intermediate steps of the conversion process. The light-driven process results in a colloid with distinctive optical properties that relates directly to the nanoprism shape of the particles. Calculations with the DDA coupled with experimental observations allowed the nanoprism dipole and quadrupole SPP bands to be assigned. Unlike the spherical particles, the nanoprisms exhibit scattering in the red region. An example for time-dependent UV–VIS spectra is given in Figure 9.52.

Aizpurua et al. [540, 541] investigated the optical response of ring-shaped gold nanoparticles with radii in the order of about 60 nm and heights of ~20 and ~40 nm, prepared by colloidal lithography. The thickness of the ring wall was estimated to be 14 ± 2 nm. Figure 9.53a shows TEM images of the prepared rings.

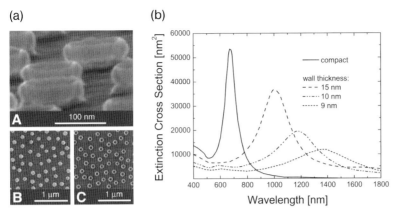

Figure 9.53 (a) Scanning electron micrographs of gold nanorings and nanodisks prepared by colloidal lithography. (A) 80° tilt image of a ring structure. The walls of the rings are thin enough for the 30 keV electrons to pass through. (B, C) Top views of disks and rings. Copyright (2003) American Physical Society. (b) Experimental extinction spectra of solid disks and rings with estimated wall thickness of the rings $d = 14 \pm 2$, 10 ± 2, and 9 ± 2 nm. The rings exhibit NIR features at larger wavelengths for thinner walls. The disks show a dipolar excitation at around 700 nm.

Compared with solid gold particles of similar size, the nanorings exhibit a redshifted localized SPP that can be tuned over an extended wavelength range by varying the ratio of the ring thickness to its radius. Exemplary spectra are shown in Figure 9.53b. The measured wavelength variation is well reproduced by numerical calculations and interpreted as originating from coupling of dipolar modes at the inner and outer surfaces of the nanorings. For the calculations, Maxwell's equations were solved using the boundary element method (BEM) [542, 543], in which the electromagnetic field is expressed in terms of charges and currents distributed on the surfaces and interfaces of the structure under consideration. The customary boundary conditions for the electromagnetic field provide a set of linear integral equations, with charges and currents as unknowns, which are solved self-consistently in the presence of the external incident light field by discretizing the integrals. The electric field associated with these SPPs exhibits uniform enhancement and polarization in the ring cavity, suggesting applications in NIR surface-enhanced spectroscopy and sensing.

The above experimental and numerical examples showed that control over not only particle size but also shape is of major interest for scientists involved in the preparation of nanostructured materials since the electronic, magnetic, and catalytic properties of these particles depend on their size and shape. A certain class of nonspherical, elongated nanoparticles that have attracted considerable attention in the last few years is *nanorods*, as testified by the many papers on this subject. We cannot refer here to all publications, but cite some selected papers [544–561]. In the following, we report on numerical and experimental studies in which the

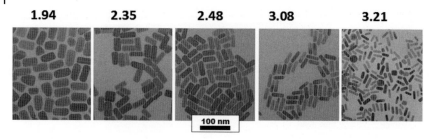

Figure 9.54 Transmission electron micrographs of nanorods with increasing aspect ratio length/width. Reproduced from [562]. Copyright 2004, by permission of Elsevier.

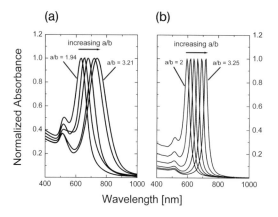

Figure 9.55 (a) Optical absorption spectra of nanorods with increasing aspect ratio. (b) BEM calculations for the same nanorods. Data courtesy of Luis M. Liz-Marzan.

above numerical methods and others were applied to simulate the optical response of nanorods.

Pérez-Juste et al. [562, 563] provided an overview of current research into the synthesis and properties of gold nanorods. Different synthetic strategies are described that have been developed to achieve decent yields and sample monodispersity. Some of the most innovative research dealing with surface modification and chemical reactivity of gold nanorods is highlighted, together with new directions such as the synthesis of core–shell particles and the interactions of gold nanorods with biomolecules. Initial results on the optical properties of nanocomposites consisting of thin films and gels with incorporated gold nanorods are reviewed. The review is concluded with a section devoted to the future perspectives for gold rods as novel materials.

Here, we present selected results for gold nanorods with increasing aspect ratio length/width = a/b. Figure 9.54 shows typical electron micrographs of the prepared nanorods.

Optical absorption spectra were recorded in the VIS–NIR spectral region from 400 to 1000 nm. They are shown in Figure 9.55a.

For nanorods of gold (and silver), a dominant SPP band corresponding to the longitudinal resonance can be observed. Its maximum position λ_{max} shifts to the red as the aspect ratio a/b increases, whereas the transverse resonance shifts slightly to shorter wavelengths. These observations are common for all nanorods of gold and silver. Similar behavior can be found for prolate spheroids, as already demonstrated in Section 9.1. However, the shape of nanorods clearly differs from that of spheroids; the latter do not have edges and corners. Therefore, simulations of the optical properties of nanorods need numerical approaches. Pérez-Juste et al. [562] used the BEM for computation of the optical absorption spectra of nanorods. The BEM is a rigorous result derived from vector diffraction theory. This method starts by expressing the electromagnetic field scattered by a nanoparticle in terms of boundary charges and currents, which, upon imposing the customary boundary conditions for the continuity of the parallel components of the electric and magnetic fields, leads to a system of surface-integral equations. This system is solved by discretizing the integrals using a set of N representative points distributed at the boundaries, so that it turns into a set of linear equations that are solved numerically by standard linear algebra techniques. The results of the calculations are depicted in Figure 9.55b. They clearly reproduce the measured spectra with respect to the peak positions. The halfwidth of the longitudinal resonance is larger in the measured spectra, resulting from the size and shape distribution. This distribution was not considered in the calculations.

A critical comparison of theoretical methods for predicting and understanding the optical response of gold nanoparticles is provided in a tutorial review by Myroshnychenko et al. [564]. It assists the reader in making a rational choice for each particular problem, while analytical models provide insights into the effects of retardation in large particles and non-locality in small particles.

Based on a previous study [565], Chau et al. [566] investigated numerically the interaction between an incident wave and an Au nanorod in air using a finite-element method. Various effects due to the variation of the aspect ratio of the rod and the polarization direction of the fields on the optical scattering and absorption cross-sections, and also the locally induced charge densities at the surfaces of the rod, are discussed in detail. Simulation results show good agreement with the experimental optical images presented in [567].

Alekseeva et al. [568] described optical monitoring of the synthesis of gold nanorods based on seed-mediated growth in the presence of the soft surfactant template cetyltrimethylammonium bromide. To separate nanorods from spheres and surfactants they fractionated samples in a density gradient of glycerol. The optical properties of the nanorods were characterized by extinction and differential light-scattering spectra (at 90°, 450–800 nm) and by the depolarization light-scattering ratio, I_{vh}/I_{vv}, measured at 90° with a helium–neon laser. Theoretical spectra and the I_{vh}/I_{vv} ratios were calculated by the T-matrix method as applied to randomly oriented nanorods, specially adapted to a new particle s-cylinder model (circular cylinders with semispherical ends). The simulated data were fitted to experimental observations by use of particle length and width as adjustable parameters, which were close to the data yielded by TEM.

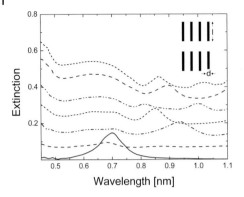

Figure 9.56 Extinction spectra for different cylinder lengths l = 91, 320, 580, 850 1100, 1340, and 1620 nm from bottom to top. For better presentation the spectra are shifted along the ordinate by 0.005. The value of d is 91 nm.

Brioude et al. [569] simulated the absorption efficiency of gold nanorods using the DDA and taking into account the real shape of the gold nanorods. The data are in good agreement with previous theoretical work based on classical electrostatic predictions and assuming that gold nanorods behave as ellipsoidal particles. From the experimental point of view, good agreement with the published data for gold nanorods was obtained.

Schaich et al. [570] studied the transmission of light along the surface normal through an air–quartz-glass interface covered with a periodic array of thin finite cylinders of gold. Optical absorption spectra were recorded over the visible to the infrared spectral range. Various structures with varying length l were observed. Figure 9.56 shows measured spectra as a function of the length l at a constant cylinder thickness d. A method of moments calculation scheme provided simulations in good quantitative agreement with the data.

10
Beyond Mie's Theory II – The Generalized Mie Theory

When nanoparticles are more or less strongly pinned to each other by interaction forces (e.g., van der Waals forces or chemical bonds), the scattering and absorption by this particle aggregate must differ from that of the primary particles which form the aggregate.

In this chapter, we consider the case of light scattering and absorption by aggregates of spheres, which is the most evolved model for interacting particles. For the simplest case of the nanoparticle dimer, a complete solution was given already in 1935 by Trinks [571], in 1965 and 1970 in the Rayleigh approximation by Olaofe [572, 573], and in 1967 by Liang and Lo [574]. Bruning and Lo [575] were the first to provide a comprehensive computationally viable solution to the problem of cooperative scattering by two large spheres.

In 1972, Langbein [465] developed an electrodynamic model for N interacting molecules to explain van der Waals attraction among molecules. The case of N spherical particles with $N > 2$ was first investigated by Borghese *et al.* [576] in 1979. Gérardy and Ausloos [577–580] presented a complete solution of the light scattering by an arbitrary assembly of N spherical particles embedded in some arbitrary homogeneous matrix material. They adopted the model of Langbein, applied it to spherical particles, extended it to arbitrary positions of the individual spheres using transformation rules for the spherical harmonics given by Jeffreys [581], and applied it to sodium nanospheres. We call this model the *generalized Mie theory* (GMT). It is currently the most accurate approach to describing spherical particle aggregates without any approximations. It includes multipoles and retardation effects within the particle in the electrodynamic case.

The first numerical evaluations of this theory applied to silver and gold nanoparticles in colloidal suspensions were carried out by Quinten *et al.* [114, 582–585].

In a series of papers, Fuller *et al.* [586–592], Hamid *et al.* [593–597], Mackowski [598, 599], Xu *et al.* [600–604], and Rouleau and Martin [605, 606] published models for the light scattering by aggregates of spheres that yielded the same results as the model of Gérardy and Ausloos. Unlike directly imposing Maxwell's boundary conditions on the surface of each particle in the aggregate, Ioannidou *et al.* [607] used the indirect mode matching (IMM) method to obtain a compact, yet exact, solution of the light scattering by aggregates. Vagov *et al.* [608] developed a technique in the quasi-static regime for the calculation of the optical response of arrays of spheres from the theory of hypercomplex variables.

10.1
Derivation of the Generalized Mie Theory

We now introduce the essentials of the theory of Gérardy and Ausloos [578]. We consider $i = 1, \ldots, N$ spherical particles with arbitrary sizes $2R_i$ lumped into an aggregate of arbitrary topography in which they are separated with center-to-center distances $d_{ij} \geq R_i + R_j$ (Figure 10.1). A plane electromagnetic wave at wavelength λ propagating in a nonabsorbing surrounding medium is scattered and absorbed by each particle to a certain amount.

The electromagnetic fields E and H outside the aggregate are assumed to result from a linear superposition of the incident fields and the scattered fields:

$$E = E_{\text{inc}} + \sum_{j=1}^{N} E_{\text{sca}}(j) \tag{10.1}$$

$$H = H_{\text{inc}} + \sum_{j=1}^{N} H_{\text{sca}}(j) \tag{10.2}$$

They are expanded to according even and odd vector spherical harmonics $M_{\substack{e \\ o}nm}$ and $N_{\substack{e \\ o}nm}$ as introduced in Equations 5.2 and 5.3 in Section 5.1.

As for N neighboring spheres the symmetry in the scattering problem is less than that for a single sphere, also vector harmonics with $m \neq 1$ contribute to the field expansions:

incident wave at particle j:

$$E_{\text{inc}}(j) = \exp(ik_M r_j) \sum_{n=1}^{\infty} \sum_{m=-n}^{n} E_{nm} \left[M_{onm}^{(1)}(k_M) - iN_{enm}^{(1)}(k_M) \right] \tag{10.3a}$$

$$H_{\text{inc}}(j) = \frac{-k_M}{\omega \mu_0} \exp(ik_M r_j) \sum_{n=1}^{\infty} \sum_{m=-n}^{n} E_{nm} \left[M_{onm}^{(1)}(k_M) - iN_{enm}^{(1)}(k_M) \right] \tag{10.3b}$$

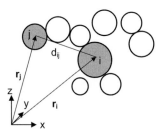

Figure 10.1 Schematic diagram of an aggregate of spherical particles.

scattered wave of particle j:

$$E_{sca}(j) = \sum_{n=1}^{\infty} \sum_{m=-n}^{n} E_{nm} \left[\beta_{nm}(j) M^{(3)}_{onm}(k_M) - i\alpha_{nm}(j) N^{(3)}_{enm}(k_M) \right] \quad (10.4a)$$

$$H_{sca}(j) = \frac{-k_M}{\omega\mu_0} \sum_{n=1}^{\infty} \sum_{m=-n}^{n} E_{nm} \left[\alpha_{nm}(j) M^{(3)}_{onm}(k_M) - i\beta_{nm}(j) N^{(3)}_{enm}(k_M) \right] \quad (10.4b)$$

inside the particle j:

$$E(j) = \sum_{n=1}^{\infty} \sum_{m=-n}^{n} E_{nm} \left[\delta_{nm}(j) M^{(3)}_{onm}(k) - i\gamma_{nm}(j) N^{(3)}_{enm}(k) \right] \quad (10.5a)$$

$$H(j) = \frac{-k}{\omega\mu_0} \sum_{n=1}^{\infty} \sum_{m=-n}^{n} E_{nm} \left[\gamma_{nm}(j) M^{(3)}_{onm}(k) - i\delta_{nm}(j) N^{(3)}_{enm}(k) \right] \quad (10.5b)$$

with the abbreviation

$$E_{nm} = \frac{1}{2} i^n \frac{2n+1}{n(n+1)} E_0 \delta_{m,\pm 1} \quad (10.6)$$

where $\delta_{m,\pm 1}$ is the Kronecker symbol. These prefactors E_{nm} given in Equation 10.6 belong to a plane wave incident along the positive z-axis with the electric field polarized along the x-axis. For other directions of incidence and polarization they can be calculated according to a transformation rule given by Gérardy and Ausloos [578] in Equation 56 of their paper. The extra phase factor $\exp(ik_M r_j)$ in Equation 10.3 takes into account that the phase of the incident wave is different at different locations r_j of the particles in the aggregate with respect to a reference frame.

As the spherical vector harmonics M_{enm}, M_{onm}, N_{enm}, and N_{onm} represent a complete set of orthogonal functions, it is possible to expand the vector functions of the scattered wave of particle j into a series of vector functions describing an incident wave at the site of particle i, using vector addition theorems (e.g., [239, 240, 609].):

$$M^{(3)}_{nm}(j) = \sum_{q=1}^{\infty} \sum_{p=-q}^{q} S_{nmqp}(i,j) M^{(1)}_{qp}(i) + T_{nmqp}(i,j) N^{(1)}_{qp}(i) \quad (10.7)$$

$$N^{(3)}_{nm}(j) = \sum_{q=1}^{\infty} \sum_{p=-q}^{q} S_{nmqp}(i,j) N^{(1)}_{qp}(i) + T_{nmqp}(i,j) M^{(1)}_{qp}(i) \quad (10.8)$$

These transformations are the essential base of the theory, and the main problem in numerical evaluation, since they introduce higher multipoles than necessary for the evaluation of the light scattering by a single particle.

Inserting Equations 10.3–10.8 in Maxwell's boundary conditions at the surface of particle i, two sets of linear equations follow for the expansion coefficients α_{nm} and β_{nm} of the scattered wave of particle i:

$$\alpha_{nm}(i) = a_n(i) \exp(ik_M r_i) + a_n(i) \sum_{\substack{j \neq i}}^{N} \sum_{q=1}^{\infty} \sum_{p=-q}^{q} \alpha_{qp}(j) S_{nmqp}(i,j)$$

$$+ \beta_{qp}(j) T_{nmqp}(i,j) \quad (10.9)$$

$$\beta_{nm}(i) = b_n(i)\exp(i\mathbf{k}_M\mathbf{r}_i) + b_n(i)\sum_{\substack{j=1 \\ j\neq i}}^{N}\sum_{q=1}^{\infty}\sum_{p=-q}^{q}\alpha_{qp}(j)T_{nmqp}(i,j)$$
$$+ \beta_{qp}(j)S_{nmqp}(i,j) \qquad (10.10)$$

The coefficients $a_n(i)$ and $b_n(i)$ are the corresponding scattering coefficients of the single isolated sphere i for the TM mode and TE mode of order n following from Mie's theory. As these coefficients enter the calculations only as a factor, the interaction among coated spheres and hollow spheres is already included.

The transformation matrices $\underline{\underline{T}}$ and $\underline{\underline{S}}$ with matrix elements $T_{nmqp}(i,j)$ and $S_{nmqp}(i,j)$ describe the transformation of the multipole (p,q) in the scattered wave of particle j into the multipole (n,m) of the incident field at the site of particle i. In this way, all multipoles of the fields of one particle are coupled to all multipoles of the fields of a second particle. (Due to their different positions, there is no orthogonality rule for the different multipoles.)

The matrix elements depend on the particle sizes R_i and R_j and the relative coordinates d_{ij}, θ_{ij} and φ_{ij} between two spheres i and j. They decrease with increasing center-to-center distance d_{ij} and so do the electrodynamic interactions, and become negligible at distances $d_{ij} \geq 5(R_i + R_j)$ [610]. Then, the following cross-sections reduce to the sum over cross-sections of N isolated spheres.

For determination of the extinction and scattering cross-sections of the whole aggregate with N spherical particles we use a conceptual sphere, the size of which is chosen so large that the aggregate is completely enclosed. Inside this conceptual sphere the fields are superposed coherently. Therefore, all special effects, as for example the weak localization and the occurrence of *hot spots*, that is, simply interference maxima of the fields at the positions of particular particles or in between them, are included. The cross-sections then follow from Poynting's law as

$$\sigma_{\text{ext}}(N) = \frac{2\pi}{k_M^2}\sum_{i=1}^{N}\sum_{n=1}^{\infty}\sum_{m=-n}^{n}\text{Re}[\alpha_{nm}(i) + \beta_{nm}(i)] \qquad (10.11)$$

$$\sigma_{\text{sca}}(N) = \frac{2\pi}{k_M^2}\sum_{i=1}^{N}\sum_{n=1}^{\infty}\sum_{m=-n}^{n}|\alpha_{nm}(i)|^2 + |\beta_{nm}(i)|^2$$
$$+ \frac{2\pi}{k_M^2}\sum_{i=1}^{N}\sum_{n=1}^{\infty}\sum_{m=-n}^{n}\text{Re}\left\{\alpha_{nm}^*(i)\left[1 - \frac{\alpha_{nm}(i)}{a_n(i)}\right]\right.$$
$$\left. + \beta_{nm}^*(i)\left[1 - \frac{\beta_{nm}(i)}{b_n(i)}\right]\right\} \qquad (10.12)$$

The corresponding phase functions or scattering intensities i_p^N and i_s^N parallel and perpendicular to the scattering plane for an N-particle aggregate are obtained as

$$i_p^N(\theta) = \left|\frac{1}{N}\sum_{i=1}^{N}\sum_{n=1}^{\infty}\sum_{m=-n}^{n}[\alpha_{nm}(i)\tau_{nm}(\theta) + \beta_{nm}(i)\pi_{nm}(\theta)]\right|^2 \qquad (10.13)$$

$$i_s^N(\theta) = \left|\frac{1}{N}\sum_{i=1}^{N}\sum_{n=1}^{\infty}\sum_{m=-n}^{n}[\alpha_{nm}(i)\pi_{nm}(\theta) + \beta_{nm}(i)\tau_{nm}(\theta)]\right|^2 \qquad (10.14)$$

The angular functions $\tau_{nm}(\theta) = \partial P_{nm}/\partial\theta$, and $\pi_{nm}(\theta) = mP_{nm}/\sin\theta$ satisfy recurrence relations given in detail in [502], and can be computed for each scattering angle θ and orders n and m. $P_{nm}(\cos\theta)$ are associated Legendre polynomials.

Most critical for the numerical evaluation of the above equations is the summation over the multipole order n, since it runs from 1 to infinity. However, analogous to the Mie theory, it can be restricted to a maximum order n_{MAX} for numerical reasons. Then, a set of $2Nn_{MAX}(n_{MAX} + 2)$ linear equations has to be solved. This computational complexity is the main shortcoming of the GMT. For large N and n_{MAX}, the solution of this system of equations requires prohibitively large computer storage and high levels of computer time.

In the long-wavelength limit, considerable simplifications can be made. Claro [611] and Claro and Fuchs [612] proposed a method based on the calculation of normal modes of electromagnetic excitations in an array of small particles. This method works well for aggregates of separated particles, but it was found [611] that the multipolar expansion used did not converge for touching particles.

10.2
Resonances

If there are MDRs or electronic resonances present in the individual particles, the electromagnetic coupling among the particles leads to new resonances, similarly to coupled mechanical oscillators. The maximum number of aggregate resonances is $2Nn_{MAX}(n_{MAX} + 2)$ because we have N oscillators with $2n_{MAX} + 1$ eigenmodes for each TM and TE polarization. The probability of excitation, however, differs strongly for all these resonances. Depending on the symmetry of the aggregate, part of them will be degenerated.

For the purpose of demonstrating the effects of electromagnetic coupling on resonances of the primary particles, we will discuss pairs and linear chains of small spheres with only dipole–dipole coupling. As these primary particles are small compared with the wavelength only electronic resonances may occur, and MDRs are excluded. In detail, we restrict considerations to surface plasmon polaritons (SPPs) in metal nanospheres; the discussion for other electronic resonances is similar. Moreover, as we want to discuss densely packed particles, we restrict considerations in this section to the *near-field zone* for the electric fields, that is, the electric field of a dipole with dipole moment p is approximately

$$E_{dip} = \frac{3r_0(r_0 \cdot p) - p}{4\pi\varepsilon_M\varepsilon_0 r^3} \qquad (10.15)$$

where the vector $r_o = r/r$ is the dimensionless unit vector in the direction of the line between the centers of both particles.

Two neighboring spherical particles and also linear chains of spherical particles can be arranged in principle along the three axes of a Cartesian coordinate system as depicted in Figure 10.2.

Assuming the incident light is propagating along the z-axis and being polarized along the x-axis, the induced dipole moments $p(1)$ and $p(2)$ are parallel to r_o in Figure 10.2a, so that

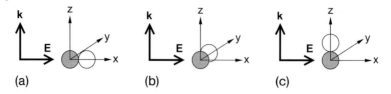

Figure 10.2 Two sphere aggregates (dimers) in the field of an electromagnetic wave with **E** parallel to **x** and **k** parallel to **z**.

$$E_{\text{dip}}(i) = \frac{2p(i)}{4\pi\varepsilon_M\varepsilon_0 d^3} \quad i=1,2 \tag{10.16}$$

In Figure 10.2b and c, however, the induced dipole moments are perpendicular to r_o, so that

$$E_{\text{dip}}(i) = -\frac{p(i)}{4\pi\varepsilon_M\varepsilon_0 d^3} \quad i=1,2 \tag{10.17}$$

where d is the center-to-center distance of the two particles. Both cases in Equations 10.16 and 10.17 can be summarized by introducing a factor F_μ:

$$F_\mu = \begin{cases} +2 & \mu=1 \\ -1 & \mu=2 \end{cases} \tag{10.18}$$

The dipole moments $p(1)$ and $p(2)$ now are coupled in two linear equations:

$$p(1) = \varepsilon_0\varepsilon_M\alpha(1)[E_{\text{inc}} + E_{\text{dip}}(2)] \tag{10.19}$$

$$p(2) = \varepsilon_0\varepsilon_M\alpha(2)[E_{\text{inc}} + E_{\text{dip}}(1)] \tag{10.20}$$

The result of this set of equations can also be expressed in terms of a dipole polarizability α_{pair} of the pair defined via

$$p_{\text{pair}} = p(1) + p(2) = \varepsilon_0\varepsilon_M\alpha_{\text{pair}}E_{\text{inc}} \tag{10.21}$$

Taking into account that the particles may be of different size, α_{pair} is given for two spherical particles of different sizes as

$$\alpha_{\text{pair}} = \frac{\alpha(1)+\alpha(2)+2\alpha(1)\alpha(2)\dfrac{F_\mu}{4\pi d^3}}{1-\alpha(1)\alpha(2)\left(\dfrac{F_\mu}{4\pi d^3}\right)^2} \tag{10.22}$$

and depends on the polarization (different results for $\mu = 1$ and $\mu = 2$).

The solution for Figure 10.2a defines the *longitudinal eigenmode* of the pair, while the solution for Figure 10.2b or c defines the two degenerated *transverse eigenmodes* of the pair. This classification into longitudinal and transverse modes can be made also for linear chains but fails for more complex aggregate topologies. For unpolarized incident light, both eigenmodes contribute to certain amounts to the extinction and scattering cross-section: 1/3 for the longitudinal eigenmode ($\mu = 1$) and, due to the degeneration, 2/3 for the transverse eigenmode ($\mu = 2$).

In the following we restrict considerations to identical spheres. Then, α_{pair} is

$$\alpha_{\text{pair}} = \frac{2\alpha}{1 - F_\mu \left(\frac{R}{d}\right)^3 \frac{\varepsilon - \varepsilon_M}{\varepsilon + 2\varepsilon_M}} \tag{10.23}$$

for each μ. It is obvious that the pair polarizability takes its maximum if the denominator vanishes, that is, if

$$\varepsilon = -2\varepsilon_M \frac{1 + \frac{1}{2} F_\mu \left(\frac{R}{d}\right)^3}{1 - F_\mu \left(\frac{R}{d}\right)^3} \tag{10.24}$$

This resonance condition differs from the condition for the single sphere, $\varepsilon = -2\varepsilon_M$, which is obtained in the case of the pair for large center-to-center distances d. For example, for $d = 2R$ the resonance of the longitudinal mode ($\mu = 1$) and the transverse mode ($\mu = 2$) of the pair occur at $\varepsilon = -3\varepsilon_M$ and at $\varepsilon = -1.66\varepsilon_M$, respectively.

In fact, there are two *in-phase* and two *opposite-phase* modes, the latter being excited by the retardation effect of the incident light. We assume – corresponding to the *quasi-static* approximation applied in this example – that retardation can be neglected and then only the *in-phase* modes are relevant for optical excitation and thus contribute to the spectrum.

Extending the above procedure to linear chains of N identical particles, one obtains for the chain polarizability for each μ

$$\alpha_{\text{chain}} = \frac{N\alpha}{1 - F_\mu(N) \left(\frac{R}{d}\right)^3 \frac{\varepsilon - \varepsilon_M}{\varepsilon + 2\varepsilon_M}} \tag{10.25}$$

with F_μ depending on N. In Table 10.1 we summarize the results for $F_\mu(N)$ for $N = 2, 3, 4, 16$, and ∞. Again, the possible excitation modes are reduced to the optically relevant *in-phase* modes. The *opposite-phase* modes are neglected here.

Another approach in the quasi-static approximation was developed by Clippe and co-workers [613, 614] that enables us to treat larger arrays of spherical particles provided that their symmetry is high (linear chains, tetrahedra, etc.) and neighboring particles are almost touching. However, instead of the explicit absorption spectra which can be derived from the cross-sections, only the positions of the absorption peaks are obtained. In this model, the particles are replaced by an undamped model harmonic oscillator with simple resonance behavior $1/(1 - \omega^2/\omega_r^2)$. The particular material properties without any resonance behavior of their own are then re-introduced by a function $M(\omega)$, so that the dipole moment p of a particle is

$$\boldsymbol{p} = e\boldsymbol{q} = 4\pi\varepsilon_0 R^3 \frac{M(\omega)}{1 - \frac{\omega^2}{\omega_r^2}} \boldsymbol{E}(\omega) \tag{10.26}$$

Table 10.1 Resonance factor $F_\mu(N)$ for linear chains of N spherical particles:

$$F_\mu(N) = \begin{Bmatrix} +2 \\ -1 \end{Bmatrix} f(N) \begin{cases} \mu = 1 \\ \mu = 2 \end{cases}$$

N	$f(N)$[a]	Reference
2	1.000	[112, 113, 578]
3	$1.478 = \dfrac{-32}{1-\sqrt{513}}$	[112, 113, 578]
4	$1.744 = \dfrac{-243}{1-\sqrt{19684}}$	[112, 578]
16	2.312	[112, 578]
∞	$2.4042 = 2\zeta(3)$	[577]

a) $\zeta(x)$ is Riemann's zeta function.

Coupling such particles to form a highly symmetric aggregate, the eigenvalues of the Hamiltonian operator of the particle system give the resonance frequencies of the coupled system. The Hamilton operator of the particle system is given by

$$H = \frac{1}{8\pi\varepsilon_0} \sum_i \frac{e_0^2}{M(\omega)R^3\omega_r^2} \left[\left(\frac{dq_i}{dt}\right)^2 + \omega_r^2 q_i^2\right] + \frac{e_0^2}{8\pi\varepsilon_0} \sum_{i,j} q_i T_{ij} q_j \quad (10.27)$$

where q_i and q_j are generalized coordinates and T_{ij} is the dipole tensor:

$$T_{ij} = \left(-\frac{\partial^2}{\partial R_{ij}^2} \frac{1}{R_{ij}}\right) \quad (10.28)$$

with eigenvalues λ_μ. The important feature here is that these eigenvalues are determined only by the aggregate geometry and are independent of the particle material properties. After solving the eigenvalue problem, the material properties are incorporated again via $M(\omega)$ to obtain the appropriate eigenfrequencies, which follow from

$$\omega_\mu = \omega_r \sqrt{1 + \lambda_\mu M(\omega_\mu)} \quad (10.29)$$

It has been proven that there is a simple relationship between the eigenvalues $\lambda_\mu(N)$ and the $F_\mu(N)$ of an aggregate with N primary particles:

$$F_\mu(N) = -8\lambda_\mu(N) \quad (10.30)$$

As an example, resonance positions for aggregates of small Au nanoparticles calculated from Clippe's theory (CELA) and for comparison with GMT restricted to dipole–dipole coupling only are shown in Figure 10.3. There are two main positions, the long-wavelength positions varying strongly with aggregate shape, and the short-wavelength positions, which are always fixed at nearly the same position close to the interband absorption edge. Clippe's more approximate theory differs

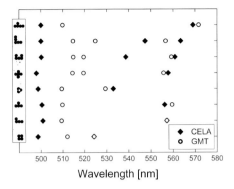

Figure 10.3 Comparison of the theory of undamped harmonic oscillators [613, 614] and the GMT restricted to dipole–dipole coupling only [113].

from the GMT for the long-wavelength positions. One essential difference is the neglect of damping in the Clippe model, which is particularly important close to the interband transition threshold.

10.3
Common Results

In the following subsections we discuss results obtained from the GMT which are common for all aggregates and particle materials. Extinction and scattering cross-sections of aggregates of mostly silver nanoparticles were calculated because the most prominent effects can be expected for primary particles exhibiting resonances. The particles are suspended either in vacuum or in water, two embedding media which are often found in nanoparticle aggregate systems. As the particles are small compared with the wavelength, n_{MAX} is restricted to $n_{MAX} = 2$ in the following calculations, except where higher multipole orders are required.

10.3.1
Influence of Shape

In Figure 10.4, computed extinction and scattering cross-section spectra are displayed for particle aggregates with $N = 5$ identical silver spheres with diameter $2R = 40\,nm$ at wavelengths ranging from 0.3 to 1 µm. For better presentation, the spectra are shifted along the ordinate by arbitrary figures added to the spectra, maintaining the important information on the shape of the spectra. Otherwise, the spectra would be clumped in a narrow region of extinction values. Particularly in the interband transition region all spectra would lie one after another. For comparison, the single particle spectra are also plotted as the bottom-most spectra. The aggregate spectra are averaged over different directions of incidence of the incident plane wave and over both polarizations of the incident wave.

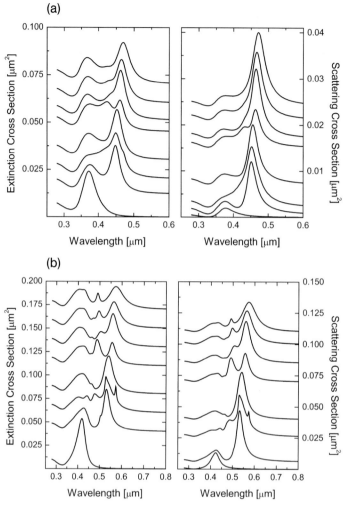

Figure 10.4 Extinction and scattering cross-section spectra of aggregates with $N = 5$ identical silver spheres of diameter $2R = 40$ nm. (a) Embedding medium vacuum; (b) embedding medium water. The next-neighbor distance d_{ij} from center-to-center of two adjacent particles is $d_{ij} = 2R$. In each case from top to bottom, the spectra belong to the following particle arrangements in the aggregate: ●●●●● ●●●⋮ ●●⋮● ⋮●⋮ ●⋮● ⋈ ✣.

The spectrum of the single silver nanoparticle exhibits the well-known SPP resonance, both in extinction and in scattering. If aggregation occurs, this resonance of the TM dipole mode splits into many new resonances, mainly contributing to extinction and scattering at longer wavelengths (low photon energies). The splitting depends sensitively on the topology of the aggregate and it turned out that the largest splitting, that is, the difference in the wavelengths at which the most prominent modes are peaked, occurs for chain-like aggregates, while the smallest peak splitting occurs for the most densely packed aggregates [610]. The number of maxima and their wavelength positions are similar in scattering and extinction spectra. However, the long-wavelength modes contribute more to the scattering than the short-wavelength modes. The latter contribute more to the absorption. The interband transition region in extinction and scattering is almost unaffected by aggregation since electron–hole excitations are less sensitive to geometries.

In detail, when evaluating the magnitudes of the cross-sections, the total extinction by the aggregates is less than that of five isolated spheres at wavelengths where the SPP is peaked. On the other hand, at long wavelengths where the SPP of the single particle has almost vanished, the extinction by the aggregates is still large, meaning an enormous enhancement of extinction and scattering in this spectral range. This enhancement effect is demonstrated in Section 10.3.4.

A general categorization of the demonstrated topology effect is impossible since the spectral features depend on the dielectric function of the particle material. For another set of optical constants the spectra may differ from those shown in Figure 10.4. Hence it also becomes impossible to invert spectra to determine the topology of the aggregate.

10.3.2
Influence of Length

On increasing the number of particles N in a chain-like aggregate, the peak splitting runs into saturation, approaching a maximum peak splitting. This can be seen in Figure 10.5, where the peak positions of the most prominent maxima in the extinction spectrum of a linear chain are plotted versus the number of particles in the chain. The peak positions are extracted from computations on chains of identical touching silver particles *in vacuo* and in water, varying N from 2 to 30. The different curves belong to particles with different single particle size $2R$. As a consequence of this effect, extinction structures at still longer wavelengths cannot be explained by interacting particles but must arise from aggregates in which coalescence led to a changed topography. This is demonstrated in Chapter 11.

10.3.3
Influence of Interparticle Distance

The peak splitting also depends on the next-neighbor distance. It decreases with increasing d_{ij}, and for $d_{ij} \geq 5(R_i + R_j)$ the corresponding spectrum hardly differs from the single particle extinction spectrum, meaning that the interaction among the particles no longer affects optical extinction. This is illustrated in Figure 10.6 for a linear chain with $N = 10$ identical silver spheres. In a regular arrangement of equally sized particles in a volume V, the distance $d_{ij} = 10R$ means a volume fraction of particulate matter or filling factor $f = 0.019$.

Experimentally, Rechberger *et al.* [615] studied the SPP excitation in pairs of identical Au nanoparticles by TEM. They produced samples by electron beam

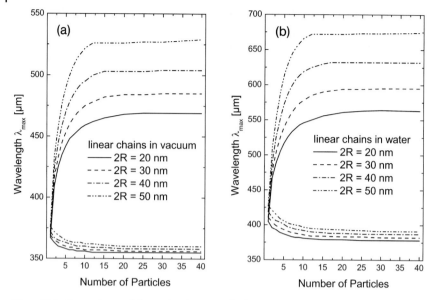

Figure 10.5 Dependence of the peak splitting on the length of the aggregate. The photon energy is plotted where the most prominent extinction maxima are peaked versus the number of particles in the chain. (a) In vacuum; (b) in water.

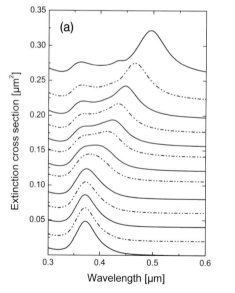
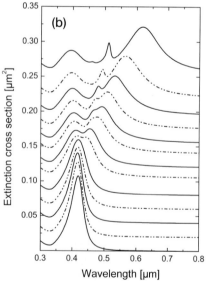

Figure 10.6 Extinction cross-section spectra of linear chains with $N = 10$ identical silver particles with increasing next-neighbor distance $R_{NN} = 2R$, 2.1R, 2.2R, 2.3R, 2.4R, 2.5R, 2.75R, 3R, 4R, 5R, and 10R from top to bottom. The bottom-most spectrum corresponds to the extinction cross-section of the single particle. (a) In vacuum; (b) in water.

lithography, obtaining 2D particle arrangements with varying interparticle distance. They found that with decreasing interparticle distance the SPP resonance shifts to longer wavelength for the polarization direction parallel to the long particle pair axis, whereas a blue shift was found for the orthogonal polarization.

10.3.4
Enhancement of Scattering and Extinction

Introducing the ratio of the extinction cross-section of an N-sphere aggregate and the cross-sections of N single particles as the *enhancement factor*, ENH:

$$ENH = \frac{\sigma_{ext,aggregate}}{N\sigma_{ext,1}} \tag{10.31}$$

comparison with the single particle is possible. Similarly, also for the scattering an enhancement factor can be defined by the cross-sections for scattering. It is evident that ENH depends on the refractive index and the size parameter. A first definition of the enhancement factor was given by Olaofe [572, 573] and Michel [616].

Many calculations on various aggregates of particles [170] showed that

1) In most cases, $ENH > 1$, meaning that the N-particle aggregate scatters and absorbs light more efficiently than N isolated spheres. A plausible explanation is that due to the multiple scattering in the near-field of each primary particle, part of the incident light is *trapped* between the particles of the aggregate and can be effectively absorbed and scattered.

2) Large values for ENH are achieved for small size parameters.

3) ENH may also be <1. This result is not unexpected because the electromagnetic coupling among the particles can also diminish the extinction at certain wavelengths and size parameters.

Here, we give an example for soot agglomerates in vacuum [617, 618]. Calculations of the extinction and scattering efficiencies were carried out for two aggregates with 16 and 25 particles shown in the inset in Figure 10.7. The single particle size was $2R = 20$ nm. In addition, an average over 266 different directions of incidence was taken at each wavelength to consider the random distribution of the aggregates in air.

The extinction efficiency spectra and the scattering efficiency spectra of both aggregates are rather similar and are distinctly larger than that of the single particle spectrum. The enhancement is between 40% and 80% for the extinction. It is caused by stronger absorption in the aggregate with respect to the single particle. However, scattering is also dramatically increased. Although the scattering is less than 10% of the extinction for both the single particle and the aggregate, the scattering by the aggregate is enhanced by a factor of 10 compared with the single particle.

330 | *10 Beyond Mie's Theory II – The Generalized Mie Theory*

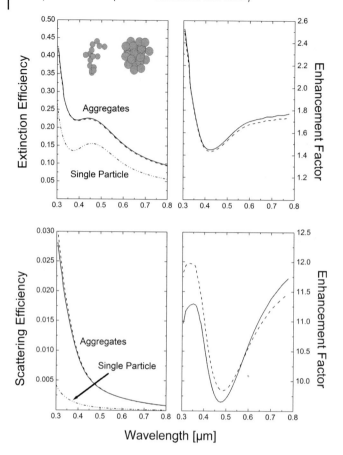

Figure 10.7 Extinction and scattering efficiencies and corresponding enhancement factors of soot aggregates with $N = 16$ and 25 particles (topologies: see inset).

One simple reason for the enhancement might be the size of the aggregate, which is larger than the single particle. However, many additional computations on densely packed aggregates the shapes of which are close to that of a larger sphere, showed that scattering of light by an aggregate cannot be explained by either a volume equivalent or a size equivalent sphere. Therefore, it is hardly possible to predict the scattering and absorption by an aggregate of arbitrary topology using a larger single spherical particle, because the size of this sphere cannot be predicted.

It should be pointed out that the enhancement in the extinction of 40–80% is valid for the high refractive index of approximately $n = 2.5 + 1.5i$ of the soot material. For a lower refractive index of $n = 2 + i$, the enhancement amounts to only 30%. This finding coincides with findings of K. A. Fuller (personal communications).

10.3.5
The Problem of Convergence

The maximum number n_{MAX} of multipoles that must be considered in the multipole expansion of the electromagnetic fields is the subject of controversial discussion in the literature. It depends on the individual particle sizes and on the density of the topological packing, that is, the smallest surface-to-surface distances, occurring in the aggregate. n_{MAX} must be at least as large as the largest order necessary in the calculation for a single sphere. For closely packed spheres, the transformation of vector harmonics of the scattered waves implies that n_{MAX} must be essentially greater. Comparison of calculated and measured spectra [114, 584, 585] showed that n_{MAX} is, under realistic conditions, not too far away from the maximum multipole order for a single sphere. On the other hand, predictions are made that especially for metal particles n_{MAX} increases to 100–1000 for almost touching spheres, which was also the finding of Ruppin [619], who used the T-matrix approach to study the effect of higher multipoles on the extinction by bispheres of Na and NaCl in the IR region.

We will discuss the dependence of calculated spectra upon the maximum number n_{max} for the example of $N = 2$ spherical particles in contact. Calculations are done for Ag and Pt, and also for nonmetallic particles of Si, Fe_2O_3, and SiO_2, and we discuss the wavelength dependence of the extinction cross-section.

We begin our discussion with two neighboring silver particles. For two particles (and also for linear chains of spherical particles) we can simply distinguish between the *longitudinal mode* of the aggregate and the *transverse mode* of the aggregate, depending on the direction of the electric field compared with the axis of the dimer (see Figure 10.8). For the longitudinal mode, the electric field is parallel to the dimer axis.

The extinction spectra of the longitudinal and transverse modes are calculated with n_{MAX} varying from 1 to 15. The resulting spectra are depicted in Figure 10.9. For comparison, the extinction spectrum of the single sphere is also plotted as the dashed-dotted-dotted line.

Obviously, the longitudinal mode does not show any convergence, in contrast to the transverse mode, for which the spectra do not differ for $n_{MAX} \geq 5$. The extinction peak shifts towards longer wavelengths with increasing n_{MAX}. This behavior of the longitudinal mode was predicted by several workers and also found in calculations by Pustovit et al. [620]. They used a method similar to the DDA to reduce

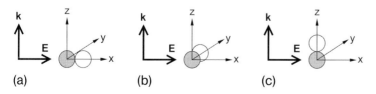

Figure 10.8 (a) Longitudinal mode and (b) and (c) transverse mode of a dimer in the field of an electromagnetic wave with **E** parallel to **x** and **k** parallel to **z**.

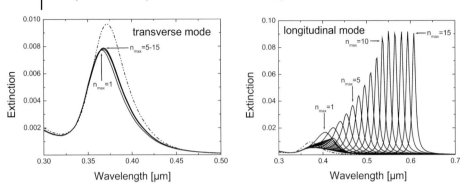

Figure 10.9 Longitudinal and transverse modes of an Ag dimer (primary particle size $2R = 40$ nm) in vacuum as a function of the maximum multipolar order n_{MAX}. The dashed-dotted-dotted line corresponds to the spectrum of the single particle.

the effort in the GMT and calculated extinction efficiency spectra for an aggregate of six Drude metal spheres. For touching spheres convergence was not achieved even for $n_{MAX} = 200$. Smith et al. [621] tried to study the convergence and to ascertain the accuracy of low multipolar orders n_{MAX} in the GMT by comparison with other theories. However, they found good agreement only if the next-neighbor distance between two gold nanoparticles was larger than $R_{NN} = 2.5R$. For touching gold particles considerable quantitative discrepancies were found. These results are in contrast to the results of many experiments on aggregation in silver and gold colloids [113, 114, 582–585]. There, TEM images showed close-lying and even touching particles in the aggregates. The optical spectra in an extended wavelength range always showed a decreasing optical extinction at long wavelengths, indicating that the calculations must converge at a low n_{MAX}. This can also be seen in detail later in Chapter 11.

Pack et al. [622] showed with calculations on aggregated silver nanoparticles that convergence of the multipole contributions can be obtained when considering also longitudinal plasmon excitations. Does this also hold true for metal nanoparticles where the SPP is damped away for the single particle, for which reason no exciting peaks appear in the spectra? To find an answer to this question, we calculated spectra of the longitudinal and transverse modes of a dimer of platinum particles. They are shown in Figure 10.10.

Again, the longitudinal mode does not converge. Apparently, the presence of free electrons is more relevant than the excitation of an SPP. Then, the model of Pack et al. should be applicable also to those metals which do not exhibit a distinct SPP in nanoparticles.

Does the longitudinal mode converge for nonmetallic particles? To obtain an answer to this question, we examined silicon, hematite (Fe_2O_3), and silica. The most interesting is the semiconductor Si. The amount of free carriers is too low to exhibit an SPP in the UV–VIS range. On the other hand, the refractive index of Si is high in the visible spectral range due to strong interband transitions in

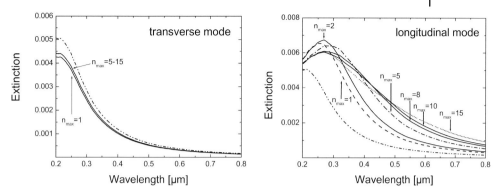

Figure 10.10 Longitudinal and transverse modes of a Pt dimer (primary particle size $2R = 40\,\text{nm}$) in vacuum as a function of the maximum multipolar order n_{MAX}. The dashed-dotted-dotted line corresponds to the spectrum of the single particle.

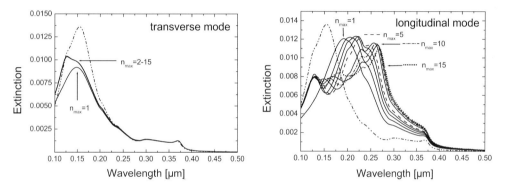

Figure 10.11 Longitudinal and transverse modes of an Si dimer (primary particle size $2R = 40\,\text{nm}$) in vacuum as a function of the maximum multipolar order n_{MAX}. The dashed-dotted-dotted line corresponds to the spectrum of the single particle.

the UV region. Therefore, Si nanoparticles are potential candidates for MDRs as already shown in Figure 6.23 in Section 6.7. We also showed there that in the vacuum ultraviolet (VUV) region at a wavelength of approximately 126 nm (9.84 eV), small Si particles exhibit a resonance due to a close-lying interband transition. This electronic resonance behaves similarly to a surface plasmon resonance when two neighboring particles interact electromagnetically.

We calculated the spectra of the Si nanoparticle dimer in the wavelength range 0.1–0.5 μm to include the electronic resonance. MDRs at longer wavelengths can be excluded for a particle size of $2R = 40\,\text{nm}$. The spectra are shown in Figure 10.11. The result is evident: also if an electronic resonance contributes to the spectrum of the single nanoparticle, the longitudinal dimer mode does not converge. In this case, however, the model of Pack *et al.* [622] cannot be applied as it needs the free electron plasma.

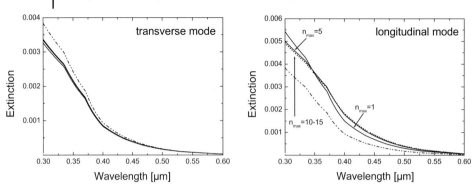

Figure 10.12 Longitudinal and transverse modes of an Fe_2O_3 dimer (primary particle size $2R = 40\,nm$) in vacuum as a function of the maximum multipolar order n_{MAX}. The dashed-dotted-dotted line corresponds to the spectrum of the single particle.

We further calculated spectra of dimers of Fe_2O_3 and SiO_2. Only the results for Fe_2O_3 are presented in Figure 10.12. In both cases, the differences in the spectra of the longitudinal mode vanish for a multipole number $n_{MAX} \geq 10$. However, this is not a general rule at all, because in the wavelength range considered both materials do not exhibit any kind of resonance. Hence it can be expected that also for these materials the convergence is lacking if one looks at wavelengths where resonances occur or for larger particles (in the case of Fe_2O_3).

What is the reason for the missing convergence? This question has not yet been answered satisfactorily. We suppose that the reason is a wrong normalization in the calculation of the cross-sections of an aggregate. The cross-section is generally defined by the ratio of the energy dissipated by absorption and scattering and the incident intensity. This holds true for all kinds of particles and incident waves. In the derivation of Equations 10.11 and 10.12 I_{inc} is assumed to be constant, which is the case of a plane wave. We expect that this assumption is wrong. Instead, one has to add up the intensities of the waves incident on each particle in the aggregate. This has, however, not yet been verified.

An interesting paper stems from Andersen et al. [623], who examined the IR optical extinction of particle aggregates of SiC, FeO, and SiO_2. These materials exhibit phonon polariton bands in the IR region which can be described by harmonic oscillators. Therefore, electronic resonances will contribute to the spectra. Andersen et al. studied the performance of two GMT and two DDA methods for calculating the extinction of aggregates. They presented results of the calculated extinction within IR absorption bands of SiC, FeO, and SiO_2 aggregates composed of 5–8 spherical particles of radius 10 nm. The aggregates had three different shapes: linear chain, semi-fractal, and simple cubic. When the real part of the refractive index is much smaller than unity, none of the four methods were able to converge. For the two DDA methods there is a strong dependence of the calculated band profiles on the exact dipole distribution within the particles. For the linear five-particle aggregate the result depends on whether the grid is a multiple of five or not, for the semi-fractal seven particle aggregate, and for the simple cubic eight-particle aggregate the dependence is on whether the grid is a multiple of three and two, respectively. Currently it does not seem possible to calculate the absorption efficiencies of clusters of spheres for materials with optical constants $n < 1$ in a reliable way. The authors assumed that the critical point is the contact geometry of the particles, which explains the importance of the resolution of the methods used. This implies that for corresponding experiments it is of great importance to study the contact points between the particles.

10.4
Extensions of the Generalized Mie Theory

An approach based upon fractal topology of coagulation aggregates was developed by Sotelo and Niklasson [624]. They constructed, in extension of the general procedure described above, large cluster arrangements by iteratively reinterpreting the small aggregates as new building units (*effective particles*) to be inserted in the GMT. It is noteworthy that such systems have indeed been found.

Al-Nimr and Arpaci [625] developed a model for the interactions among particles in the Rayleigh approximation. It is very similar to the exact comprehensive solutions, but more similar to the DDA, except that it is intended for the interactions among particles and not the interactions of dipoles as elements of a single (nonspherical) particle.

A direct extension of the GMT to absorbing surrounding media was given by Lebedev [260] and was applied to silver nanoparticle aggregates by Lebedev and Stenzel [626].

10.4.1
Incident Beam

Here, we first discuss the difference between plane wave excitation and *evanescent wave* excitation of aggregates. For that purpose, we present exemplary results for the extinction efficiency spectrum of a pair of silver nanoparticles with $2R = 40$ nm on a glass substrate where the evanescent wave can be generated by total internal reflection. The spectra in Figure 10.13 are compared with spectra obtained for a plane wave with grazing incidence at the substrate. For the plane wave the electric field vector is always perpendicular to the pair axis due to the grazing incidence. Then, the spectra for s- and p-polarization must coincide, showing the spectrum of the transverse mode with retardation of the incident wave at the site of the second particle. The peak position is at $\lambda = 366$ nm, close to the peak position of the unretarded transverse mode. For the evanescent wave, the spectrum in s-polarization is rather similar to that for the plane wave and is peaked at the same wavelength. In contrast, the spectrum in p-polarization shows a resonance peaked at $\lambda = 367$ nm that belongs to the transverse mode and in addition a peak at $\lambda = 422$ nm that can be assigned to the longitudinal mode. This resonance is caused by the p-polarized electric field which is rotating in the plane of incidence for evanescent waves due to the complex phase shift associated with total internal

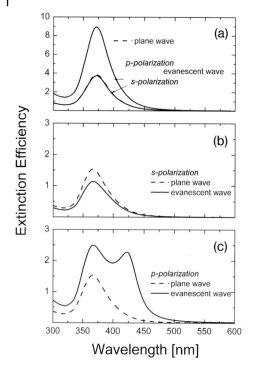

Figure 10.13 Comparison of plane wave excitation and evanescent wave excitation in the GMT. (a) Excitation of the single particle with a plane wave and an evanescent wave. (b) Dimer excited with s-polarized light. (c) Dimer excited with p-polarized light.

reflection. Then, the electric field vector can be decomposed into components parallel and perpendicular to the pair axis. The parallel component corresponds to the case of a longitudinal electromagnetic wave. The efficiency in p-polarization is at least a factor of two larger than that in s-polarization.

Barton et al. [627] used their previously derived single spherical particle–arbitrary beam interaction theory to develop an iterative procedure for the determination of the electromagnetic field for a beam incident on two adjacent spherical particles. Exemplary calculations of internal and near-field normalized source function distributions were presented. Also presented were calculations demonstrating the effect of the relative positioning of the second adjacent particle on far-field scattering patterns.

10.4.2
Nonspherical Particles

Solutions which take into account electromagnetic coupling are also known for spheroids [628–630] and infinite cylinders [631–637] and cubes [638]. It should be mentioned that this list of references is not complete; however, it covers the most relevant papers concerning this subject.

10.4 Extensions of the Generalized Mie Theory

In the following we briefly introduce the extension of the GMT to the coupling of infinitely long cylinders. The theory developed by Bever and Allebach [633] is based on interaction potentials whereas those of Felbacq et al. [635] and Lee and Grzesik [636] are based on vector harmonics expansions. On neglecting reflection of the incident and scattered waves at the surface of the substrate in the model of Lee and Grzesik, all solutions yield the same results. For convenience we restrict considerations to the solution without reflection.

Consider a set of N parallel cylinders of diameter $2R_i$, $i = 1, ..., N$, in a transparent matrix, being illuminated perpendicularly by a plane wave. The incident wave becomes partly scattered and partly absorbed by each cylinder to a certain extent. The corresponding electromagnetic fields can be expressed as the sum of TM and TE partial waves. In contrast to vector harmonics, Bever and Allebach used scalar potential functions u and v. This formulation of the scattering problem with scalar potentials is completely adequate for the formulation with vector harmonics. The potentials of the incident wave in the coordinate system of the jth cylinder are

$$u_{\text{inc}}^{(j)} = -\frac{\exp(ik_M r_j \cos\beta)}{k_M} \sum_{n=-\infty}^{\infty} i^{n+1} J_n(k_M r_j) \exp(-in\theta_j) \tag{10.32}$$

$$v_{\text{inc}}^{(j)} = \frac{\exp(ik_M r_j \cos\beta)}{k_M} \sum_{n=-\infty}^{\infty} i^{n+1} J_n(k_M r_j) \exp(-in\theta_j) \tag{10.33}$$

where $J_n(k_M r_j)$ is the Bessel function of order n in the reference system of particle j. The exponential $\exp(ik_M r_j \cos\beta)$ takes into account the phase of the incident wave at the site of the jth cylinder with respect to a reference frame. k_M is the wavenumber of the incident wave in the surrounding medium. Similarly, the TM and TE potentials of the scattered wave of the jth cylinder are

$$u_{\text{sca}}^{(j)} = \frac{1}{k_M} \sum_{n=-\infty}^{\infty} i^{n+1} \alpha_n(j) H_n(k_M r_j) \exp(-in\theta_j) \tag{10.34}$$

$$v_{\text{sca}}^{(j)} = -\frac{1}{k_M} \sum_{n=-\infty}^{\infty} i^{n+1} \beta_n(j) H_n(k_M r_j) \exp(-in\theta_j) \tag{10.35}$$

where $\alpha_n(j)$ and $\beta_n(j)$ are the scattering coefficients to be determined from the boundary conditions. $H_n(k_M r_j)$ is the Hankel function of the first kind of order n in the reference system of particle j. Inside each cylinder the transmitted waves are described by the potentials

$$u_{\text{int}}^{(j)} = \frac{1}{k} \sum_{n=-\infty}^{\infty} i^{n+1} \gamma_n(j) J_n(kr_j) \exp(-in\theta_j) \tag{10.36}$$

$$v_{\text{int}}^{(j)} = -\frac{1}{k} \sum_{n=-\infty}^{\infty} i^{n+1} \delta_n(j) J_n(kr_j) \exp(-in\theta_j) \tag{10.37}$$

where $\gamma_n(j)$ and $\delta_n(j)$ are coefficients of the transmitted wave also to be determined from Maxwell's boundary conditions and k is the in general complex wavenumber inside the cylinder.

The potentials u and v must satisfy the boundary conditions that u, $\varepsilon(\partial u/\partial r)$, v, and $\varepsilon(\partial v/\partial r)$ must be continuous across the boundaries at $r_j = a_j$ with $j = 1, ..., N$. For a set of parallel cylinders the scattered fields of all cylinders $i \neq j$ are expressed as incident fields at the jth cylinder. This is accomplished by means of the Graf addition theorem [84] as follows:

$$\exp(-in\theta_i)H_n(k_M r_i) = \begin{cases} \sum_{q=-\infty}^{\infty} (-1)^q H_{n+q}(k_M r_{ij}) J_q(k_M r_j) \exp(iq\theta_j) & j > i \\ (-1)^n \sum_{q=-\infty}^{\infty} H_{n+q}(k_M r_{ij}) J_q(k_M r_j) \exp(iq\theta_j) & j < i \end{cases} \quad (10.38)$$

where r_{ij} is the center-to-center distance between the ith and jth cylinders. Then, two independent sets of linear equations for the coefficients $\alpha_n(j)$ and $\beta_n(j)$ result from applying Maxwell's boundary conditions:

$$\alpha_n(j) = a_n(j)\left[\exp(ik_M r_j \cos\beta) + i^{n+1}\exp(in\beta)\sum_{\substack{i=1 \\ i\neq j}}^{N}\sum_{n=-\infty}^{\infty} A_{ij}^{-n}\right] \quad (10.39)$$

$$\beta_n(j) = b_n(j)\left[\exp(ik_M r_j \cos\beta) + i^{n+1}\exp(in\beta)\sum_{\substack{i=1 \\ i\neq j}}^{N}\sum_{n=-\infty}^{\infty} B_{ij}^{-n}\right] \quad (10.40)$$

where $a_n(j)$ and $b_n(j)$ again are the single-cylinder scattering coefficients for the TM and TE case. The transformation matrix elements A_{ij}^{-n} and B_{ij}^{-n} contain the information about the relative position of the ith cylinder with respect to the jth cylinder and can be calculated from

$$A_{ij}^{-n} = \begin{cases} \sum_{q=-\infty}^{\infty} (-1)^q i^{q+1} \alpha_q(i) H_{n+q}(k_M r_{ij}) \exp(-iq\beta) & j < i \\ (-1)^n \sum_{q=-\infty}^{\infty} i^{q+1} \alpha_q(i) H_{n+q}(k_M r_{ij}) \exp(-iq\beta) & j > i \end{cases} \quad (10.41)$$

$$B_{ij}^{-n} = \begin{cases} \sum_{q=-\infty}^{\infty} (-1)^q i^{q+1} \beta_q(i) H_{n+q}(k_M r_{ij}) \exp(-iq\beta) & j < i \\ (-1)^n \sum_{q=-\infty}^{\infty} i^{q+1} \beta_q(i) H_{n+q}(k_M r_{ij}) \exp(-iq\beta) & j > i \end{cases} \quad (10.42)$$

Having solved these linear equations, the extinction and scattering efficiencies for the array of N parallel cylinders can be calculated from Poynting's law as

$$Q_{\text{ext}}(N) = Q_{\text{ext}}^{\text{TM}}(N) + Q_{\text{ext}}^{\text{TE}}(N) = \frac{2}{(k_M R)^2}\sum_{i=1}^{N}\sum_{n=-\infty}^{\infty} \text{Re}[\alpha_n(i) + \beta_n(i)] \quad (10.43)$$

$$Q_{sca}(N) = Q_{sca}^{TM}(N) + Q_{sca}^{TE} = \frac{2}{(k_M R)^2} \sum_{i=1}^{N} \sum_{n=-\infty}^{\infty} |\alpha_n(i)|^2 + |\beta_n(i)|^2$$

$$+ \frac{2}{(k_M R)^2} \sum_{i=1}^{N} \sum_{n=-\infty}^{\infty} \text{Re}\left\{\alpha_n^*(i)\left[1 - \frac{\alpha_n(i)}{a_n(i)}\right] + \beta_n^*(i)\left[1 - \frac{\beta_n(i)}{b_n(i)}\right]\right\} \quad (10.44)$$

Kottmann and Martin [639, 640] investigated the SPP resonances of interacting silver nanowires of 50 nm diameter. Whereas the individual cylinders exhibited a single plasmon resonance, much more complex spectra of resonances were observed for interacting structures. The number and magnitude of the different resonances depend on the illumination direction and on the distance between the particles. For very small separations, a dramatic field enhancement between the particles was observed, where the electric field amplitude reached 100 times the illumination.

Byun *et al.* [641] investigated arrays of metallic nanowires on dielectric substrates for use as biosensors. Calculations were carried out using rigorous coupled-wave analysis [642, 643]. Silver or gold nanowires that are periodic on the *x*-axis were assumed to be aligned on the *y*-axis. The nanowires with a complex dielectric function were regularly patterned on a glass substrate with $n_{sub} = 1.515$. The nanowire period was considered in the range 250–400 nm such that far-field dipolar interactions dominated. TM-polarized light, the electric field of which oscillated in parallel to the nanowire grating vector, was assumed to be normally incident. The results showed that the resonance spectrum depended strongly on the nanowire period and profile. For nanowire periods in the far-field coupling, dipole interactions between metallic nanowires resulted in a blue shift of the SPP resonance as the grating period varied from 400 to 250 nm. The results indicated that the extinction spectra of the SPP resonance sensor based on metallic nanowires are fairly linear and significantly sensitive to changes in the refractive indices of dielectric binding media if design parameters are properly optimized.

Tsuei and Barber [644] considered the multiple scattering problem for two parallel infinite dielectric cylinders for plane wave illumination perpendicular to the cylinder axes. Numerical results showed the coupling effect with respect to cylinder size, separation, and orientation of the cylinder axes with respect to the incident wave. The coupling effect was illustrated by calculations of the internal and near-field scattering intensities (angular-dependent scattering) for end-on and broadside incidence.

11
The Generalized Mie Theory Applied to Different Systems

In principle, the generalized Mie theory (GMT) can be applied in all cases where the distance among spherical particles is less than about five times the mean diameter. For larger distances, the interaction among the particles becomes negligible. In this chapter, we present numerical results on aggregated systems for numerous particle materials and wavelength regions, obtained by applying the GMT rigorously. The calculations are partly supplemented by experimental results where the filling factor f is large enough to take into account the electromagnetic interaction but less than in dense-packed systems. The latter will be treated in Chapter 12. A first review on several applications was given in [645].

We point out that in most of the following calculated spectra the surrounding is assumed to have a constant refractive index $n_M = 1$ (vacuum), whereas in the experiments it may be larger. Nevertheless, provided that the refractive index n_M remains real valued, that is, the surrounding medium is not absorbing, comparison with the calculations is possible.

Similarly to the single spherical and nonspherical particles in Chapters 6–9, we consider various categories for the primary particles:

- metal particles
- semimetal and semiconductor particles
- nonabsorbing dielectric particles
- carbonaceous particles
- particles with phonon polaritons
- miscellaneous particles.

In the calculations we always averaged the aggregate spectra over 72 directions of incidence of the incident plane wave and over both polarizations of the incident wave. The optical constants used are referred to locally.

Calculations are carried out for (a) particle aggregates with $N = 5$ identical spheres with diameter $2R = 40$ nm, (b) linear chains with $N = 2$–10 identical particles with diameter $2R = 40$ nm, and (c) a special icosadeltahedral aggregate of 55 primary particles with $2R = 20$ nm, called CLU55 in the following. For better presentation, the spectra of (a) are shifted along the ordinate by arbitrary figures added to the spectra, maintaining the important information on the

Optical Properties of Nanoparticle Systems: Mie and Beyond. Michael Quinten
Copyright © 2011 WILEY-VCH Verlag GmbH & Co. KGaA, Weinheim
ISBN: 978-3-527-41043-9

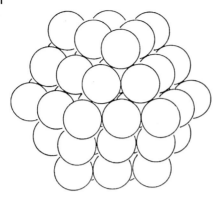

Figure 11.1 Icosadeltahedral aggregate CLU55 of 55 identical spheres.

shape of the spectra. For the same reason, the spectra of (b) are multiplied by the factors 1, 2, 4, 6, 8, 10, 12, 14, and 16 and a logarithmic scale is used. Otherwise, the spectra would be clumped in a narrow region of extinction values. For comparison, the spectrum of two single particles is also plotted as the bottom-most spectrum, except for (c) where the spectrum of 55 single particles is plotted.

The specific features of the icosadeltahedral aggregate as sketched in Figure 11.1 are its compactness and the nearly spherical shape. Nevertheless, since it is built up from spheres its surface is rough, making it comparable to a larger nonspherical particle. Hence it is well suited for comparison with approximations for nonspherical particles. We will compare the extinction cross-section spectrum of the icosadeltahedral aggregate with the cross-sections of the primary particles and the volume-equivalent sphere with $2R_{ev} = (55)^{1/3} \times 2R = 76\,\text{nm}$.

11.1
Metal Particles

11.1.1
Calculations

As already seen and discussed in Chapter 10, the most prominent effects of the electromagnetic interaction among neighboring particles in an aggregate are obtained if the single particle exhibit resonances. For metallic nanoparticles, this means that mainly primary particles with surface plasmon polariton (SPP) resonances will show considerable changes in the optical extinction due to the electromagnetic interaction. Here, we will present results for aluminum, sodium, potassium, silver, gold, and platinum.

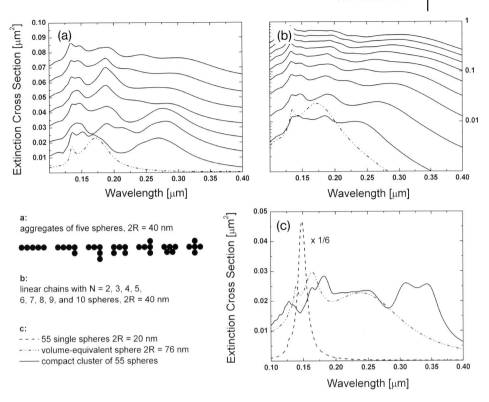

Figure 11.2 Extinction cross-section spectra of (a) aggregates with $N = 5$ identical aluminum spheres of $2R = 40$ nm, (b) linear chains of Al particles, and (c) an icosadeltahedral aggregate of 55 particles with $2R = 20$ nm in comparison with the volume-equivalent sphere with $2R_{ev} = 76$ nm.

We start with the presentation and discussion of the spectra of aluminum, sodium, and potassium nanoparticle aggregates in Figures 11.2–11.4. The spectrum of the single aluminum, sodium, or potassium nanoparticle exhibits the well-known SPP resonance in the corresponding wavelength range. If aggregation occurs, the SPP splits into many new resonances, contributing to increased extinction (and scattering) at both longer wavelengths (low photon energies) and lower wavelengths (high photon energies) than the position of SPP resonance. Interband transitions at low wavelengths are missing for these metals, for which reason also the resonances at short wavelengths can be clearly resolved. The splitting depends sensitively on the topology of the aggregate, as already seen in Chapter 10. The largest peak splitting, that is, the difference in the wavelengths at which the most prominent modes are peaked, occurs again for chain-like aggregates. Aggregates of aluminum, sodium, and potassium nanoparticles exhibit the highest number of resonances and the most pronounced effects due to aggregation.

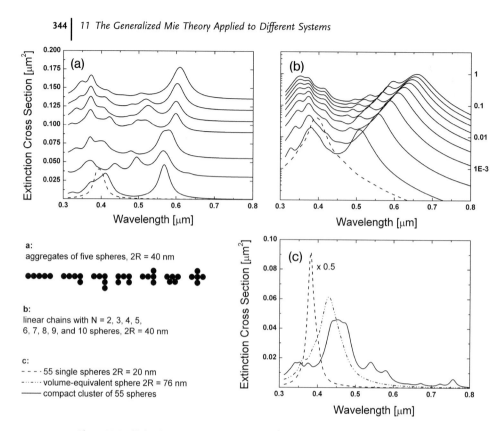

Figure 11.3 Extinction cross-section spectra of (a) aggregates with $N = 5$ identical sodium spheres of $2R = 40$ nm, (b) linear chains of Na particles, and (c) an icosadeltahedral aggregate of 55 particles with $2R = 20$ nm in comparison with the volume-equivalent sphere with $2R_{ev} = 76$ nm.

Comparing the spectrum of the cluster CLU55 with the single particle spectrum of 55 single particles and that of the volume-equivalent sphere with $2R_{ev} = 76$ nm, it is clearly dominated by many resonances that contribute over the whole spectral region. Due to these resonances, it also clearly deviates from the spectrum of the volume-equivalent sphere, so that for compact aggregates of aluminum, sodium, or potassium the comparison with a volume-equivalent sphere becomes invalid (Figures 11.5 and 11.6).

Also for silver and gold the nanoparticles exhibit the well-known SPP resonance in the corresponding wavelength range. If aggregation occurs, this resonance splits into many new resonances, contributing to increased extinction (and scattering) mainly at longer wavelengths (low photon energies) because the close-lying interband transitions prevent resonances at lower wavelengths (high photon energies), as before for aluminum, sodium, and potassium. The splitting again depends sensitively on the topology of the aggregate as already seen in Chapter 10. The largest peak splitting, that is, the difference in the wavelengths at which the most prominent modes are peaked, occurs for chain-like aggregates. Note that the

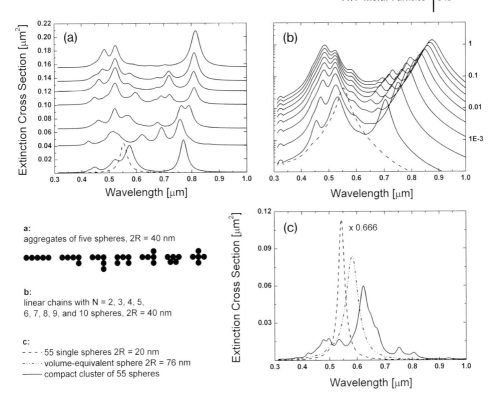

Figure 11.4 Extinction cross-section spectra of (a) aggregates with $N = 5$ identical potassium spheres of $2R = 40$ nm, (b) linear chains of K particles, and (c) an icosadeltahedral aggregate of 55 particles with $2R = 20$ nm in comparison with the volume-equivalent sphere with $2R_{ev} = 76$ nm.

spectra for Au nanoparticle aggregates were calculated with water as surrounding medium. For vacuum as surrounding material, the peak splitting is too low and the resonances of the aggregates cannot be resolved.

Comparing the spectrum of the cluster CLU55 with the single particle spectrum of 55 single particles and that of the volume-equivalent sphere with $2R_{ev} = 76$ nm, it is also clearly dominated by many resonances that contribute over the whole spectral region. Due to these resonances, it also clearly deviates from the spectrum of the volume-equivalent sphere, so that for compact aggregates of silver and gold the comparison with a volume-equivalent sphere becomes invalid.

The similarity of the spectra of CLU55 for gold and for silver is striking. Except for the interband transition regions, they exhibit the same features, but in different wavelength ranges. The spectra appear to be a *fingerprint* of the icosahedral structure, provided that the optical constants for gold and silver are similar. The similarity is not so significant for the CLU55 formed by sodium and potassium nanoparticles, but is also recognizable.

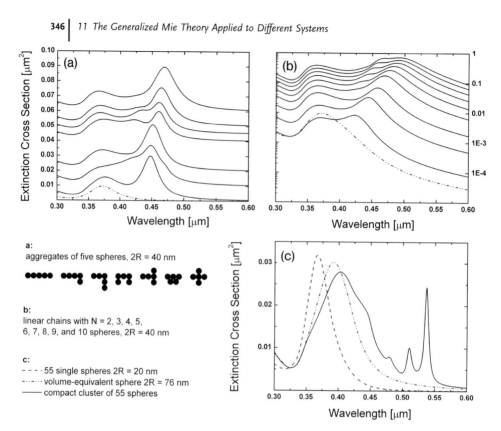

Figure 11.5 Extinction cross-section spectra of (a) aggregates with $N = 5$ identical silver spheres of $2R = 40$ nm, (b) linear chains of Ag particles, and (c) an icosadeltahedral aggregate of 55 particles with $2R = 20$ nm in comparison with the volume-equivalent sphere with $2R_{ev} = 76$ nm.

The last example is for platinum (Figure 11.7). For this metal and also for Pd, Rh, Ru, Fe, Ni, Co, and many others, neither the single particle with $2R = 20$ nm nor the single particle with $2R = 40$ nm exhibits an SPP, but only a monotonic decrease of the extinction proportional $1/\lambda$ from short to long wavelengths. In consequence, the aggregation merely influences the extinction. It only increases with the number of particles in the aggregate. Different shapes of aggregates with the same number of primary particles have almost the same spectrum. Comparing the aggregate spectrum of CLU55 with the volume-equivalent sphere, the extinction of CLU55 is only slightly increased at wavelengths below 600 nm.

11.1.2
Experimental Results

11.1.2.1 Extinction of Light in Colloidal Gold and Silver Systems

Extinction of light by aggregated small silver and gold particles in aqueous suspensions has been examined by various workers [113, 114, 582–585, 646–649].

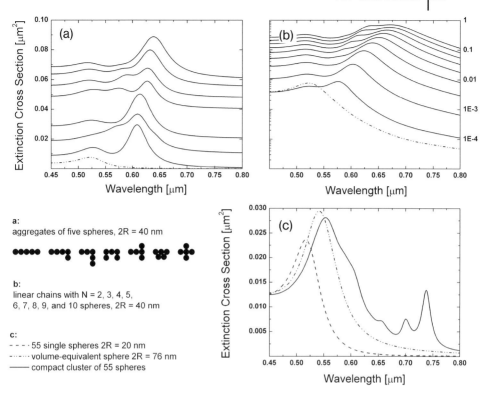

Figure 11.6 Extinction cross-section spectra of (a) aggregates with $N = 5$ identical gold spheres of $2R = 40$ nm in water, (b) linear chains of Au particles, and (c) an icosadeltahedral aggregate of 55 particles with $2R = 20$ nm in comparison with the volume-equivalent sphere with $2R_{ev} = 76$ nm.

Aggregation was induced by addition of small amount of $CuSO_4$, $CaCl_2$, or Na_2CO_3 solution. Then, the repulsive electrostatic force among the particles becomes weaker and controlled aggregation into aggregates of various mean sizes and topologies can be induced. The aggregation process is diffusion limited and can be stopped by addition of a solution containing macromolecules, for example, gelatin solution. Gelatin forms mechanically protecting shells around the aggregates and particles. When using $CuSO_4$ solution, the concentration of copper is low in the suspensions as neither copper particles are generated nor copper shells formed around the colloidal silver or gold particles. Furthermore, the absorption of the colloidal suspension remains unaffected by the added $CuSO_4$.

After adding a constant amount of a salt solution stronger than a threshold concentration (depending on the salt) to the colloid, the aggregation starts and the aggregates begin to grow. After a time t of a few minutes a sample is extracted and is stabilized with gelatin from a gelatin solution. With increasing time various samples with increasing amounts of larger aggregates are obtained.

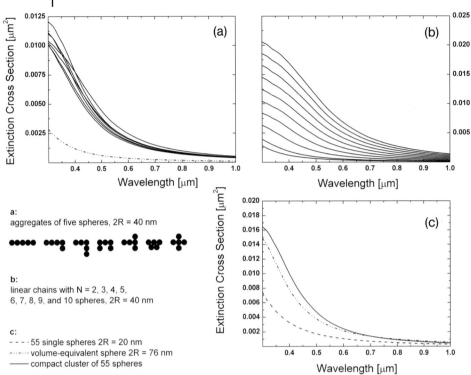

Figure 11.7 Extinction cross-section spectra of (a) aggregates with $N = 5$ identical platinum spheres of $2R = 40$ nm, (b) linear chains of Pt particles, and (c) an icosadeltahedral aggregate of 55 particles with $2R = 20$ nm in comparison with the volume-equivalent sphere with $2R_{ev} = 76$ nm.

Figure 11.8 provides the spectra of two exemplary aggregate series for gold and silver, showing the main features in the measured extinction spectra. For the sake of better presentation, a semilogarithmic plot of the extinction data versus the wavelength is used and the curves are shifted along the ordinate by arbitrary factors. The extinction spectra of the isolated particles (open circles) are plotted for comparison at the bottom of each graph. The samples were obtained by addition of small amounts of 10^{-4} M $CaCl_2$ solution and stabilization of the aggregated samples at different time points with small amounts of gelatin solution.

If aggregation of particles occurs, the extinction increases at longer wavelengths (low photon energies) and decreases at the wavelength where the SPP of the corresponding isolated particles is peaked. The interband transition region is almost unaffected. These results are in agreement with computed spectra of aggregates of various topologies, meaning that both the decrease in the extinction at lower wavelengths and the increase at longer wavelengths result from the electromagnetic coupling among the particles in the aggregates. The random distribution of different aggregates of various topologies in the samples, however, makes a simple assignment of the additional peaks to a certain kind of aggregates almost

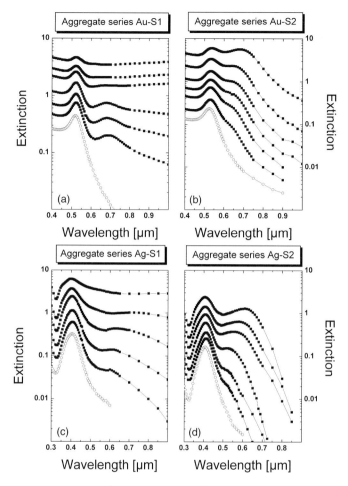

Figure 11.8 Measured extinction spectra of aggregated samples of (a) and (b) silver particles with $2R = 20.0$ and 27.4 nm and (c) and (d) gold particles with $2R = 30.2$ and 38 nm. The samples contain a random distribution of aggregates of various topologies, but with increasing state of aggregation from bottom to top. For better presentation the spectra are shifted along the y-axis by arbitrary factors.

impossible. Additionally, many isolated particles are still present and contribute to the extinction at short wavelengths. Hence, in the measured spectra the short-wavelength maximum is higher than the long-wavelength extinction maximum, which is in contrast to calculated spectra in Section 11.1.1. For a quantitative comparison of measured extinction spectra and computed spectra, an extended analysis of TEM images of aggregated samples must be carried out.

TEM applied on the samples showed that the single particles lumped into densely packed aggregates, the topology of which is random with the primary particles almost touching. Typical TEM images of various samples are presented in Figure 11.9. With the growth of larger aggregates, the number of single particles

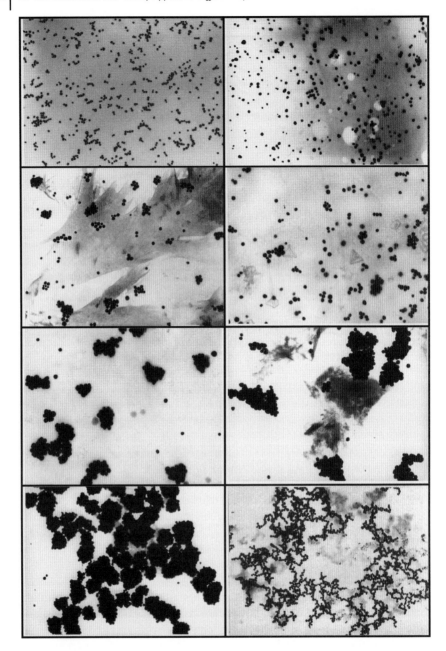

Figure 11.9 Collection of typical transmission electron micrographs of aggregated gold and silver particles.

in the samples decreases. In some cases, the aggregation process leads to more chain-like structures which grow to fractal-like samples. An example can be seen in the bottom-most micrograph. A detailed discussion of such fractal-like samples is given in Section 11.1.6.

The changes in the extinction spectra can also be recognized from the color of the samples. For silver, the isolated particle sample appears yellow due to the SPP which is peaked in the violet light. The color of the aggregated samples changes to orange, brown, and green with increasing amount of larger aggregates in the sample. For gold, the red color of the isolated particle sample changes for the aggregated samples to violet and blue with increasing amount of larger aggregates in the sample. For illustration, color pictures of samples with increasing amounts of larger aggregates are shown in Figure 11.10 for a silver and a gold sample series. Additionally, in Figure 11.11 the color coordinates for aggregated gold and silver colloids are summarized in a chromaticity diagram. They were obtained by evaluation of the extinction spectra.

A quantitative comparison of measured extinction spectra and computed spectra was performed for two samples. For that purpose, an aggregate topology histogram

Figure 11.10 Color pictures of samples with isolated and aggregated silver and gold particles. A color version of this figure can be found in the color plates at the end of the book.

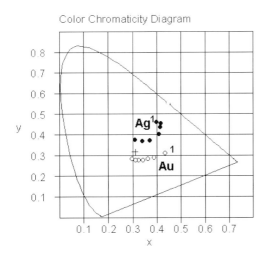

Figure 11.11 Chromaticity diagram of aggregated gold and silver colloidal suspensions.

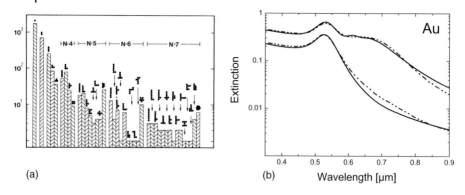

Figure 11.12 (a) Number distribution of aggregates of certain topologies; (b) measured and calculated extinction spectra of the sample and the single particle gold colloid.

was constructed. Model aggregates with simple topology were introduced and the aggregates in the samples were assigned to the corresponding most similar model aggregate.

The first example is an aggregated gold colloid sample. The aggregation was obtained by addition of a small amount of 0.2 M Na_2CO_3 solution and addition of gelatin solution for stabilization. The aggregate topology histogram and the measured and calculated spectra are shown in Figure 11.12.

For each aggregate, the corresponding extinction spectrum was calculated and weighted with the number from the histogram. The aggregate spectra were averaged over different directions of incidence of the incident plane wave and over both polarizations of the incident wave. For calculation, optical constants from [40] were used.

The measured and calculated spectra are in excellent agreement. For this agreement, the maximum number of multipoles n_{MAX} must be assumed to be $n_{MAX} = 3$.

The second example is an aggregated silver colloid sample, obtained by addition of a small amount of 10^{-4} M $CaCl_2$ solution and stabilizing with gelatin solution. The aggregate topology histogram and the measured and calculated spectra are shown in Figure 11.13.

Again, for each aggregate the corresponding extinction spectrum was calculated and weighted with the number from the histogram. The aggregate spectra were averaged over different directions of incidence of the incident plane wave and over both polarizations of the incident wave. For calculation, optical constants from [40] were used.

The measured and calculated spectra are in fairly good agreement. The calculated spectrum exhibits two narrow peaks at longer wavelengths instead of a broad single peak as in the measured spectrum. The reason may be the less extensive statistics in the aggregate topology histogram. To smear out these peaks, more aggregates of $N = 3, 4$, and 5 particles with more compact topology are necessary. For this agreement, the maximum number of multipoles n_{MAX} must again be assumed to be $n_{MAX} = 3$.

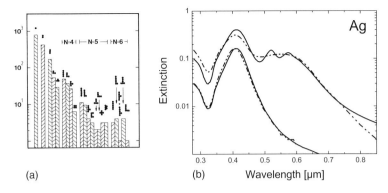

Figure 11.13 (a) Number distribution of aggregates of certain topologies; (b) measured and calculated extinction spectra of the sample and the single particle silver colloid.

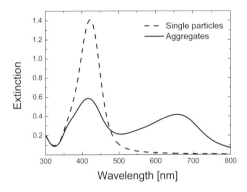

Figure 11.14 Extinction spectrum of an aggregated Ag colloid containing mainly linear chains of particles. For comparison, the spectrum of the single particle colloid is also shown.

The third example demonstrates the effect of aggregation on the apparent color of the colloid. Silver nanoparticles with mean size $2R = 45$ nm were produced in a colloidal solution by chemical reduction [650]. Aggregation was induced by addition of a certain amount of $CuSO_4$ to the colloidal solution and stabilization of the colloid after a few seconds by adding gelatin solution. TEM analysis showed that mainly smaller linear chains have formed in this sample. Therefore, the extinction spectrum in Figure 11.14 exhibits two clearly separated extinction maxima at wavelength 416 and 662 nm caused by these aggregates. The gap in the green spectral region with low absorption leads to a green color in transmission.

11.1.2.2 Total Scattering of Light by Aggregates

Total scattering of aggregated samples $[\sigma_{sca}(\lambda)]$ was measured at wavelengths between 350 and 700 nm using the modified integrating-sphere spectrometer shown in Figure 7.32 in Section 7.5.

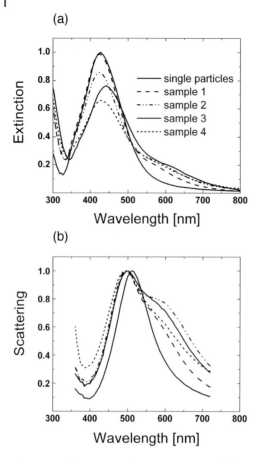

Figure 11.15 Extinction and scattering spectra of silver particles with diameter $2R = 36\,\text{nm}$ (dashed lines) and four aggregated samples, obtained by addition of various small amounts of $10^{-6}\,\text{M}$ CuSO$_4$ solution.

As the sample is inside the integrating sphere, problems arise with multiple absorption and scattering of light scattered by and transmitted through the sample. Consequently, the photomultiplier does not receive a signal which is proportional to the corresponding cross-section, but is influenced by the reabsorption of light, resulting in modified extinction and scattering spectra. Figure 11.15 provides measured spectra of samples containing isolated and aggregated silver spheres [269, 651].

The aggregates were obtained by adding various small amounts of $10^{-6}\,\text{M}$ CuSO$_4$ solution to the single particle colloid with mean diameter $2R = 36\,\text{nm}$. The particles formed aggregates with different topologies. The extinction by the aggregated samples is increased at longer wavelengths in accordance with the results of the GMT. However, in the scattering spectra the peak positions of the short-wavelength modes are shifted from about $\lambda = 420–430\,\mathrm{nm}$ to about $\lambda = 500–510\,\mathrm{nm}$. Additionally, the intensity at $\lambda = 420–430\,\mathrm{nm}$ is decreased. At wavelengths longer than about $\lambda = 530\,\mathrm{nm}$, the scattering spectra of the aggregated samples are comparable to the corresponding extinction spectra.

The shift of the short-wavelength peaks in the scattering spectra to longer wavelengths is dominated by absorption of the scattered light. This has already been discussed in Section 7.5 for the spectra of single particles. The reason is that for a single aggregate the surrounding medium containing all other aggregates and single particles is absorptive. Then, the Mie theory extended to absorbing embedding media could be applied. For aggregates, an appropriate model is still lacking. At long wavelengths where the scattering is the dominant contribution to the extinction of the aggregates, the spectra are less influenced by the reabsorption and the peaks in the scattering remain almost unshifted.

11.1.2.3 Angle-Resolved Light Scattering by Nanoparticle Aggregates

Angle-resolved light scattering is an efficient method to distinguish spherical particles from nonspherical particles. Therefore, it is suitable also for distinguishing between aggregates and single particles.

Kahlau et al. [650, 652] measured relative scattering intensities $i_p(\theta)$ for p-polarization (in the scattering plane) and $i_s(\theta)$ for s-polarization (perpendicular to the scattering plane) of silver nanoparticle colloidal solution samples containing aggregates. This was done with a photogoniometer set-up where the aqueous suspension is contained in a cylindrical quartz cell (diameter $\varnothing = 10\,\mathrm{mm}$, refractive index $n = 1.46$), which is concentric with a larger glass cell ($\varnothing = 50\,\mathrm{mm}$, $n = 1.52$). The space between the two cells is filled with doubly distilled water. The monochromatic light from an He–Ne laser is scattered by the sample and the scattered light is detected with a photomultiplier with high sensitivity for photon counting at wavelengths between 185 and 830 nm. It is mounted on a goniometer arm and is moved in a circle around the sample. This set-up allows measurements for scattering angles ranging from 15° to 160°.

Figure 11.16 provides exemplarily the measured extinction spectra and the angle-resolved light scattering for samples with isolated silver particles with mean diameter $2R = 49.7\,\mathrm{nm}$ and three aggregated samples. Aggregation was induced by addition of small amounts of $10^{-4}\,\mathrm{M}$ $CuSO_4$ solution to the colloid and stabilizing the samples with a gelatin solution at different time points. The extinction spectra start with the isolated particle spectrum and extend to systems with increasing number of aggregated nanoparticles. The extinction is increased over a wide wavelength range at longer wavelengths as expected for aggregated silver nanoparticles. The relative scattering intensities $i_p(\theta)$ and $i_s(\theta)$ are plotted versus the scattering angle from 20° to 140°. The small silver particles behave like Rayleigh scatterers, showing $\cos^2\theta$ behavior in p-polarization and constant scattering in s-polarization. The size distribution leads to broadening of the minimum at $\theta = 90°$ in $i_p(\theta)$. For the aggregates, the minimum at $\theta = 90°$ in angle-resolved light scattering is smeared out compared with the isolated particles. Furthermore, the scattering in the forward direction is increased with respect to the backward direction.

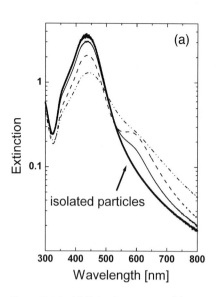
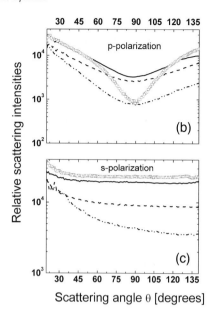

Figure 11.16 (a) Extinction spectra of the isolated silver particles with $2R = 49.7$ nm and three aggregated samples. (b, c) Relative scattering intensities $i_p(\theta)$ and $i_s(\theta)$ of the same samples with sample 3 at the top of each plot. The corresponding single particle spectra are plotted as squares.

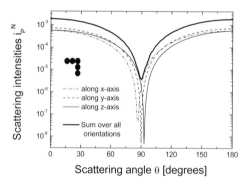

Figure 11.17 Angular distribution function i_p^N of a five-particle aggregate (see inset) for three orientations of the aggregate with respect to the incident wave. The bold solid curve gives the sum over all three directions.

For an explanation of the broadening of the minimum in $i_p(\theta)$ at $\theta = 90°$ due to aggregation, we show in Figure 11.17 the calculated scattering intensities for a five-particle aggregate arranged along the x-, y-, and z-axes. The sum over all three orientations corresponds to the scattering intensity which can be expected for a statistical distribution of the aggregate in space. It can be clearly recognized that the minimum appears at different angles for the three different orientations, 86°, 90°, and 93°. In consequence, the sum is broadened around the minimum, which appears in that case at $\theta = 90°$. For a broad variety of aggregates, this effect may lead to smearing out of the minimum.

For a quantitative interpretation of the angular distribution of light scattered by aggregates, the scattering intensities at the angles 40°, 90°, and 140° were examined, and two new quantities were introduced to characterize all samples [650, 652]. These are the *relative depth* of the minimum for p-polarized light, defined as $I(140°)/I(90°)$ (backscattering to side scattering), and the *asymmetry* in scattering, defined as the ratio $I(40°)/I(140°)$ (forward scattering to backscattering). For the examples presented the parameters are between 2.38 and 25.3 for the relative depth and between 1.04 and 3.25 for the asymmetry.

In many other isolated particle systems examined, the measured asymmetry parameter was between 1 and 1.3, meaning nearly symmetric scattering of the particles in the forward and backward scattering directions. Larger parameters, between 1.3 and 3.5, were obtained for samples containing aggregated particles. The larger the aggregates were, the larger was this parameter. Values of 1.3–1.6 were obtained for the samples with small aggregates. The relative depth is infinite in the ideal case of a point dipole, but under experimental conditions the finite aperture of the detector and the size distribution of the nanoparticles lead to finite values. All isolated silver particle systems examined had parameters between 10 and 25. In contrast, the samples containing aggregates had parameters between 3 and 7. This is a clear difference compared with the single particle systems, indicating a minimum that is smeared out not only due to the size distribution of the primary spheres and the finite detector aperture: a further reason is that the angle-resolved scattering depends on the orientation of the aggregate with respect to the polarization of the incident field. Averaging over all orientations and various aggregates leads to a flattened minimum at $\theta = 90°$.

The observed differences allow the samples to be classified according to the introduced parameters. Figure 11.18 shows the asymmetry ratio plotted versus the relative depth for all samples examined. In this plot, samples with isolated nanoparticles are clearly separated from aggregated samples. Furthermore, in the domain of the isolated particles, one can learn from the position of any arbitrary sample along with the abscissa about the present size distribution of the corresponding sample. Samples with a narrow size distribution have large relative depths; for example, for a system of monodisperse polystyrene particles with $2R = 82 \pm 4$ nm a relative depth of 2500 was measured with this method. On the other hand, the asymmetry parameter gives information about the size of the primary sphere. Larger spheres have larger asymmetry parameters because of increased forward scattering.

In addition to the asymmetry parameter and the relative depth, the depolarization ratio was determined at $\theta = 90°$. It is equal to zero for an exactly spherical particle and assumes larger values with increasing deviation from a spherical shape. Aggregates also deviate from a spherical shape and depolarization is nonzero. Figure 11.19 depicts the depolarization ratio of all silver particle aggregate

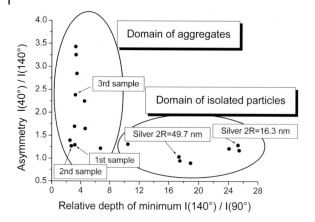

Figure 11.18 Plot of the asymmetry parameter versus the relative depth for several silver samples.

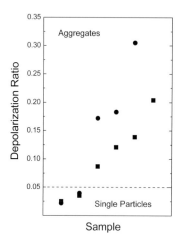

Figure 11.19 Depolarization ratio for single particles and aggregates of Ag- nanoparticles.

samples investigated. The samples containing isolated nanoparticles have depolarization ratios less than 0.05.

11.1.2.4 PTOBD on Aggregated Gold and Silver Nanocomposites

Photothermal optical beam deflection (PTOBD) was applied to solid samples containing either isolated Au or Ag nanoparticles or aggregated nanoparticles to determine the absorption by these particles [653].

The PTOBD method was developed by Boccara *et al.* [654]. It is applicable to systems containing strongly absorbing particles. A sample with absorption coefficient α is mounted on a transparent substrate ($\alpha_s = 0$) for a heating light beam. The background behind the substrate is also assumed to be nonabsorbing ($\alpha_{bg} = 0$).

The sample is heated periodically by a chopped pump beam. The periodic absorption signal is distinguished from that caused by static heating using a lock-in amplifier. The intensity $I(r,t)$ on the sample surface is homogeneous with

$$I(r,t) = \begin{cases} I_0 \cos(2\pi vt) & r \leq R \\ 0 & r > R \end{cases}, \tag{11.1}$$

where v is the chopper frequency, whereas in the sample ($-d < z < 0$), the intensity is attenuated according to the Lambert–Beer law:

$$I(z,r,t) = I(r,t)\exp(-\alpha z) \tag{11.2}$$

leading to the heating of the sample. The heat transfer in the system air–sample–(substrate plus background) is given by the solution of the heat transfer equation in each region:

$$\frac{\partial}{\partial t}(\rho c T) - \nabla(\kappa \nabla T) = Q(r,t) \tag{11.3}$$

where convection is neglected and $Q(r,t)$ are heat sources at site **r**. From continuity of temperature and heat flux, both at $z = 0$ and $z = -d$, finally the temperature distribution in the air in front of the absorbing sample can be obtained.

The dependence of the temperature on the absorption coefficient α is in general complicated and is treated in more detail in fundamental papers on photoacoustics and photothermal beam deflection [654–659]. It follows that only for thermally thick opaque samples and for transparent samples is it possible to obtain information on the absorption coefficient of the sample from the deflection of a laser beam. In these cases, the temperature gradient dT/dz leads to a refractive index gradient dn/dz in the air above the sample, which causes the deflection of the laser beam. The deflection angle ϑ is proportional to the absorption coefficient α. This effect is called the *mirage effect*.

Absorption measurement using PTOBD is only possible with reference to a reference sample, which consists here of a gelatin film (absorption coefficient $\alpha_G \approx 0$) on a glass substrate ($\alpha_{gl} = 0$) coated with black lacquer. This combination shows absorption which is independent of the wavelength.

In Figure 11.20, measured extinction spectra and absorption spectra (PTOBD) are shown for a series of aggregated silver and gold nanoparticles. For the sake of better presentation, a semilogarithmic plot of the extinction and absorption data versus the wavelength is used. The spectra are shifted along the ordinate by arbitrary factors. The curves at the bottom correspond to the extinction and absorption by the isolated particles. From bottom to top the spectra of the aggregate samples are plotted with increasing state of aggregation. The factors for shifting the absorption spectra are properly chosen to plot the spectra on the corresponding extinction spectra.

The silver particle samples were obtained from a silver colloid containing particles with mean particle size $2R = 35.5$ nm. The gold particle systems were obtained from a colloidal gold particle suspension with mean particle size $2R = 16.9$ nm. Aggregation was induced in both suspensions by addition of different very small

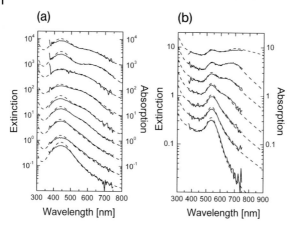

Figure 11.20 Extinction (dashed lines) and absorption (solid lines) spectra of (a) silver particle aggregates (2R = 35.5 nm) and (b) gold particle aggregates (2R = 16.9 nm).

amounts (some tens of microliters) of 0.1 M CuSO$_4$ solution. The samples contained aggregates of different sizes and shapes. The aggregate size increased with increasing sample number, which corresponds to an increased amount of added CuSO$_4$ solution. The samples were prepared by mixing the colloidal suspensions containing the particles and aggregates with a high-concentration gelatin solution (56 g l^{-1}) and heating to reduce the water content. The samples were filled in moulds to prepare solid samples with a smooth surface. The volume fraction f in the solid samples was increased up to $f = 10^{-4}$.

The absorption spectra are fairly similar to the extinction spectra in the whole spectral region. The influence of the aggregates can be clearly seen: extinction and absorption are increased at longer wavelengths with respect to the single particle spectrum due to the aggregates. However, quantitative determination of the absorption was not possible, because the calibration of the photothermal optical beam deflection is complicated and did not allow quantitative determination of the absorption to be carried out.

11.1.2.5 Light-Induced van der Waals Attraction

Electromagnetic coupling among nanoparticles not only results in splitting of resonances and therefore additional extinction bands in the long-wavelength region, but also in additional electromagnetic forces among the particles. Preliminary calculations of Maxwell's tensor of tension yielded very low forces that can be attributed to light-induced additional van der Waals forces. A theoretical formulation of van der Waals forces for nanoparticles was published in [660]. These dispersion forces are possible in metal nanoparticles upon excitations of SPPs and their electromagnetic multipolar interactions.

In experiments with 10 nm Au nanoparticles in hydrosols with slow spontaneous aggregation, a drastic acceleration of the aggregation process was directly observed during 10 min of illumination with an intense laser beam (argon ion

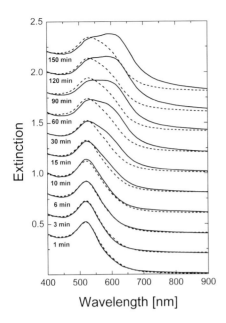

Figure 11.21 Light-induced aggregation in Au hydrosols [661, 662]. Comparison of the optical extinction spectra of the irradiated system (solid lines) with the nonirradiated system (dashed lines).

laser, 0.5 W, 514.5 nm) by comparing the extinction spectra of both illuminated and nonilluminated nanoparticles [661, 662]. The recorded extinction spectra are summarized in Figure 11.21.

During the first 15 min, the spectra of the irradiated and the nonirradiated sample are almost identical. After 30 min, the changes caused by the irradiation can clearly be recognized. Similar results have been reported [663, 664] for Au nanoparticles immersed in 2-propanol. Beside coagulation aggregation, coalescence and strong Ostwald ripening were detected there, pointing to photochemical effects.

Light absorption, scattering, and scattering of a probe beam were studied numerically with Newton's equations and the coupled dipole equations for penta-particle aggregates by Perminov et al. [665]. The relative changes in optical responses were large compared with the linear, low-intensity limit and relatively fast with characteristic times of nanoseconds. Time and intensity dependences were shown to be sensitive to the initial potential of the aggregation forces.

11.1.2.6 Coalescence of Nanoparticles

Once aggregation is initiated in many-particle systems, coalescence of particles is difficult to prevent. Optical properties of coalescence aggregates, however, are much more difficult to describe theoretically than those of coagulation aggregates. One possibility is to use numerical methods such as the DDA or the T-matrix approach.

In the following experimental example, we illustrate the transition from nano-structured material with coagulation aggregates to coalescence aggregates by the

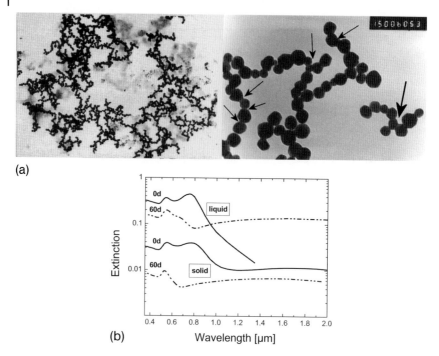

Figure 11.22 (a) Transmission electron micrographs showing quasi-fractal Au nanoparticle (2R = 38 nm) coagulation aggregates. When coalescence occurred after 60 days characteristic links had formed between adjacent particles (see arrows). (b) Extinction spectra of the coagulated and coalesced samples.

corresponding optical extinction spectra [114, 666]. The samples were aqueous colloidal systems and the same colloidal systems partly dried on quartz substrates to obtain solid samples. In a first step, coagulation aggregates of chain-like quasi-fractal topology were created in the colloidal suspensions (see the TEM images in Figure 11.22). The corresponding optical extinction spectra in Figure 11.22 in aqueous solution and on a quartz substrate show a typical two-peak structure descending at long wavelengths. After a time of 60 days, coalescence could be observed in both systems by formation of extended grain boundaries on the one hand and by the changes in the optical spectra on the other. In contrast to the coagulated systems, the long-wavelength peak decreased in magnitude and a new, broad extinction feature arose at wavelengths longer than 1200 nm. This extinction can be assigned to the newly formed larger and irregularly shaped coalescence aggregates. Collective excitations of the conduction electrons now refer to these new, larger units as a whole. As an approximation, their shape can be roughly approximated by ellipsoids. The resulting eccentricity points to about 10 particle diameters for the long axis, a result which was confirmed by TEM analysis.

In a study on the coalescence of nanoparticles, Hellmers *et al.* [667] investigated the sintering process of silver nanoparticles by light scattering calculations. First the scattering behavior of two separate silver spheres was calculated. Then the authors assumed sintering between them, leading to a single particle with a concave, peanut-like shape, approximated by a Cassini oval. For light scattering studies, an advanced T-matrix algorithm – the nullfield method with discrete sources – was used. Four different particle systems were considered. In the first step the spheres were separated, then they touched each other, and finally sintering started and a single particle was obtained, which was modeled by a Cassini oval approach and for two different stages of the process. The cross-sections for the two separate spheres exhibit one high peak which can be assigned to the longitudinal mode of a pair of spheres. If the particles were sintered, the new non-spherical silver particle exhibited two well-separated resonances for which the peak splitting increased with increasing sintering.

11.1.2.7 Further Experiments with Gold and Silver Nanoparticles

Based on the coupled-dipole method, Karpov et al. [668] proposed a theory for the optical properties of fractal clusters for the problem of adequately describing the evolution of optical spectra of any polydisperse silver colloid with particles falling within the range of most characteristic sizes (5–30 nm). The dipole–dipole interactions were shown to be a key factor in determining the broadening of the sol absorption spectra during the course of fractal aggregation. There exists a clear and strong correlation between the degree of particle aggregation and the aggregate structure on the one hand and the shape of the optical spectra on the other. This also means a clear confirmation of the results for gold and silver nanoparticles obtained with the GMT.

More recently, Ou et al. [669, 670] prepared Ag–Si_3N_4 composite films with metal fractions f of 20 and 33% by magnetron sputtering. The Ag particles inside the composite films were of radius 1 nm for $f = 20\%$ and 3 nm for $f = 33\%$. The optical absorption and near-field enhancement properties of the composite films were measured. Extra absorption due to interaction among the Ag nanoparticles was observed, the intensity of which depended on the particle size and metal fraction. The composite films with higher extra absorption at longer wavelengths had better near-field enhancement properties. In the second paper, the authors used the DDA method to calculate interactions among two neighboring Ag particles to interpret the extra absorption at longer wavelengths. The calculated results revealed that the interaction between metal particles actually induces this extra absorption and improves the near-field enhancement property of metal–dielectric composite films. This finding is in full agreement with calculations according to the GMT.

Akamatsu et al. [671] characterized polymer thin films containing silver particles and silver sulfide nanoparticles. They prepared particles with mean size ranging from 4.5 to 9.1 nm. The optical absorption spectra exhibited a clear red shift (to longer wavelengths) of the SPP peak, which was explained by the authors as resulting not only from the increasing size but mainly from interactions among the particles, because the volume fraction of the metal in the polymer film increased when preparing larger particles. Similar results were obtained by the authors for Au nanoparticles [672], where the effect of the interaction on the peak position could be clearly demonstrated.

11.2
Semimetal and Semiconductor Particles

Metal nanoparticles which interact strongly with visible light are useful for molecular detection and biosensing. To facilitate the useful application of these nanoparticles, it is important to design and fabricate the particles at a desired frequency. Khlebtsov and co-workers [673–675] therefore used colloidal gold nanoparticle aggregates as nanosensors in biological systems.

The optical extinction spectra of small nanoparticles of the half-metal ZrN in vacuum exhibit an extinction maximum at wavelengths around $\lambda = 435$ nm that can be interpreted as collective excitation of the free electrons – the SPP [101]. It can be expected that this peak splits into new resonances when the particles are lumped into aggregates. This is demonstrated in Figure 11.23, where computed

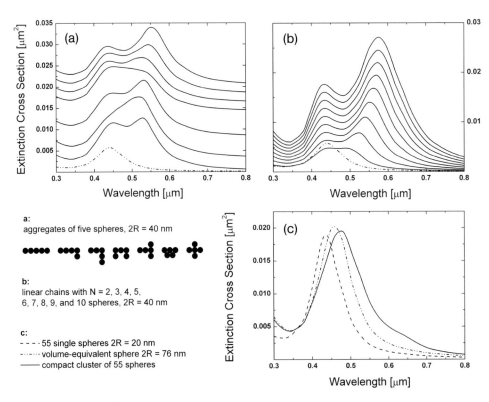

Figure 11.23 Extinction cross-section spectra of (a) aggregates with $N = 5$ identical zirconium nitride spheres of $2R = 40$ nm, (b) linear chains of ZrN particles, and (c) an icosadeltahedral aggregate of 55 particles with $2R = 20$ nm in comparison with the volume-equivalent sphere with $2R_{ev} = 76$ nm.

extinction and scattering cross-section spectra for ZrN aggregates in vacuum are displayed.

Indeed, if aggregation occurs, this resonance splits into new resonances, contributing to increased extinction (and scattering) mainly at longer wavelengths (low photon energies). The splitting depends on the topology of the aggregate, but is less pronounced than for the metals Al, Na, K, Ag, and Au considered earlier. The largest peak splitting occurs for chain-like aggregates.

Comparing the spectrum of the cluster CLU55 with the single particle spectrum of 55 single particles and that of the volume-equivalent sphere with $2R_{ev} = 76\,nm$, it is slightly shifted to longer wavelengths and at approximately 650 nm a weak resonance can be recognized.

Analogous calculations for TiN showed that for this material the aggregation hardly influences the spectra, similarly to gold in vacuum. Embedding in a surrounding medium with higher refractive index, however, leads to a clear splitting of the aggregate modes in the spectrum. Then, the experimental results of W. Hoheisel (personal communication) presented in Figure 11.24 can also be interpreted. The measured optical density spectra in Figure 11.24c exhibit a broad maximum at wavelengths between 0.73 and 0.77 μm and a minimum at around 0.5 μm. The measured spectra can be explained by a pot-pourri of calculated spectra in Figure 11.24a and b, resulting in a broad maximum at longer wavelengths.

Silicon exhibits strong interband transitions in the near-UV region close to the visible spectral region. Hence it is a good candidate for electronic resonances in the interband transition region. If aggregation occurs, these resonances must also

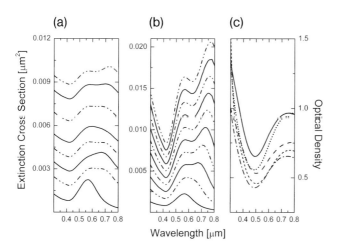

Figure 11.24 Extinction cross-section spectra of (a) aggregates with $N = 5$ identical titanium nitride spheres of $2R = 40\,nm$ and (b) linear chains of TiN particles in water. (c) Optical density of various particle ensembles of aggregated TiN nanoparticles. Data courtesy of W. Hoheisel.

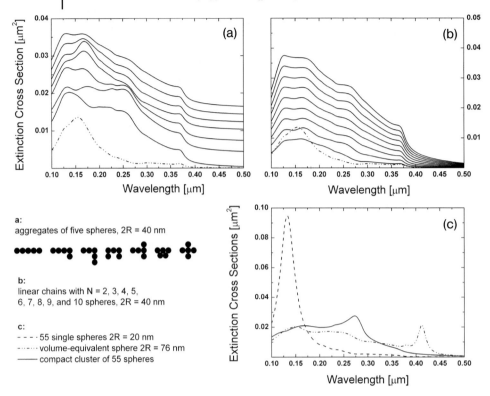

Figure 11.25 Extinction cross-section spectra of (a) aggregates with $N = 5$ identical silicon spheres of $2R = 40$ nm, (b) linear chains of Si particles, and (c) an icosadeltahedral aggregate of 55 particles with $2R = 20$ nm in comparison with the volume-equivalent sphere with $2R_{ev} = 76$ nm.

undergo a splitting into new resonances. This can be followed from the spectra in Figure 11.25.

Obviously, the influence of the shape is stronger than the influence of the size of the aggregates. The linear chain further does not have the largest splitting. Moreover, the length of the linear chain is not so important for the spectrum. The reason for this behavior of silicon nanoparticle aggregates in contrast to metal nanoparticle aggregates considered earlier is that in the spectral region of the electronic resonances of silicon the resonance conditions cannot be fulfilled for all possible resonances. In metals, the free electron contribution to the dielectric function of the corresponding metal leads to a negative real part of the dielectric function, which further decreases with increasing wavelength. In the spectral region close to a harmonic oscillator (interband transition), the real part of the dielectric function becomes negative only in a small spectral region. Then, not all resonance conditions of the aggregate can be fulfilled. This effect has already been discussed for the metal yttrium [676], where the dielectric function can be described in the wavelength range 0.2–1.9 μm by a harmonic oscillator contribution and a free electron contribution. For yttrium, aggregation mainly led only to a broadening of the single particle electronic resonance because all eigenmodes of the aggregates contribute only in the wavelength range between 0.3 and 0.5 μm.

Comparing the spectrum of the cluster CLU55 with the single particle spectrum of 55 single particles and that of the volume-equivalent sphere with $2R_{ev} = 76$ nm, they exhibit a resonance at different wavelengths (CLU55 at ~270 nm, volume-equivalent sphere at ~420 nm) so that they cannot be compared at all. Compact aggregates of silicon therefore cannot be approximated with a volume-equivalent sphere.

11.3
Nonabsorbing Dielectrics

At first glance, for nonabsorbing dielectric nanoparticles the formation of aggregates should not yield significant differences with respect to the single isolated particle. This can be seen from the plots in Figure 11.26, showing the extinction cross-section spectra of silica particles, forming aggregates with $N = 5$ identical particles in Figure 11.26a, and linear chains with $N = 2, 3, 5, 7, 10$, and 20 in Figure 11.26b. The aggregate spectra in Figure 11.26a are almost identical and deviate only slightly from each other at short wavelengths. Deviations from the scattering of a single particle can be best seen from the slopes of the extinction spectra of the linear chains in Figure 11.26b. For this purpose, the spectra are plotted in a double-logarithmic plot. With increasing chain length, the slope increases from -4 (Rayleigh scattering) for the single particle to -3.24 for the chain with $N = 20$ particles. This means that the scattering by these aggregates changes from $1/\lambda^4$ (Rayleigh scattering) to $1/\lambda^n$ with $n < 4$.

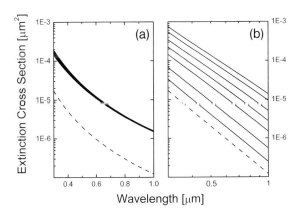

Figure 11.26 (a) Extinction (= scattering) spectra of aggregates of $N = 5$ identical silica spheres with $2R = 40$ nm and (b) extinction spectra of linear chains with $N = 2, 5, 10, 15$, and 20 identical SiO_2 spheres in a double-logarithmic plot. The dashed line corresponds to the single particle spectrum.

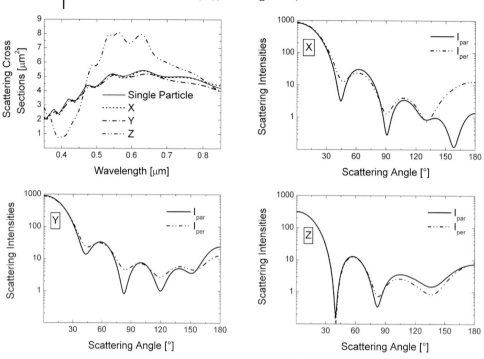

Figure 11.27 Extinction (= scattering) efficiencies and relative scattering intensities for a single, isolated silica sphere with $2R = 900$ nm and a dimer ($N = 2$) of silica spheres.

If the primary dielectric spheres are sufficiently large for MDRs to occur, electromagnetic coupling among the particles in the aggregate must also lead to new MDRs. The principal effect is obvious: due to the mutual scattering fields the amplitudes and phases of the waves within each particle are changed, and hence the interference effects among them. The effects can be followed from Figure 11.27, where extinction efficiency spectra and the scattering intensities i_p^N and i_s^N of an isolated SiO_2 particle with $2R = 900$ nm and of a dimer ($N = 2$) are plotted. The scattering intensities are computed at the wavelength $\lambda = 514$ nm with the maximum multipolar order $n_{MAX} = 20$. The three different curves belong to three different orientations of the dimer with respect to the incident wave. The incident wave propagates along the z-axis and is polarized along with the x-axis.

The efficiencies of the dimer differ only slightly from that of the single sphere if the dimer is arranged along the x- or y-axis. Only slight shifts of the resonance positions can be observed. For the dimer arranged along the z-axis, the retardation of the incident field and the interferences lead to a clearly different efficiency of the dimer. In unpolarized light, all three contributions determine the efficiency of the dimer which will also clearly differ from the efficiency of the single particles. Considering the scattering intensities of the dimer parallel and perpendicular to the scattering plane, it is obvious that the orientation of the dimer with respect to the incident wave plays a key role. All three cases result in a different angular scattering by the dimer.

11.4 Carbonaceous Particles

The optical properties of particles that consist of carbonaceous material are of interest in many fields of research such as aerosol science and astrophysics. While in astrophysics carbon is recognized as an element forming solid particles which are an important component of interstellar dust, the carbonaceous particles from combustion of oil products, coal, and carbon-containing organics are of increasing interest for the radiative balance of the Earth. In both astrophysical systems and terrestrial atmospheric aerosols they directly influence the energy balance of the system by absorption and re-emission of radiation.

One main and important part of dust in space or soot from exhaust are fractal-like and fluffy aggregates of polycrystalline carbonaceous spherical particles formed by nearly spherical primary particles. Their optical properties can be modeled by clusters of spheres within the GMT. Since it is exact (depending on the number of multipoles treated) and quickly converging, it allows the calculation of extinction spectra also in wavelength regions where the primary particle size is already comparable to the wavelength.

From the astrophysical point of view, the UV region is extremely important since the π–π* electronic transitions in such particles are observed in the interstellar extinction curve [162, 163]. Clustering effects on these absorption bands have been studied [677], but only for synthetic dielectric constants (except for graphite). Further, an experimental study of clustering effects in this wavelength region has been published [167].

The effect of clustering on the π–π* electronic transitions is depicted in Figure 11.28, where the optical extinction cross-section spectra of various aggregates of $N = 5$ particles with $2R = 40$ nm are presented, using optical constants of graphite parallel to the crystalline c-axis (C_{par}) and of graphite perpendicular to the c-axis (C_{per}). The interstellar extinction feature at wavelength 217.5 nm only appears for C_{per}. The aggregation mainly broadens the peak caused by the π–π* electronic transitions. For the second peak at around 80 nm, mainly the peak position is affected by the aggregation. For C_{par} the aggregation leads to a slight red shift of the long-wavelength peak and narrowing of the short-wavelength peak.

Quinten et al. [678] studied the wavelength-dependent extinction by absorbing carbonaceous particles over a wide spectral region. This requires the use of optical constants of a carbonaceous material relevant for both cosmic dust and terrestrial aerosols. As cosmic and terrestrial carbon particles are considered to be of a highly disordered structure, the data published by Jäger et al. [166] and Schnaiter et al. [167] were used for amorphous carbons produced by pyrolysis of cellulose. These data have the advantage that they cover a range of different microstructures (sp^2/sp^3 hybridization ratios) and, therefore, electrical conductivity and optical

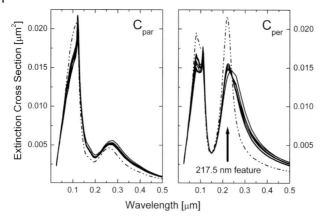

Figure 11.28 Extinction cross-section spectra of aggregates with $N = 5$ identical graphite spheres of $2R = 40$ nm for graphite parallel to the c-axis (C_{par}) and perpendicular to the c-axis (C_{per}).

properties. In the UV spectral range, pure graphite was also considered. Some detailed calculations on the influence of the inner structure of carbonaceous grains on the optical properties were made by Michel et al. [679] using also other computational techniques [680].

All spectra presented in Figure 11.29 show an increase in the extinction cross-section towards the UV with absorption maxima or shoulders in the 200–300 nm wavelength region. These features are related to transitions of the π-electrons in aromatic units of sp^2-hybridized carbon atoms from the valence band (π-band) to the π^* (conduction) band. The spectrum of the *graphite sphere* in Figure 11.29a exhibits a maximum in the extinction at $\lambda = 226$ nm, which led to assignments of the strong UV maximum of the interstellar extinction curve at 217.5 nm to extinction by graphite particles. Nowadays, it is assumed that cosmic carbon particles are structurally disordered, consisting of graphitic microcrystallites [basic structural units (BSUs)] with sizes of the order of about 1 nm. These BSUs are formed by sp^2-hybridized atoms and are embedded in sp^3-hybridized material which does not contribute to the π–π^* band. Frequently, also bent *graphitic* units, which can be described by mixtures of sp^2- and sp^3-hybridized atoms (see also fullerenes and nanotubes), can be found in disordered carbonaceous particles [681].

As an analog for the carrier of the *interstellar UV bump*, hydrogenated amorphous carbon particles produced by evaporation of carbon in a hydrogen–argon atmosphere have been proposed [167]. Spherical particles composed of this material called *H50* show the extinction maximum at exactly the correct wavelength position and with the correct width (Figure 11.29b). In contrast, the disordered carbonaceous materials prepared by Jäger et al. [166] (CEL400, CEL1000) are characterized by a very broad and much weaker band with the maximum at wavelengths of 280–300 nm, indicating a higher degree of disorder.

From the comparison of Figure 11.29a–d, it is evident that the significance of shape and clustering effects is related to the strength of the UV bands. Whereas the differences between the spectra calculated for the different particle geometries

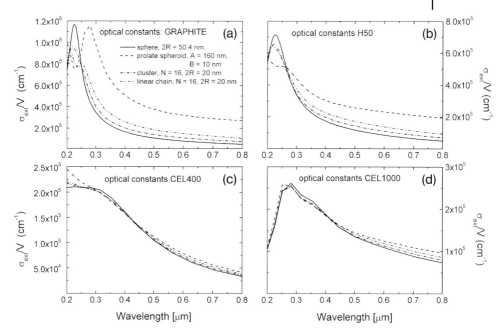

Figure 11.29 Optical extinction spectra of a compact aggregate and a linear chain with N = 16 primary spheres with sizes 2R = 20 nm, a volume-equivalent sphere, and a volume-equivalent prolate spheroid in the wavelength range 0.2–0.8 μm. Panels (a)–(d) correspond to different sets of optical constants.

are fairly small for CEL400 and CEL1000 (Figure 11.29c–d), significant differences between these spectra are found for graphite and H50. Already for the compact clusters, the π-electron band is broadened (Figure 11.29a and b), shifted (Figure 11.29b), or becomes asymmetric (Figure 11.29a) compared with that of the volume-equivalent sphere. The dependence on the material properties can be understood by considering the scattering behavior of the single spheres constituting the aggregates. The stronger the transition of the π-electrons, the more their absorption is dominated by resonances of the polarizability of the sphere. They are very sensitive to all changes in the particle geometry and also to the electromagnetic coupling among primary particles via the scattered waves.

More clearly, these effects can be observed in the spectra of the linear chains, where a splitting of the single-sphere resonance is evident in the case of the stronger π–π* transitions. One of the new resonances is shifted towards smaller wavelengths (e.g., for graphite to $\lambda = 212$ nm) and the other one is shifted towards longer wavelengths ($\lambda = 252$ nm) compared with the single particle band ($\lambda = 226$ nm). In Figure 11.29b, the splitting is not that distinct but is visible as an asymmetry of the extinction band. Coalescence of the linear aggregates to prolate ellipsoids of revolution enhances the resonance splitting considerably (see Figure 11.29a and b). Here, it is so strong that for the blue-shifted resonance only the long-wavelength decline is visible in the available wavelength range.

11.5
Particles with Phonon Polaritons

Interaction of light with almost all ionic crystals is determined by the coupling of IR radiation with TO phonons. This leads to a significant change in the dielectric function of the material. This coupling can be interpreted in the framework of harmonic oscillators. As already seen and discussed for spherical and nonspherical magnesia (MgO) and sodium chloride (NaCl) single nanoparticles, one large resonance peak is obtained for spherical particles and at least two clearly separated resonances for nonspherical particles.

If MgO or NaCl spherical nanoparticles build aggregates, the single particle peak splits into new resonances contributing to increased extinction (and scattering) at longer wavelengths and also shorter wavelengths than the position of the electronic resonance. The splitting depends sensitively on the topology of the aggregate. The largest peak splitting occurs for chain-like aggregates. This can be seen from Figure 11.30 for MgO and Figure 11.31 for NaCl.

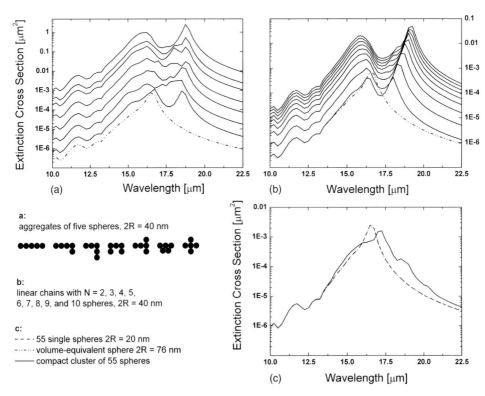

Figure 11.30 Extinction cross-section spectra of (a) aggregates with $N = 5$ identical magnesia spheres of $2R = 40$ nm, (b) linear chains of MgO particles, and (c) an icosadeltahedral aggregate of 55 particles with $2R = 20$ nm in comparison with the volume-equivalent sphere with $2R_{ev} = 76$ nm.

a:
aggregates of five spheres, $2R = 40$ nm

b:
linear chains with $N = 2, 3, 4, 5, 6, 7, 8, 9$, and 10 spheres, $2R = 40$ nm

c:
- - - 55 single spheres $2R = 20$ nm
······ volume-equivalent sphere $2R = 76$ nm
——— compact cluster of 55 spheres

11.5 Particles with Phonon Polaritons

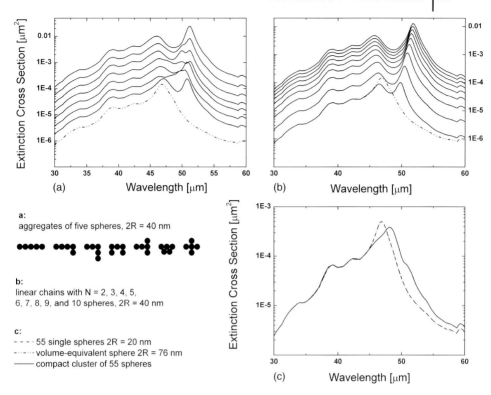

Figure 11.31 Extinction cross-section spectra of (a) aggregates with $N = 5$ identical sodium chloride spheres of $2R = 40$ nm, (b) linear chains of NaCl particles, and (c) an icosadeltahedral aggregate of 55 particles with $2R = 20$ nm in comparison with the volume-equivalent sphere with $2R_{ev} = 76$ nm.

The spectrum of the cluster CLU55 differs from the volume-equivalent sphere with $2R_{ev} = 76$ nm for both materials mainly in the long-wavelength range. The peak is shifted to longer wavelengths and some additional broad resonances appear in this spectral range for CLU55.

Fuchs [463] not only developed a model for the optical properties of ionic crystal cubes but also compared it with measured spectra. An example is given in Figure 11.32. Whereas the calculated spectrum (dashed line) fits fairly well the measured spectrum for MgO (solid line), the difference is larger for NaCl. Obviously, if one compares it with the aggregate spectra above, the measured spectrum seems to be determined by smaller aggregates that lead to a red shift of the resonance peak and to additional long-wavelength extinction.

In a second paper, Fuchs calculated the IR absorption by cube pairs [682] with the DDA. This resulted in additional extinction peaks mainly at longer wavelengths similar to the spectra of aggregates of spherical NaCl nanoparticles. These results better explain the measured spectrum in Figure 11.32.

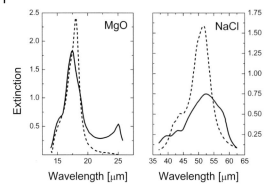

Figure 11.32 Calculated (dashed line) and measured (solid line) extinction of cubic MgO and NaCl particles in the infrared spectral region. After [463].

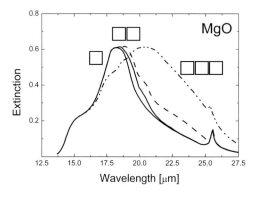

Figure 11.33 Measured extinction of several samples containing cubic MgO nanoparticles in the infrared spectral region. Data courtesy of M. Essig.

M. Essig (personal communication) examined cubic MgO nanoparticles in the infrared and measured the spectra of several samples. In addition transmission electron microscopy was applied to characterize the samples. Figure 11.33 provides his measured spectra of several samples containing MgO nanoparticles. The spectra clearly differ mainly at longer wavelengths. From electron microscopy, Essig was able to assign the single isolated particles and also aggregates of two and three cubic particles to the spectra. From comparison with the calculated aggregate spectra of spherical nanoparticles above, this assignment appears to be plausible, because aggregates with only a few particles have similar spectra. For better agreement, however, we believe that a mixture of different aggregates and the nonspherical shape of the primary particle must be taken into account.

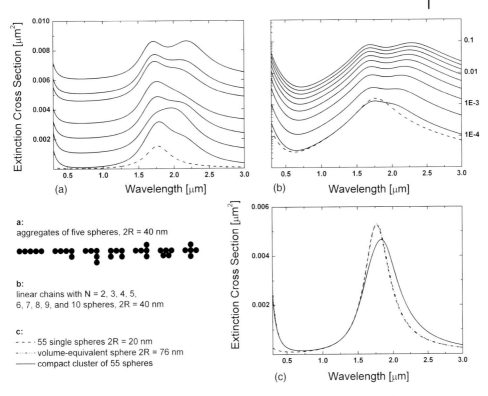

Figure 11.34 Extinction cross-section spectra of (a) aggregates with $N = 5$ identical ITO spheres of $2R = 40$ nm, (b) linear chains of ITO particles, and (c) an icosadeltahedral aggregate of 55 particles with $2R = 20$ nm in comparison with the volume-equivalent sphere with $2R_{ev} = 76$ nm.

11.6 Miscellaneous Particles

Indium-doped tin oxide (ITO) nanoparticles and lanthanum hexaboride (LaB$_6$) nanoparticles have been already discussed as particles for IR absorbers. The reason is that these particles exhibit an SPP resonance in the NIR region. Therefore, we can expect splitting of this resonance if nanoparticles are lumped into aggregates.

In fact, as follows from Figures 11.34 and 11.35, aggregates of nanoparticles of both materials exhibit new resonances that contribute to the extinction spectra mainly at longer wavelengths. The splitting depends on the topology of the aggregate, but is less pronounced than for the metals. The largest peak splitting occurs for chain-like aggregates.

The spectrum of the cluster CLU55 is very similar to that of a single nanoparticle. Compared with the volume-equivalent sphere with $2R_{ev} = 76$ nm, it is slightly

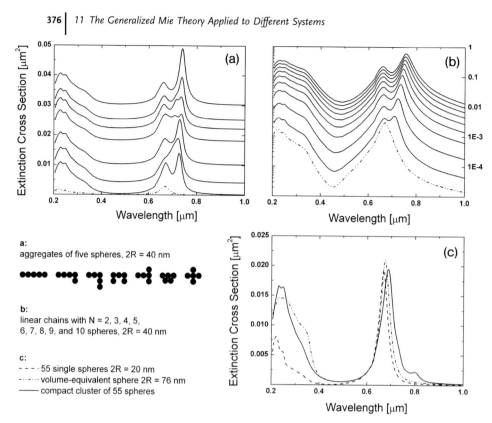

Figure 11.35 Extinction cross-section spectra of (a) aggregates with $N = 5$ identical lanthanum hexaboride spheres of $2R = 40$ nm, (b) linear chains of LaB$_6$ particles, and (c) an icosadeltahedral aggregate of 55 particles with $2R = 20$ nm in comparison with the volume-equivalent sphere with $2R_{ev} = 76$ nm.

red shifted and broadened. For LaB$_6$ even a small resonance at $\lambda = 900$ nm appears which is absent in the calculations for a single spherical particle.

11.7
Aggregates of Nanoparticles of Different Materials

In some applications, the dielectric functions of the nanoparticles may differ. This is the case, for example, in *scanning near-field optical microscopy* (SNOM), where the scanning tip and particles on a substrate usually have different optical material functions. Three classes of particle materials – metals, dielectrics and semiconductors – must be considered. The dielectrics must be further divided into absorbing and nonabsorbing dielectrics. We restrict considerations here to metal–metal interactions; for other examples, see [683]. We consider the interaction of a silver nanoparticle with diameter $2R = 40$ nm with another Ag, Au, Pt, Pd, or Fe metal particle of the same size and show computed extinction efficiency spectra in Figure 11.36.

11.7 Aggregates of Nanoparticles of Different Materials | 377

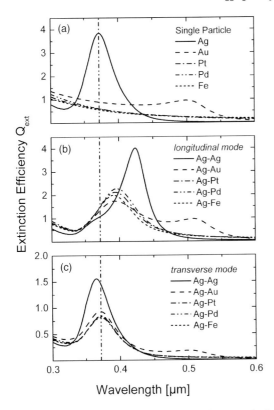

Figure 11.36 Extinction efficiency spectra of (a) isolated metal clusters, (b) the longitudinal mode of pairs of metal–silver nanoparticles, and (c) the transverse mode of pairs of metal–silver nanoparticles. The vertical line represents the peak position of the single silver nanoparticle plasmon polariton resonance.

In Figure 11.36a the efficiency spectra of the single nanoparticles are summarized. The SPP resonance for the silver particle, peaked at $\lambda = 390$ nm, and the gold particle, peaked at $\lambda = 524$ nm, can be clearly recognized. All other metal nanoparticles do not exhibit such an SPP. Figure 11.36b and c show the efficiencies of a pair of particles in interaction for the longitudinal and the transverse mode, respectively.

The coupling among two adjacent silver particles leads to the largest splitting of the SPP resonance into the longitudinal and transverse modes of the particle pair. An interesting case is the coupling among a silver and a gold nanoparticle, because both particles exhibit an SPP. However, the SPPs are peaked at different, well-separated wavelengths. Indeed, in the spectra of the silver–gold pair both SPPs contribute to the spectrum, but are red shifted for the longitudinal mode and are blue shifted for the transverse mode. Moreover, the peak magnitudes are decreased with respect to the isolated particles.

The coupling of the silver particle with another, plasmon-less metal particle obviously also results in a splitting of the silver nanoparticle SPP. However, the red shift of the longitudinal pair resonance is less than for the coupling between two silver particles. For the transverse mode the peak position is even the same as for the single particle silver SPP. Hence, for an assembly of randomly oriented nanoparticle pairs, only one broad peak will be observed due to superposition of the two peaks. In all spectra a decrease in the magnitudes and broadening of the pair resonances can be recognized, except for the coupling between two silver particles. The magnitude of the corresponding longitudinal pair resonance decreases by a factor of 2 and the transverse pair resonance peak decreases by a factor of approximately 4 with respect to the single silver nanoparticle SPP. For the silver pair the transverse mode peak is approximately one-third of the single nanoparticle peak, whereas the longitudinal peak is almost as high as the single nanoparticle SPP.

In Figure 11.37 we summarize peak positions and halfwidths of the longitudinal and transverse modes for all investigated pairs of nanoparticles. The weakest effect on the SPP of the silver particle occurs for silica, or more generally for nonabsorbing dielectrics. For these particle materials, for the second particle the splitting of the resonance is almost negligible and cannot be resolved for unpolarized illumi-

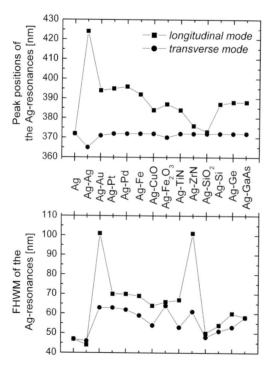

Figure 11.37 Peak positions and halfwidths of the longitudinal and transverse modes of a pair of nanoparticles with different optical constants ($2R = 40$ nm, touching spheres).

nation. Obviously, the full halfwidth of the transverse and the longitudinal modes becomes rather large if the second particle also exhibits an SPP like gold and titanium nitride (TiN). The reason is that the SPP resonance wavelength of these nanoparticles is close to that of the silver particle and, hence, both resonances overlap.

11.8
Optical Particle Sizing

Optical particle sizing is based on light scattering by single particles in a small sensing volume. The scattered light intensity depends on the size, shape, and refractive index of the particle. Assuming spherical particles, it is possible to determine the size distribution from evaluating the scattered light intensity and comparing it with calibration curves calculated from Mie theory. However, for nonspherical particles, a correct assignment to the particle size is more difficult since the scattering of light by nonspherical particles is different from that of spheres. For example, in a study on the calibration of optical particle counters, Friehmelt and Heidenreich [684] showed that for nonspherical quartz particles the scattered light intensity increases compared with spherical particles of the same bulk material. This increase in scattered light intensity was observed for particles with equivalent volume diameters larger than approximately $0.7\,\mu m$. For smaller particles no difference between spheres and nonspherical particles could be detected. Although these results are not very surprising, so far no mathematical model is able to predict the scattering behavior of nonspherical particles in such a way that detailed experimental calibration procedures become unnecessary. Here, we model nonspherical particles by means of aggregates of closely packed spherical particles. We have developed a mathematical model that calculates light scattering intensities by particle aggregates for a specific white light/90° optical particle counter (OPC) of the type described first by Umhauer [685]. The useful size range of this OPC is $2R = 0.5–30\,\mu m$ with a resolution of $\Delta\log(2R) = 0.1423$.

In order to study the effect of aggregates on the scattered intensity measured with an OPC, the following computational procedure was applied [686]:

Step 1: Computation of $i_s(\theta,N)$ and $i_p(\theta,N)$ in the range $[\theta_0 - \vartheta, \theta_0 + \vartheta]$ at a fixed wavelength λ_i using GMT. This is the main task in our computation procedure.

Step 2: Integration of the data obtained in step 1 over the finite cylindrical aperture using the model of Heyder and Gebhart [687]:

$$S(\lambda) = \frac{\lambda^2}{4\pi^2} \int_{\theta_0-\vartheta}^{\theta_0+\vartheta} \left\{ [i_s(\theta) + i_p(\theta)]\cos^{-1}\left(\frac{\cos\vartheta - \cos\theta\cos\theta_0}{\sin\theta\sin\theta_0}\right) \right.$$
$$\left. + \frac{1}{2}[i_s(\theta) - i_p(\theta)]\sin\left[2\cos^{-1}\left(\frac{\cos\vartheta - \cos\theta\cos\theta_0}{\sin\theta\sin\theta_0}\right)\right] \right\} \sin\theta\,d\theta \quad (11.4)$$

Step 3: Repetition of steps 1 and 2 for all wavelengths λ_i provided by the light source.

Step 4: Calculation of the total scattered light, S_{total}, taking into account the spectral distribution of the light source convoluted with the detector sensitivity $L(\lambda)$:

$$S_{total} = \frac{\int_{\lambda_1}^{\lambda_2} S(\lambda)L(\lambda)d\lambda}{\int_{\lambda_1}^{\lambda_2} L(\lambda)d\lambda} \tag{11.5}$$

This computation procedure has to be repeated for each particle size in order to obtain an intensity curve versus particle diameter.

We computed intensity curves for exemplary closely packed aggregates of spheres of varying primary particle size. The incident light was assumed to propagate along the positive z-axis of a reference frame, being unpolarized. The examined aggregates of N = 2, 3, or 4 spheres are given as insets in Figures 11.38–11.40. The refractive index of the primary particles is assumed to be n = 1.6, which is the refractive index of polystyrene at wavelengths in the visible spectral region.

In Figure 11.38, the calculated scattered intensity curves as a function of the particle size for a pair of identical particles are summarized. In principle, the dimer may be arranged along the x-axis (dimer 1) or along the z-axis (dimer 2) in the reference frame given in the plot. Then, different intensities result mainly for larger particles. In comparison with the scattered intensity of a single spherical particle, given as the solid line, the intensity curves of the two dimers do not differ from this reference curve for equivalent volume diameters less than about 0.126 μm, which corresponds to a primary particle size of 0.1 μm. This means that the OPC cannot distinguish between the aggregates and single particles for such small

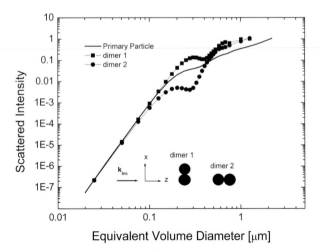

Figure 11.38 Calculated scattered intensity in the OPC for the aggregates dimer 1 and dimer 2 of two identical spheres in comparison with the calculated intensity curve for a single spherical particle. The refractive index is n = 1.6.

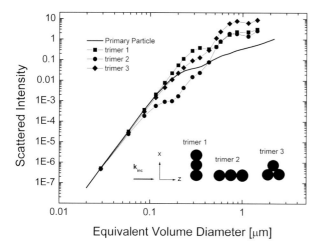

Figure 11.39 Calculated scattered intensity in the OPC for the aggregates trimer 1, trimer 2 and trimer 3 of three identical spheres in comparison with the calculated intensity curve for a single spherical particle. The refractive index is $n = 1.6$.

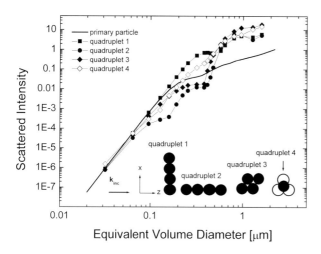

Figure 11.40 Calculated scattered intensity in the OPC for the aggregates quadruplet 1, quadruplet 2, quadruplet 3 and quadruplet 4 of four identical spheres in comparison with the calculated intensity curve for a single spherical particle. The refractive index is $n = 1.6$.

particles. For larger sizes the intensity scattered by dimer 1 is larger than that of a single sphere of equivalent volume by a factor of up to 5. In contrast, the intensity curve of dimer 2 lies under the reference curve for a single sphere in the size range between 0.126 and 0.4788 µm (0.1 and 0.32 µm primary particle size). The intensity is almost constant between 0.126 and 0.3 µm and then increases rapidly, approaching the single particle curve for 0.4788 µm. In this case, the aggregate will be assigned to a single sphere of 0.126 μm as long as the equivalent volume diameter is between 0.126 and 0.3 μm. For sizes larger than 0.4788 μm (0.32 μm primary particle size), the intensity scattered by dimer 2 is larger than that of the volume-equivalent sphere by a factor of up to 5 and approaches the intensity curve for dimer 1. The differences in the scattering by dimers 1 and 2 mainly result from a retardation in the excitation of the particles by the incident light. For dimer 1, the incident light arrives simultaneously at both particles, whereas for dimer 2 the incident wave is phase shifted on arriving at the second particle. This phase shift increases with increasing primary particle size. For particles smaller than about 0.126 μm, the phase shift is as small as the retarded excitation of the second particle does not influence the scattering by the dimer remarkably. Then, the scattered intensity curves of the two dimers are almost identical.

In Figure 11.39, the corresponding results for the scattered intensity in the OPC are depicted for three different aggregates with $N = 3$ primary particles. The intensities scattered by trimers 1 and 2 can be discussed similarly to the dimers in Figure 11.38. Again, for the cluster along the z-axis (trimer 2), retardation in the excitation of the particles in the aggregate leads to significant deviations in scattering compared with the aggregate trimer 1. The deviations appear already at smaller primary particle sizes of about 0.086 μm (0.06 μm primary particle size) due to the increased length of the aggregate. For particles larger than 0.1442 μm (0.1 μm primary particle size), the scattering by trimer 1 is larger than that of the volume-equivalent sphere by a factor of up to 5. For the aggregate trimer 3 the more compact topography of the aggregate leads to an intensity curve that is close to the reference intensity curve of a single particle for particle sizes less than 0.2 μm (0.14 μm primary particle size). For larger sizes the intensity is enhanced by a factor of up to 9.

With increasing number of particles in the aggregate, the number of different aggregate topologies increases. Then, the intensity curves also become more complicated due to various retardation effects. In Figure 11.40, a final example is given for four aggregates of $N = 4$ primary particles. As becomes obvious from this figure, the total length of an aggregate also leads to deviations of the intensity curve from the reference intensity curve. This can be followed from the intensity curves plotted in Figure 11.40, in which already for particle sizes larger than 0.1 μm (0.063 μm primary particle size) the scattered intensity differs from that of a single sphere. It follows that the corresponding intensity curve is closer to the reference curve of the single particle the more compact the aggregate is. This holds true for sizes up to 0.22 μm (0.14 μm primary particle size). For larger primary particle sizes, all aggregates scatter more efficiently than the single particle.

11.9
Stochastically Distributed Spheres

The general – and most realistic – case of nanocomposites consists of more or less densely packed single particles plus aggregates with various neighbor distances,

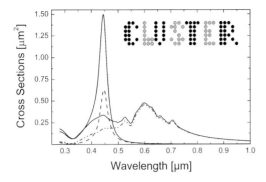

Figure 11.41 Extinction and scattering spectra (solid and dashed lines) of Ag nanoparticles (2R = 40 nm) and seven aggregates with a total of 91 nanoparticles in glass (N-BK7). Optical constants of Ag from [40].

coordination numbers, sizes, and shapes, which for their part again are packed more or less closely together. Mostly, they are additionally brought up on substrates. In principle, this general case can be treated with the GMT, because many-aggregate topologies are admitted and the particle coordinates are the only input parameters. Consequently, the GMT is not limited to one particle aggregate alone, but allows the computation of several neighboring aggregates simultaneously. Once the computer facilities permit the choice of a sufficiently extended portion of the sample which reflects the main features of the topology of the whole sample, the GMT will be the appropriate tool to treat arbitrary nanoparticle samples as it uses the *coherent superposition of electromagnetic fields* instead of the incoherent superposition of intensities in radiative transfer models. In fact, this goal is not that unreachable even today, as can be seen from Figure 11.41, where we present both the extinction and the scattering spectra of seven close-lying different Ag particle aggregates forming the word CLUSTER.

Note that the spectra of this CLUSTER aggregate published in the book by Kreibig and Vollmer [11] were calculated by Quinten using optical constants of Johnson and Christy [37] for silver and a constant refractive index $n_M = 1.71$ for the surrounding medium.

Less spectacular than the above CLUSTER, but more realistic, are samples with particles seeded on a substrate. In the following, we show some examples for nanoparticles of Au and Ag nanoparticles.

In the first example [419], 78 Au nanoparticles with distributed sizes are spread out on the substrate as shown in the inset of Figure 11.42. The corresponding extinction cross-section spectrum, calculated with the GMT, is shown in comparison with the spectrum of all 78 nanoparticles without electromagnetic interaction. In both calculations (Mie and GMT), the different particle sizes are taken fully into account.

Obviously, the SPP resonance becomes red shifted when taking into account electromagnetic interaction among the particles. The shift amounts to $\Delta = 31$ nm. No splitting into at least a short- and a long-wavelength resonance is observed.

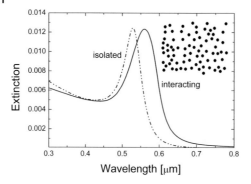

Figure 11.42 Statistical distribution of 78 seeded Au nanoparticles (different sizes) and extinction cross-section spectra of the isolated particles (dashed line) and the coupled particles (solid line).

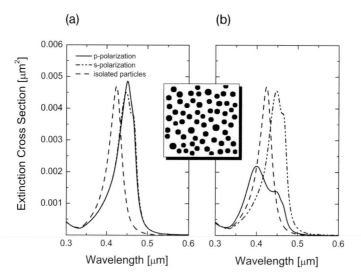

Figure 11.43 Polarization-dependent extinction cross-section spectra of 64 seeded Ag nanoparticles ($2R = 6$ nm) for angle of incidence (a) $\theta_i = 0°$ and (b) $\theta_i = 60°$. The spectra are compared with those of 64 noninteracting Ag nanoparticles (dashed line).

The reason is that the mean interparticle distance is still high as two clearly separated peaks cannot result from the interaction but are close together, forming one shifted and broadened peak. Its peak position is dominated by the long-wavelength resonance.

A similar result is obtained in the second example for Ag nanoparticles [170]. Here, we calculated the extinction cross-section spectra of 64 Ag nanoparticles with mean particle size $2R = 6$ nm. They are spread out on the substrate as shown in the inset of Figure 11.43. Again, the spectra of the coupled particles are compared with the spectra of noninteracting particles. The computed spectra are additionally resolved according to the polarization and the angle of incidence is $\theta_i = 0°$ in Figure 11.43a and 60° in Figure 11.43b.

Again, the interaction among the particles leads to a red shift of the plasmon polariton resonance, amounting to $\Delta = 27$ nm for perpendicular incidence. The spectra for the two polarizations do not clearly differ. Only a slight difference between the polarizations [the electric field along the x-axis (s-polarization) and along the y-axis (p-polarization)] can be observed, meaning that the arrangement of the particles on the substrate is not completely symmetric. In the preceding example for Au nanoparticles, this polarization dependence was not extra resolved. If the angle of incidence has changed to $\theta = 60°$, a clear dependence on the polarization can be observed. In s-polarization, the spectrum is the same as for perpendicular incidence. In p-polarization of the incident wave, however, two peaks can be recognized: one short-wavelength peak, which is clearly resolved, and one long-wavelength peak, which appears only as shoulder. The reason for this splitting is that p-polarized light with an angle of incidence $\theta_i \neq 0°$ can be divided into two parts, one part being perpendicular to the substrate surface and the other being along the substrate surface. The polarization dependence and the splitting in p-polarization become more obvious with increasing single particle size. A comprehensive discussion of such metal island films and their optical properties during the growth of the nanoparticles was presented by Lebedev et al. [689] in 1999.

Metal nanoparticle systems were prepared by Ag ion implantation into perfect single crystalline SiO_2 by Liu et al. [690, 691]. The samples can be divided into two regions: the shallower implanted layer contains noninteracting small Ag nanoparticles and the deeper layer contains interacting large nanoparticles. In the deeper layer, a broadening and red shift of the SPP resonance of the silver spheres can be observed at low doses. For doses exceeding approximately 5×10^{16} Ag^+ ions cm^{-2}, even a splitting into two well-separated extinction bands can be observed in the spectral absorption. Obviously, here even chain-like aggregates of certain length form as in the case of the colloidal suspension of Kahlau [650] presented in Section 11.1.2.1.

Although effective medium models were used by Pivin et al. [692] to explain the spectra of Ag nanoparticles implanted in SiO_2, this is also an example of the interaction among nanoparticles in films and substrates. SiO_2–Ag colloid films were prepared by ion implantation and ion beam mixing. Whereas for low metal concentrations the shape of the spectra is determined by the nanoparticle size, at higher concentrations it depends largely on their mutual interaction. This interaction induces a broadening of the resonance in implanted films and on the contrary a narrowing in ion beam mixed films. In Figure 11.44 we show a typical TEM image of larger Ag nanoparticles in SiO_2 and their distribution in the matrix, and the optical density spectra of four samples obtained for different ion fluences. The SPP peak of the silver nanoparticles is clearly broadened and shows a red shift due to the interaction among the more or less closely lying particles.

Taleb et al. [693] and Pileni [694] described the collective properties of silver nanoparticles organized in two-dimensional superlattices. The particles were found in the form of either a well-organized two-dimensional array of isolated

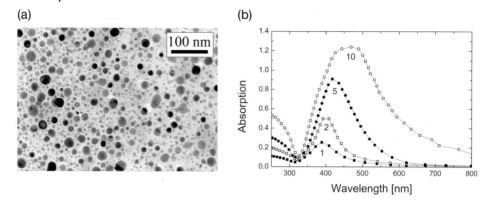

Figure 11.44 (a) Typical transmission electron micrograph of silver particles from ion implantation in silica. (b) Optical density spectra of the SiO_2–Ag systems at different ion fluences. Data and TEM image courtesy of H. Hofmeister.

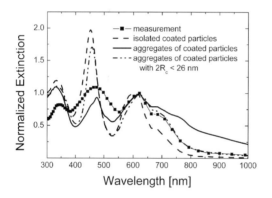

Figure 11.45 Optical properties of a copper phthalocyanine film with inclusions of Ag nanospheres and aggregates.

particles or disordered and coalesced particles distributed more or less randomly on the surface. The optical spectra are compared with both polarized and unpolarized light. When particles are arranged in a hexagonal array, an asymmetric and broad peak is observed. Under p-polarized light, the low-wavelength peak could be clearly resolved. With a disordered and coalesced system, the low-wavelength peak disappears whereas a peak towards long wavelengths is observed. This is attributed to coalesced particles.

Quinten et al. [695] studied the linear optical properties of copper phthalocyanine (CuPc) films with inclusions of copper, silver, or gold nanoparticles. For calculation of the optical properties of these films, the size distribution of the particles and the interaction among the particles were taken into account. As the CuPc films are strongly absorbing in several spectral regions, including the visible region, the single particles were simulated as coated spheres, consisting of a copper, silver, or gold core with a CuPc overlayer. Figure 11.45 provides an example of Ag nanoparticles in a CuPc film. Most of the nanoparticles have core sizes with $2R_c < 26$ nm with a mean shell thickness $d = 16.1$ nm. The volume fraction of suspended particles is $f = 0.2$. With these parameters, extinction cross-section spectra were computed, assuming a pot-pourri of isolated heterogeneous particles and also a pot-pourri of planar aggregates of 5×5 coated particles.

The pot-pourri of aggregates fits the measured spectrum better than the pot-pourri of isolated particles. In particular, the absorption at wavelengths above 620 nm is better fitted because the absorption by the single particles is already decreased in this spectral region. However, also for the aggregates there are some differences in this spectral region. Obviously, the amount of particles with $2R_c > 26$ nm is overestimated, leading to stronger absorption than in the measurement. If the particle size is restricted to particles with maximum $2R_c = 26$ nm (dashed-dotted line), rather good agreement with the measured spectrum is obtained without loss of information at shorter wavelengths.

Fucile et al. [237, 238] developed in 1997 a rigorous solution for the scattering and absorption by a single spherical particle on a surface. In 1999, the same group introduced a theoretical approach suitable for exact calculations of the scattering pattern from a dielectric substrate sparsely seeded either with single spheres or with aggregates of spheres [696, 697]. The main idea was to consider also the incident light reflected by the surface and the light scattered by each particle and reflected at the surface in the matrix equations of the coupled spheres. The theory correctly describes effects such as the depolarization and the shadow effect that are expected to occur, on both theoretical and experimental grounds, because of the anisotropy of the deposited particles.

11.10
Aggregates of Spheres and Numerical Methods

The electromagnetic interaction among spherical particles can also be treated within the framework of the common numerical methods discrete dipole approximation (DDA) or volume integral equation formulation (VIEF), T-matrix or extended boundary condition method (EBCM), finite difference time domain (FDTD) method, and generalized multipole technique. In the following we give some examples and a list of relevant publications in this field. This list does not claim to be complete but gives a good overview of relevant studies.

Many of these publications are concerned with the optical properties of fractal-like particles. They play a distinct role in atmospheric science where carbonaceous species form soot particles. In contrast, more fluffy aggregates of spheres can be found in interstellar matter and are of interest in astrophysics.

11.10.1
Applications of the Discrete Dipole Approximation

Félidj et al. [698] applied the discrete dipole approximation to simulate aggregates of gold nanoparticles in colloidal suspensions. Aggregation in the colloidal suspension led to a broad band strongly shifted towards long wavelengths, similarly to the experiments presented in Section 11.1.2. The observed extinction spectrum was simulated as a suitably weighted sum of extinction cross-sections of various objects of any size and shape. Various aggregates were assumed and the extinction spectrum was calculated with DDA. Then, the weight of each aggregate was adjusted so that fairly good agreement of calculated and measured spectrum resulted. This work is rather similar to that presented for gold and silver colloids in Section 11.1.2, where the calculations were carried out with the GMT and the weights of the aggregates were taken from TEM images.

Xu and Gustafson [699] used the DDA for calculations on two-sphere aggregates of carbonaceous particles. Scattering intensities at a fixed wavelength and cross-sections as a function of the size parameter were calculated. They compared the calculations with the GMT and with laboratory experiments. The DDA works well for small volume structures; its validity is, however, challenged on larger structures.

In a comprehensive project, Andersen et al. [700] studied the extinction by compact and fractal polycrystalline graphitic clusters consisting of touching identical spheres. Three general methods for computing the extinction of the clusters in the wavelength range 0.1–100 μm, namely the GMT and two different discrete-dipole approximation methods. Clusters of $N = 4, 7, 8, 27, 32, 49, 108$, and 343 particles with radii of either 10 or 50 nm were considered in three different geometries: open fractal (fractal dimension $D = 1.77$), simple cubic, and face-centered cubic. The fractal and compact clusters display an extinction at long wavelengths, of the same order of magnitude as computed with the GMT. It seems that the DDA approximations grossly overestimate the long-wavelength extinction of small fractal structures. The DDA codes should therefore be used with caution for this type of problem. The GMT computations were compared with the observed interstellar extinction curve. It was found that the results for small and medium-sized (fewer than 50 particles) fractal clusters are in fair agreement with observational constraints, whereas those of compact simple cubic and face-centered cubic clusters are not.

Iskander et al. [701–703] used the volume integral equation formulation (VIEF) to make intensive calculations on five smoke agglomerates having a widely varying number of particles ranging from 12 to 372 and an almost constant fractal dimension in the range 1.7–1.9. From these calculations, they derived expansion coefficients for an empirical equation. The equation satisfies the frequency dependence of the absorption at lower frequencies and up to the resonance frequency of the agglomerate. Numerical results obtained from the empirical equation are generally in good agreement with those calculated by the VIEF for soot particles with a magnitude of complex permittivity in the range 1–7. The equation provides a simple and inexpensive method for calculating the optical absorption of aerosol agglomerates, which otherwise require complicated and expensive methods of calculation.

The shape anisotropy of gold and silver nanorods gives rise to two distinct orientational modes for the primary particle. In regular arrangements nanorods can be assembled end-to-end and side-by-side. Optical absorption spectra of gold nanorod assemblies have earlier been observed to show a red shift of the longitudinal plasmon band for the end-to-end linkage of nanorods, resulting from the plasmon coupling between neighboring nanoparticles, similar to the assembly of gold nanospheres. Jain *et al.* [704] observed, however, that side-by-side linkage of nanorods in solution shows a blue shift of the longitudinal plasmon band and a red shift of the transverse plasmon band. Optical spectra calculated using the DDA method were used to simulate plasmon coupling in assembled nanorod dimers. The longitudinal plasmon band was found to shift to longer wavelengths for end-to-end assembly, but a shift to shorter wavelengths was found for the side-by-side orientation, in agreement with the optical absorption experiments. The strength of plasmon coupling was seen to increase with decreasing inter-nanorod distance and an increase in the number of interacting nanorods. For both side-by-side and end-to-end assemblies, the strength of the longitudinal plasmon coupling increases with increasing nanorod aspect ratio as a result of the increasing dipole moment of the longitudinal plasmon. For both the side-by-side and end-to-end orientations, the simulation of a dimer of nanorods having dissimilar aspect ratios showed a longitudinal plasmon resonance with both a blue- and a red-shifted component, as a result of symmetry breaking. A similar result was observed for a pair of similar aspect ratio nanorods assembled in a nonparallel orientation.

11.10.2
Applications of the T-Matrix approach

Beginning in 1991, the group of Mishchenko and Mackowski used the T-matrix approach to simulate the optical properties of aggregates of axially symmetric particles [705–710]. However, they mainly studied only the components of the differential scattering cross-section, that is, the angular distribution of the scattered light, for dielectric particles at one fixed wavelength.

Lou and Charalampopoulos [711] developed an exact form of the internal-field equation for an assembly of Rayleigh size spherical particles irradiated by an electromagnetic wave on the basis of the T-matrix approach. Multiple-scattering effects are accounted for by the T-matrix. The general exact solution for the internal electric field can be reduced to the form of the equation obtained by the DDA.

Wang *et al.* used a modified T-matrix for calculation of the elastic light scattering by a finite domain of distributed spherical scatterers [712] and spheroidal scatterers [713]. The modification comprised different exterior spaces for the incident wave and the scattered wave and included multiple scattering effects.

11.10.3
Other Methods

Light scattering by individual Ag nanoparticles and structures was studied spectroscopically by Tamaru *et al.* [714]. Individual particles were selected and manipulated with a micromanipulator installed inside a scanning electron microscope.

With typical particle dimensions of a few hundred nanometers, the plasma resonances of particles and the coupled modes of particle pairs were observed in the visible region. The polarization dependence of the resonance frequencies strongly reflects the shape anisotropy. With a simple approximation to take the glass substrate into account, the results are in good agreement with the analytical calculations by Mie scattering, and with numerical calculations by the FDTD method. The calculations with the FDTD method also confirm the GMT.

Drolen and Tien [715] examined the scattering and absorption of agglomerated soot particles. The primary particles were assumed to be Rayleigh scatterers, so that electromagnetic interactions among the primaries could be taken into account by the model of Jones [716, 717]. Clusters of up to 136 primaries were considered. The analytical results indicated that the soot particle aggregates can be approximated with a volume-equivalent sphere of corresponding radius. The refractive index of this volume-equivalent sphere was obtained by the effective medium model of Garnett (Maxwell-Garnett model; effective medium models are discussed in detail later in Chapter 14).

In a similar way, Stognienko et al. [718] studied the optical properties of coagulated particles in astrophysics. Rather than using the GMT, the DDA and effective medium models were used.

Pustovit et al. [620] utilized a simplified approach to compute the optical properties of aggregates in the long-wavelength limit. Considering only pair interactions, a very large number of multipoles could be taken into account in the computations for aggregates with $N = 6$ particles. The method is in good agreement with the GMT for nontouching sparse particle aggregates when the relative distances between the two particles in each pair are not too large. Even in the case of touching particles, acceptable agreement in the region of the main extinction peaks was observed. At long wavelengths, the convergence of the multipolar expansion is poor in the case of touching spheres.

The absorption cross-section of a system of two spheres was derived by an analytic method based on the separability of the Laplace equation in bispherical coordinates by Ruppin [719]. The method is applicable in the long-wavelength limit. The evolution of the absorption spectrum with decreasing distance between the spheres is demonstrated by numerical calculations. The implications for the interpretation of experimental infrared absorption spectra of small crystallites are discussed.

A scale-invariant theory of resonant Rayleigh scattering by fractal clusters was developed by Shalaev et al. [720]. The main result was that the scattering cross-section is greatly enhanced. Simulations dealing with two examples of fractal structures, namely random walk and cluster–cluster aggregates, were presented. The numerical results confirmed the theoretical predictions of the scaling behavior for both absorption and scattering, and allowed the corresponding exponents to be obtained.

An analysis of radiative absorption and scattering by aggregates of spheres in the Rayleigh limit was developed by Mackowski [721] with an electrostatics analysis. The electric fields incident upon and scattered by the aggregate were represented by the gradient of a potential that satisfies Laplace's equation. The analytical solution for the potential that exactly satisfies the boundary conditions at the surfaces of the spheres was obtained with a coupled spherical harmonics method. Calculations of the polarizability tensor and the absorption, scattering, and depolarization factors were performed on fractal-like clusters of spheres, with refractive index values that are characteristic for carbonaceous soot in the visible and IR regions. The results indicated an enhancement of the absorption cross-sections of fractal soot aggregates at the mid-IR wavelengths. The predicted spectral variation of soot absorption at visible and the mid-IR wavelengths was shown to agree well with experimental measurements.

A computationally profitable theoretical approach based on the first-order Born dipole approximation was applied by di Stasio and Massoli [722] to evaluate the scattering/extinction properties of randomly oriented filamentary soot agglomerates composed of Rayleigh particles. Four main morphologies, that is, the straight and the zigzag chain and the random and the fractal cluster, were considered. One major result was that the influence of the multiple scattering on the extinction factor is not negligible and it is quantifiable below 10% for primary diameters not larger than 40 nm.

With the numerically exact superposition T-matrix method, Liu *et al.* [723] carried out extensive computations of scattering and absorption properties of soot aggregates with varying states of compactness and size. The fractal dimension, D_f, was used to quantify the geometric mass dispersion of the clusters. It was shown that for smaller values of the monomer radius, the absorption cross-section tends to be relatively constant when $D_f < 2$, but increases rapidly when $D_f > 2$. The scattering cross-section and single-scattering albedo increased monotonically as fractals evolve from chain-like to more densely packed morphologies. Overall, the results for soot fractals differed profoundly from those calculated for the respective volume-equivalent soot spheres and also for the respective external mixtures of soot monomers under the assumption that there are no electromagnetic interactions between the monomers.

Vargas *et al.* [724] used a T-matrix formalism to calculate local electric fields around aggregates of prolate spheroids in the long-wavelength regime. The calculations were conducted as a function of interparticle distance and also angle of orientation. The observed red shifts in the resonant wavelengths of the characteristic peaks were shown to obey an exponential relationship as a function of interparticle separation and a sinusoidal relationship as a function of angle of rotation of the spheroid. The behavior of the cluster is discussed and the two effects of separation and rotation are compared.

Ludwig [725] used the generalized multipole technique to study the scattering by aggregates of two or three spheres.

12
Densely Packed Systems

The electromagnetic response of a densely packed particle assembly may also be obtained from coherent superposition of electromagnetic fields as in the *generalized Mie theory* (GMT). The GMT is applicable to all particle systems containing spherical particles. However, available computer capacities restrict its applicability to a maximum of a few hundred particles. This is not sufficient to calculate the optical properties of many realistic densely packed samples, for example, pigment layers (lacquers), but it is sufficient to calculate the response of closely packed particle aggregates.

Unlike the GMT, *radiative transfer* models use the incoherent superposition of intensities. Therefore, they are capable of treating huge amounts of particles such as water droplets in clouds or nanoparticles in lacquers. A very wide variety of multiple-scattering models and techniques have been developed over the long history of the field, particularly in astrophysical and atmospheric contexts. These methods have been reviewed and tabulated in the comprehensive book by van de Hulst [726].

Radiative transfer considers the stationary distribution of intensities in space caused by incoherent interaction of radiation with matter, in which the geometry, the radiation source distribution, and boundary conditions are well known. The radiative transfer models take into account the energy conservation law, and also principles of thermodynamics, electrodynamics, and geometric optics.

According to the rules of radiative transfer [727], the change in the intensity of a light beam along a path s in a volume of scatterers and absorbers is given in the stationary (time independent) case as

$$\frac{\partial I(s,\theta,\phi)}{\partial s} = -KI(s,\theta,\phi) - SI(s,\theta,\phi)$$
$$+ (K+S)\iint w(\Omega,\Omega')I(s,\theta,\phi,\theta',\phi')\mathrm{d}\Omega' \qquad (12.1)$$

where the quantities K and S account for the absorption and scattering processes per unit length. Note that S only contains backscattering. The quantity $w(\Omega,\Omega')$, the phase function, stands for the probability of scattering of photons coming from the solid angle Ω into the solid angle Ω'. The terms with K and S diminish the intensity, whereas the integral term increases the intensity (source term). The phase function is usually represented as a function of the scattering angle θ and

Optical Properties of Nanoparticle Systems: Mie and Beyond. Michael Quinten
Copyright © 2011 WILEY-VCH Verlag GmbH & Co. KGaA, Weinheim
ISBN: 978-3-527-41043-9

the asymmetry parameter g. A common approach is the Henyey–Greenstein function:

$$w(\Omega, \Omega') = w(\cos\theta) = 0.5(1-g^2)(1-g^2+2\cos\theta)^{1.5} \tag{12.2}$$

For vanishing source term, the remaining differential equation yields the Lambert–Beer law:

$$\frac{\partial I(s,\theta,\phi)}{\partial s} = -(K+S)I(s,\theta,\phi) \tag{12.3}$$

For nonvanishing source term the calculation of the optical response of the nanoparticle composite is more complicated.

For further reading, we refer to the book by Mishchenko et al. [728], which provides substantial knowledge about classical and modern developments in the theory of multiple light scattering in media composed of randomly positioned particles.

12.1
The Two-Flux Theory of Kubelka and Munk

In fact, radiative transfer is applicable only if the phases of the scattered light are statistically distributed and the radiation fields are assumed to be independent, thereby permitting the addition of the field intensities rather than amplitudes. Therefore, it is applicable only to diluted systems, or to systems where the particles are larger than the wavelength, as in atmospheric research. Its applicability to submicron- and nanometer-sized particles must be considered to be critical. In concentrated particulate dispersions such as paints and coatings, the assumptions of radiative transfer are not fulfilled.

However, there have been successfully established *many-flux theories*, which assume certain numbers of photon fluxes in the (+x) direction and the (−x) direction as an approach to the radiative transfer through a particle assembly. Among the various flux models (six-, four-, two-flux), the two-flux model of Kubelka and Munk [18] is the most popular and attained distinct importance, since it is simple and frequently yields fairly accurate results. It is briefly described below. For a more detailed discussion of the Kubelka–Munk theory we refer to the book by Kortüm [729]. Extensions have been derived for rough surfaces [730] and for glazing surfaces [731].

Consider two fluxes across the x-direction through a thin layer of thickness Δx in a sample of total thickness d (see Figure 12.1). The illumination of the sample is assumed to be diffuse.

The flux $I_+(x)$ in the positive x-direction is decreased by absorption and backscattering inside the layer and is increased by backscattering of the intensity $I_-(x)$ in the negative x-direction:

$$dI_+(x) = -(K+S)I_+(x)dx + SI_-(x)dx \tag{12.4}$$

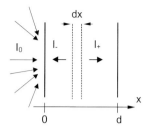

Figure 12.1 The principle of the two-flux radiation transfer of Kubelka and Munk.

Vice versa, the flux $I_-(x)$ in the negative x-direction is decreased by absorption and backscattering inside the layer and is increased by backscattering of the intensity $I_+(x)$ in the positive x-direction:

$$-dI_-(x) = -(K+S)I_-(x)dx + SI_+(x)dx \quad (12.5)$$

The minus sign on the left-hand side of Equation 12.5 is due to the convention that $I_-(x)$ is counted in the negative x-direction.

The above differential equations can be solved only for the ratio $I_-(x)/I_+(x)$. The results for $x = 0$ (front side of the sample) and $x = d$ (rear side of the sample) are of special interest. The ratio $I_-(x=0)/I_+(x=0)$ at $x = 0$ defines the diffuse reflectance R_{KM} of the sample:

$$R_{KM} = \frac{(1-\gamma)(1+\gamma)[\exp(Ad) - \exp(-Ad)]}{(1+\gamma)^2 \exp(Ad) - (1-\gamma)^2 \exp(-Ad)} \quad (12.6)$$

Analogously, the ratio $I_-(x=d)/I_+(x=d)$ at $x = d$ can be used to define the diffuse transmittance T_{KM} of the sample:

$$T_{KM} = \frac{4\gamma}{(1+\gamma)^2 \exp(Ad) - (1-\gamma)^2 \exp(-Ad)} \quad (12.7)$$

where γ and A are abbreviations representing K and S as

$$\gamma = \sqrt{\frac{K}{K+2S}} \quad (12.8)$$

$$A = \sqrt{K(K+2S)} \quad (12.9)$$

For large values of Ad the transmittance becomes zero. The reflectance is then the reflectance of an opaque layer:

$$R_\infty = 1 + \frac{K}{S} - \sqrt{\left(\frac{K}{S}\right)^2 + 2\frac{K}{S}} \quad (12.10)$$

The ratio K/S is also known as the *Kubelka–Munk unit*.

In practice, K is not available from measurements. Instead, R_∞ and S are measured for opaque systems and Equation 12.10 is resolved for K.

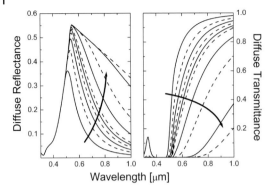

Figure 12.2 Computed diffuse reflectance and transmittance spectra of silver pigment films with $2R = 50$ nm and variable film thickness $d = 50$, 100, 150, 200, 250, 300, 500, 1000, 3000, and 5000 nm. The arrows indicate the direction of increasing film thickness.

The quantities K and S can be related to the cross-sections for extinction and scattering, σ_{ext} and σ_{sca}, the asymmetry parameter g (amount of backscattered and forward scattered light), and the particle volume V_p of a single nanoparticle via [732–734]

$$K = \frac{\sigma_{ext} - \sigma_{sca}}{V_p} = \frac{\sigma_{abs}}{V_p} \quad (12.11)$$

$$S = \frac{3}{4}\frac{\sigma_{sca}(1-g)}{V_p} \quad (12.12)$$

These relations allow the calculation of R_{KM} and T_{KM} for all dense particle ensembles as a function of the thickness d of the sample. As an example, Figure 12.2 shows the diffuse reflectance and transmittance through a system with spherical silver particles with diameter $2R = 50$ nm. The film thickness is increasing, amounting to $d = 50$ nm (one particle layer), 100, 150, 200, 250, 300, 500, 1000, 3000, and 5000 nm. The diffuse reflectance R_{KM} exhibits a maximum lying in the wavelength range 510–550 nm, depending on the film thickness. Its magnitude increases from $R_{KM} = 0.36$ to 0.55. Moreover, R_{KM} increases at longer wavelengths with increasing sample thickness (which is a multiple of the single particle diameter). In the diffuse transmittance T_{KM} a spectral band with transmittance $T_{KM} \approx 0$ between 390 m and 470 nm can clearly be recognized. At longer wavelengths the transmittance is high for thin films but decreases with increasing film thickness.

The changes in diffuse reflectance and transmittance also result in changes in the color of the nanoparticle system. The color resulting from the transmittance spectra is red for all thicknesses. The corresponding color coordinates in Figure 12.3 lie along a line in the color chromaticity diagram. Note that the transmittance in the visible spectral region, however, decreases strongly with increasing

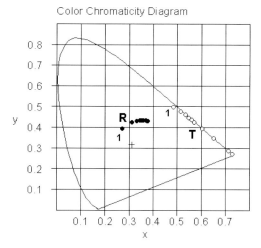

Figure 12.3 Chromaticity diagram showing the color coordinates of Ag particle films with $2R = 50$ nm and increasing thickness d, obtained from the reflectance and transmittance spectra in Figure 12.2.

thickness so that actually the sample becomes opaque. This is the case at a film thickness of approximately $d = 500$ nm. The corresponding apparent colors in the reflectance start with green for the lowest thickness and turn to yellow for thickness $d \geq 500$ nm, due to the increase in reflectance at long wavelengths with increasing film thickness. For a thickness of $d \geq 3000$ nm the color coordinates do not change further.

12.2
Applications of the Kubelka–Munk Theory

In this section, the Kubelka–Munk theory is applied to various densely packed particle systems. In all calculations we assume a free-standing particle layer system, which means that the influence of a supporting substrate is not considered. In principle, this is possible with correspondingly changed equations for R_{KM} and T_{KM}. For more details we refer to [729]. Moreover, we assume the surrounding medium to have the refractive index $n_M = 1$.

The most interesting cases are particle systems with particles that are known as color pigments. Hence we will treat here systems of Au, Ag, Cr_2O_3, Fe_2O_3, Cu_2O, TiN, and ZrN nanoparticles. In addition, SiO_2 and TiO_2 nanoparticles are considered as representatives of all transparent particles used as white pigments. We turn our attention also to Si, ITO and LaB_6 nanoparticle systems because of their distinct properties in the UV and NIR spectral regions. Unlike in earlier chapters, gold and silver particle systems will be treated last.

12.2.1
Dense Systems of Color Pigments: Cr_2O_3, Fe_2O_3, and Cu_2O

The calculations in Chapter 6 showed that small spheres of Cr_2O_3, Fe_2O_3, or Cu_2O are well suited as color pigments in diluted systems due to characteristic absorption bands in the visible spectral range. Not all particle sizes are favorable for that purpose. In lacquers the same particles are now densely packed and multiple scattering among the particles influences the coloring properties of the single, isolated particles.

Indeed, when looking at the collective properties of a dense system of Cr_2O_3 nanoparticles, the calculations according to the Kubelka–Munk theory yield a dark blue–green for the color obtained from R_{KM} for particles with $2R \approx 100\,nm$ that slowly turns into a bluish green for particles with $2R = 260–280\,nm$. In Figure 12.4a we present the diffuse reflectance for a particle system with $2R = 270\,nm$ for a monolayer and the opaque system.

The diffuse reflectance of the chromium oxide nanoparticle monolayer exhibits two maxima, one smaller maximum at 420 nm and one broader maximum at 540 nm with a width of about 100 nm. The latter maximum determines the color resulting from this reflectance spectrum. A third broad maximum appears at 860 nm in the NIR region. This maximum, however, does not affect the apparent color and vanishes with increasing thickness of the particle layer. Finally, for the opaque system, the reflectance increases continuously with increasing wavelength in the NIR spectral region.

For iron oxide (Fe_2O_3), already for small particles with $2R = 50\,nm$ a color impression is obtained. In Table 12.1 we summarize the color impression from calculated diffuse reflectance spectra of dense Fe_2O_3 nanoparticle systems as a function of the particle diameter.

The calculations of R_{KM} for dense systems of Fe_2O_3 nanoparticles showed that a red color can be obtained with particles of $2R = 240–270\,nm$. Figure 12.4b depicts

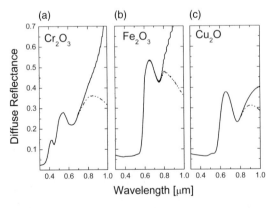

Figure 12.4 Computed diffuse reflectance spectra for a monolayer (dashed-dotted-dotted) and an opaque layer of (a) Cr_2O_3 nanoparticles with $2R = 270\,nm$, (b) Fe_2O_3 nanoparticles with $2R = 250\,nm$, and (c) Cu_2O nanoparticles with $2R = 250\,nm$.

Table 12.1 Color impression and color coordinates x, y, Y of dense Fe_2O_3 nanoparticle systems obtained from calculated diffuse reflectance spectra. The norm light source used is D65 with 2° observer.

2R (nm)	x	y	Y	Color impression
50	0.4514	0.3923	5.574	Dark brown
100	0.4401	0.3892	18.61	Brown
150	0.4355	0.3456	18.48	Reddish brown
200	0.4089	0.4179	26.74	Dark yellow
250	0.5087	0.3544	18.65	Red
300	0.4429	0.3637	12.49	Reddish brown

the diffuse reflectance for a particle system with $2R = 250$ nm for a monolayer and the opaque system. The monolayer shows an almost constant reflectance at wavelengths below 540 nm. Then, the reflectance increases rapidly to approach a maximum at 655 nm. The steep increase with edge wavelength 590 nm and the maximum at 655 nm lead to a red color for this system. In the NIR region the behavior is similar to that of the Cr_2O_3 nanoparticle system.

Analogously, for dense Cu_2O nanoparticle systems we find a red color preferably for particles with $2R = 240$–260 nm. Figure 12.4c depicts diffuse reflectance spectra for a particle system with $2R = 250$ nm for a monolayer and the opaque system. On looking at the reflectance spectra, this system appears to be almost a copy of the iron oxide particle system, except that the magnitude of the reflectance is less. The monolayer shows an almost constant reflectance at wavelengths below 540 nm. Then, the reflectance increases rapidly to approach a maximum at 655 nm. The steep increase with edge wavelength 590 nm and the maximum at 655 nm lead to a red color for this system.

12.2.2
Dense Systems of White Pigments: SiO_2 and TiO_2

Silica (SiO_2) has been established as the workhorse nanoparticle for a long time. However, the seemingly endless functionalization combinations attract ever increasing attention to SiO_2 nanoparticles. TiO_2 pigments are used in large amounts in almost all composites, that is, in lacquers, colors, plastics, and synthetic fibers. They are intended to be used as UV absorbers and as white pigments. Titania (TiO_2) is strongly absorbing in the near-UV region due to the onset of direct interband transitions. These close-lying interband transitions cause a rather high refractive index in the visible spectral region.

Here, we present representative calculated diffuse reflectance spectra for particles with diameter $2R = 100$ nm for both silica and titania. The spectra are summarized in Figure 12.5. For other sizes the principle effects shown with these spectra are similar.

400 | *12 Densely Packed Systems*

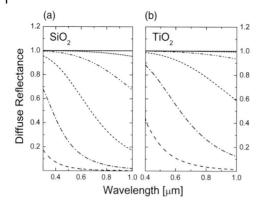

Figure 12.5 Computed diffuse reflectance spectra of white pigment films with $2R = 100$ nm and variable film thickness $d = 200$ nm, 1 μm, 10 μm, 100 μm, and 1 mm, and the opaque layer. (a) SiO_2; (b) TiO_2.

As the single particles of both materials are transparent in the visible spectral region, the diffuse reflectance clearly shows a strong dependence on the particle density, that is, the thickness of the particle layer. For a monolayer of particles the reflectance is high at lower wavelengths of violet and blue light. Vice versa, the transmittance through such a monolayer is also high, but for longer wavelengths. With increasing film thickness the reflectance increases at longer wavelengths. A complete 100% reflectance is obtained for very thick films of ~5 mm for titania and ~30 mm for silica. For larger particles (typically $2R = 200\text{–}350$ nm for titania as white pigment), 100% reflectance is obtained for thinner films.

A spectral projected gradient method (SPGM) was applied by Curiel *et al.* [735] to invert measured diffuse reflectance spectra of thick titanium dioxide pigmented coatings on black substrates illuminated with semi-isotropic diffuse radiation. The Lorenz–Mie theory and Fresnel relations were used to obtain the first approximation required by the SPGM. The spectral dependence of the retrieved values of the SPGM agrees fairly well above the absorption edge of the pigment with the values that correspond to the first approximation. The authors concluded that some of the differences between the retrieved and the first approximation values in this region of the spectrum might be clarified if an effective refractive index and a better treatment of the absorption of the matrix are included.

12.2.3
Dense Systems of ZrN and TiN Nanoparticles

TiN and ZrN belong to a larger group of compounds with extreme stability (hardness, chemical inertness, high melting point, high Young's modulus) with a

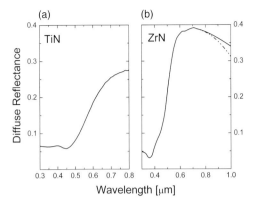

Figure 12.6 Computed diffuse reflectance spectra for a monolayer (dashed-dotted-dotted) and an opaque layer of (a) TiN nanoparticles and (b) ZrN nanoparticles with $2R = 200$ nm.

ceramic behavior with respect to hardness and inertness, and a *metallic* behavior with high electrical and thermal conductivity and free electron-like optical behavior. From the calculations in Chapter 6, we know that ZrN and TiN nanoparticles exhibit rather similar behavior to gold nanoparticles. Indeed, bulk TiN and bulk gold also have similar reflectivity and appear to be golden yellow.

The calculations according to Kubelka and Munk for dense systems of ZrN and TiN nanoparticles result in similar behavior of both materials. For opaque systems with particle diameters $2R = 160–240$ nm the color impression from the diffuse reflectance is golden yellow. Figure 12.6 depicts exemplary spectra for a single particle size of $2R = 200$ nm for one monolayer and the opaque system. Obviously, for both materials the reflectance of the opaque layer is almost identical with the spectrum of the monolayer. Only slight differences appear in the NIR spectral region for ZrN. The steep increase in the reflectance at wavelengths between 550 m and 600 nm leads to a golden yellow color of these coatings, similar to the regular reflection of bulk gold. Hence dense opaque systems of ZrN and TiN can replace bulk gold coatings with the advantage of being a hard coating.

12.2.4
Dense Systems of Silicon Nanoparticles

Silicon exhibits strong interband transitions in the near-UV close to the visible spectral region. They cause a high refractive index in the visible spectral region whereas the absorption index becomes smaller with increasing wavelength. As we showed in Chapter 6, silicon nanoparticles are therefore good candidates for electronic resonances in the interband transition region and for MDRs in the visible and NIR region. These resonances should also influence the diffuse reflectance of dense Si nanoparticle systems. Actually, the calculations according to the Kubelka–Munk theory showed the following:

- In the UV and visible spectral region characteristic spectral features appear in the diffuse reflectance, which must result from these single particle resonances.
- The spectral features depend on the particle size, but not on the thickness of the particle layer. The diffuse reflectance increases only in the NIR region with increasing layer thickness.
- The spectral features in the visible spectral region lead to a characteristic size-dependent color in the reflectance of the particle layer.

In Table 12.2 we summarize the color impression from calculated diffuse reflectance spectra of dense Si nanoparticle systems as a function of the particle diameter. Obviously, the color changes within a narrow range of particle sizes. This is caused by the size-sensitive resonances of the nanoparticles enhanced by the high refractive index of silicon.

The calculated spectra of the diffuse reflectance R_{KM} in Figure 12.7 show the reason for the color of the dense Si nanoparticle systems. A steep increase in the

Table 12.2 Color impression and color coordinates x, y, Y of dense Si nanoparticle systems obtained from calculated diffuse reflectance spectra. The norm light source used is D65 with 2° observer.

2R (nm)	x	y	Y	Color impression
130	0.3280	0.3885	34.58	Blue–green
140	0.3377	0.4630	42.42	Green
150	0.3695	0.4703	48.76	Green
160	0.4025	0.4386	50.08	Yellow
180	0.4465	0.3623	38.42	Red
200	0.4050	0.3188	26.9	Violet

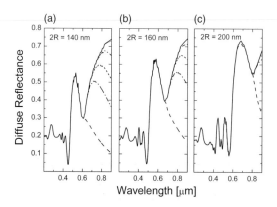

Figure 12.7 Computed diffuse reflectance spectra for dense silicon nanoparticle films with (a) $2R = 140$ nm, (b) $2R = 160$ nm, and (c) $2R = 200$ nm. The film thickness is $d = 2R$, 10R, 20R, 40R, and 100R, and the opaque layer.

reflectance to a distinct reflectance maximum appearing in the visible spectral region determines the color impression of the whole film. The spectral position of this band shifts with increasing size to longer wavelengths (peak positions: 525, 565, 680 nm). In addition, two further fairly narrow reflectance maxima appear in the UV region and shift into the visible spectral range with increasing single particle size. For each particle size the corresponding features remain unchanged on increasing the film thickness. In contrast, the reflectance increases in the NIR region with increasing film thickness, without affecting the apparent color. At still longer wavelengths above 1200 nm (not shown here) where silicon becomes transparent, the opaque film finally exhibits 100% reflectance, similarly to SiO_2 and TiO_2.

12.2.5
Dense Systems of IR Absorbers: ITO and LaB_6

ITO films are often used in applications where an electrical current is used for switching a process, and where in addition the film must be transparent, for example, in flat panel displays or OLEDs (organic light emitting diodes) or as a TCO (transparent conductive oxide) in thin-film photovoltaics. The first materials used for NIR absorbers based on nanoparticles were also the conductive oxides antimony-doped tin oxide (ATO) and indium-doped tin oxide ITO). Another interesting group of materials for IR absorbers are the rare earth hexaborides, particularly lanthanum hexaboride (LaB_6). LaB_6 has been attracting the most attention from this group, because it is of interest also for other applications.

The various calculations on densely packed ITO nanoparticle systems showed that the diffuse reflectance R_{KM} does not exhibit extraordinary features in the UV and NIR regions. Systems with small particles exhibit some smaller features in the visible spectral region that lead to an apparent blue color, but for larger particles this turns to gray without distinct features in the spectral diffuse reflectance.

In contrast, systems with LaB_6 nanoparticles do exhibit characteristic features in the diffuse reflectance that lead to a color in reflectance. For all particle sizes the opaque systems clearly exhibit a high reflectance for blue and violet light and a deep minimum around 600 nm, followed by an increase in the reflectance again to longer wavelengths. The reflectance in the red and NIR and also in the blue and violet spectral region depends on the particle size and increases with increasing particle size. This can be seen from Figure 12.8a, where reflectance spectra for opaque systems are plotted as a function of the particle size. The color resulting from these reflectance spectra is blue. Figure 12.8b additionally shows the dependence on the film thickness for a film with particles of $2R = 150$ nm. Obviously already the monolayer exhibits the characteristic spectral features. Only in the NIR region does the reflectance still increase slightly for the opaque system.

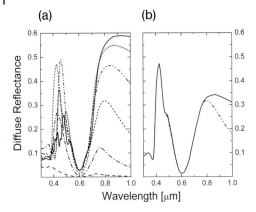

Figure 12.8 Computed diffuse reflectance spectra for dense lanthanum hexaboride (LaB$_6$) nanoparticle systems. (a) Opaque layers with particle sizes $2R$ = 50, 100, 150, 200, 250, and 300 nm; (b) a monolayer (dashed-dotted-dotted) and the opaque layer (solid) with $2R$ = 150 nm.

12.2.6
Dense Systems of Noble Metals: Ag and Au

The use of colloidal gold and silver for coloring, especially of glass, is old and reaches back to the Ancient World. One should keep in mind, however, that it was known how to use the colloids for coloring glasses but it was unknown that nanoparticles are the reason for these coloring effects. First descriptions of how to obtain ruby gold glasses can be found in the bibliography of Ashurbanipal in Nineve (seventh century BC). Another famous example from the Romans is the Lycurgus cup (fourth century AD, British Museum, London) and other Roman glasses. The Lycurgus cup is treated in more detail in the next subsection. In view of the applications in the following two subsections, we consider here the diffuse reflectance of dense particle systems of gold and silver with the refractive index of the surrounding medium assumed to be n_M = 1.5. This is a fairly good approximation for soda-lime glass. Furthermore, the particle size was fixed at $2R$ = 70 nm. In many further calculations we also examined smaller particles but with similar results.

The diffuse reflectance and transmittance spectra are presented in Figure 12.9 for Ag and Figure 12.10 for Au. The layer thickness was varied and amounts to d = 70 (one monolayer), 140, 350, 700, 1400, 2800, 7000, and 14 000 nm, and the opaque layer. The spectra for the silver nanoparticle systems are similar to that presented in Figure 12.2. Due to the higher refractive index of the surrounding medium and the larger single particle size, all features are shifted to longer wavelengths. Hence the apparent color in reflectance turns to yellow already for two monolayers. Only the monolayer exhibits a yellow–green color. In transmittance the color is red for all layer thicknesses. The system rapidly becomes opaque. Similarly, the dense Au nanoparticle system exhibits a maximum in the diffuse

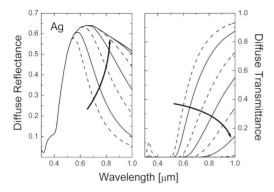

Figure 12.9 Computed diffuse reflectance and transmittance spectra of silver pigment films with $2R = 70$ nm and variable film thickness $d = 70, 140, 350, 700, 1400, 2800,$ 7000, and 14 000 nm, and the opaque system. The arrows indicate the direction of increasing film thickness. The refractive index of the surrounding is $n_M = 1.5$.

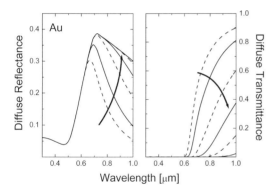

Figure 12.10 Computed diffuse reflectance and transmittance spectra of gold pigment films with $2R = 70$ nm and variable film thickness $d = 70, 140, 350, 700, 1400, 2800,$ 7000, and 14 000 nm, and the opaque system. The arrows indicate the direction of increasing film thickness. The refractive index of the surrounding is $n_M = 1.5$.

reflectance R_{KM} lying in the wavelength range 660–720 nm, depending on the film thickness. R_{KM} increases at longer wavelengths with increasing sample thickness (which is a multiple of the single particle diameter). In the diffuse transmittance T_{KM} a spectral band with transmittance $T_{KM} \approx 0$ between 300 and 600 nm can clearly be recognized. At longer wavelengths the transmittance is high for thin films but decreases rapidly with increasing film thickness. The apparent color in reflectance is yellow–brown and in transmittance red. Similar results are obtained for dense Au nanoparticle systems with smaller particle size. Moreover, the spectral features shift to shorter wavelengths for a smaller refractive index n_M of the surrounding matrix material.

12.2.7
The Lycurgus Cup

The Lycurgus cup is the only complete example of a very special type of glass because its color changes when held up to the light. The opaque green cup turns to a glowing translucent red when light is shone through it (see Figure 12.11). It has been proven that tiny amounts of Ag and Au are contained in the glass. The different colors in reflectance and transmittance are therefore often ascribed to gold and silver, which are believed to contribute to the apparent colors as nanoparticles of size approximately $2R = 70$ nm. Additionally, manganese is present that also colors the glass purple–red.

The scene on the cup depicts an episode from the myth of Lycurgus, a king of the Thracians (around 800 BC). A man of violent temper, he attacked Dionysos and one of his maenads, Ambrosia. Ambrosia called out to Mother Earth, who transformed her into a vine. She then coiled herself about the king, and held him captive. The cup shows this moment when Lycurgus is entrapped by the branches of the vine, while Dionysos, a Pan, and a satyr torment him for his evil behavior.

Experiments on stained glasses with silver and gold nanoparticles (see Chapter 6) showed that low concentrations of gold nanoparticles lead to a purple–red and low concentrations of silver nanoparticles lead to a yellow color, in transmittance and reflectance. Therefore, we suppose that higher concentrations of the particles or electromagnetic interactions among the particles must be assumed to explain the different apparent colors. Indeed, as seen in Chapter 11, formation of chain-like aggregates can actually result in green and red colors, but again the same color in transmittance and reflectance.

A reasonable explanation for the color change from reflectance to transmittance can only be obtained with densely packed nanoparticles in a thin layer. This hypothesis is supported by experiments on ion-implanted copper, silver, and gold glasses. The nanoparticles formed here are distributed in a thin surface layer

Figure 12.11 The Lycurgus cup. A color version of this figure can be found in the color plates at the end of the book. Copyright: The Trustees of the British Museum.

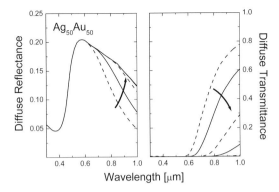

Figure 12.12 Computed diffuse reflectance and transmittance spectra of Ag$_{50}$Au$_{50}$ alloy pigment films with $2R = 70$ nm and variable film thickness $d = 70, 140, 350, 700, 1400,$ 2800, 7000, and 14 000 nm, and the opaque system. The arrows indicate the direction of increasing film thickness. The refractive index of the surrounding is $n_M = 1.5$.

region. The previous calculations on dense gold and silver nanoparticle systems with the Kubelka–Munk theory showed that monolayers or double layers exhibit different colors in reflectance and transmittance. However, neither gold nor silver layers actually result in an opaque green color in reflectance and a red color in transmittance. To resolve this dilemma, we also considered the possibility of gold–silver alloy particles Ag$_x$Au$_{1-x}$ with $x \leq 1$. From a large series of calculations on dense gold–silver alloy nanoparticle systems we found that an opaque green color in reflectance and a red color in transmittance are obtained for $x = 0.5$ and a maximum of two layers of particles with size $2R = 70$ nm in a surrounding medium with refractive index $n_M = 1.5$. This result coincides with observation since one has to take into account that the manganese also contributes to the color. In Figure 12.12 we show the corresponding calculated diffuse reflectance and transmittance spectra, again with layer thicknesses $d = 70$ (one monolayer), 140, 350, 700, 1400, 2800, 7000, and 14 000 nm, and the opaque layer. All particle layers show a maximum in the reflectance at wavelength 580 nm. The reflectance only increases at longer wavelengths with increasing layer thickness and rapidly runs into its final reflectance curve for the opaque system. The peak at 580 nm and the steep increase in the reflectance at shorter wavelengths remain unaffected by the increase in the layer thickness. In contrast, the transmittance decreases rapidly so that only for a few layers does the dense system remain transparent. The apparent color in transmittance is red for the monolayer.

12.3
Improvements of the Kubelka–Munk Theory

The derivation of Equations 12.6 and 12.7 assumes that the light incident upon the slab, and also the light in every slice within the slab, is totally diffuse, that is, it has no angular dependence. The more realistic case of a collimated light source is addressed by the so-called extended Kubelka–Munk models [736, 737]. As the Kubelka–Munk model assumes also diffuse radiation within each slice of the slab, which is not the case for a collimated light source, some extended models incorporate more than two radiation fluxes, and allow the scattering and absorption parameters to vary with depth. Early three-flux models of Ryde [738] and Reichman [739] included a collimated component in the forward flux. More elaborate four-flux [732, 740, 741] and many-flux [732, 742, 743] models have been developed, and also models that take into account variations in scattering with depth [737, 744, 745] by taking into account the different pathlengths for collimated and diffuse components of the radiation field. For large optical depth and low absorption, the photon diffusion theory [746–750] has proved useful.

The most versatile four-flux model seems to be that developed by Maheu et al. [740]. From this theory, explicit relations for specular and diffuse components of reflectance and transmittance are available in terms of particle concentration, volumetric scattering and absorption cross-sections of the particles, film thickness, and forward scattering ratios. It has been compared with numerical solutions of the equation of radiative transfer [751] and also with highly accurate Monte Carlo simulations [752, 753]. A generalization of Maheu et al.'s solution was given by Vargas [754].

A new method was proposed for solving the equations for the multiple scattering of light by Elias and Elias [755]. It was shown that the introduction of an auxiliary function in the diffusion equation together with its expansion on the spherical harmonics allows one to compute the exact angular distribution of the light scattered by a diffusing medium. In contrast to the case of N-flux methods, the result is expressed as a function of both angles θ and φ. Although the equations to be solved remain complex at first sight, they are actually well suited to an efficient numerical resolution.

In the following, we want to improve the Kubelka–Munk model using the GMT. The examples in Chapter 11 show that the GMT also applies to nanoparticle composites with filling factors $f > 10^{-2}$, presuming that the inclusions are spherical. In fact, however, the available computer facilities restrict considerations to particle numbers of a few hundred. This may be sufficient for the demonstrated examples of particles seeded on substrates, but on going to three-dimensional densely packed nanocomposites, for example pigment films and powders, this extrapolation is limited. Admittedly, as we have shown in Chapter 10, the interaction runs into saturation with increase in the number of particles in the aggregate (for linear chains about $N = 30$). This allows the GMT to be combined with the Kubelka–Munk theory.

In the Kubelka–Munk theory, the absorption constant K and the scattering constant S per unit length of the dense composite material are related to the cross-sections for extinction and scattering σ_{ext} and σ_{sca}, the asymmetry parameter g and the particle volume V_p Equations 12.11 and 12.12. These relations are normally used with single particle cross-sections and asymmetry parameters. Combination of the Kubelka–Munk approach with the GMT is now simple: use the cross-sections

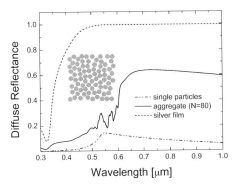

Figure 12.13 Diffuse reflectance of two films according to the Kubelka–Munk theory. The solid line is the reflectance obtained when using K and S of the 80-particle aggregate shown at the right side. The refractive index of the surrounding is $n_M = 1.5$.

and the asymmetry parameter of an aggregate in the calculation of the quantities K and S instead of single particle properties and calculate the reflectance of an opaque layer. This has been done for a film of 80 closely packed silver particles with $2R = 20$ nm. The values of K and S obtained for this film with the GMT were inserted in Equations 12.6–12.9 to obtain the diffuse reflectance of the opaque layer. In Figure 12.13 we show the reflectance spectra of this opaque film in comparison with the reflectance of an opaque film where K and S stem from single particle properties. In both cases the surrounding medium for the particles was assumed to have a constant refractive index of $n = 1.5$.

Obviously, films consisting of aggregates exhibit a higher reflectance by a factor of 2–3 than films of single particles. Particularly at long wavelengths the difference is clear: the reflectance of the film with single particles decreases by a factor of 2 with increase in wavelength from 550 to 1000 nm, whereas that of the film with aggregates decreases only slightly in this wavelength interval. In the wavelength interval between 480 and 610 nm, however, some striking features appear in the diffuse reflectance of the aggregate layer system. They are the result of the multiple resonances in the extinction and scattering cross-sections of the aggregate.

The clearly increased diffuse reflectance of the dense particle system with K and S from the calculation according to the GMT encourages the use of this approach for explanation of the reflectance of thin gold particle films. Dusemund et al. [34, 756] and Kreibig [757] carefully prepared densely packed monolayers of gold nanoparticles with mean size $2R = 20$ nm on quartz substrates. The particles are ligand stabilized so that coalescence of the particles was prevented. Optical absorption (transmittance) and reflectance spectra were recorded. Figure 12.14 shows the spectra and a TEM image of the sample. Obviously the electromagnetic interaction among the particles leads to a broad absorption band with its peak shifted to longer wavelengths. At short wavelengths the reflectance spectrum is comparable to the reflectance of bulk gold, but decreases again with increasing wavelength. A reasonable explanation for the reflectance of the monolayer of densely packed gold

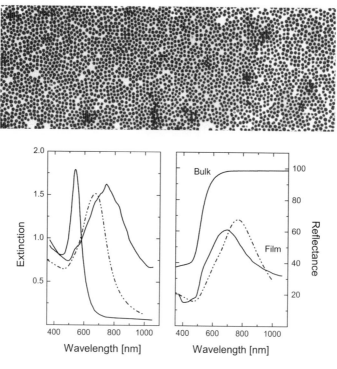

Figure 12.14 Transmission electron micrograph, extinction, and reflectance of a thin layer of densely packed ligand-stabilized gold nanoparticles with $2R = 20$ nm in comparison with the calculated extinction and the diffuse reflectance of an aggregate of $N = 80$ particles. The refractive index of the surrounding is $n_M = 1.6$.

nanoparticles is given by the Kubelka–Munk approach extended by the GMT. The calculated diffuse reflectance spectrum exhibits similar behavior to the measured reflectance spectrum. However, although this spectrum reflects the magnitude of the reflectance of the measured film and the decrease at longer wavelength, the peak position is clearly shifted to longer wavelengths by approximately $\Delta\lambda = 80$ nm. Improvements may be possible when using values for the refractive index of the surrounding of $n_M < 1.6$, as used in the present calculations.

13
Near-Field and SERS

For a plane electromagnetic wave a distinction between far-field and near-field is meaningless as long as the wave travels in free space. It only becomes meaningful when the wave hits a boundary between two different media. Then, the fields in the vicinity of this surface clearly differ from those far away from the surface. A classical example is a metal sphere in a homogeneous static electric field. As the field trajectories must end at the surface of the metal sphere, the curvature of the field lines clearly deviates from the straight line far from this sphere. We already discussed in Section 5.5 in detail that the electromagnetic fields scattered by a small spherical particle are strongly enhanced compared with the far-field. This effect, mostly reported for gold and silver nanospheres, has been shown

- To be independent of the particle material. For a silica sphere and a silver sphere of same size, the enhancement factor in the same!
- To be additionally increased for strongly absorbing particles such as metal particles. The scattering cross-section of a silver nanoparticle out of (the SPP) resonance at a wavelength $\lambda = 514$ nm is about 25 times larger than that of a silica sphere of same size.
- To be further increased for metal nanoparticles if they are excited in resonance. The scattering cross-section of a silver nanosphere in resonance is about 100 times larger than that out of resonance.
- To be strongly size dependent. Whereas the near-field enhancement is in the order of 10 for particles with a radius $R = 100$ nm, it increases to 10^7 for a radius $R = 1$ nm. This finding corresponds to the behavior of the electrical and magnetic field of the scattered wave close to the particle surface for which the near-field cross-section is approximately proportional to $1/R^{3.54}$ at a distance $r = R$.

These locally enhanced fields in the vicinity of a small particle play an important role

- in waveguiding along particle chains
- in near-field optical microscopy
- in surface-enhanced Raman scattering (SERS).

Additionally, they are important in some cases of optical nonlinearities of small particles. In this chapter, we concentrate on the above-mentioned applications.

13.1
Waveguiding Along Particle Chains

A fundamental problem of integrated optics is how to transport light energy in matter structures of transverse dimensions which are considerably smaller than the wavelength of light. The main problem for studying such subwavelength-sized light guiding principles is that the size of transmission lines within an integrated optical device is a limiting factor for further miniaturization towards a future nanooptics. The smallest beam diameter possible within a given conventional, dielectric optical waveguide is determined by the effective wavelength of the beam in the core material. In cases of materials with a negative real part of the dielectric function, such as metals, such a cut-off dimension does not exist, as in that case the corresponding transversal wavevector components are imaginary. Thus waveguiding systems consisting of metal structures can overcome the limits valid for conventional dielectric wave optics. On the other hand, absorption along the traveling path can rapidly diminish the power of the transported light.

Takahara *et al.* [758] theoretically analyzed systems based on signal propagation along cylindrical metal structures. Inspired by their results, Quinten *et al.* [759] looked for an alternative possibility for light guiding in subwavelength metal structures, namely light transport via the electrodynamic interaction between a sequence of closely spaced metal nanoparticles. The principle was analyzed theoretically by model calculations for Ag spheres with respect to transmission loss as a function of particle size and interparticle spacings.

For the model calculations of the light energy transport properties, an infinitely long linear chain of identical Ag spheres with constant spacing between the spheres was used in a modified GMT. The GMT was modified so that only the first sphere in the chain was assumed to be irradiated by the light field; all other spheres acquired their plasma oscillation energy by coupling. As the particle line is infinitely long, it behaves like a transmission line of constant impedance. The energy flows in only one direction, and the intensity distribution along the line decreases constantly as it is not subject to interference by reflections. For computational reasons, however, the number of spheres had to be final and was restricted on $N = 50$ particles.

As we were interested in the intensity distribution in the near-field of the aggregates, we restricted our computations to the electromagnetic fields and intensities in the near-field according to the following procedure:

Step 1: Solving the sets of linear equations (10.9) and (10.10) for a chain of N identical spheres separated with a center-to-center distance d, and a plane wave propagating along the positive z-axis being polarized either parallel or normal to the chain axis (x-axis) incident on the first particle.

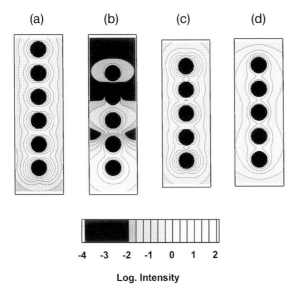

Figure 13.1 x–y electric field intensity distribution along a chain of 25 nm radius Ag nanospheres for the two polarization directions: (a) normal to the chain axis and (b) parallel to the chain axis. Panels (c) and (d) show the corresponding field distribution for a chain of final particle number N = 5. Only the first particle at the bottom is assumed to be irradiated by light.

Step 2: Calculation of the scattered electromagnetic fields of the particles in an arbitrary point $r = (x,y,z)^T$ outside the spheres. We restricted our calculations here to $z = 0$, meaning the equatorial plane of the spheres.

Step 3: Coherent superposition of the scattered fields of all particles $j = 1, \ldots, N$ by adding $E_{sca}(r,j)$ and $H_{sca}(r,j)$ to $E_{sca}(r)$ and $H_{sca}(r)$.

Step 4: Calculation of the intensity $I(r) = \sqrt{(\varepsilon_0/\mu_0)}|E_{sca}(r)|^2$

where ε_0 and μ_0 are the permittivity and permeability of free space, respectively.

To obtain a complete map of the intensity in the x–y plane, the components x and y of vector r were varied continuously in a limited range. For each new set of x,y coordinates steps 2–4 were repeated. The model calculations were carried out for Ag as this metal has the lowest optical damping properties in the visible region. We used metal optical constants of Ag from [40]. To obtain the resonance of a single particle of $2R = 50$ nm at wavelength 488 nm (argon ion laser line) we assumed a dielectric shell around the particles.

Figure 13.1 shows a typical spatial distribution of the calculated field intensity at the beginning of an N = 50 spheres long particle chain, for irradiating light polarization (a) parallel and (b) normal to the chain axis. Note the logarithmic scale in gray levels! As can be seen from the picture, with normal excitation the guiding is strongly damped, whereas with parallel excitation we observe efficient guiding. Figure 13.1c and d show for comparison the corresponding field distributions at a particle array of final particle number of N = 5. In this case the field intensity is distributed in a complicated way along the line, as the signal reflected at the end of the line causes interference effects.

For determination of the transmission loss along a chain of metal spheres, a cross-section of the two-dimensional information shown in Figure 13.1 was extracted. Figure 13.2 shows the corresponding data obtained by cutting parallel to the particle chain axis at zero distance to each particle surface (see inset). Due to the large number of particles in the chain, the signal reflected from the end of the line can be neglected, as argued above. The field intensity decays continuously with increasing distance, with some oscillations superimposed at the beginning and at the end of the chain due to the impedance misfit. Note the starting values larger than 1 due to the resonantly enhanced field of the first particle. Transmission losses were extracted from Figure 13.2 by a least-squares fit of an exponential decay applied to the cases $d/R = 2.0$, 2.4, and 3.0, shown in Table 13.1. The minimum transmission loss is observed for a center-to-center distance/radius ratio d/R of approximately 3. At zero interparticle distance ($d/R = 2.0$) the transmission loss is a factor of 2.3 higher than at the optimum distance and at larger distances the loss also increases strongly, as expected. In the optimum case $d/R = 3.0$ for the exponential decay law $I = I_0 \exp(-\gamma x)$ a signal damping constant of $\gamma = 0.0083\,\text{nm}^{-1}$, corresponding to a 1/e damping length of approximately 0.9 µm, was evaluated. At higher interparticle distances than three times the particle radius the logarithmic signal versus distance curve deviates strongly from a straight line, indicating essential nonexponentiality, with a faster decay at the input side.

In conclusion, we have found an optimum particle–particle distance and attenuation factors in the 0.1 µm^{-1} range, a value comparable to the results obtained by Takahara et al. [758] for cylindrical waveguide geometries.

After this initial paper and a paper by Brongersma et al. [760], this type of guiding due to near-field coupling was first demonstrated experimentally in

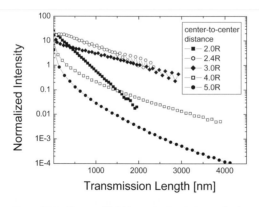

Figure 13.2 Decay of field intensity in chain axis direction. Values are taken at identical positions of each particle as shown by the open circles in the inset (section of Figure 13.1 parallel to the axis). The parameter d/R is the ratio of interparticle (center-to-center) distance to particle radius. The intensities are normalized to the irradiated light field intensity. Note the strong nonexponential decay for $d/R > 3$.

Table 13.1 Signal damping coefficient γ for the exponential part of the slope: per particle and per nanometer for different ratios d/R of particle radius to interparticle distance[a].

d/R	γ (per particle)	γ (per nm)	$I_{1\mu m}/I_{irrad}$
2.0	0.19	0.0037	0.60
2.4	0.11	0.0018	6.2
3.0	0.083	0.0011	1.3
4.0			0.21
5.0			0.029

a) The cases with $d/R = 4.0$ and 5.0 show strong nonexponential decay, therefore a damping coefficient cannot be given. The last column shows the signal intensity after 1 µm guiding length relative to the irradiated light intensity. All data for $2R = 50$ nm.

macroscopic structures operating in the microwave regime by Maier et al. [761]. The propagation loss in a straight array is 3 dB per 8 cm. Routing of energy around 90° corners is possible with a power loss of 3–4 dB. In a series of further publications, Maier et al. [762–768] also examined nanometer-sized structures and showed that coupling between adjacent particles sets up coupled plasmon modes that give rise to coherent propagation of energy along the array. A point dipole analysis predicts group velocities of energy transport that exceed $0.1c$ along straight arrays and shows that energy transmission and switching through chain networks such as corners and tee structures are possible at high efficiencies. Plasmon waveguides can be used to build nanoscale optical devices with a lateral mode size well below the diffraction limit. Radiation losses into the far-field are expected to be negligible due to the near-field nature of the coupling, and resistive heating leads to transmission losses of about 6 dB µm^{-1} for gold and silver particles.

Salerno et al. [769] reported in 2001 on the near-field optical response of a small square grating of 333 gold nanoparticles tailored by electron-beam lithography. Photon scanning tunneling microscopy (PSTM) was applied to acquire near-field optical images. Two different incident wavelengths were used to characterize the intensity and the spatial localization of the electromagnetic near-field both in and out of resonance for the excitation of particle plasmons. Indeed, particle plasmons appeared as a clear increase in the light field enhancement when properly excited. In the PSTM images of the 333 particle grating they found a spatial intensity profile clearly different from that of the single particle. The number of interference maxima within the grating depends on the ratio of the grating constant to the light wavelength. There is obviously no crucial influence of the relative particle position and therefore of the number of nearest neighboring particles – within the array or not – on the spatial intensity profile.

Salerno et al. [770] reported on scanning near-field optical microscopy (SNOM) of one-dimensional chains of gold nanoparticles which are optically excited at their particle plasmon frequency under total internal reflection. The shape and mutual distance of the nanoparticles are tailored by lithographic fabrication. It was shown that the optical near-field profiles are determined by two contributions: the particle plasmon near-field and interference patterns due to the interaction of the exciting light wave and diffracted radiative grating orders.

Girard and Quidant [771] presented a theoretical analysis of the transport of optical energy along narrow chains of metal particles. Under permanent and localized illumination, they pointed out a transmission band where the optical near-field becomes commensurate with the metal arrangement. Transverse modes are found to lead to a higher transmittance than longitudinal modes. Within the framework of the dipolar approximation, the dispersion relation of the chain was derived, allowing the energy of both longitudinal and transverse modes to be compared. However, for larger and nonspherical particles lithographically designed at the surface of a sample, the point-like description becomes questionable and more elaborate models are highly desirable.

Gray and Kupka [772] used the finite-difference time-domain (FDTD) method to study a variety of arrays of silver cylinders with nanometer-scale diameters with the aim of assessing the possibility of nanoscale confinement and propagation of radiation. A funnel-like configuration was found to be one interesting possibility for achieving propagation of light through features confined in one dimension to less than 100 nm.

13.2
Scanning Near-Field Optical Microscopy

With the development of the optical near-field microscopy [773–778], shortly after the invention of the scanning tunneling microscope in 1982 a scanning technique was established that allows spatial resolution beyond the diffractive limit with optical means and at the same time uses the versatile methods of optical spectroscopy. The idea of the scanning near-field optical microscope traces back to Synge [779].

The high resolution is obtained by the detection of short-range electromagnetic interactions among the specimen and probe in the near-field. The decisive components are localized evanescent electromagnetic fields such as those obtained by total internal reflection. In common scanning near-field optical microscopes these fields are generated by a metallized and sharpened fiber with a small aperture of typically 50–100 nm. The sample disturbs the near-field of the probe and light is emitted from the sample that can be detected in reflection or transmission. Unlike this aperture SNOM we will discuss a modified version of the apertureless near-field optical microscope, employing scattering of an evanescent wave at a nanometer-sized silicon tip of a standard cantilever interacting with the sample [780, 781], which has yielded spatial resolution in the 1 nm range. Instead of focusing the exciting light beam on the sample, we consider evanescent waves from total internal reflection. They propagate along a plane surface and are scattered by the nonradiating probe and by particles seeded on the surface.

In a first approach to the problem of calculating the power scattered from the tip–sample region, we modeled the probe tip by a spherical particle of radius R_0

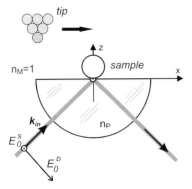

Figure 13.3 Sketch of the scattering geometry for the scan of an Si cluster across a surface with a single spherical particle in the center of the reference frame.

and particle material with complex refractive index n_0 and assumed single spheres of radius R_1 or aggregates of identical particles of radius R_1 with complex refractive index n_1 as the sample. In order to understand the basic properties of evanescent-wave scattering by such a multi-particle complex, we further assumed that we could neglect multiple scattering effects involving the surface. It could be expected that the influence of the surface in this scattering problem is similar to that in the case of a single particle on a substrate.

In the following, we restrict considerations to the example of an x–y scan of a tip nanoparticle of Si ($2R_0 = 20$ nm) over a surface with (a) a single Ag nanoparticle ($2R_1 = 40$ nm) and (b) a linear chain of identical silver nanoparticles [782]. Figure 13.3 shows a sketch of the geometry in this scattering problem. In all calculations it was assumed that the evanescent wave is generated by total internal reflection at the glass–air interface with an angle of incidence of $\theta_i = 60°$. The corresponding refractive index of the glass prism was assumed to be $n_P = 1.5$ at all wavelengths under consideration. In all cases, the scan range was 400×400 nm in the x–y-plane, and the calculations were carried out for 200×200 positions of the scanning probe particle. The silicon particle was scanned across the silver sphere at height $z = 50$ nm of its center relative to the glass substrate, so allowing contact of the tip with the particles on the surface. The scan was carried out at wavelength $\lambda = 514$ nm.

Figure 13.4 displays the scattered intensity obtained in the scan of the silicon particle across the surface with a single Ag nanoparticle for both polarizations. The difference between s- and p-polarization is evident from the figure. In p-polarization, a single maximum of the scattered intensity is observed in the center of the scan range, when the silicon particle is in contact with the silver particle. In s-polarization, two maxima appear, shifted along the y-axis, that is, in the direction of the electric field. The qualitative features of the image obtained in s-polarization can be explained as follows. Along the x-axis at $y = 0$ the incidence is always in-plane with the pair axis of the probe–sample pair. Then, only the

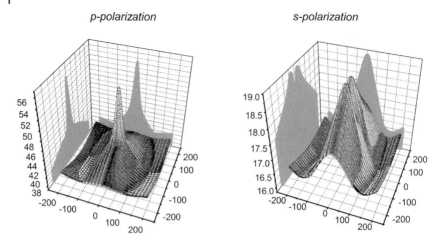

Figure 13.4 Three-dimensional plot of the scattered intensity in p- and s-polarization of an incident evanescent wave for the scan of an Si nanoparticle ($2R_0 = 20$ nm) across a surface with a single spherical Ag nanoparticle ($2R_1 = 40$ nm).

transverse mode is excited. For all other y-positions, the incidence is out-of-plane with the pair axis and the electric field vector can be decomposed into two components parallel and perpendicular to the pair axis. Then, the more efficient longitudinal mode is also excited by a certain amount. Hence the scattered intensity increases and approaches maxima close to $x = \pm 35$ nm but in the direction of the incident electric field. In p-polarization the scattered intensity is always dominated by the longitudinal mode and becomes maximum at $x = 0$ and $y = 0$ where only the longitudinal mode contributes to the intensity pattern. At large distances of the probe particle the intensity tends to a constant value which is the sum of the intensities scattered by the isolated silicon and silver cluster. This value is different in s- and p-polarization as the scattering of evanescent waves by single nanoparticles is polarization dependent.

In many further calculations we examined the dependence of scan results on particle sizes, particle materials, and the wavelength of the incident wave [782]. The results are summarized in the following.

- **Particle Sizes**

Depending on the particle size the total scattered intensity increases with increasing size. However, the basic features of the scans remain almost unchanged.

- **Particle Materials**

As the probe material Si and Si_3N_4 were used, which are commonly used materials in near-field microscopy. For the sample material we used Ag, GaAs, and SiO_2 as representative of metals, semiconductors, and isolators. Again, the basic features of the scans remain almost unchanged; however, the total intensity depends on the material. In s-polarization the minimum at $x = 0$ has already vanished for the nanoparticle pairs Si–GaAs and Si–SiO$_2$ and is replaced by a narrow maximum.

- **Wavelength**

For comparison of the scattered intensity at different wavelengths, we calculated scans of a Si probe ($2R_0 = 20$ nm) over a surface with a silver nanoparticle ($2R_1 = 40$ nm) at wavelengths $\lambda_1 = 514$ nm and $\lambda_2 = 633$ nm, which are the most often used laser wavelengths in common scattering experiments. The ratio $I(\lambda_1)/I(\lambda_2)$ of the scattered intensities approximately amounts to 2.94 for p-polarization and 2.83 for s-polarization. These ratios are close to the ratio $(\lambda_1/\lambda_2)^4 = 2.3$ that is expected from single particle scattering in the Rayleigh limit. The difference is caused by the electromagnetic coupling among the nanoparticles that already in the Rayleigh limit affects scattering and absorption.

Figure 13.5 displays the scattered intensity for both polarizations obtained in the scan of the silicon particle across the surface with a linear chain of $N = 5$ Ag nanoparticles. Along the x-axis all scans are slightly asymmetric on going from $-x$ to x. The asymmetry is caused by retardation in the excitation of the particles in the chain: the evanescent wave is phase shifted at each particle site when propagating along the x-axis. Six intensity maxima with five intensity minima in between can clearly be resolved. The number of minima corresponds to the number of

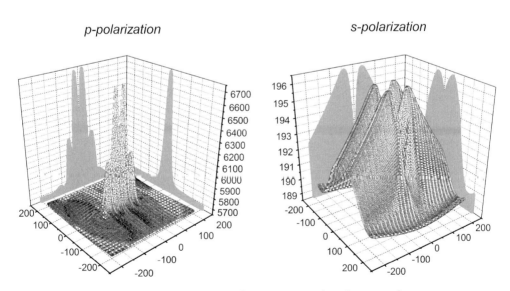

Figure 13.5 Three-dimensional plot of the scattered intensity in p- and s-polarization of an incident evanescent wave for the scan of a Si nanoparticle ($2R_0 = 20$ nm) across a surface with a chain of $N = 5$ spherical Ag nanoparticles ($2R_1 = 40$ nm).

particles in the chain, while the number of maxima is increased by 1. This result was also obtained in further line scans along linear chains with varying number N of clusters, provided that N is below 8. For larger N the influence of further particles in the chain decreases with increasing N, resulting in deviations from the above rule. The distance of the peak positions of the outermost maxima $\delta x = 196$ nm approximately corresponds to the length of the aggregate in the x-direction (200 nm), although the positions of these maxima are slightly shifted compared with the physical ends of the aggregate. This result was not confirmed by further calculations on linear chains with varying N. For example, for a linear chain of $N = 2$ spheres with $2R_1 = 40$ nm the distance of the outermost peak positions amounts to $\delta x = 56$ nm instead of 80 nm, and for a linear chain with $N = 6$ spheres it amounts to $\delta x = 250$ nm instead of 240 nm.

Doicu et al. [783] developed numerical schemes on the basis of the discrete sources method for calculation of the evanescent wave scattering by a sensor tip near a plane surface. With these algorithms they calculated, however, only the angular resolved scattering and not the power scattered by the tip as a function of x, y, z.

Wurtz et al. [784] reported an investigation of the field scattered by isolated silver nanoparticles on a glass substrate observed by apertureless near-field optical microscopy with an illumination wavelength of 404 nm. They observed an unexpected spatial extension of the scattered field. The near-field contrast (both distribution and intensity) was shown to be strongly sensitive to polarization of the incident light. A large field enhancement observed in polarization was interpreted as the result of the contribution of the particle plasmon resonance to the diffracted field. The results were further discussed in terms of the particle shape and the experimental configuration used.

Xiao [785] developed a self-consistent procedure to calculate the induced fields at the probe and at the surface microscopic features in scattering SNOM. Two incident fields, total internal reflection and external reflection, were examined numerically. The results were used to discuss the resolution of the device, the position of the far-field detector, and the probe material. The position of the far-field detector and the illuminating method proved to be the most important parameters for the image obtained in the far-field.

Krenn et al. [786] used a near-field optical microscope to probe the surface plasmon fields on a smooth silver film with superimposed specifically designed nanoscale surface structures, tailored by means of electron beam lithography. Interference patterns resulting from scattering of surface plasmons were observed, together with enhanced near-field intensities localized in subwavelength areas.

13.3
SERS with Aggregates

Raman scattering by molecules has been proven to be enhanced if the molecules are deposited on metallic nanoparticles. This has been reported by several groups in the period from 1970 to 1985. We refer the reader to a list of publications and reviews that appeared in this period (note that this list does not claim to be complete) [787–811].

In some of these papers based on Mie's theory [805, 806], the relation between SERS and the resonant optical absorption by Cu, Ag, or Au nanoparticles was emphasized. In the quasi-static approximation this relation can roughly be derived. For that purpose, the differences between near-field and far-field are neglected.

Let us consider one molecule at the interface of a spherical metal particle which is very small compared with the wavelength of light. The Raman radiation is emitted in a two step process:

1) Excitation of the molecule. The exciting field can exceed the incident field on account of the resonance behavior of the inner field in the particle; its increase is then given by the enhancement factor $3\varepsilon(\omega_i)/[\varepsilon(\omega_i) + 2\varepsilon_M(\omega_i)]$.

2) Emission from the molecule. The emitted field is, again, enhanced by the metal particle resonance; the factor is now $3\varepsilon(\omega_R)/[\varepsilon(\omega_R) + 2\varepsilon_M(\omega_R)]$, where ω_i and ω_R are the frequencies of the incident and the Raman-scattered light, respectively.

Hence the total enhancement factor for the scattered intensity in view of the electromagnetic resonance of the small metal particle is given by

$$G = \left| \frac{E_{NF}(\omega_i)}{E_{inc}(\omega_i)} \frac{E_{NF}(\omega_i)}{E_{inc}(\omega_i)} \right|^2 \propto \left| \frac{\varepsilon(\omega_i)}{\varepsilon(\omega_i) + 2\varepsilon_M(\omega_i)} \frac{\varepsilon(\omega_R)}{\varepsilon(\omega_R) + 2\varepsilon_M(\omega_R)} \right|^2 \quad (13.1)$$

As $\omega_i - \omega_R \ll \omega_i, \omega_R$ we substitute $\varepsilon(\omega_R) \approx \varepsilon(\omega_i)$ to give

$$G \propto \left| \frac{\varepsilon(\omega_i)}{\varepsilon(\omega_i) + 2\varepsilon_M(\omega_i)} \right|^4. \quad (13.2)$$

The resonance behavior of Equation 13.2 at the minimum of the denominator is reflected in the optical extinction cross-section of such spherical metallic particles (without the Raman-active molecule), and hence $G(\omega) \propto \sigma_{ext}^4(\omega)$.

From this equation, it appears necessary to have Cu, Ag, or Au nanoparticles, as these nanoparticles exhibit a resonance due to the collective excitation of the free electron plasma. On the other hand, we showed in Chapter 6 that resonances may also appear due to electronic interband transitions that can be described by a harmonic oscillator model. Indeed, SERS was found not only for Cu, Ag, and Au nanoparticles, but also for Li, Ni, Pd, Pt, Cd, and Hg nanoparticles. Otto et al. [812, 813] pointed to the fact that roughness is not the only origin of SERS. In addition, SERS also incorporates a short-range *chemical effect* which is restricted to localized adsorbate-surface interaction. The electromagnetic approach described above is therefore only one possible explanation for SERS. In the following, however, we will further consider the electromagnetic approach.

In fact, on turning again to nanoparticles, the above estimation is rather rough, because it neglects radiation damping, surface scattering, and retardation effects. Moreover, in practice of SERS the particles are typically 50–80 nm in size and, therefore, beyond Rayleigh's approximation, and the wavelengths commonly used in SERS are $\lambda = 780$ and 830 nm, which are far from the resonance position of any single particle. From this, it follows that we need a new model which takes into account the larger size and the wavelengths in the NIR region of the exciting radiation. Further, the new model must yield increased scattering, since the near-field of the scattered wave is important rather than the inner field of the particle. As the tangential component of the inner field must fit the tangential components of the fields outside the particle, a resonance of the inner field also can be recognized in the scattered field close to the surface of the particle. A high potential for SERS, therefore, is shown by elongated nonspherical particles (ellipsoids, nanorods, finite cylinders) and aggregates of spherical particles. We will consider in the following only the aggregates of spheres. Potential particle materials are again Ag and Au, but also ZrN, as has turned out from all the earlier discussions on spherical and nonspherical particles.

Blatchford et al. [802] and Wokaun and co-workers [814, 815] showed experimentally that aggregates of spherical particles are most advantageous for a large Raman enhancement. Blatchford et al. found additional extinction bands at wavelengths between 650 and 750 nm and increased scattering at wavelengths between 650 and 800 nm caused by the aggregates. A few first theoretical treatments [613, 614] and some more recent studies [816–820] already gave qualitative agreement with experimental results and also yielded *hot* spots depending on the size and topography of the aggregates and on the polarization of the incident light [819, 820]. However, these studies were based on the coupling of dipoles, that is, the primary particles in the aggregates were assumed to be small compared with the wavelength of light and also the radiated fields of the particles were approximated by only dipole radiation. This is a serious restriction compared with the analytical solution of light scattering by aggregates presented in Chapter 10. Hence a quantitative determination of Raman intensities remains difficult from the coupled dipole approximation. Markel et al. [820] gave an estimation also for the near-field intensity in the hot spots, showing that it can exceed a factor of 10^5. They concluded that with these local fields, SERS enhancements of 10^{10} can be achieved. A more quantitative analysis, however, is lacking.

We report here on numerical results for planar silver and gold nanoparticle aggregates with $N = 9$, 13, and 16 primary particles with diameter $2R = 50$ nm, as illustrated in the insets in Figures 13.6 and 13.7, where efficiency spectra and the scattering enhancement $Q_{sca}(N)/[NQ_{sca}(1)]$ are presented for silver in Figure 13.7 and for gold in Figure 13.8. The spectra were calculated for unpolarized incident light. In addition, near-field intensities at wavelength $\lambda = 830$ nm are given for the gold nanoparticle particle aggregate with $N = 16$ primary particles in Figure 13.8.

The effect of aggregation of silver nanoparticles can again be clearly recognized: the *N*-particle aggregate exhibits additional extinction features at longer wavelengths where the surface plasmon polariton of the single particle is strongly decreased, due to the coupling among the primary particles. In the spectral region of the SPP, the extinction by the aggregate is decreased. Moreover, at long wavelengths the scattering is the main contributor to the extinction, also for the single particles, whereas at short wavelengths the absorption dominates the extinction.

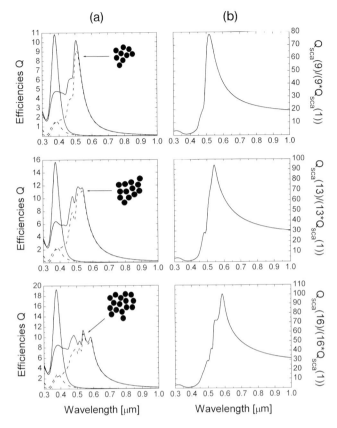

Figure 13.6 (a) Efficiency spectra and (b) scattering enhancement of planar aggregates of silver nanoparticles with diameter $2R = 50$ nm and $N = 9$, 13, and 16 and topologies shown in the insets.

In Figure 13.6b, the scattering enhancement $Q_{sca}(N)/[NQ_{sca}(1)]$ is shown. The scattering enhancement for the considered aggregates is a minimum of 20 ($N = 9$) and is peaked at a wavelength λ_0 that is dependent on the number N of primary particles. For $N = 9$ it is $\lambda_0 = 519$ nm and the enhancement is 78, for $N = 13$ it is $\lambda_0 = 545$ nm and the enhancement is 94, and for $N = 16$ it is $\lambda_0 = 587$ nm and the enhancement is 100. This means that already in the far-field zone the scattering of light by these N-particle aggregates exceeds the scattering by N isolated particles by a factor of 80–100 at λ_0. Compared with a single particle, the enhancement is still larger by the factor N; for example, it is 1600 for the aggregate with $N = 16$ primary particles. The maximum enhancement shifts to longer wavelengths with increasing aggregate size. Nevertheless, we point again to the fact that this shift will run into saturation for large aggregates.

For aggregates of gold nanoparticles, the extinction spectra in Figure 13.7 are similar to those of silver aggregates in Figure 13.6 except that the resonances of

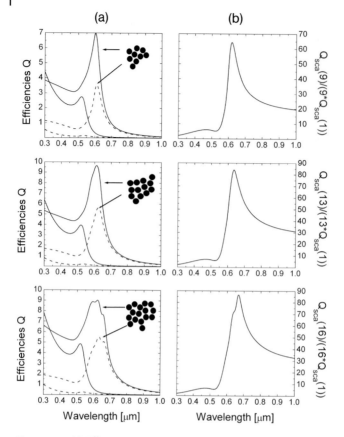

Figure 13.7 (a) Efficiency spectra and (b) scattering enhancement of planar aggregates of gold nanoparticles with diameter $2R = 50$ nm and $N = 9$, 13, and 16 and topologies shown in the insets.

the aggregates are less resolved than for silver. Only one broadened extinction peak can be observed that is shifted to longer wavelengths. The peak position, and in consequence also the peak in the scattering enhancement, shift to longer wavelengths with increasing aggregate size. This result confirms the experimental findings of Blatchford et al. [802] who measured the absorption and scattering by aggregated gold colloids. Moreover, they found that the Raman scattering is maximum where the aggregates exhibit a maximum in scattering. The corresponding values for λ_0 and the peak magnitude are $\lambda_0 = 619$ nm with peak magnitude 64 for $N = 9$, $\lambda_0 = 634$ nm and peak magnitude 84 for $N = 13$, and $\lambda_0 = 667$ nm with peak magnitude 87 for $N = 16$. The minimum enhancement amounts to 20 compared with N primary particles.

The discussion of the spatial distribution of the near-field intensities is restricted to the aggregate with $N = 16$ gold particles. For this aggregate the intensity of the

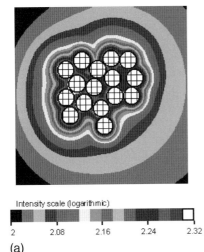
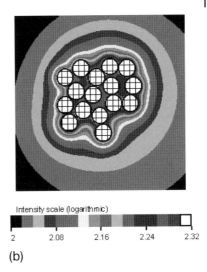

Figure 13.8 Near-field intensities in the equatorial plane (x–y plane) of a planar aggregate with $N = 16$ Au particles for an incident wave propagating along the z-axis, being polarized along (a) the x-axis and (b) the y-axis. A color version of this figure can be found in the color plates at the end of the book.

scattered light in the vicinity of the 16 primary particles is calculated at $\lambda = 830$ nm in the equatorial plane, which in the present case is the x–y plane. The results are depicted in Figure 13.8 for a plane incident wave propagating along the positive z-axis of a reference frame centered in one particle of the aggregate and being polarized either along the x-axis (Figure 13.8a) or along the y-axis (Figure 13.8b). The hatched spherical regions correspond to the nanoparticles. The intensities are plotted with a logarithmic scale. On comparison with the near-field scattering by a single particle, it can be recognized that the aggregate as a whole just behaves similarly to a larger single particle whose size is still small compared with the wavelength $\lambda = 830$ nm as the dipole radiation dominates the scattering intensities. In between the particles, however, the constructive interference of all scattered fields yields distinct regions where the near-field intensities are particularly large. The location of these hot spots depends on the polarization of the incident light, which can be recognized on comparing Figure 13.8a with Figure 13.8b. For the electric field parallel to the x-axis the hot spot includes the seven left-most particles in the aggregate, whereas for the electric field parallel to the y-axis the hot spot is centered in the aggregate and includes six particles. Comparing the intensity in the hot spots with the near-field intensities at the surface of a single particle with $2R = 50$ nm (not presented here), the intensity in the hot spot is increased by a factor of 37, which exceeds the factor of 16 coming from the number of primary particles in the aggregate.

In summary, we find from our calculations that at wavelengths in common SERS experiments (780 and 830 nm), the enhancement of the far-field intensity of the elastically scattered light of the aggregates considered is from about 180 ($N = 9$) to 640 ($N = 16$) compared with a single particle for silver or gold particle aggregates with $2R = 50$ nm. From the analysis of the near-field intensities we obtain a further enhancement by a factor of 37 in the hot spots for the aggregate with $N = 16$ primary particles. This means that we obtain a total enhancement of the intensity in elastic light scattering amounting to 24 000 in the hot spots. This is only a factor of 4 smaller than the estimation in [819], but also shows that enhancements of at least 10^8 can be obtained in SERS when using the appropriate single particles and aggregates. The advantage of our approach is that we can calculate near-field and far-field intensities of aggregates beyond the coupled dipole approach, that is, also for larger primary particles.

14
Effective Medium Theories

Consider a composite of at least two nonmixable components: well-separated nanoparticles statistically distributed in a nonabsorbing homogeneous matrix (see Figure 14.1). Its optical response is determined both by the particles and by the matrix material and is difficult to predict in general. However, if it is possible to replace the inhomogeneous composite by a homogeneous material of one common dielectric function ε_{eff}, the reflectance, transmittance, and absorbance of this medium can be calculated as a linear response. For that purpose, a model for the dielectric function ε_{eff} of this *effective medium* must be established as a function of the particle properties and the surrounding matrix and the concentration of particles in the composite.

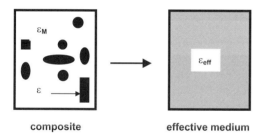

composite   effective medium

Figure 14.1 Scheme of the effective medium. The realistic composite is replaced by a homogeneous effective medium.

The fundamental question is how to obtain this effective dielectric function ε_{eff} in terms of the dielectric function ε of the nanoparticles and ε_M of the matrix and of suitably chosen topology parameters. A considerable number of equations for the effective dielectric function are available today based upon different approximative models. The most important will be reviewed and discussed in this chapter.

The most limiting assumption of all *effective medium theories* is the assumption that scattering by the particles can be neglected, for which the nanoparticles must be very small compared with the wavelength of light. Hence their application to

Table 14.1 Values of the dielectric constant and the scattering cross-section of very small particles in the Rayleigh limit and at the same wavelength $\lambda = 514$ nm.

Material	Dielectric constant ε	$\sigma_{sca} \propto \left|\dfrac{\varepsilon - \varepsilon_M}{\varepsilon + 2\varepsilon_M}\right|^2$
Ag	$-10.3 + i0.205$	1.853
GaAs	$17.66 + i3.207$	0.725
Si	$17.89 + i0.525$	0.72
Si_3N_4	$4.15 + i0$	0.263
SiO_2	$2.13 + i0$	0.075

composites with transparent, purely scattering particles is questionable. However, also for strongly absorbing particles the question may arise of whether they can be applied because those particles usually scatter the light more strongly than transparent particles. This can be recognized from Table 14.1, where the scattering cross-sections of small spherical particles are calculated in the Rayleigh approximation for a fixed size $2R$ and at wavelength $\lambda = 514$ nm. Compared with a silica sphere, the silver sphere of same size scatters the light about 25 times more strongly. This is the result for a silver sphere off resonance. In resonance the scattering cross-section of the silver sphere is additionally increased by a factor of 3–4 due to the SPP resonance. As a consequence of this stronger scattering, the failure of effective medium models must become obvious earlier for metal nanoparticles than for dielectric nanoparticles.

A further restricting assumption is that the particles must be well separated. Electromagnetic particle–particle interactions are usually omitted.

The existing effective medium models essentially differ in the way in which an average is calculated from the dielectric functions ε and ε_M of the embedded nanoparticles and the matrix, respectively.

The first effective medium concept goes back to Newton (see [821]), later modified by Beer [822], Gladstone and Dale [823], Landau and Lifschitz [824], and Looyenga [825]. It is based on simply averaging certain powers of the dielectric functions of the two mixed components, weighted with the filling factor f:

$$\varepsilon_{\text{eff}} = f\varepsilon + (1-f)\varepsilon_M \quad \text{Newton} \tag{14.1}$$

$$\varepsilon_{\text{eff}}^{\frac{1}{2}} = f\varepsilon^{\frac{1}{2}} + (1-f)\varepsilon_M^{\frac{1}{2}} \quad \text{Beer, Gladstone and Dale} \tag{14.2}$$

$$\varepsilon_{\text{eff}}^{\frac{1}{3}} = f\varepsilon^{\frac{1}{3}} + (1-f)\varepsilon_M^{\frac{1}{3}} \quad \text{Landau and Lifschitz, Looyenga} \tag{14.3}$$

By applying upper and lower bounds, Lichtenecker [821] obtained the *logarithmic mixture rule*, which was later used in careful investigations of effective medium theories by Niklasson et al. [826]:

$$\log(\varepsilon_{\text{eff}}) = f\log(\varepsilon) + (1-f)\log(\varepsilon_M) \tag{14.4}$$

The simplest approach to an effective medium that explicitly considers also the shape of the particles stems from J. C. Maxwell Garnett[1] [827] in 1904 for spherical inclusions:

$$\varepsilon_{\text{eff}} = \varepsilon_M \frac{(\varepsilon + 2\varepsilon_M) + 2f(\varepsilon - \varepsilon_M)}{(\varepsilon + 2\varepsilon_M) - f(\varepsilon - \varepsilon_M)} \quad \text{or} \quad \frac{\varepsilon_{\text{eff}} - \varepsilon_M}{\varepsilon_{\text{eff}} + 2\varepsilon_M} = f \frac{\varepsilon - \varepsilon_M}{\varepsilon + 2\varepsilon_M} \quad (14.5)$$

Using the abbreviation

$$\Lambda = \frac{\varepsilon - \varepsilon_M}{\varepsilon + 2\varepsilon_M} \quad (14.6)$$

the Maxwell-Garnett equation can be rewritten as

$$\varepsilon_{\text{eff}} = \varepsilon_M \frac{1 + 2f\Lambda}{1 - f\Lambda} \quad (14.7)$$

The advantage of this equation becomes obvious in the following. For spherical particles that are small compared with the wavelength, the dipole polarizability α is (see also Section 4.7)

$$\alpha = 3V_p \frac{\varepsilon - \varepsilon_M}{\varepsilon + 2\varepsilon_M} = 3V_p \Lambda \quad (14.8)$$

This relation between Λ and α now allows the Maxwell-Garnett equation to be extended to nonspherical particles. Expressions for α for ellipsoids, spheroids, and cubes were given in Chapter 9.

In the Bruggeman ansatz [828], the dielectric constant is given as

$$\varepsilon_{\text{eff}} = E \pm \sqrt{E^2 + \frac{\varepsilon \varepsilon_M}{2}} \quad (14.9)$$

with

$$E = \frac{\varepsilon_M (2 - 3f) - \varepsilon (1 - 3f)}{4} \quad (14.10)$$

Usually, the positive square root is used. However, if the imaginary part of ε_{eff} becomes negative, the negative square root has to be used to obtain the correct dielectric constant.

Bergman [829–831] developed a *generalized effective medium* model. The starting point of the Bergman theory is the expression for the orientation averaged polarizability of one nanoparticle of variable shape:

$$\alpha = V_p \sum_{i=1}^{N} \frac{C_i (\varepsilon - \varepsilon_M)}{\varepsilon_M + G_i (\varepsilon - \varepsilon_M)} \quad (14.11)$$

---

1) His approach is well known as the Maxwell-Garnett model, although Maxwell is in fact the third Christian name of James Clerk Maxwell Garnett.

The next step is to introduce the variables $t = \varepsilon_M/(\varepsilon_M - \varepsilon)$ and $g_i = C_i/\varepsilon_M$. On going formally to a macroscopic sample with large numbers of arbitrarily irregular inclusions, the number of modes N will increase sufficiently to extend N formally to infinity. Then, g_i becomes a continuously variable quantity denoted β between the boundaries of zero and unity. The summation over i can be replaced by integration over β and we formally obtain

$$\varepsilon_{\text{eff}} = \varepsilon_M \left[1 - f \int_0^1 \frac{g(\beta)}{t - \beta} d\beta \right] \tag{14.12}$$

where the function $g(\beta)$ is called the *spectral density*, which reflects the topology of the sample.

This is the *Bergman equation* (or *Bergman theorem*) [829–831]. Here, ε_{eff} is defined from the assumption that the volume average of the energy density W of the electric field in a capacitor filled with inhomogeneous material is invariant if the inhomogeneous sample is replaced by a fictitious homogeneous effective medium of ε_{eff}. In this concept, the volume fraction f just represents a scaling factor without a direct influence on topology. The dielectric properties of particle and matrix materials enclosed merely in the variable t occur *separately* from the topology in the Bergman equation in a transparent but formal way.

The explicit consideration of structural effects and electrodynamic interaction was introduced in effective medium approaches for the first time by Lamb et al. [832]. They used a multiple-scattering approach that yields an effective propagation wavevector. Thus lowest-order corrections of the filling factor and weak scattering are included.

Felderhof and Jones [833–836] introduced the electrodynamic influences of neighboring particles in statistically inhomogeneous samples as a correction of the Maxwell-Garnett equation:

$$\varepsilon_{\text{eff}} = \varepsilon_M \left(1 + \frac{3f}{1 - 3t - f} \right) \tag{14.13}$$

with respect to two-cluster pair correlations by a correction term $C(t)$:

$$\varepsilon_{\text{eff}} = \varepsilon_M \left[1 + \frac{3f}{1 - 3t - f - C(t)} \right] \tag{14.14}$$

where the variable t is defined as $t = \varepsilon_M/(\varepsilon_M - \varepsilon)$. $C(t)$ was computed for particle pairs from the direct dipolar coupling. The result is

$$\varepsilon_{\text{eff}}^{\text{pair}} = \frac{(4\pi N\alpha)^2 V_T}{3} \left[\int_{2R}^{\infty} dr \cdot r \left[\frac{2(\varepsilon - \varepsilon_M)/r^2}{\varepsilon(r^3 - 2R^3) + 2\varepsilon_M(r^3 + R^3)} \right] \right.$$
$$\left. + \int_{2R}^{\infty} dr \cdot r \left[\frac{2(\varepsilon - \varepsilon_M)/r^2}{\varepsilon(r^3 + R^3) + \varepsilon_M(2r^3 - R^3)} \right] \right] \tag{14.15}$$

where r is center-to-center distance of the nanoparticles, R the nanoparticle radius, α the single particle polarizability, N the number of particles in the sample volume, and V_T the single particle volume.

All these results for ε_{eff} can be used to calculate the reflectance, transmittance, and absorbance of the composite using Maxwell's relation:

$$n + i\kappa = \sqrt{\varepsilon_1 + i\varepsilon_2} \tag{14.16}$$

between the dielectric function and the refractive index, Fresnel's equations for reflectance and transmittance, and

$$\gamma_a(\lambda) = \frac{4\pi}{\lambda}\kappa \tag{14.17}$$

for the absorption constant.

Extensions of existing effective medium theories have regularly been made by different authors. Mainly the Maxwell-Garnett and the Bruggeman equations have been extended to obtain better agreement between experiment and theory. We will come back to these extensions later in Section 14.3. Here, we will treat only the first and most prominent extension by Gans and Happel outlined in the information box.

14.1
Theoretical Results for Dielectric Nanoparticle Composites

The similarity of effective medium models with the molar refraction used to calculate the refractive index of mixed glasses makes them attractive for calculation of the refractive index of dielectric nanoparticle–dielectric matrix composites. In Table 14.2 we give results obtained for Al_2O_3, TiO_2, ZrO_2, and SiO_2 nanoparticles embedded in the polymer poly(methyl methacrylate) (PMMA) using the Maxwell-Garnett equation. The idea is to increase or to lower the refractive index of PMMA by oxide nanoparticle inclusions.

Actually, the inclusion of the highly refracting Al_2O_3, TiO_2, and ZrO_2 nanoparticles results in an increased refractive index. Vice versa, the inclusion of SiO_2

Table 14.2 Refractive indices of PMMA with several oxide nanoparticle inclusions.

Material	f	n		
		486 nm	587 nm	656 nm
PMMA		1.4977	1.4918	1.4895
Al_2O_3 in PMMA	0.01	1.5005	1.4945	1.4919
	0.10	1.5246	1.5185	1.5159
TiO_2 in PMMA	0.01	1.5094	1.5026	1.4998
	0.10	1.6152	1.6015	1.5958
ZrO_2 in PMMA	0.01	1.5042	1.4982	1.4956
	0.10	1.5629	1.5564	1.5531
SiO_2 in PMMA	0.01	1.4974	1.4915	1.4889
	0.10	1.4943	1.4884	1.4859

INFO – Mie Theory-Based Extension of Maxwell-Garnett Theory

An extension of Maxwell-Garnett's equation to the full expressions for the electric partial modes a_n and the magnetic partial modes b_n according to Mie's theory instead of the simple approximative Λ was made by Gans and Happel [837] in 1909 and later by Doyle [838] and Ruppin [839].

Gans and Happel obtained the following effective dielectric function (for the electric field) and the effective magnetic permeability (for the magnetic field):

$$\varepsilon_{\text{eff}} = \varepsilon_m \frac{1 + 2f\Lambda_{\text{el}}^{\text{GH}}}{1 - f\Lambda_{\text{el}}^{\text{GH}}}; \quad \mu_{\text{eff}} = \mu_m \frac{1 + 2f\Lambda_{\text{magn}}^{\text{GH}}}{1 - f\Lambda_{\text{magn}}^{\text{GH}}} \quad (14.18)$$

with

$$\Lambda_{\text{el}}^{\text{GH}} = \frac{3i}{2x^3}(a_1 - a_2); \quad \Lambda_{\text{magn}}^{\text{GH}} = \frac{3i}{2x^3}b_1 \quad \text{with} \quad x = \frac{2\pi R}{\lambda}\sqrt{\varepsilon_M} \quad (14.19)$$

The derivation of ε_{eff} and μ_{eff} by Gans and Happel is correct only up to a certain point. Then, the authors made the erroneous assumption that the electric partial modes a_n represent *only* the electric field and the magnetic partial modes b_n represent *only* the magnetic field. This assumption is wrong! As we have already shown in Section 5.1, the electric or transverse magnetic modes a_n stand for both (!) E^{TM} and H^{TM}. Vice versa, the magnetic or transverse electric modes b_n stand for both E^{TE} and H^{TE}. Hence it is incorrect to assign the a_n only to the electric field and the b_n only to the magnetic field.

In a correct derivation, only ε_{eff} can be deduced, but polarization dependent:

$$\varepsilon_{\text{eff}}^{\text{TM,TE}} = \varepsilon_m \frac{1 + 2f\Lambda^{\text{TM,TE}}}{1 - f\Lambda^{\text{TM,TE}}} \quad (14.20)$$

with

$$\Lambda^{\text{TM}} = \frac{3i}{2x^3}\left(3a_1 + \frac{100}{x^2}a_2\right); \quad \Lambda^{\text{TE}} = \frac{3i}{2x^3}b_1 \quad \text{with} \quad x = \frac{2\pi R}{\lambda}\sqrt{\varepsilon_M} \quad (14.21)$$

An effective magnetic permeability cannot be deduced. Hence the use of $\Lambda_{\text{magn}}^{\text{GH}}$ may lead to confusion, since according to Equation 14.18 for nonmagnetic materials with ε, ε_M, and $\mu = \mu_M = 1$, a magnetization M of the effective medium may result for large enough filling factors. For instance, for glass nanoparticles ($\varepsilon = 2.25$) in air ($\varepsilon_M = 1$) and a filling factor $f = 0.3$, the effective permeability would amount to $\mu_{\text{eff}} \approx 1.001$, and for silver nanoparticles in air it would even amount to $\mu_{\text{eff}} \approx 1.1$, both at wavelength $\lambda = 633$ nm.

Doyle's result is included in Λ^{TM} since he only considered the TM dipole mode a_1. A conceptual inconsequence of this Mie-based extension is, however, that this more general particle polarizability makes sense only if the higher multipoles are essentially excited, that is, particle sizes are beyond the Rayleigh limit, automatically admitting the possibility of light scattering.

nanoparticles with a refractive index lower than that of PMMA also lowers the refractive index of the composite. The effects of the inclusions depend on the volume fraction f.

The Maxwell-Garnett approach seems to be helpful here to predict the refractive index of dielectric–dielectric composites. Nevertheless, we want to point to the fact that these nanoparticles still scatter the light. A well-suited measure of the influence of the scattering is the haze of such composites introduced in Section 2.3. If Al_2O_3, TiO_2, or ZrO_2 nanoparticles with $2R = 10$ nm or larger are suspended in a PMMA plate of 1 mm thickness, the mean haze becomes intolerable for filling factors $f \geq 0.01$, according to the standard ASTM D 1003-97: *Test Method for Haze and Luminous Transmittance of Transparent Plastics*.

Another application of effective medium approaches for dielectric nanoparticle composites is the prediction of optical properties of transparent dielectric particles with absorbing inclusions. This is an important problem in atmospheric science, astronomy, and optical particle sizing. Hence theories have been developed to calculate the scattering by particles composed of more than one distinct refractive index. For examples, we refer to [840–842].

14.2
Theoretical Results for Metal Nanoparticle Composites

The original work of Garnett dealt with metal nanoparticle inclusions in a transparent medium to explain in a simple way the optical properties of such systems, like colloidal suspensions or colored glasses. They were published even before Gustav Mie developed his theory on light scattering by spherical particles. Later, Bruggeman recognized that the Maxwell-Garnett approach is incomplete in the sense that percolation, that is, conductive paths through the sample, at higher volume fractions is not included. He developed a new approach for ε_{eff} (see Equation 14.9) that includes percolation.

If the filling factor f is less than 0.001, the results for ε_{eff} of the most prominent Maxwell-Garnett and Bruggeman effective medium models and the mixing rule of Looyenga do not differ. This changes drastically for higher volume fractions. Figure 14.2 shows a comparison of these three models for Ag inclusions ($f = 0.1$) in a transparent matrix with a constant refractive index $n_M = 1.5$ (approximately valid for various crown glasses).

For the Maxwell-Garnett model, the refractive index n_{eff} of this composite exhibits features that indicate the presence of a harmonic oscillator with a resonance wavelength of $\lambda = 435$ nm, which is close to the wavelength position of Ag spheres with $2R = 2$ nm in a medium with $n_M = 1.5$ ($\lambda = 421$ nm). In consequence, when using the corresponding κ_{eff} in calculation for the absorption constant $\gamma_a(\lambda) = 4\pi \kappa_{eff}(\lambda)/\lambda$ of a planar homogeneous medium, almost the same absorption peak can be observed as in the extinction by single isolated Ag nanoparticles, but shifted to longer wavelengths (red shift). This can be recognized from Figure 14.3, where the absorption coefficient calculated with the above κ_{eff} is calculated. This

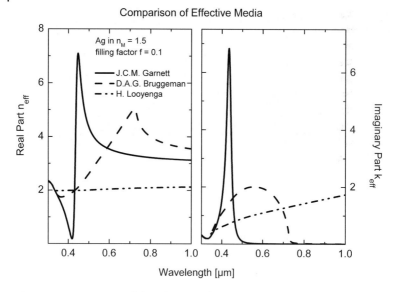

Figure 14.2 Comparison of the refractive index $n_{eff} + i\kappa_{eff}$ from the Maxwell-Garnett, Bruggeman and Looyenga models for Ag inclusions ($f = 0.1$) in a transparent matrix with a constant refractive index $n_M = 1.5$.

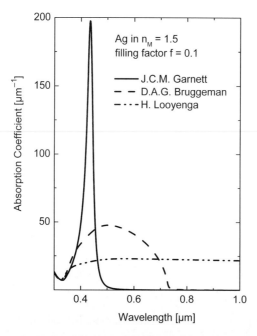

Figure 14.3 Comparison of the absorption coefficient resulting from κ_{eff} from the Maxwell-Garnett, Bruggeman and Looyenga models for Ag inclusions ($f = 0.1$) in a transparent matrix with a constant refractive index $n_M = 1.5$.

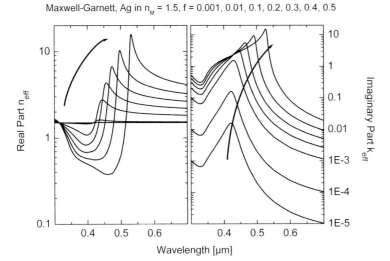

Figure 14.4 Refractive index $n_{eff} + i\kappa_{ef}$ of a transparent matrix with a constant refractive index $n_M = 1.5$ with Ag inclusions as a function of the volume fraction of inclusions according to the Maxwell-Garnett model.

is not so for the other two models. The Bruggeman n_{eff} appears to be composed of a series of close-lying harmonic oscillators with different oscillator strengths. However, the curvatures of n_{eff} and κ_{eff} are unexpectedly strange and cannot really be explained by a sum over harmonic oscillators. Completely unexpected is the behavior of the Looyenga n_{eff}. Whereas the real part is almost constant, the imaginary part, and hence the absorption in the composite, increases continuously with increasing wavelength. Correspondingly, the absorption coefficient is also modified for the Bruggeman model. For the Looyenga mixing rule, the absorption even appears to be almost identical with the absorption coefficient of a silver film.

Next, we consider the influence of the volume fraction on the refractive index of the effective medium. We only give results for the Maxwell-Garnett model in Figure 14.4. The volume fraction varies and takes the values $f = 0.001$, 0.01, 0.1, 0.2, 0.3, 0.4, and 0.5.

Despite the logarithmic scale used, the changes in the real part of n_{eff} cannot be resolved for $f = 0.001$. The imaginary part κ_{eff} clearly shows the behavior of a harmonic oscillator. The harmonic oscillator behavior is maintained up to volume fractions of $f = 0.1$. For higher volume fractions, clear deviations can be observed that cannot be assigned to any realistic model for dielectric functions or refractive indices. This trend can also be observed in the absorption coefficient displayed in Figure 14.5.

Finally, we consider the effect of the particle shape on the refractive index of the composite. Again, we use Ag inclusions in a matrix with constant refractive index $n_M = 1.5$. The filling factor is $f = 0.1$, for which the Maxwell-Garnett approach still results in a harmonic oscillator behavior of the refractive index n_{eff}. In all cases presented in Figure 14.6 the refractive index n_{eff} and the absorption index κ_{eff}

436 | *14 Effective Medium Theories*

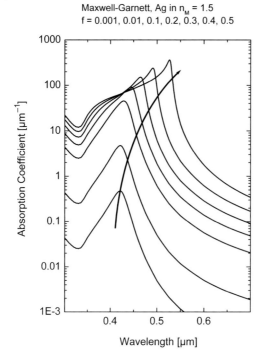

Figure 14.5 Absorption coefficient of a transparent matrix with a constant refractive index $n_M = 1.5$ with Ag inclusions as a function of the volume fraction of inclusions according to the Maxwell-Garnett model.

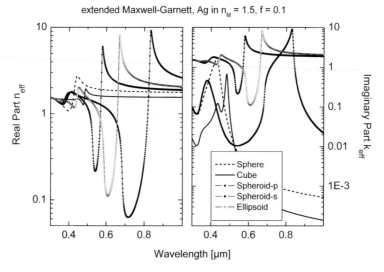

Figure 14.6 Refractive index $n_{eff} + i\kappa_{eff}$ of a transparent matrix with a constant refractive index $n_M = 1.5$ with Ag inclusions as a function of the shape of the inclusions. Calculations according to the Maxwell-Garnett model for $f = 0.1$.

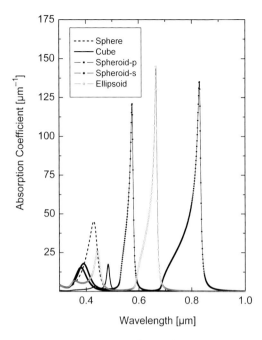

Figure 14.7 Absorption coefficient of a transparent matrix with a constant refractive index $n_M = 1.5$ with Ag inclusions as a function of the shape of the inclusions. Calculations according to the Maxwell-Garnett model for $f = 0.1$.

exhibit features that are similar to those of a harmonic oscillator. However, the curvature, especially at the lower wavelength side, do not really correspond to a harmonic oscillator behavior. The position of the *resonance wavelength* depends on the particle shape. For ellipsoids and spheroids, clearly two *resonances* can be identified, according to the geometry factors (depolarization factors). These resonances can be recognized also in the absorption coefficient in Figure 14.7. Except for the sphere, all other particles lead to peaks in the absorption coefficient with a shoulder at short wavelengths and an abrupt decrease for longer wavelengths.

14.3
Experimental Examples

In this last section, we give an overview of several experiments where effective medium models were applied to calculate optical properties of the composite. In most examples the original approach (Maxwell-Garnett, Bruggeman) was not sufficient and was extended to take into account shapes, aggregation, sizes and shape distribution.

Abelès and Gittleman [843] examined the optical properties of various metal–dielectric and semiconductor–dielectric composite systems. The results are compared with the predictions of the Maxwell-Garnett effective medium model and its extension to an ellipsoidal shape. Only the original Maxwell-Garnett theory predicts the characteristic optical features of granular metals (here Ag). In the case of the semiconductor–dielectric composites (here Si–SiC), the observed red shift in the transverse optical phonon frequency in the far-infrared region is very small and does not allow one to discriminate between the two models. In the case of the Ge–Al_2O_3 films, both models are in good agreement with the experimental results for the optical constants near the absorption edge.

Granqvist and Hunderi [844] prepared composites containing Au nanoparticles with $2R = 3$–4 nm. For interpretation of the measured optical transmittance they used the Maxwell-Garnett and Bruggeman models and an extension of effective medium models by Hunderi. For low filling factors the models are indistinguishable. For higher filling factors, the authors additionally considered deviations from the spherical shape, the size distribution, and dielectric coatings on the particles, but without experimental evidence that they affect the optical properties significantly. The authors also considered dipole–dipole coupling, which is accounted for by a set of effective depolarization factors, that is, they assumed well-defined geometric configurations of spheres. With this extension, they obtained in excellent agreement with the experimental data.

Norrman et al. [845] continued the experiments of Granqvist and Hunderi on gold particles in a dielectric matrix. For interpretation of the optical transmittance spectra they used the Maxwell-Garnett formalism extended to prolate spheroids. Interactions were taken into account according to Bedeaux and Vlieger [846]. They achieved in excellent agreement with the experimental results.

An overview of cermets (metal–ceramic composites) with application of the Maxwell-Garnett and Bruggeman approaches was given by Granqvist [847].

Optical properties of granular tin films were investigated by Truong et al. [848] in the spectral range 0.22–1.0 µm. Reflectance was calculated using the Maxwell-Garnett approach. An extension of this theory – the revised dipole theory that includes frequency-dependent terms neglected in the static approximation – was also used to characterize the granular films with a better description of the optical behavior when the particle size exceeds 10 nm.

Electron beam coevaporated Al–Si composite films were produced by Niklasson and Craighead [849] consisting of a mixture of aluminum and silicon particles. The microstructure suggests the use of the Bruggeman theory to account for the optical properties of the films, since the microgeometry underlying this theory is that of a random mixture of spheres of the two components. Indeed, very good agreement between the experimental reflectance and predictions by the Bruggeman theory was found for volume fractions less than 0.33. Furthermore, the agreement was obtained without the introduction of fitting parameters.

The Maxwell-Garnett theory extended to include the shape factor and the size of the metal particles embedded in a dielectric matrix was used by Thériault and Boivin [850] to explain the observed optical constants of a Cu–PbI_2 cermet material. Fairly good agreement was observed for a volume fraction ranging from $f = 0$ to 0.12.

Experimental spectral transmission curves of four cermets, Cu–PbCl$_2$, Cu–NaF, Ag–PbCl$_2$, and Ag–NaF, were compared with calculated curves obtained from the generalized Maxwell- Garnett theory by Chandonnet and Boivin [851]. The results showed that the microstructure of the metal particles in the dielectric matrix is independent of the nature of the dielectric, as predicted by the generalized Maxwell-Garnett theory. However, measurements also indicated that the microstructure is the same for both copper and silver cermets, which cannot be explained within the context of the Maxwell-Garnett theory.

Heilmann et al. [852–854] investigated plasma–polymer–silver composites prepared by simultaneous or alternating plasma polymerization and metal evaporation. Transmission spectra were recorded and interpreted with the Maxwell-Garnett and Bruggeman models. Fairly good agreement was obtained for the two-dimensional distribution of the particles using the Maxwell-Garnett approach, whereas for three-dimensional particle distributions in the matrix the Bruggeman approach yielded better results. A comparison with the Bergman effective medium model was also made [854].

Leitner et al. [855] examined the optical properties of a metal island film close to a smooth metal surface. The silver island film was positioned close to a silver mirror with a quartz interlayer in between. The interlayer thickness could be changed stepwise. A dramatic change in the optical reflection behavior of this mirror system was observed depending on the thickness of the interlayer. At a low interlayer thickness the reflectivity loss is high in a broad spectral band, so the mirror appears to be nearly black. At greater distances the mirror becomes colored: as the spacer layer thickness is increased, the color varies from bright blue to yellow, orange, and violet. The phenomenon was analyzed in terms of an effective medium theory by using TEM data and an electromagnetic model for the optical constants of the metal island film. For island films with a sufficiently high absorbance, the spectra are characterized by two sharp minima where the reflectivity drops to values below 10^{-3}.

The optical properties of discontinuous copper films on quartz substrates at volume fractions ranging from 0.19 to 1.0 were investigated by Dobierzewska-Mozrzymas and Bieganski [856]. Three phases could be distinguished: (1) the dielectric phase for $f < 0.54$, (2) the percolation phase for $f = 0.54 \pm 0.05$, and (3) the metallic phase for $f > 0.54$. A modified Maxwell-Garnett theory (shape factor, size effect, surface oxide layer) was successfully used to interpret the experimental data for films with low volume fraction ($f \approx 0.19$–0.26).

Nagendra and Lamb [857] prepared germanium–silver composite thin films having different concentrations of Ag, ranging from 7 to 40%, by DC cosputtering of Ge and Ag. The morphology was determined to be island-like, that is, Ag particles distributed in Ge. The optical constants of the composites showed a semiconductor behavior, even for 40% Ag in Ge. Comparison of the n and k data with the Maxwell-Garnett and Bruggeman theories showed that both theories have limited scope in predicting the optical properties of semiconductor–metal composite films in the infrared region.

Based on the research of Niklasson et al. [826], the Maxwell–Garnett and Bruggeman theories for two-component systems were directly extended to three-component systems by Luo [858]. Two theories were obtained, of which one had already been obtained by Wood and Ashcroft [859] and Stroud and Pan [860] by different or similar methods. The improved theories were employed in the calculation of the spectral reflectance of three samples of electrolytic colored anodic aluminum oxide film, Au–Cu–Al$_2$O$_3$ and Ag–Cu–Al$_2$O$_3$, and the results were compared with the experimental data.

Dalacu and Martinu [353] examined composites of Au nanoparticles with varying size in SiO$_2$ in the wavelength range 300–850 nm. They produced gold particles by plasma deposition in an SiO$_2$ film. Analysis was performed using the Maxwell-Garnett effective medium approach. From comparison of the measured and calculated spectra, they determined the optical constants of the Au nanoparticles with an additional evaluation of the interband edge.

Kürbitz et al. [861] doped a melt of a commercially manufactured glass with copper compounds to prepare copper nanoparticles in glass. The glass obtained was opaque black at the usual thickness and looked dark red in bulbs of incandescent lamps due to the high absorption caused by the particles. The pseudo-optical constants of this material were determined as a function of wavelength in the range 350–700 nm by ellipsometric measurements. They could be reproduced very well by a model that consisted of a roughness layer situated on a substrate of glass containing spherical copper particles with a Gaussian size distribution with $2R = 6.5$ nm and $\sigma = 1.6$ nm and a volume concentration of 2.4×10^{-3}. For this modeling, the dielectric function of the roughness layer was approximated by the Bruggeman effective-medium theory and that of the copper-containing glass substrate was calculated on the basis of the theory of Gans and Happel. The results were verified by TEM investigations.

Mandal et al. [862] prepared silver–silica nanocomposite thin films by a high-pressure DC sputtering technique. Films deposited at lower substrate temperature showed a narrow distribution of nanoparticles with nearly spherical shape. An increase in substrate temperature resulted in films with a non-uniform size and shape due to the agglomeration of the nanoparticles. This size and shape distribution affects the optical absorbance spectra and results in a broad and asymmetric surface plasmon band. A shape distribution introduced in the Maxwell-Garnett or Bruggeman effective medium theory was found to give a reasonable description of the experimentally observed optical absorption spectra.

Takeda et al. [863] investigated the linear and nonlinear optical properties of Cu nanoparticle composites with high Cu concentrations. Negative Cu ions of 60 keV were implanted into amorphous SiO$_2$ at a fixed flux and various fluences. Optical absorption increased with the Cu volume fraction but the SPP peak was maximized around a fraction of $f = 0.1$ and attenuated beyond that fraction. The best agreement with the absorption spectra was found with the Maxwell-Garnett model.

References

1. Ostwald, W. (1914) *Die Welt der vernachlässigten Dimensionen. Eine Einführung in die moderne Kolloidchemie mit besonderer Berücksichtigung ihrer Anwendungen*, Theodor Steinkopff, Dresden.
2. Kunckel, J. (1689) *Ars Vitraria Experimentalis oder Vollkommene Glasmacherkunst*, Frankfurt.
3. Born, M., and Wolf, E. (1975) *Principles of Optics*, Pergamon Press, Oxford.
4. Stratton, J.A. (1941) *Electrodynamic Theory*, McGraw-Hill, New York.
5. Jackson, J.D. (1981) *Classical Electrodynamics*, 2nd edn, John Wiley & Sons, Inc., New York.
6. van de Hulst, H.C. (1981) *Light Scattering by Small Particles*, Dover Publications, New York.
7. Kerker, M. (1969) *The Scattering of Light*, Academic Press, San Diego, CA.
8. Bohren, C.F., and Huffman, D.R. (1983) *Absorption and Scattering of Light by Small Particles*, John Wiley & Sons, Inc., New York.
9. Schuerman, D.W. (ed.) (1980) *Light Scattering by Irregularly Shaped Particles*, Plenum Press, New York.
10. Barber, P.W., and Chang, R.K. (eds) (1988) *Optical Effects Associated with Small Particles*, World Scientific, Singapore.
11. Kreibig, U., and Vollmer, M. (1995) *Optical Properties of Metal Clusters*, Springer Series in Materials Science, vol. **25**, Springer, Berlin.
12. Mishchenko, M.I., Hovenier, J.W., and Travis, L.D. (eds) (2000) *Light Scattering by Nonspherical Particles*, Academic Press, San Diego, CA.
13. Shalaev, V.M. (ed.) (2002) *Optical Properties of Nanostructured Random Media*, Topics in Applied Physics, vol. **82**, Springer, Berlin.
14. Borghese, F., Denti, P., and Saija, R. (2002) *Scattering from Model Nonspherical Particles*, Springer, Berlin.
15. Mishchenko, M.I., Travis, L.D., and Lacis, A.A. (2002) *Scattering, Absorption, and Emission of Light by Small Particles*, Cambridge University Press, Cambridge.
16. Babenko, V.A., and Astafyeva, L.G. (2003) *Electromagnetic Scattering in Disperse Media*, Springer, Berlin.
17. Mie, G. (1908) Beiträge zur Optik trüber Medien, speziell kolloidaler Metallösungen (Contributions to the optics of turbid media, especially colloidal metal suspensions). *Ann. Phys. (Leipzig)*, **25**, 377–445.
18. Kubelka, P., and Munk, F. (1931) Ein Beitrag zur Optik der Farbanstriche. *Z. Tech. Phys.*, **12**, 593–601.
19. Schönauer, D., and Kreibig, U. (1985) Topography of samples with variably aggregated metal particles. *Surf. Sci.*, **156**, 100–111.
20. Jullien, R., and Botet, R. (1987) *Aggregation and Fractal Aggregates*, World Scientific, Singapore.
21. Meakin, P. (1988) Fractal aggregates. *Adv. Colloid Interface Sci*, **28**, 249–331.
22. Botet, R., and Jullien, R. (1990) Fractal aggregates of particles. *Phase Transitions*, **24**, 691–736.
23. Derjaguin, B. (1940) On the repulsive forces between charged colloidal particles and on the theory of slow

Optical Properties of Nanoparticle Systems: Mie and Beyond. Michael Quinten
Copyright © 2011 WILEY-VCH Verlag GmbH & Co. KGaA, Weinheim
ISBN: 978-3-527-41043-9

coagulation and stability of lyophobe sols. *Trans. Faraday Soc.*, **36**, 203–211.
24 Verwey, E.J.W., and Overbeek, J.Th.G. (1948) *Theory of the Stability of Lyophobic Colloids*, Elsevier, Amsterdam.
25 McCartney, L.N., and Levine, S. (1969) An improvement on Derjaguin's expression at small potentials for the double layer interaction energy of two spherical colloidal particles. *J. Colloid Interface Sci.*, **30**, 345–354.
26 Hofmeister, F. (1888) Zur Lehre von der Wirkung der Salze. *Arch. Exp. Path. Pharmakol.*, **24**, 247–260.
27 Schmickler, W. (1985) Die Elektrochemie im Umbruch. *Nachr. Chem. Tech. Lab.*, **33**, 872–878.
28 Lang, N.D., and Kohn, W. (1970) Theory of metal surfaces: charge density and surface energy. *Phys. Rev. B*, **1**, 4555–4568.
29 von Smoluchowski, M. (1916) Drei Vorträge über Diffusion, Brownsche Molekularbewegung und Koagulation von Kolloidteilchen. *Phys. Z.*, **17**, 557–599.
30 Ostwald, W. (1900) Über die vermeintliche Isomerie des roten und gelben Quecksilberoxyds und die Oberflächenspannung fester Körper. *Z. Phys. Chem.*, **34**, 495–503.
31 Bedeaux, D., and Vlieger, J. (1974) A statistical theory of the dielectric properties of thin island film. I. The surface material coefficients. *Physica*, **73**, 287–311.
32 Vlieger, J., and Bedeaux, D. (1980) A statistical theory for the dielectric properties of thin island films. *Thin Solid Films*, **69**, 107–130.
33 Bedeaux, D., and Vlieger, J. (1983) A statistical theory for the dielectric properties of thin island films: application and comparison with experimental results. *Thin Solid Films*, **102**, 265–281.
34 Dusemund, D., Hoffmann, A., Salzmann, T., Kreibig, U., and Schmid, G. (1991) Cluster matter: the transition of optical elastic scattering to regular reflection. *Z. Phys. D*, **20**, 305–308.
35 Wolf, P.E., and Maret, G. (1985) Weak localization, coherent backscattering of photons in disordered media. *Phys. Rev. Lett.*, **55**, 2696–2699.
36 Palik, E.D. (ed.) (1985) *Handbook of Optical Constants of Solids I*, Academic Press, San Diego, CA.; Palik, E.D. (ed.) (1991) *Handbook of Optical Constants of Solids II*, Academic Press, San Diego, CA.
37 Johnson, P.B., and Christy, R.W. (1972) Optical constants of the noble metals. *Phys. Rev. B*, **6**, 4370–4379.
38 Paquet, V. (1966) Optische Eigenschaften von Silber-Gold-Misch-Schichten. Diploma thesis. University of Saarland, Saarbrücken.
39 Hagemann, H.-J., Gudat, W., and Kunz, C. (1974) DESY Report SR-74/7, Deutsches Elektronen Synchrotron, Hamburg.
40 Quinten, M. (1996) Optical constants of gold and silver clusters in the spectral range between 1.5 eV and 4.5 eV. *Z. Phys. B*, **101**, 211–217.
41 Granqvist, C.G., and Buhrman, R.A. (1976) Ultrafine metal particles. *J. Appl. Phys.*, **47**, 2200–2219.
42 Chýlek, P., and Ramaswamy, V. (1982) Lower and upper bounds on extinction cross sections of arbitrarily shaped strongly absorbing or strongly reflecting nonspherical particles. *Appl. Opt.*, **21**, 4339–4344.
43 Brendel, R., and Bormann, D. (1992) An infrared dielectric function model for amorphous solids. *J. Appl. Phys.*, **71**, 1–6.
44 Kim, C.C., Garland, J.W., Abad, H., and Raccah, P.M. (1992) Modeling the optical dielectric function of semiconductors: extension of the critical-point parabolic-band approximation. *Phys. Rev. B*, **45**, 11749–11767.
45 Ashcroft, N.W., and Mermin, N.D. (1976) *Solid State Physics*, Holt, Rinehardt & Winston, New York.
46 Ehrenreich, H., and Philipp, H.R. (1962) Optical properties of Ag and Cu. *Phys. Rev.*, **128**, 1622–1629.
47 Feibelmann, P.J. (1982) Surface electromagnetic fields. *Prog. Surf. Sci.*, **12**, 287–407.
48 Sturm, K. (1992) Die Wechselwirkung von Licht mit Materie, in *Synchrotronstrahlung zur Erforschung kondensierter Materie*, vol. 23 (ed. Forschungszentrum Jülich GmbH), IFF-Ferienkurs Forschungszentrum Jülich, Chapter 3, pp. 3.1–3.60.

49 Ehrenreich, H., and Cohen, M.H. (1959) Self-consistent field approach to the many-electron problem. *Phys. Rev.*, **115**, 786–790.

50 Hohenberg, P., and Kohn, W. (1964) Inhomogeneous electron gas. *Phys. Rev.*, **136** (3B), B864–B870.

51 Kohn, W., and Sham, L.J. (1965) Self-consistent equations including exchange and correlation effects. *Phys. Rev.*, **140** (4A), A1133–A1138.

52 Ceperley, D.M., and Alder, B.J. (1980) Ground state of the electron gas by a stochastic method. *Phys. Rev. Lett.*, **45**, 566–569.

53 Hubbard, J. (1957) The description of collective motions in terms of many-body perturbation theory. II. The correlation energy of a free-electron gas. *Proc. R. Soc. London Ser. A*, **243**, 336–352.

54 Adler, S.L. (1962) Quantum theory of the dielectric constant in real solids. *Phys. Rev.*, **126**, 413–420.

55 Wiser, N. (1963) Dielectric constant with local field effects included. *Phys. Rev.*, **129**, 62–69.

56 Roaf, D.J. (1962) The Fermi surfaces of copper, silver and gold II. Calculation of the Fermi surfaces. *Philos. Trans. R. Soc. London Ser. A*, **255**, 135–152.

57 Bassani, F., and Parravicini, G.P. (1975) *Electronic States, Optical Transitions in Solids*, Pergamon Press, Oxford.

58 Jellison, G.E., Jr., and Modine, F.A. (1996) Parametrization of the optical functions of amorphous materials in the interband region. *Appl. Phys. Lett.*, **69**, 371–373.

59 O'Leary, S.K., Johnson, S.R., and Lim, P.K. (1997) The relationship between the distribution of electronic states, the optical absorption spectrum of an amorphous semiconductor: an empirical analysis. *J. Appl. Phys.*, **82**, 3334–3340.

60 Karlsson, B., and Ribbing, C.G. (1982) Optical constants and spectral selectivity of stainless steel and its oxides. *J. Appl. Phys.*, **53**, 6340–6346.

61 Tyndall, J. (1869) On the blue colour of the sky, the polarization of skylight, and on the polarization of light by cloudy matter generally. *Proc. R. Soc. London*, **17**, 223–233.

62 Lord Rayleigh (Strutt, J.W.) (1871) On the scattering of light by small particles. *Philos. Mag.*, **41**, 447–454.

63 Lord Rayleigh (Strutt, J.W.) (1871) On the light from the sky, its polarization and colour. *Philos. Mag.*, **41**, 107–110. 274–279.

64 Lord Rayleigh (Strutt, J.W.) (1881) On the electromagnetic theory of light. *Philos. Mag.*, **12**, 81–101.

65 Lorenz, L. (1890) On the light reflected and refracted by a transparent sphere. *K. Dan. Vidensk. Selsk. Skr.*, **6**, 1–62.

66 von Ignatowski, W. (1905) Reflexion elektromagnetischer Wellen an einem dünnen Draht. *Ann. Phys. (Leipzig)*, **18**, 495–522.

67 Seitz, W. (1906) Die Beugung des Lichtes an einem dünnen, zylindrischen Drahte. *Ann. Phys. (Leipzig)*, **21**, 1013–1029.

68 Wriedt, T. (1998) A review of elastic light scattering theories. *Part. Part. Syst. Charact.*, **15**, 67–74.

69 Bergmann, L., and Schäfer, C. (1993) *Lehrbuch der Experimentalphysik*, vol. 3, Optik, Neunte Auflage, Walter de Gruyter, Berlin.

70 Hertz, H. (1888) Die Kräfte elektrischer Schwingungen, behandelt nach der Maxwellschen Theorie. *Ann. Phys.*, **36**, 1–22.

71 Righi, A. (1901) Sui campi elettromagnetici e particolarmente su quelli creati da cariche elettriche o da poli magnetici in movimento. *Nuovo Cimento*, **2**, 104–121.

72 Sauter, F. (1967) Der Einfluss von Plasmawellen auf das Reflexionsvermögen von Metallen (I). *Z. Phys.*, **203**, 488–494.

73 Forstmann, F. (1967) Der Einfluss von Plasmawellen auf das Reflexionsvermögen von Metallen (II). *Z. Phys.*, **203**, 495–514.

74 Melnyk, A.R., and Harrison, M.J. (1970) Theory of optical excitation of plasmons in metals. *Phys. Rev. B*, **2**, 835–857.

75 Clanget, R. (1972) Der Einfluss von Plasmawellen auf die optischen Eigenschaften leitender Schichten und metallischer Kolloide. *Optik*, **35**, 180–194.

76 Ruppin, R. (1973) Optical properties of a plasma sphere. *Phys. Rev. Lett.*, **31**, 1434–1437.

77 Ruppin, R. (1975) Optical properties of small metal spheres. *Phys. Rev. B*, **11**, 2871–2876.

78 Ruppin, R. (1982) Spherical and cylindrical surface polaritons in solids, in *Electromagnetic Surface Modes* (ed. A.D. Boardman), John Wiley & Sons, Inc., New York, pp. 345–398.

79 Möglich, F. (1933) Beugungserscheinungen an kleinen Kugeln in der Nähe von Brennpunkten konvergenter Kugelwellen. *Ann. Phys. (5th Ser.)*, **17**, 825–862.

80 Gans, R. (1925) Strahlungsdiagramme ultramikroskopischer Teilchen. *Ann. Phys. (Leipzig)*, **76**, 29–38.

81 Pedersen, J.S. (1970) Analysis of small-angle scattering data from colloids and polymer solutions: modeling and least-squares fitting. *Adv. Colloid Interface Sci.*, **70**, 171–210.

82 Steubing, W. (1908) Über die optischen Eigenschaften kolloidaler Goldlösungen. *Ann. Phys. (Leipzig)*, **26**, 329–371.

83 Abramovitz, M., and Stegun, I. (eds) (1970) *Handbook of Mathematical Functions*, Dover Publications, New York.

84 Wiscombe, W.J. (1979) Mie scattering calculations: advances in techniques and fast vector-speed computer codes. NCAR Technical Note, NCAR_TN-140_STR, National Center for Atmospheric Research, Boulder, CO.

85 Nichols, E.F., and Hull, G.F. (1903) The pressure due to radiation. *Phys. Rev.*, **17**, 26–50.

86 Debye, P. (1909) Der Lichtdruck auf Kugeln von beliebigem Material. *Ann. Phys. (Leipzig)*, **30**, 57–136.

87 Lord Rayleigh (1910) The problem of the whispering gallery. *Philos. Mag.*, **20**, 1001–1004.

88 Ashkin, A., and Dziedzic, J.M. (1971) Optical levitation of micron sized spheres. *Appl. Phys. Lett.*, **19**, 283–285.

89 Ashkin, A., and Dziedzic, J.M. (1977) Feedback stabilization of optically levitated particles. *Appl. Phys. Lett.*, **30**, 202–204.

90 Ashkin, A., and Dziedzic, J.M. (1980) Observation of light scattering from nonspherical particles using optical levitation. *Appl. Opt.*, **19**, 660–668.

91 Ashkin, A., and Dziedzic, J.M. (1981) Observation of optical resonances of dielectric spheres by light scattering. *Appl. Opt.*, **20**, 1803–1814.

92 Smith, N.V. (1969) Optical constants of sodium and potassium from 0.5 to 4.0 eV by split-beam ellipsometry. *Phys. Rev.*, **183**, 634–644.

93 Kreibig, U. (1976) Small silver particles in photosensitive glass: their nucleation and growth. *Appl. Phys.*, **10**, 255–264.

94 Kittel, C. (1971) *Introduction to Solid State Physics*, 4th edn. John Wiley & Sons, Inc., New York.

95 Inagaki, T., Emerson, L.C., Arakawa, E.T., and Wiliams, M.W. (1976) Optical properties of solid Na and Li between 0.6 and 3.8 eV. *Phys. Rev. B*, **13**, 2305–2313.

96 Cohen, H. (1958) Optical constants, heat capacity and the Fermi surface. *Philos. Mag.*, **3**, 762–775.

97 Rasigni, M., and Rasigni, G. (1977) Optical constants of lithium deposits as determined from the Kramers–Kronig analysis. *J. Opt. Soc. Am.*, **67**, 54–59.

98 Raether, H. (1980) *Excitation of Plasmons and Interband Transitions by Electrons*, Springer Tracts in Modern Physics, vol. 111, Springer, Berlin.

98a Raether, H. (1988) *Surface Plasmons on Smooth and Rough surfaces and on Gratings*, Springer Tracts in Modern Physics, vol. 111, Springer, Berlin.

99 Otter, M. (1961) Temperaturabhängigkeit der optischen Konstanten massiver Metalle. *Z. Phys.*, **161**, 539–549.

100 Choyke, W.J., Vosko, S.H., and O'Keeffe, T.W. (1971) A comparison of the optical properties of single crystal and liquid mercury in the energy range 0.5 to 8.25 eV. *Solid State Commun.*, **9**, 361–367.

101 Hibbins, A.P., Sambles, J.R., and Lawrence, C.R. (1998) Surface plasmon-polariton study of the optical dielectric function of titanium nitride. *J. Mod. Opt.*, **45**, 2051–2062.

102 Veszelei, M., Anderson, K., Ribbing, C.G., Järrendahl, K., and Arwin, H. (1994) Optical constants and Drude analysis of sputtered zirconium nitride films. *Appl. Opt.*, **33**, 1993–2001.

103 Weaver, J.H., and Olson, C.G. (1977) Optical absorption of hcp yttrium. *Phys. Rev. B*, **15**, 590–594.

104 Underwood, S., and Mulvaney, P. (1994) Effect of the solution refractive index on the color of gold colloids. *Langmuir*, **10**, 3427–3430.

105 Messinger, B.J., van Raben, K.U., Chang, R.K., and Barber, P.W. (1981) Local fields at the surface of noble-metal microspheres. *Phys. Rev. B*, **24**, 649–657.

106 Quinten, M. (1995) Local fields and Poynting vectors in the vicinity of the surface of small spherical particles. *Z. Phys. D*, **35**, 217–224.

107 Quinten, M. (2001) Local fields close to the surface of nanoparticles and aggregates of nanoparticles. *Appl. Phys. B*, **73**, 245–255.

108 Xu, H. (2004) Electromagnetic energy flow near nanoparticles–I: single spheres. *J. Quant. Spectrosc. Radiat. Transfer*, **87**, 53–67.

109 Kreibig, U., and Quinten, M. (1994) Electromagnetic excitations in large clusters, in *Clusters of Atoms, Molecules* (ed. H. Haberland), Springer Series in Chemical Physics, vol. **56**, Springer, New York, Chapter 3.5, pp. 321–359.

110 Hecht, J. (1980) Optical properties of K smoke particles from 480–620 nm. *J. Opt. Soc. Am.*, **70**, 694–697.

111 Kreibig, U. (1971) Untersuchungen zum Einfluß der Größe von Edelmetall-Mikrokristalliten auf die optischen Materialkonstanten im Bereich der Kugel-Plasma-Resonanz. PhD thesis. University of Saarland, Saarbrücken.

112 Schönauer, D. (1983) Optische Eigenschaften von Systemen wechselwirkender kleiner Goldteilchen. Diploma thesis. University of Saarland, Saarbrücken.

113 Quinten, M. (1985) Der Einfluß der Bildung von Teilchenaggregaten auf die optischen Eigenschaften kolloidaler Goldsysteme. Diploma thesis. University of Saarland, Saarbrücken.

114 Quinten, M. (1989) Analyse der optischen Eigenschaften inhomogener Materie am Beispiel aggregierter, kolloidaler Edelmetall-Systeme. PhD thesis. University of Saarland, Saarbrücken.

115 Kreibig, U., and von Fragstein, C. (1969) The limitation of electron mean free path in small silver particles. *Z. Phys.*, **224**, 307–323.

116 Kreibig, U. (1977) Anomalous frequency and temperature dependence of the optical absorption of small gold particles. *J. Phys. IV, Colloq. C2*, **38**, 97–103.

117 Kreibig, U. (2010) Copper nanoparticles: optical plasmon- and bandstructure excitations. In preparation.

118 Uchida, K., Kaneko, S., Omi, S., Hata, C., Tanji, H., Asahara, Y., Ikushima, A.J., Tokizaki, T., and Nakamura, A. (1994) Optical nonlinearities of a high concentration of small metal particles dispersed in glass: copper and silver particles. *J. Opt. Soc. Am. A*, **11**, 1236–1243.

119 Magruder, R.H., III, Haglund, R.F., Jr, Yang, L., Wittig, J.E., and Zuhr, R.A. (1994) Physical, optical properties of Cu nanoclusters fabricated by ion implantation in fused silica. *J. Appl. Phys.*, **76**, 708–715.

120 Magruder, R.H., III, Wittig, J.E., and Zuhr, R.A. (1993) Wavelength tunability of the surface plasmon resonance of nanosize colloids in glass. *J. Non-Cryst. Solids*, **163**, 162–168.

121 Takeda, Y., Lu, J., Plaksin, O.A., Amekura, H., Kono, K., and Kishimoto, N. (2004) Optical properties of dense Cu nanoparticle composites fabricated by negative ion implantation. *Nucl. Instrum. Methods Phys. Res. B*, **219–220**, 737–741.

122 Kishimoto, N., Gritsyna, V.T., Kono, K., Amekura, H., and Saito, T. (1997) Negative copper ion implantation into silica glasses at high dose rates and optical measurements. *Nucl. Instrum. Methods Phys. Res. B*, **127–128**, 579–582.

123 Kishimoto, N., Gritsyna, V.T., Kono, K., Amekura, H., and Saito, T. (1997) High current implantation of negative copper ions into silica glasses. *Mater. Res. Soc. Symp. Proc.*, **438**, 435–440.

124 Radtke, U. (1987) Kolloide Silber-Goldlegierungen in photosensitivem Glas. Diploma thesis. University of Saarland, Saarbrücken.

125 Pal, A., Shah, S., and Devi, S. (2007) Preparation of stable silver and silver–gold bimetallic nanoparticle in W/O microemulsion containing Triton X-100. *Afr. Phys. Rev.*, **1**, Special Issue (Microfluidics):0001, 1–3.

126 Torigoe, K., Nakajima, Y., and Esumi, K. (1993) Preparation, characterization of colloidal silver–platinum alloys. *J. Phys. Chem.*, **97**, 8304–8309.

127 White, J.U. (1942) Long optical paths of large aperture. *J. Opt. Soc. Am.*, **32**, 285–288.

128 White, J.U. (1976) Very long optical paths in air. *J. Opt. Soc. Am.*, **66**, 411–416.

129 Graff, A. (1995) Präparation und Charakterisierung von nanometergroßen Aerosolpartikeln. Diploma thesis. RWTH Aachen, Aachen.

130 Oslender, F. (1995) Aufbau einer Meßapparatur zur spektroskopischen Analyse der Extinktion von nanometergroßen Aerosolen geringer Teilchenanzahlkonzentration. Diploma thesis. RWTH Aachen, Aachen.

131 Graff, A., Oslender, F., and Quinten, M. (1995) Spectroscopic characterization of ultrafine aerosol particles in the visible spectral region. In Proceedings of the 4th Optical Particle Sizing, Nürnberg, pp. 493–502.

132 Graff, A., Oslender, F., and Quinten, M. (1996) Determination of the layer thickness of submicron coated aerosol particles from optical extinction measurements. *J. Aerosol Sci.*, **27**, 313–324.

133 Russell, B.K., Mantovani, J.G., Anderson, V.E., Warmack, R.J., and Ferrell, T.L. (1987) Experimental test of the Mie theory for microlithographically produced silver spheres. *Phys. Rev. B*, **35**, 2151–2154.

134 Banerjee, S., and Chakravorty, D. (1998) Optical absorption of composites of nanocrystalline silver prepared by electrodeposition. *Appl. Phys. Lett.*, **72**, 1027–1029.

135 Cai, W., Hofmeister, H., Rainer, T., and Chen, W. (2001) Optical properties of Ag and Au nanoparticles dispersed within the pores of monolithic mesoporous silica. *J. Nanopart. Res.*, **3**, 443–453.

136 Wilk, N.R., Jr and Schreiber, H.D. (1998) Optical properties of gold in acetate glasses. *J. Non-Cryst. Solids*, **239**, 192–196.

137 Sönnichsen, C., Franzl, T., Wilk, T., von Plessen, G., and Feldmann, J. (2002) Plasmon resonances in large noble-metal clusters. *New J. Phys.*, **4**, 93.1–93.8.

138 Papavassilion, G.C., and Kokkinakis, T. (1974) Optical absorption spectra of surface plasmons in small copper particles. *J. Phys. F*, **4**, L67–L68.

139 Doremus, R.H., Kao, S.-C., and Garcia, R. (1992) Optical absorption of small copper particles and the optical properties of copper. *Appl. Opt.*, **31**, 5773–5778.

140 Peña, O., Rodríguez-Fernández, L., Rodríguez-Iglesias, V., Kellermann, G., Crespo-Sosa, A., Cheang-Wong, J.C., Silva-Pereyra, H.G., Arenas-Alatorre, J., and Oliver, A. (2009) Determination of the size distribution of metallic nanoparticles by optical extinction spectroscopy. *Appl. Opt.*, **48**, 566–572.

141 Ila, D., Williams, E.K., Sarkisov, S., Smith, C.C., Poker, D.B., and Hensley, D.K. (1998) Formation of metallic nanoclusters in silica by ion implantation. *Nucl. Instrum. Methods Phys. Res. B*, **141**, 289–293.

142 Umeda, N., Kishimoto, N., Takeda, Y., Lee, C.G., and Gritsyna, V.T. (2000) Thermal stability of nanoparticles in silica glass implanted with high-flux Cu^- ions. *Nucl. Instrum. Methods Phys. Res. B*, **166–167**, 864–870.

143 Plaksin, O.A., Takeda, Y., Amekura, H., Kono, K., Umeda, N., and Kishimoto, N. (2005) Optical responses of negative-copper-ion implanted Al_2O_3. *Trans. Mater. Res. Soc. Jpn.*, **30**, 753–756.

144 Amekura, H., Umeda, N., Kono, K., Takeda, Y., Kishimoto, N., Buchal, C., and Mantl, S. (2007) Dual surface plasmon resonances in Zn nanoparticles in SiO_2: an experimental study based on optical absorption and thermal stability. *Nanotechnology*, **18**, 395707, 1–6.

145 Turkevich, J., and Kim, G. (1970) Palladium: preparation and catalytic properties of particles of uniform size. *Science*, **169**, 873–879.

146 Rostalski, J. (1995) Optische Untersuchungen an Metallclustern in nichtkonventionellen Einbettmedien. Diploma thesis. RWTH Aachen, Aachen.

147 Johnson, P.B., and Christy, R.W. (1974) Optical constants of transition metals: Ti, V, Cr, Mn, Fe, Co, Ni, and Pd. *Phys. Rev. B*, **9**, 5056–5070.

148 Amekura, H., Kitazawa, H., Umeda, N., Takeda, Y., and Kishimoto, N. (2004) Nickel nanoparticles in silica glass fabricated by 60 keV negative-ion implantation. *Nucl. Instrum. Methods Phys. Res. B*, **222**, 114–122.

149 Behrens, H., and Ebel, G. (eds) (1981) *Physics Data: Optical Properties of Metals, Parts 1 and 2*, Fachinformationszentrum Energie, Physik, Mathematik GmbH, Eggenstein-Leopoldshafen.

150 Gartz, M. (2001) *Clusterphysik mit LUCAS, einer neuartigen Laser-Nanoteilchen-Quelle mit höchster Leistung*, Aachener Beiträge zur Physik kondensierter Materie 31, Wissenschaftsverlag Mainz, Aachen.

151 Takeda, Y., and Kishimoto, N. (2003) Nonlinear optical properties of metal nanoparticle composites for optical applications. *Nucl. Instrum. Methods Phys. Res. B*, **206**, 620–623.

152 Reinholdt, A., Pecenka, R., Pinchuk, A., Runte, S., Stepanov, A., Weirich, T., and Kreibig, U. (2004) Structural, compositional, optical and colorimetric characterization of TiN-nanoparticles. *Eur. Phys. J. D*, **31**, 69–76.

153 Reinholdt, A. (2005) Untersuchungen zur Anwendbarkeit von Titannitrid- und Zirkonnitrid-Nanopartikeln als Farbpigmente und zum magnetischen Verhalten von Nickel/Nickeloxid-Kern-Hülle-Nanopartikeln. PhD thesis. RWTH Aachen, Aachen.

154 Cortie, M.B., Giddings, J., and Dowd, A. (2010) Optical properties and plasmon resonances of titanium nitride nanostructures. *Nanotechnology*, **21**, 115201.

155 Tomczak, N., Janzewski, D., Han, M., and Vancso, G.J. (2009) Designer polymer–quantum dot architectures. *Prog. Polym. Sci.*, **34**, 393–430.

156 Toshifumi, T., Adachi, S., Nakanishi, H., and Ohtsuka, K. (1993) Optical constants of $Zn_{1-x}Cd_xTe$ ternary alloys: experiment and modeling. *Jpn. J. Appl. Phys.*, **32** (Part 1), 3496–3501.

157 Ozaki, S., and Adachi, S. (1993) Optical constants of $ZnSe_xTe_{1-x}$ ternary alloys. *Jpn. J. Appl. Phys.*, **32**, 2620–2625.

158 Katari, J.E.B., Colvin, V.L., and Alivisatos, A.P. (1994) X-ray photoelectron spectroscopy of CdSe nanocrystals with applications to studies of the nanocrystal surface. *J. Phys. Chem.*, **98**, 4109–4117.

159 Colvin, V.L., Schlamp, M.C., and Alivisatos, A.P. (1994) Light-emitting diodes made from cadmium selenide nanocrystals and a semiconducting polymer. *Nature*, **370**, 354–357.

160 Alivisatos, A.P. (1995) Semiconductor nanocrystals. *MRS Bull.*, **8/1995**, 23–32.

161 Robel, I., Subramanian, V., Kuno, M., and Kamat, P.V. (2006) Quantum dot solar cells. Harvesting light energy with CdSe nanocrystals molecularly linked to mesoscopic TiO_2 films. *J. Am. Chem. Soc.*, **128**, 2385–2393.

162 Stecher, T.P. (1965) Interstellar extinction in the ultraviolet. *Astrophys. J.*, **142**, 1683–1684.

163 Fitzpatrick, E.L., and Massa, D. (1986) An analysis of the shapes of ultraviolet extinction curves I–The 2175 Å bump. *Astrophys. J.*, **307**, 286–294.

164 Draine, B.T., and Lee, H.M. (1984) Optical properties of interstellar graphite and silicate grains. *Astrophys. J.*, **285**, 89–108.

165 Draine, B.T. (1985) Tabulated optical properties of graphite and silicate grains. *Astrophys. J. Suppl. Ser.*, **57**, 587–594.

166 Jäger, C., Mutschke, H., and Henning, Th. (1998) Optical properties of carbonaceous dust analogues. *Astron. Astrophys.*, **332**, 291–299.

167 Schnaiter, M., Mutschke, H., Dorschner, J., Henning, Th., and Salama, F. (1998) Matrix-isolated nanosized carbon grains as an analog for the 217.5 nm feature carrier. *Astrophys. J.*, **498**, 486–496.

168 Li, A., and Greenberg, J.M. (1997) A unified model of interstellar dust. *Astron. Astrophys.*, **323**, 566–584.

169 Matijevic, E., and Scheiner, P. (1978) Ferric hydrous oxide sols. III. Preparation of uniform particles by hydrolysis of Fe(III)-chloride, -nitrate and -perchlorate solutions. *J. Colloid Interface Sci.*, **63**, 509–524.

170 Quinten, M. (1998) *Physical and Chemical Properties of Surfaces in Nano-Particle Systems*, Aachener Beiträge zur Physik kondensierter Materie 24, Wissenschaftsverlag Mainz, Aachen.

171 Hsu, W.P., and Matijévic, E. (1985) Optical properties of monodispersed hematite hydrosols. *Appl. Opt.*, **24**, 1623–1630.

172 Guo, S., Arwin, H., Jacobsen, S.N., Järrendahl, K., and Helmersson, U. (1995) A spectroscopic ellipsometry study of cerium dioxide thin films grown on sapphire by rf magnetron sputtering. *J. Appl. Phys.*, **77**, 5369–5376.

173 Measured by SOPRA SA, with an ellipsometer. Data available at www.sopra-sa.com.

174 Gerfin, T., and Graetzel, M. (1996) Optical properties of tin-doped indium oxide determined by spectroscopic ellipsometry. *J. Appl. Phys.*, **79**, 1722–1729.

175 Gmelin-Institut fur anorganische Chemie und Grenzgebiete (Ed). (1974) *Gmelin Handbuch der Anorganischen Chemie, Part C1: Rare Earth Elements – Sc, Y, La und Lanthanide – Hydrides and Oxides*, 8th edn, Springer, Berlin.

176 Aden, A.L., and Kerker, M. (1951) Scattering of electromagnetic waves from two concentric spheres. *J. Appl. Phys.*, **22**, 1242–1246.

177 Güttler, A. (1952) Die Miesche Theorie der Beugung durch dielektrische Kugeln mit absorbierendem Kern und ihre Bedeutung für Probleme der interstellaren Materie und des atmosphärischen Aerosols. *Ann. Phys. (Leipzig)*, **11**, 65–98.

178 Wait, J.R. (1963) Electromagnetic scattering from a radially inhomogeneous sphere. *Appl. Sci. Res. B*, **10**, 441–450.

179 Mikulski, J.J., and Murphy, E.L. (1963) The computation of electromagnetic scattering from concentric spherical structures. *IEEE Trans. Antennas Propag.*, **AP-11**, 169–177.

180 Bhandari, R. (1985) Scattering coefficients for a multilayered sphere: analytic expressions and algorithms. *Appl. Opt.*, **24**, 1960–1967.

181 Sinzig, J. (1993) Optische Untersuchungen an Hüllen-Clustern. Diploma thesis. RWTH Aachen, Aachen.

182 Sinzig, J., and Quinten, M. (1994) Scattering and absorption by spherical multilayer particles. *Appl. Phys. A*, **58**, 157–162.

183 Kai, L., and Massoli, P. (1994) Scattering of electromagnetic-plane waves by radially inhomogeneous spheres: a finely stratified sphere model. *Appl. Opt.*, **33**, 501–511.

184 Johnson, B.R. (1996) Light scattering by a multilayer sphere. *Appl. Opt.*, **35**, 3286–3296.

185 Yang, W. (2003) Improved recursive algorithm for light scattering by a multilayered sphere. *Appl. Opt.*, **42**, 1710–1720.

186 Wannemacher, R., Pack, A., and Quinten, M. (1999) Resonant absorption and scattering in evanescent fields. *Appl. Phys. B*, **68**, 225–232.

187 Edgar, A. (1999) Light scattering from copper-coated dielectric particles in fluorozirconate glass. *J. Non-Cryst. Solids*, **256–257**, 323–327.

188 Morriss, R.H., and Collins, L.F. (1964) Optical properties of multilayer colloids. *J. Chem. Phys.*, **41**, 3357–3363.

189 Sinzig, J., Radtke, U., Quinten, M., and Kreibig, U. (1993) Binary clusters: homogeneous alloys and nucleus–shell structures. *Z. Phys. D*, **26**, 242–245.

190 Mulvaney, P., Giersig, M., and Henglein, A. (1993) Electrochemistry of multilayer colloids: preparation and absorption spectrum of gold-coated silver particles. *J. Phys. Chem.*, **97**, 7061–7064.

191 Moskovits, M., Smová-Sloufová, I., and Vlcková, B. (2002) Bimetallic Ag–Au nanoparticles: extracting meaningful optical constants from the surface-

plasmon extinction spectrum. *J. Chem. Phys.*, **116**, 10435–10446.
192 Liz-Marzan, L.M., Correa-Duarte, M.A., Pastoriza-Santos, I., Mulvaney, P., Ung, T., Giersig, M., and Kotov, N.A. (2001) Core–shell nanoparticles and assemblies thereof, in *Handbook of Surfaces and Interfaces of Materials*, vol. 3 (ed. H. Nalwa), Academic Press, San Diego, CA, pp. 190–232.
193 Rodriguez-Gonzalez, B., Burrows, A., Watanabe, M., Kiely, C.J., and Liz-Marzan, L.M. (2005) Multishell bimetallic Au/Ag nanoparticles: synthesis, structure and optical properties. *J. Mater. Chem.*, **15**, 1755–1759.
194 Leung, K.M. (1986) Optical bistability in the scattering and absorption of light from nonlinear microparticles. *Phys. Rev. A*, **33**, 2461–2464.
195 Uchida, K., Kaneko, S., Omi, S., Hata, C., Tanji, H., Asahara, Y., Ikushima, A.J., Tokizaki, T., and Nakamura, A. (1994) Optical nonlinearities of a high concentration of small metal particles dispersed in glass: copper and silver particles. *J. Opt. Soc. Am. A*, **11**, 1236–1243.
196 Haus, J.W., Kalyaniwalla, N., Inguva, R., and Bowden, C.M. (1989) Optical bistability in small metallic composites. *J. Appl. Phys.*, **65**, 1420–1423.
197 Kalyaniwalla, N., Haus, J.W., Inguva, R., and Birnboim, M.H. (1990) Intrinsic optical bistability for coated spheroidal particles. *Phys. Rev. A*, **42**, 5613–5621.
198 Bergman, D.J., Levy, O., and Stroud, D. (1994) Theory of optical bistability in a weakly nonlinear composite medium. *Phys. Rev. B*, **49**, 129–134.
199 Levy-Nathanson, R., and Bergman, D.J. (1995) Electrical resonances and optical bistability in periodic composite materials. *J. Appl. Phys.*, **77**, 4263–4273.
200 Neuendorf, R. (1995) Intrinsische optische Bistabilität edelmetallbeschichteter Halbleiter-Cluster. Diploma thesis. RWTH Aachen, Aachen.
201 Neuendorf, R., Quinten, M., and Kreibig, U. (1996) Optical bistability in small heterogeneous clusters. *J. Chem. Phys.*, **104**, 6348–6354.
202 McCall, S.L., and Hahn, E.L. (1969) Self-induced transparency. *Phys. Rev.*, **183**, 457–485.
203 Oldenburg, S.J., Jackson, J.B., Westcott, S.L., and Halas, N.J. (1999) Infrared extinction properties of gold nanoshells. *Appl. Phys. Lett.*, **75**, 2897–2899.
204 Sershen, S.R., Westcott, S.L., West, J.L., and Halas, N.J. (2001) An opto-mechanical nanoshell-polymer composite. *Appl. Phys. B*, **73**, 379–381.
205 Jackson, J.B., and Halas, N.J. (2001) Silver nanoshells: variations in morphologies and optical properties. *J. Phys. Chem. B*, **105**, 2743–2746.
206 Zsigmondy, R. (1925) *Kolloidchemie I, Allgemeiner Teil*, Spamer, Leipzig; Zsigmondy, R. (1927) *Kolloidchemie II, Spezieller Teil*, Spamer, Leipzig.
207 Barnickel, P., and Wokaun, A. (1989) Silver coated latex spheres – optical absorption spectra and surface enhanced Raman scattering. *Mol. Phys.*, **67**, 1355–1372.
208 Mayer, A.B.R., Grebner, W., and Wannemacher, R. (2000) Preparation of silver–latex composites. *J. Phys. Chem. B*, **104**, 7278–7285.
209 Kan, C., Cai, W., Li, C., Zhang, L., and Hofmeister, H. (2003) Ultrasonic synthesis and optical properties of Au/Pd bimetallic nanoparticles in ethylene glycol. *J. Phys. D*, **36**, 1609–1614.
210 Liz-Marzan, L.M., Giersig, M., and Mulvaney, P. (1996) Synthesis of nanosized gold–silica core shell particles. *Langmuir*, **12**, 4329–4335.
211 Tuo, L., Jooho, M., Augusto, A.A., Mecholsky, J.J., Talham, D.R., and Adair, J.A. (1999) Preparation of Ag/SiO$_2$ nanosize composites by a reverse micelle and sol–gel technique. *Langmuir*, **15**, 4328–4334.
212 Santos, I.P., Koktysh, D.S., Mamedov, A.A., Giersig, M., Kotov, N.A., and Liz-Marzan, L.M. (2000) One-pot synthesis of Ag@TiO$_2$ nanoparticles and their layer-by-layer assembly. *Langmuir*, **16**, 2731–2735.
213 Eswaranand, V., and Pradeep, T. (2002) Zirconia protected silver clusters through functionalised monolayers. *J. Mater. Chem.*, **12**, 2421–2425.

214 Nair, A.S., Tom, R.T., Suryanarayanan, V., and Pradeep, T. (2003) ZrO_2 bubbles from core shell nanoparticles. *J. Mater. Chem.*, **13**, 297–300.

215 Wind, M.M., Vlieger, J., and Bedeaux, D. (1987) The polarizability of a truncated sphere on a substrate I. *Physica A*, **141**, 33–57.

216 Yamaguchi, T., Yoshida, S., and Kinbara, A. (1974) Optical effect of the substrate on the anomalous absorption of aggregated silver films. *Thin Solid Films*, **21**, 173–187.

217 Ruppin, R. (1983) Surface modes and optical absorption of a small sphere above a substrate. *Surf. Sci.*, **127**, 108–118.

218 Lindell, I.V., Sihvola, A.H., Muinonen, K.O., and Barber, P.W. (1991) Scattering by a small object close to an interface. I. Exact-image theory formulation. *J. Opt. Soc. Am. A*, **8**, 472–476.

219 Muinonen, K.O., Sihvola, A.H., Lindell, I.V., and Lumme, K.A. (1991) Scattering by a small object close to an interface. II. Study of backscattering. *J. Opt. Soc. Am. A*, **8**, 477–482.

220 Bobbert, P.A., and Vlieger, J. (1986) Light scattering by a sphere on a substrate. *Physica A*, **137**, 209–241.

221 Bobbert, P.A., Vlieger, J., and Greef, R. (1986) Light reflection from a substrate sparsely seeded with spheres – comparison with an ellipsometric experiment. *Physica A*, **137**, 243–257.

222 Nahm, K.B., and Wolfe, W.L. (1987) Light-scattering models for spheres on a conducting plane: comparison with experiment. *Appl. Opt.*, **26**, 2995–2999.

223 Videen, G. (1991) Light scattering from a sphere on or near a surface. *J. Opt. Soc. Am. A*, **8**, 483–489.

224 Videen, G. (1992) Light scattering from a sphere on or near a surface: errata. *J. Opt. Soc. Am. A*, **9**, 844–845.

225 Videen, G. (1993) Light scattering from a sphere behind a surface. *J. Opt. Soc. Am. A*, **10**, 110–117.

226 Videen, G., Turner, M.G., Iafelice, V.J., Bickel, W.S., and Wolfe, W.L. (1993) Scattering from a small sphere near a surface. *J. Opt. Soc. Am. A*, **10**, 118–126.

227 Videen, G. (1995) Light scattering from a particle on or near a perfectly conducting surface. *Opt. Commun.*, **115**, 1–7.

228 Videen, G. (1996) Light scattering from an irregular particle behind a plane interface. *Opt. Commun.*, **128**, 81–90.

229 Videen, G. (1997) Polarized light scattering from surface contaminants. *Opt. Commun.*, **143**, 173–178.

230 Bosi, G. (1992) Retarded treatment of substrate-related effects on granular films. *Physica A*, **190**, 375–392.

231 Ruppin, R. (1992) Optical absorption by a small sphere above a substrate with inclusion of nonlocal effects. *Phys. Rev. B*, **45**, 11209–11215.

232 Johnson, B.R. (1992) Light scattering from a spherical particle on a conducting plane: I. Normal incidence. *J. Opt. Soc. Am. A*, **9**, 1341–1351.

233 Johnson, B.R. (1993) Light scattering from a spherical particle on a conducting plane: I. Normal incidence: errata. *J. Opt. Soc. Am. A*, **10**, 766.

234 Johnson, B.R. (1994) Morphology-dependent resonances of a dielectric sphere on a conducting plane. *J. Opt. Soc. Am. A*, **11**, 2055–2064.

235 Johnson, B.R. (1996) Calculation of light scattering from a spherical particle on a surface by the multipole expansion method. *J. Opt. Soc. Am. A*, **13**, 326–337.

236 Eremin, Y., and Orlov, N.V. (1996) Simulation of light scattering from a particle upon a wafer surface. *Appl. Opt.*, **35**, 6599–6604.

237 Fucile, E., Denti, P., Borghese, F., Saija, R., and Sindoni, O.I. (1997) Optical properties of a sphere in the vicinity of a plane surface. *J. Opt. Soc. Am. A*, **14**, 1505–1514.

238 Fucile, E., Borghese, F., Denti, P., Saija, R., and Sindoni, O.I. (1997) General reflection rule for electromagnetic. Multipole fields on a plane interface. *IEEE Trans. Antennas Propag.*, **AP-45**, 868–875.

239 Cruzan, O.R. (1962) Translational addition theorems for spherical vector wave functions. *Q. Appl. Math.*, **20**, 33–40.

240 Felderhof, B.U., and Jones, R.B. (1987) Addition theorems for spherical wave solutions of the vector Helmholtz equation. *J. Math. Phys.*, **28**, 836–839.

241 Liu, C., Weigel, T., and Schweiger, G. (2000) Structural resonances in a dielectric sphere on a dielectric surface illuminated by an evanescent wave. *Opt. Commun.*, **185**, 249–261.

242 Kim, J.H., Ehrman, S.H., Mulholland, G.W., and Germer, T.A. (2002) Polarized light scattering by dielectric and metallic spheres on silicon wafers. *Appl. Opt.*, **41**, 5405–5412.

243 Kim, J.H., Ehrman, S.H., Mulholland, G.W., and Germer, T.A. (2004) Polarized light scattering by dielectric and metallic spheres on oxidized silicon surfaces. *Appl. Opt.*, **43**, 585–591.

244 Bosi, G. (1994) Optical response of a thin film of spherical particles on a dielectric substrate: retarded multipolar treatment. *J. Opt. Soc. Am. B*, **11**, 1073–1083.

245 Bosi, G. (1996) Optical response of a thin film of spherical particles upon a dielectric substrate: retarded multipolar treatment including multiple reflections. *J. Opt. Soc. Am. B*, **13**, 1691–1696.

246 Linden, S., Kuhl, J., and Giessen, H. (2001) Controlling the interaction between light and gold nanoparticles: selective suppression of extinction. *Phys. Rev. Lett.*, **86**, 4688–4691.

247 Linden, S., Christ, A., Kuhl, J., and Giessen, H. (2001) Selective suppression of extinction within the plasmon resonance of gold nanoparticles. *Appl. Phys. B*, **73**, 1–6.

248 Girard, C. (1992) Plasmon resonances and near-field optical microscopy: a self-consistent theoretical model. *Appl. Opt.*, **31**, 5380–5387.

249 Bohren, C.F., and Hunt, A.J. (1977) Scattering of electromagnetic waves by a charged sphere. *Can. J. Phys.*, **55**, 1930–1935.

250 Rostaski, J., and Quinten, M. (1996) Effect of a surface charge on the halfwidth and peak position of cluster plasmons in colloidal metal particles. *Colloid Polym. Sci.*, **274**, 648–653.

251 Monzon, J.C. (1989) Three-dimensional field expansion in the most general rotationally symmetric anisotropic material: application to scattering by a sphere. *IEEE Trans. Antennas Propag.*, **37**, 728–735.

252 Lacoste, D., van Tiggelen, B.A., Rikken, G.L., and Sparenberg, A. (1998) Optics of a Faraday-active Mie sphere. *J. Opt. Soc. Am. A*, **15**, 1636–1642.

253 Casperson, L.W. (1981) Electromagnetic modes of an inhomogeneous sphere. *Appl. Opt.*, **20**, 2738–2741.

254 Perelman, A.Y. (1996) Scattering by particles with radially variable refractive indices. *Appl. Opt.*, **35**, 5452–5460.

255 Bohren, C.F. (1974) Light scattering by an optically active sphere. *Chem. Phys. Lett.*, **29**, 458–462.

256 Mundy, W.C., Roux, J.A., and Smith, A.M. (1974) Mie scattering by spheres in an absorbing medium. *J. Opt. Soc. Am.*, **64**, 1593–1597.

257 Chýlek, P. (1977) Light scattering by small particles in an absorbing medium. *J. Opt. Soc. Am.*, **67**, 561–563.

258 Bohren, C.F., and Gilra, D.P. (1979) Extinction by a spherical particle in an absorbing medium. *J. Colloid Interface Sci.*, **72**, 215–221.

259 Quinten, M., and Rostaski, J. (1996) Lorenz–Mie theory for spheres immersed in an absorbing host medium. *Part. Part. Syst. Charact.*, **13**, 89–96.

260 Lebedev, A.N. (2000) Theoretical description of the optical response of heterogeneous absorbing materials. PhD thesis. Technical University Chemnitz, Chemnitz.

261 Sudiarta, I.W., and Chýlek, P. (2001) Mie-scattering formalism for spherical particles embedded in an absorbing medium. *J. Opt. Soc. Am. A*, **18**, 1275–1278.

262 Sudiarta, I.W., and Chýlek, P. (2002) Mie scattering by a spherical particle in an absorbing medium. *Appl. Opt.*, **41**, 3545–3546.

263 Sudiarta, I.W., and Chýlek, P. (2001) Mie scattering efficiency of a large spherical particle embedded in an absorbing medium. *J. Quant. Spectrosc. Radiat. Transfer*, **70**, 709–714.

264 Fu, Q., and Sun, W. (2001) Mie theory for light scattering by a spherical particle in an absorbing medium. *Appl. Opt.*, **40**, 1354–1361.

265 Videen, G., and Sun, W. (2003) Yet another look at light scattering from

particles in absorbing media. *Appl. Opt.*, **42**, 6724–6727.

266 Ruppin, R. (2002) Extinction due to surface modes near a spherical inclusion in a dispersive medium. *Phys. Status Solidi B*, **233**, 331–338.

267 Salzmann, T. (1991) Vielteilcheneffekte in der optischen Absorption, Streuung und Extinktion von Metallclustersystemen. Diploma thesis. University of Saarland, Saarbrücken.

268 Hoffmann, A. (1991) Extinktion höherer Ordnung als Kollektiveffekt in Vielteilchensystemen am Beispiel von Edelmetallkolloiden. Diploma thesis. University of Saarland, Saarbrücken.

269 Stier, J. (1993) Optische Untersuchungen mit einem Ulbricht–Kugel Spektrometer: Streuung, Absorption und Extinktion von Silberclustern unterschiedlicher Topologie. Diploma thesis. RWTH Aachen, Aachen.

270 Garnett, J.C.M. (1904) Colours in metal glasses and in metallic films. *Philos. Trans. R. Soc. London*, **203**, 385–420.

271 Gartz, M. (1994) C60-Metall-Mischschichten. Diploma thesis. RWTH Aachen, Aachen.

272 Lebedev, A.N., Gartz, M., Kreibig, U., and Stenzel, O. (1999) Optical extinction by spherical particles in an absorbing medium: application to composite absorbing films. *J. Eur. Phys. D*, **6**, 365–373.

273 Hövel, H. (1995) Grenzflächeneigenschaften von Metallclustern. PhD thesis. RWTH Aachen, Aachen.

274 Li, J., and Chýlek, P. (1993) Resonances of a dielectric sphere illuminated by two counterpropagating plane waves. *J. Opt. Soc. Am. A*, **10**, 687–692.

275 Casperson, L.W., Yeh, C., and Yeung, W.F. (1977) Single particle scattering with focused laser beams. *Appl. Opt.*, **16**, 1104–1107.

276 Colak, S., Yeh, C., and Casperson, L.W. (1979) Scattering of focused beams by tenuous particles. *Appl. Opt.*, **18**, 294–302.

277 Yeh, C., Colak, S., and Barber, P.W. (1982) Scattering of sharply focused beams by arbitrarily shaped dielectric particles: an exact solution. *Appl. Opt.*, **21**, 4426–4433.

278 Gouesbet, G., and Gréhan, G. (1982) Sur la généralisation de la théorie de Lorenz–Mie. *J. Opt. (Paris)*, **13**, 97–103.

279 Gouesbet, G., Gréhan, G., and Maheu, B. (1985) Scattering of a Gaussian beam by a Mie scatter center, using a Bromwich formulation. *J. Opt. (Paris)*, **16**, 83–93.

280 Gouesbet, G., Gréhan, G., and Maheu, B. (1985) The order of approximation in a theory of the scattering of a Gaussian beam by Mie scatter center. *J. Opt. (Paris)*, **16**, 239–247.

281 Gréhan, G., Maheu, B., and Gouesbet, G. (1986) Scattering of laser beams by Mie scatter senders: numerical results using a localised approximation. *Appl. Opt.*, **25**, 3539–3548.

282 Maheu, B., Gréhan, G., and Gouesbet, G. (1987) Generalized Lorenz–Mie theory: first exact values and comparisons with the localized approximation. *Appl. Opt.*, **26**, 23–25.

283 Gouesbet, G., Gréhan, G., and Maheu, B. (1988) Expressions to compute the coefficients g_n^m in the generalized Lorenz–Mie theory using finite series. *J. Opt. (Paris)*, **19**, 35–48.

284 Gouesbet, G., Gréhan, G., and Maheu, B. (1988) Computations of the g_n^m coefficients in the generalized Lorenz–Mie theory using three different methods. *Appl. Opt.*, **27**, 4874–4883.

285 Gouesbet, G., Maheu, B., and Gréhan, G. (1988) Light scattering from a sphere arbitrarily located in a Gaussian beam, using a Bromwich formulation. *J. Opt. Soc. Am. A*, **5**, 1427–1443.

286 Gouesbet, G., Gréhan, G., and Maheu, B. (1990) Localized interpretation to compute all the coefficients g_n^m in the generalized Lorenz–Mie theory. *J. Opt. Soc. Am. A*, **7**, 998–1007.

287 Lock, J.A., and Gouesbet, G. (1994) Rigorous justification of the localized approximation to the beam-shape coefficients in generalized Lorenz–Mie theory. I. On-axis beams. *J. Opt. Soc. Am. A*, **11**, 2503–2515.

288 Gouesbet, G., and Lock, J.A. (1994) Rigorous justification of the localized approximation to the beam-shape coefficients in generalized Lorenz–Mie

theory. II. Off-axis beams. *J. Opt. Soc. Am. A*, **11**, 2516–2525.

289 Barton, J.P., Alexander, D.R., and Schaub, S.A. (1988) Internal, near-surface electromagnetic fields for a spherical particle irradiated by a focused laser beam. *J. Appl. Phys.*, **64**, 1632–1639.

290 Khaled, E.E.M., Hill, S.C., Barber, P.W., and Chowdhury, D.Q. (1992) Near-resonance excitation of dielectric spheres with plane waves and off-axis Gaussian beams. *Appl. Opt.*, **31**, 1166–1169.

291 Khaled, E.E.M., Hill, S.C., and Barber, P.W. (1993) Scattered and internal intensity of a sphere illuminated with a Gaussian beam. *IEEE Trans. Antennas Propag.*, **41**, 295–303.

292 Lock, J.A. (1995) Interpretation of extinction in Gaussian-beam scattering. *J. Opt. Soc. Am. A*, **12**, 929–938.

293 Lock, J.A., Hodges, J.T., and Gouesbet, G. (1995) Failure of the optical theorem for Gaussian-beam scattering by a spherical particle. *J. Opt. Soc. Am. A*, **12**, 2708–2715.

294 Gouesbet, G., Letellier, C., Gréhan, G., and Hodges, J.T. (1996) Generalized optical theorem for on-axis Gaussian-beam. *Opt. Commun.*, **125**, 137–157.

295 Lock, J.A. (1995) Improved Gaussian beam-scattering algorithm. *Appl. Opt.*, **34**, 559–570.

296 Doicu, A., and Wriedt, T. (1997) Computation of the beam-shape coefficients in the generalized Lorenz–Mie theory by using the translational addition theorem for spherical vector wave functions. *Appl. Opt.*, **36**, 2971–2978.

297 Khaled, E.E.M., Hill, S.C., and Barber, P.W. (1994) Light scattering by coated sphere illuminated with a Gaussian beam. *Appl. Opt.*, **33**, 3308–3314.

298 Onofri, F., Gréhan, G., and Gouesbet, G. (1995) Electromagnetic scattering from a multilayered sphere located in an arbitrary beam. *Appl. Opt.*, **34**, 7113–7124.

299 Wu, Z.S., Guo, L.X., Ren, K.F., Gouesbet, G., and Gréhan, G. (1997) Improved algorithm for electromagnetic scattering of plane waves and shaped beams by multilayered spheres. *Appl. Opt.*, **36**, 5188–5198.

300 Barton, J.P. (1997) Electromagnetic-field calculations for a sphere illuminated by a higher-order Gaussian beam. I. Internal and near-field effects. *Appl. Opt.*, **36**, 1303–1311.

301 Barton, J.P. (1998) Electromagnetic field calculations for a sphere illuminated by a higher-order Gaussian beam. II. Far-field scattering. *Appl. Opt.*, **37**, 3339–3344.

302 Shifrin, K.S., and Zolotov, I.G. (1994) Quasi-stationary scattering of electromagnetic pulses by spherical particles. *Appl. Opt.*, **33**, 7798–7804.

303 Chew, H., Wang, D.S., and Kerker, M. (1979) Elastic scattering of evanescent electromagnetic waves. *Appl. Opt.*, **18**, 2679–2687.

304 Barchiesi, D., and van Labeke, D. (1993) Application of Mie scattering of evanescent waves to scanning tunneling optical microscopy theory. *J. Mod. Opt.*, **40**, 1239–1254.

305 Liu, C., Kaiser, T., Lange, S., and Schweiger, G. (1995) Structural resonances in a dielectric sphere illuminated by an evanescent wave. *Opt. Commun.*, **117**, 521–531.

306 Liu, C. (1999) Strukturresonanzen der Lichtstreuung eines sphärischen Mikropartikels im Feld einer evaneszenten elektromagnetischen Welle. PhD thesis. Ruhr-University Bochum, Bochum.

307 Quinten, M., Pack, A., and Wannemacher, R. (1999) Scattering and extinction of evanescent waves by small particles. *Appl. Phys. B*, **68**, 87–92.

308 Liebsch, A. (1993) Surface-plasmon dispersion and size dependence of Mie resonance: silver versus simple metals. *Phys. Rev. B*, **48**, 11317–11328.

309 Kreibig, U. (1974) Electronic properties of small silver particles: the optical constants and their temperature dependence. *J. Phys. F*, **4**, 999–1014.

310 David, E. (1939) Deutung der Anomalien der optischen Konstanten dünner Metallschichten. *Z. Phys.*, **114**, 389–406.

311 Euler, J. (1954) Infrared properties of metals and the mean free paths of conduction electrons. *Z. Phys.*, **137**, 318–332.

312 von Fragstein, C., and Römer, H. (1958) Über die Anomalie der optischen Konstanten. *Z. Phys.*, **151**, 54–71.

313 Hampe, W. (1958) Beitrag zur Deutung der anomalen optischen Eigenschaften feinstteiliger Metallkolloide in großer Konzentration. I. Bestimmung des Füllfaktors dünner Schichten eines Kolloids Gold–SiO$_2$. *Z. Phys.*, **152**, 470–475.

314 Hampe, W. (1958) Beitrag zur Deutung der anomalen optischen Eigenschaften feinstteiliger Metallkolloide in großer Konzentration. II. Experimentelle Ermittlung der Absorptionskurve und Deutung des Absorptionsmechanismus des Systemes Gold–SiO$_2$. *Z. Phys.*, **152**, 476–494.

315 Doyle, W.T. (1958) Absorption of light by colloids in alkali halide crystals. *Phys. Rev.*, **111**, 1067–1062.

316 Doremus, R.H. (1964) Optical properties of small gold particles. *J. Chem. Phys.*, **40**, 2389–2396.

317 Doremus, R.H. (1965) Optical properties of small silver particles. *J. Chem. Phys.*, **42**, 414–417.

318 von Fragstein, C., and Schönes, F.J. (1967) Absorptionskoeffizient und Brechungsindex feinster Goldkugeln im nahen Ultrarot. *Z. Phys.*, **198**, 477–493.

319 Kleemann, W. (1968) Absorption of colloidal silver in KCl. *Z. Phys.*, **215**, 113–120.

320 Skillman, D., and Berry, C. (1968) Effect of particle shape on spectral absorption of colloidal silver in gelatin. *J. Chem. Phys.*, **48**, 3297–3304.

321 Dickey, I.M., and Paskin, A. (1968) Phonon spectrum changes in small particles and their implications for superconductivity. *Phys. Rev. Lett.*, **21**, 1441–1143.

322 Kreibig, U. (1978) The transition cluster–solid state in small gold particles. *Solid State Commun.*, **28**, 767–769.

323 Kreibig, U. (1978) Lattice defects in small metallic particles and their influence on size effects. *Z. Phys. B*, **31**, 39–47.

324 Dubiel, M., Hofmeister, H., and Schurig, E. (1998) Interface effects at nanosized silver particles in glass. *Recent Res. Dev. Appl. Phys.*, **1**, 69–77.

325 Cai, W., Hofmeister, H., and Dubiel, M. (2001) Importance of lattice contraction in surface plasmon resonance shift for free and embedded silver particles. *Eur. Phys. J. D*, **13**, 245–253.

326 Pines, D. (1964) *Elementary Excitations in Solids*, Benjamin, New York.

327 Fröhlich, H. (1937) Die spezifische Wärme der Elektronen kleiner Metallteilchen bei tiefen Temperaturen. *Physica*, **4**, 406–412.

328 Kawabata, A., and Kubo, R. (1966) Electronic properties of fine metallic particles. II. Plasma resonance absorption. *J. Phys. Soc. Jpn.*, **21**, 1765–1772.

329 Kubo, R. (1962) Electronic properties of metallic fine particles I. *J. Phys. Soc. Jpn.*, **17**, 975–986.

330 Gor'kov, L.P., and Eliashberg, G.M. (1965) Minute metallic particles in an electromagnetic field. *Sov. Phys. JETP*, **21**, 940–947.

331 Sander, L. (1968) Quantum theory of perpendicular electrical conductivity in a thin metallic film. *J. Phys. Chem. Solids*, **29**, 291–294.

332 Kubo, R. (1969) Electrons in small metal particles, in *Polarization, Matière et Rayonnement* (ed. Société Francaise de Physique), Presse Université de France, Paris, pp. 325–339.

333 Kawabata, A. (1970) Electronic properties of fine metallic particles. III. E.S.R absorption line shape. *J. Phys. Soc. Jpn.*, **29**, 902–911.

334 Rice, M.J., Schneider, W.R., and Strässler, S. (1973) Electronic polarizabilities of very small metallic particles and thin films. *Phys. Rev. B*, **8**, 474–482.

335 Cini, M., and Ascarelli, P. (1974) Quantum size effects in metal particles and thin films by an extended R. P. A. *J. Phys. F*, **4**, 1998–2008.

336 Ascarelli, P., and Cini, M. (1975) "Red shift" of the surface plasmon resonance absorption by fine metal particles. *Solid State Commun.*, **18**, 385–388.

337 Lushnikov, A.A., and Simonov, A.J. (1974) Surface plasmons in small metal particles. *Z. Phys.*, **270**, 17–24.

338 Genzel, L., Martin, T.P., and Kreibig, U. (1975) Dielectric function and plasma

resonances of small metal particles. *Z. Phys. B*, **21**, 339–346.

339 Ruppin, R., and Yatom, H. (1976) Size and shape effects on the broadening of the plasma resonance absorption in metals. *Phys. Status Solidi B*, **74**, 647–654.

340 Genzel, L., and Kreibig, U. (1980) Dielectric function and infrared absorption of small metal particles. *Z. Phys. B*, **37**, 93–101.

341 Wood, D.M., and Ashcroft, N.W. (1982) Quantum size effects in the optical properties of small metallic particles. *Phys. Rev. B*, **25**, 6255–6274.

342 Apell, P., Monreal, R., and Flores, F. (1984) Effective relaxation time in small spheres: diffuse surface scattering. *Solid State Commun.*, **52**, 971–973.

343 Monreal, R., Giraldo, J., Flores, F., and Apell, P. (1985) Far-infrared optical absorption due to surface phonon excitations in small metal particles. *Solid State Commun.*, **54**, 661–663.

344 Hache, F., Ricard, D., and Flytzanis, C. (1986) Optical nonlinearities of small metal particles: surface-mediated resonance and quantum size effects. *J. Opt. Soc. Am. B*, **3**, 1647–1655.

345 Barma, M., and Subrahmanyam, V. (1989) Optical absorption in small metal particles. *J. Phys. Condens. Matter*, **1**, 7681–7688.

346 Jortner, J., and Pullman, B. (eds) (1987) *Small Finite Systems*, Reidel, Dordrecht.

347 Jortner, J., Even, U., Ben-Horin, N., Scharf, D., Barnett, R.N., and Landman, U. (1989) Dynamic and quantum size effects in molecular clusters. *Z. Phys. D*, **12**, 167–171.

348 Jortner, J. (1992) Clusters as a key to the understanding of properties as a function of size and dimensionality, in *The Physics and Chemistry of Finite Systems: from Clusters to Crystals* (eds P. Jena, S. Khanna, and B. Rao), Kluwer, Dordrecht, pp. 1–18.

349 Jortner, J. (1992) Cluster size effects. *Z. Phys. D*, **24**, 247–275.

350 Yannouleas, C., and Broglia, R.A. (1992) Landau damping and wall dissipation in large metal clusters. *Ann. Phys.*, **217**, 105–141.

351 Zaremba, E., and Persson, B.N.J. (1987) Dynamic polarizability of small metal particles. *Phys. Rev. B*, **35**, 596–606.

352 Kreibig, U. (1970) Kramers–Kronig analysis of the optical constants of small silver particles. *Z. Phys.*, **234**, 307–318.

353 Dalacu, D., and Martinu, L. (2000) Spectroellipsometric characterisation of plasmon-deposited Au/SiO$_2$ nanocomposite films. *J. Appl. Phys.*, **87**, 228–235.

354 Charlé, K.-P., Frank, F., and Schulze, W. (1984) The optical properties of silver microcrystallites in dependence on size and the influence of the matrix environment. *Ber. Bunsenges. Phys. Chem.*, **88**, 350–354.

355 Charlé, K.-P., Schulze, W., and Winter, B. (1989) The size dependent shift of the surface plasmon absorption band of small spherical metal particles. *Z. Phys. D*, **12**, 471–475.

356 Persson, B.N.J. (1993) Polarizability of small spherical metal particles: influence of the matrix environment. *Surf. Sci.*, **281**, 153–162.

357 Hövel, H., Fritz, S., Hilger, A., Kreibig, U., and Vollmer, M. (1993) Width of cluster plasmon resonances: bulk dielectric functions and chemical interface damping. *Phys. Rev. B*, **48**, 18178–18188.

358 Kreibig, U., Hilger, A., Hövel, H., and Quinten, M. (1996) Optical properties of free and embedded metal clusters: recent results, in *Large Clusters of Atoms and Molecules* (ed. P. Martin), Kluwer, Dordrecht, pp. 475–493.

359 Kreibig, U., Gartz, M., and Hilger, A. (1997) Mie resonances: sensors for physical and chemical cluster interface properties. *Ber. Bunsenges. Phys. Chem.*, **101**, 1593–1604.

360 Kreibig, U. (1997) Optics of nanosized metals, in *Handbook of Optical Properties Vol. II: Optics of Small Particles, Interfaces and Surfaces* (eds R. Hummel and P. Wissmann), CRC Press, Boca Raton, FL, pp. 145–190.

361 Kreibig, U., Gartz, M., Hilger, A., and Hövel, H. (1998) Optical investigations of surfaces and interfaces of metal clusters, in *Advances in Metal*

Semiconductor Clusters, vol. **4** (ed. M. Duncan), JAI Press, Stamford, CT, pp. 345–393.

362 Hilger, A., Cüppers, N., Tenfelde, M., and Kreibig, U. (2000) Surface and interface effects in the optical properties of silver nanoparticles. *Eur. Phys. J. D*, **10**, 115–118.

363 Hilger, A. (2001) Grenzflächenanalyse durch Mie-Plasmon-Spektroskopie an Edelmetallclustern. PhD thesis. RWTH Aachen, Aachen.

364 Bönnemann, H., Hormes, J., and Kreibig, U. (2001) Nanostructured metal clusters and colloids, in *Handbook of Surfaces and Interfaces of Materials*, vol. **3** (ed. H. Nalwa), Academic Press, San Diego, CA, pp. 1–87.

365 Kreibig, U., and Quinten, M. (2004) Heterogeneous materials, in *Encyclopedia of Modern Optics*, vol. **3** (eds B. Guenther, D. Steel, and L. Bayvel), Elsevier Academic Press, Amsterdam, pp. 446–460.

366 Pinchuk, A., Hilger, A., von Plessen, G., and Kreibig, U. (2004) Substrate effect on the optical response of silver nanoparticles. *Nanotechnology*, **15**, 1890–1896.

367 Hilger, A., von Hofe, Th., and Kreibig, U. (2005) Recent investigations of size and interface effects in nanoparticle composites. *Nova Acta Leopold.*, **NF92**, 9–19.

368 Kreibig, U., Hilger, A., Hövel, H., Quinten, M., Wagner, D., and Ditlbacher, H. (2006) A short survey of optical properties of metal nanostructures, in *Functional Properties of Nanostructured Materials* (ed. R. Kassing), Springer, Heidelberg, pp. 75–110.

369 Kreibig, U. (2008) Interface-induced dephasing of Mie-plasmon polaritons. *Appl. Phys. B*, **93**, 79–89.

370 Yeh, C. (1964) Perturbation approach to the diffraction of electromagnetic waves by arbitrarily shaped dielectric obstacles. *Phys. Rev.*, **135**, A1193–A1201.

371 Yeh, C. (1965) Perturbation method in the diffraction of electromagnetic waves by arbitrarily shaped penetrable obstacles. *J. Math. Phys.*, **6**, 2008–2013.

372 Erma, V.A. (1968) An exact solution for the scattering of electromagnetic waves from conductors of arbitrary shape. I. Case of cylindrical symmetry. *Phys. Rev.*, **173**, 1243–1257.

373 Erma, V.A. (1968) Exact solution for the scattering of electromagnetic waves from conductors of arbitrary shape. II. General case. *Phys. Rev.*, **176**, 1544–1553.

374 Erma, V.A. (1968) Exact solution for the scattering of electromagnetic waves from bodies of arbitrary shape. III. Obstacles with arbitrary electromagnetic properties. *Phys. Rev.*, **179**, 1238–1246.

375 Martin, R.J. (1993) Mie scattering formulae for non-spherical particles. *J. Mod. Opt.*, **40**, 2467–2494.

376 Asano, S., and Yamamoto, G. (1975) Light scattering by a spheroidal particle. *Appl. Opt.*, **14**, 29–49.

377 Onaka, T. (1980) Light scattering by spheroidal grains. *Ann. Tokyo Astron. Obs.*, **18**, 1–54.

378 Sinha, B.P., and MacPhie, R.H. (1975) Electromagnetic scattering from prolate spheroids for axial incidence. *IEEE Trans. Antennas Propag.*, **AP-23**, 676–679.

379 Sinha, B.P., and MacPhie, R.H. (1977) Electromagnetic scattering by prolate spheroids for plane waves with arbitrary polarization and angle of incidence. *Radio Sci.*, **12**, 171–184.

380 Kurtz, V., and Salib, S. (1993) Scattering and absorption of electromagnetic radiation by spheroidally shaped particles: computation of the scattering properties. *J. Imag. Sci. Technol.*, **37**, 43–60.

381 Voshchinnikov, N.V., and Farafonov, V.G. (1993) Optical properties of spheroidal particles. *Astrophys. Space Sci.*, **204**, 19–86.

382 Voshchinnikov, N.V. (1996) Electromagnetic scattering by homogeneous and coated spheroids: calculations using the separation of variables method. *J. Quant. Spectrosc. Radiat. Transfer*, **55**, 627–636.

383 Hodge, D.B. (1970) Eigenvalues and eigenfunctions of the spheroidal wave equation. *J. Math. Phys.*, **11**, 2308–2312.

384 Flammer, C. (1957) *Spheroidal Wave Functions*, Stanford University Press, Stanford, CA.

385 Siegel, K.M., Schultz, F.V., Gere, B.H., and Sleator, F.B. (1956) The theoretical and numerical determination of the radar cross section of a prolate spheroid. *IEEE Trans. Antennas Propag.*, **AP-4**, 266–275.

386 Wait, J.R. (1966) Theories of prolate spheroidal antennas. *Radio Sci.*, **1**, 475–513.

387 Porstendorfer, S. (1997) Numerische Berechnung von Extinktions- und Streuspektren sphäroidaler Metallpartikel beliebiger Größe in dielektrischer Matrix. PhD thesis. Martin-Luther-University Halle, Halle.

388 Porstendorfer, S., Berg, K.-J., and Berg, G. (1999) Calculation of extinction and scattering spectra of large spheroidal gold particles embedded in a glass matrix. *J. Quant. Spectrosc. Radiat. Transfer*, **63**, 479–486.

389 Asano, S. (1979) Light scattering properties of spheroidal particles. *Appl. Opt.*, **18**, 712–723.

390 Asano, S., and Sato, S. (1980) Light scattering by randomly oriented spheroidal particles. *Appl. Opt.*, **19**, 962–974.

391 Voshchinnikov, N.V., Il'in, V.B., Henning, Th., Michel, B., and Farafonov, V.G. (2000) Extinction and polarization of radiation by absorbing spheroids: shape/size effects and benchmark results. *J. Quant. Spectrosc. Radiat. Transfer*, **65**, 877–893.

392 Cooray, M.F.R., and Ciric, I.R. (1992) Scattering of electromagnetic waves by a coated dielectric spheroid. *J. Electromagn. Waves Appl.*, **6**, 1491–1507.

393 Farafonov, V.G., Voshchinnikov, N.V., and Somsikov, V. (1996) Light scattering by a core–mantle spheroidal particle. *Appl. Opt.*, **35**, 5412–5426.

394 Gurwich, I., Kleiman, M., Shiloah, N., and Cohen, A. (2000) Scattering of electromagnetic radiation by multilayered spheroidal particles: recursive procedure. *Appl. Opt.*, **39**, 470–477.

395 Han, Y., and Wu, Z. (2001) Scattering of a spheroidal particle illuminated by a Gaussian beam. *Appl. Opt.*, **40**, 2501–2509.

396 Han, Y., and Wu, Z. (2002) Absorption and scattering by an oblate particle. *J. Opt.*, **4**, 74–77.

397 Barton, J.P. (1995) Internal and near-surface electromagnetics fields for a spheroidal particle with arbitrary illumination. *Appl. Opt.*, **34**, 8472–8473.

398 Barton, JP (2001) Internal, near-surface, and scattered electromagnetic fields for a layered spheroid with arbitrary illumination. *Appl. Opt.*, **40**, 3598–3607.

399 Han, Y., Gréhan, G., and Gouesbet, G. (2003) Generalized Lorenz–Mie theory for a spheroidal particle with off-axis Gaussian-beam illumination. *Appl. Opt.*, **42**, 6621–6629.

400 Bobbert, P.A., and Vlieger, J. (1987) The polarizability of a spheroidal particle on a substrate. *Physica A*, **147**, 115–140.

401 Roman-Velazquez, C.E., Noguez, C., and Barrera, R.G. (2000) Substrate effects on the optical properties of spheroidal nanoparticles. *Phys. Rev. B*, **61**, 10427–10436.

402 Cooray, M.F.R., and Ciric, I.R. (1993) Wave scattering by a chiral spheroid. *J. Opt. Soc. Am. A*, **10**, 1197–1203.

403 Gans, R. (1912) Über die Form ultramikroskopischer Teilchen. *Ann. Phys. (Leipzig)*, **37**, 881–900.

404 Roessler, D.M., Wang, D.-S.Y., and Kerker, M. (1983) Optical absorption by randomly oriented carbon spheroids. *Appl. Opt.*, **22**, 3648–3659.

405 Stookey, S.D., and Araujo, R.J. (1968) Selective polarization of light due to absorption by small elongated silver particles. *Appl. Opt.*, **7**, 777–779.

406 Weyl, W.A. (1951) *Coloured Glass*, Society of Glass Technology, Sheffield.

407 Mennig, M., Berg, K.-J., and Fröhlich, F. (1985) On production and optical properties of soda lime silicate glass containing nonspherical silver clusters in surface near region. In Conference Proceedings of the 6th International Symposium on High Purity Materials in Science and Technology, Dresden, pp. 139–144.

408 Mennig, M., and Berg, K.-J. (1991) Determination of size, shape and concentration of spheroidal silver

colloids embedded in glass by VIS-spectroscopy. *Mater. Sci. Eng. B*, **9**, 421–424.

409 Drost, W.G. (1991) Rotationsellipsoidförmige Silberkolloide in Glasoberflächen–Herstellung, Nachweis und Anwendungsmöglichkeiten. PhD thesis. University of Halle-Wittenberg, Halle.

410 Seifert, G., Kaempfe, M., Berg, K.-J., and Graener, H. (2001) Production of "dichroitic" diffraction gratings in glasses containing silver nanoparticles via particle deformation with ultrashort laser pulses. *Appl. Phys. B*, **73**, 355–359.

411 Berg, K.-J., Berg, G., and Drost, W.G. (1994) Farbig-mikrostrukturierte Glaspolarisation. *Glastech. Ber. Glass Sci. Technol.*, **67**, 2–6.

412 Berg, K.-J., Dehmel, A., and Berg, G. (1995) Colour-microstructured glass polarizers. *Glastech. Ber. Glass Sci. Technol.*, **68** (C1), 554–559.

413 Hofmeister, H., Drost, W.G., and Berger, A. (1999) Oriented prolate silver particles in glass–characteristics of novel dichroic polarizers. *Nanostruct. Mater.*, **12** (1999), 207–210.

414 Heilmann, A., Quinten, M., and Werner, J. (1998) Optical response of thin plasma-polymer films with non-spherical silver nanoparticles. *Eur. Phys. J. B*, **3**, 455–461.

415 Heilmann, A., Quinten, M., and Kiesow, A. (1998) Optical properties of nonspherical silver nanoparticles embedded in a plasma polymer thin film matrix. *MRS Proc.*, **501**, 73–78.

416 Quinten, M., Heilmann, A., and Kiesow, A. (1999) Refined interpretation of optical absorption spectra of nanoparticles in plasma-polymer films. *Appl. Phys. B*, **68**, 707–712.

417 Heilmann, A., and Hamann, C. (1991) Deposition, structure, and properties of plasma polymer metal composite films. *Prog. Colloid Polym. Sci.*, **85**, 102–110.

418 Götz, T., Vollmer, M., and Träger, F. (1993) Desorption of metal atoms with laser light of different polarization. *Appl. Phys. A*, **57**, 101–104.

419 Götz, T., Hoheisel, W., Vollmer, M., and Träger, F. (1995) Characterization of large supported metal clusters by optical spectroscopy. *Z. Phys. D*, **33**, 133–145.

420 Hövel, H., Hilger, A., Nusch, I., and Kreibig, U. (1997) Experimental determination of deposition induced cluster deformation. *Z. Phys. D*, **42**, 203–208.

421 Hilger, A., Tenfelde, M., and Kreibig, U. (2001) Silver nanoparticles deposited on dielectric surfaces. *Appl. Phys. B*, **73**, 361–372.

422 Hanarp, P., Käll, M., and Sutherland, D.S. (2003) Optical properties of short range ordered arrays of nanometer gold disks prepared by colloidal lithography. *J. Phys. Chem. B*, **107**, 5768–5772.

423 Lord Rayleigh (Strutt, J.W.) (1918) The dispersal of light by a dielectric cylinder. *Philos. Mag.*, **36**, 365–376.

424 Wait, J.R. (1955) Scattering of a plane wave from a circular dielectric cylinder at oblique incidence. *Can. J. Phys.*, **33**, 189–195.

425 Lind, A.C., and Greenberg, J.M. (1966) Electromagnetic scattering by obliquely oriented cylinders. *J. Appl. Phys.*, **37**, 3195–3203.

426 Cohen, A., and Acquista, C. (1982) Light scattering by tilted cylinders: properties of partial wave coefficients. *J. Opt. Soc. Am.*, **72**, 531–534.

427 Adey, A.J. (1956) Scattering of electromagnetic waves by coaxial cylinders. *Can. J. Phys.*, **34**, 510–520.

428 Kerker, M., and Matijevic, E. (1961) Scattering of electromagnetic waves from concentric infinite cylinders. *J. Opt. Soc. Am.*, **51**, 506–508.

429 Evans, L.B., Chen, J.C., and Churchill, S.W. (1964) Scattering of electromagnetic radiation by infinitely long, hollow and coated cylinders. *J. Opt. Soc. Am.*, **54**, 1004–1007.

430 Samaddar, S.N. (1970) Scattering of plane electromagnetic waves by radially inhomogeneous infinite cylinders. *Nuovo Cimento*, **66**, 33–50.

431 Kai, L., and D'Alessio, A. (1995) Finely stratified cylinder model for radially inhomogeneous cylinders normally irradiated by electromagnetic plane waves. *Appl. Opt.*, **34**, 5520–5530.

432 Gurwich, I., Shiloah, N., and Kleiman, M. (1999) The recursive algorithm for

electromagnetic scattering by tilted infinite circular multilayered cylinder. *J. Quant. Spectrosc. Radiat. Transfer*, **63**, 217–229.

433 Ruppin, R. (2001) Extinction properties of thin metallic nanowires. *Opt. Commun.*, **190**, 205–209.

434 Boustimi, M., Baudon, J., Féron, P., and Robert, J. (2003) Optical properties of metallic nanowires. *Opt. Commun.*, **220**, 377–381.

435 Ruppin, R. (2002) Extinction by a circular cylinder in an absorbing medium. *Opt. Commun.*, **211**, 335–340.

436 Sun, W., Loeb, N.G., and Lin, B. (2005) Light scattering by an infinite circular cylinder immersed in an absorbing medium. *Appl. Opt.*, **44**, 2338–2342.

437 Arias-Gonzalez, J.R., and Nieto-Vesperinas, M. (2000) Near-field distributions of resonant modes in small dielectric objects on flat surfaces. *Opt. Lett.*, **25**, 782–784.

438 Arias-Gonzalez, J.R., and Nieto-Vesperinas, M. (2001) Resonant near-field eigenmodes of nanocylinders on flat surfaces under both homogeneous and inhomogeneous lightwave excitation. *J. Opt. Soc. Am. A*, **18**, 657–665.

439 Videen, G., and Ngo, D. (1997) Light scattering from a cylinder near a plane interface: theory and comparison with experimental data. *J. Opt. Soc. Am. A*, **14**, 70–78.

440 Madrazo, A., and Nieto-Vesperinas, M. (1995) Scattering of electromagnetic waves from a cylinder in front of a conducting plane. *J. Opt. Soc. Am. A*, **12**, 1298–1309.

441 Kozaki, S. (1982) A new expression for the scattering of a Gaussian beam by a conducting cylinder. *IEEE Trans. Antennas Propag.*, **AP-30**, 881–887.

442 Kozaki, S. (1982) Scattering of a Gaussian beam by a homogeneous dielectric cylinder. *J. Appl. Phys.*, **51**, 7195–7200.

443 Gouesbet, G., and Mees, L. (1999) Generalized Lorenz–Mie theory for infinitely long elliptical cylinders. *J. Opt. Soc. Am. A*, **16**, 1333–1341.

444 Gouesbet, G., and Gréhan, G. (1994) Interaction between shaped beams and an infinite cylinder, including a discussion of Gaussian beams. *Part. Part. Syst. Charact.*, **11**, 299–308.

445 Gouesbet, G. (1995) Interaction between Gaussian beams and infinite cylinders, by using the theory of distributions. *J. Opt. (Paris)*, **26**, 225–239.

446 Gouesbet, G. (1995) Scattering of a first-order Gaussian beam by an infinite cylinder with arbitrary location and arbitrary orientation. *Part. Part. Syst. Charact.*, **12**, 242–256.

447 Gouesbet, G. (1997) Scattering of higher-order Gaussian beams by an infinite cylinder. *J. Opt. (Paris)*, **28**, 45–65.

448 Gouesbet, G. (1997) Interaction between an infinite cylinder and an arbitrary-shaped beam. *Appl. Opt.*, **36**, 4292–4304.

449 Ren, K.F., Gréhan, G., and Gouesbet, G. (1997) Scattering of a Gaussian beam by an infinite cylinder in the framework of generalized Lorenz–Mie theory: formulation and numerical results. *J. Opt. Soc. Am. A*, **14**, 3014–3025.

450 Lock, J.A. (1997) Scattering of a diagonally incident focused Gaussian beam by an infinitely long homogeneous circular cylinder. *J. Opt. Soc. Am. A*, **14**, 640–652.

451 Lock, J.A. (1997) Morphology-dependent resonances of an infinitely long circular cylinder illuminated by a diagonally incident plane wave or a focused Gaussian beam. *J. Opt. Soc. Am. A*, **14**, 653–661.

452 Barton, J.P. (1999) Internal and near-surface electromagnetic fields for an infinite cylinder illuminated by an arbitrary focused beam. *J. Opt. Soc. Am. A*, **16**, 160–166.

453 Mroczka, J., and Wysoczanski, D. (2000) Plane-wave and Gaussian-beam scattering on an infinite cylinder. *Opt. Eng.*, **39**, 763–770.

454 Bohren, C.F. (1978) Scattering of electromagnetic waves by an optically active cylinder. *J. Colloid Interface Sci*, **66**, 105–109.

455 Kluskens, M.S., and Newman, E.H. (1991) Scattering by a multilayer chiral

cylinder. *IEEE Trans. Antennas Propag.*, **39**, 91–96.

456 Cohen, L.D., Haracz, R.D., Cohen, A., and Acquista, C. (1983) Scattering of light from arbitrarily oriented finite cylinders. *Appl. Opt.*, **22**, 742–748.

457 Waterman, P.C., and Pedersen, J.C. (1992) Scattering by finite wires. *J. Appl. Phys.*, **72**, 349–359.

458 Waterman, P.C., and Pedersen, J.C. (1995) Electromagnetic scattering and absorption by finite wires. *J. Appl. Phys.*, **78**, 656–667.

459 Waterman, P.C., and Pedersen, J.C. (1998) Scattering by finite wires of arbitrary E, n, O. *J. Opt. Soc. Am. A*, **15**, 174–184.

460 Wang, R.T., and van de Hulst, H.C. (1995) Application of the exact solution for scattering by an infinite cylinder to the estimation of scattering by a finite cylinder. *Appl. Opt.*, **34**, 2811–2821.

461 Mishchenko, M.I., Travis, L., and Macke, A. (1996) Scattering of light by polydisperse, randomly oriented, finite circular cylinders. *Appl. Opt.*, **35**, 4927–4940.

462 Napper, D.H. (1967) Light scattering by polyhedral particles in the Rayleigh–Gans domain. *Kolloid Z. Z. Polym.*, **223**, 141–145.

463 Fuchs, R. (1975) Theory of optical properties of ionic crystal cubes. *Phys. Rev. B*, **11**, 1732–1740.

464 Ruppin, R. (1996) Plasmon frequencies of cube shaped metal clusters. *Z. Phys. D*, **36**, 69–71.

465 Langbein, D. (1972) *Van der Waals Attraction*, Springer Tracts in Modern Physics, vol. **72**, Springer, Berlin.

466 Purcell, E.M., and Pennypacker, C.R. (1973) Scattering and absorption of light by nonspherical dielectric grains. *Astrophys. J.*, **186**, 705–714.

467 Draine, B.T. (1988) The discrete-dipole approximation and its application to interstellar graphite grains. *Astrophys. J.*, **333**, 848–872.

468 Draine, B.T., and Flatau, P.J. (1994) Discrete-dipole approximation for scattering calculations. *J. Opt. Soc. Am. A*, **11**, 1491–1499.

469 Barber, P.W., and Hill, S.C. (eds) (1990) *Light Scattering by Particles: Computational Methods*, World Scientific, Singapore.

470 Singham, S.B., and Bohren, C.F. (1988) Light scattering by an arbitrary particle: the scattering-order formulation of the coupled-dipole method. *J. Opt. Soc. Am. A*, **5**, 1867–1872.

471 Mulholland, G.W., Bohren, C.F., and Fuller, K.A. (1994) Light scattering by agglomerates: coupled electric and magnetic dipole method. *Langmuir*, **10**, 2533–2546.

472 Singham, S.B., and Bohren, C.F. (1989) Hybrid method in light scattering by an arbitrary particle. *Appl. Opt.*, **28**, 517–522.

473 Bourrely, C., Chiappetta, P., Lemaire, T., and Torrésani, B. (1992) Multidipole formulation of the coupled dipole method for electromagnetic scattering by an arbitrary particle. *J. Opt. Soc. Am. A*, **9**, 1336–1340.

474 Clausius, R. (1879) *Abhandlungen über die mechanische Wärmetheorie*, vol. **2**, Vieweg, Braunschweig.

475 Mossotti, O.F. (1850) Discussione analitica sull'influenza che l'azione di un mezzo dielettrico ha sulla distribuzione dell'elettricità alla superficie di più corpi elettrici disseminati in esso. *Mem. Soc. Sci. Modena*, **24**, 49–74.

476 Dungey, C.E., and Bohren, C.F. (1991) Light scattering by nonspherical particles: a refinement to the coupled-dipole method. *J. Opt. Soc. Am. A*, **8**, 81–87.

477 Singham, S.B., and Bohren, C.F. (1987) Light scattering by an arbitrary particle: a physical reformulation of the coupled dipole method. *Opt. Lett.*, **12**, 10–12.

478 Chiappetta, P. (1980) A new model for scattering by irregular absorbing particles. *Astron. Astrophys.*, **83**, 348–353.

479 Chiappetta, P. (1980) Multiple scattering approach to light scattering by arbitrarily shaped particles. *J. Phys. A*, **13**, 2101–2108.

480 Goedecke, G.H., and O'Brien, S. (1988) Scattering by irregular inhomogeneous particles via the digitized Green's function algorithm. *Appl. Opt.*, **27**, 2431–2438.

481 Livesay, D.E., and Chen, K.M. (1974) Electromagnetic fields induced in arbitrarily biological bodies. *IEEE Trans. Microwave Theory Tech.*, **MTT-22**, 1273–1280.

482 Hage, J.I., and Greenberg, J.M. (1990) A model for the optical properties of porous grains. *Astrophys. J.*, **361**, 251–259.

483 Hage, J.I., Greenberg, J.M., and Wang, R.T. (1991) Scattering from arbitrarily shaped particles: theory and experiment. *Appl. Opt.*, **30**, 1141–1152.

484 Piller, N.B. (1997) Improved coupled-dipole approximations for EM scattering, in *Electromagnetic and Light Scattering – Theory and Applications* (eds Y. Eremin and T. Wriedt), Moscow Lomonosov State University, Moscow, pp. 79–83.

485 Piller, N.B. (1997) Influence of the edge meshes on the accuracy of the coupled-dipole approximation. *Opt. Lett.*, **22**, 1674–1676.

486 Piller, N.B., and Martin, O.J.F. (1998) Extension of the generalized multipole technique to anisotropic media. *Opt. Commun.*, **150**, 1–6.

487 Gay-Balmaz, P., and Martin, O.J.F. (2001) Electromagnetic scattering of high-permittivity particles on a substrate. *Appl. Opt.*, **40**, 4562–4569.

488 Piller, N.B., and Martin, O.J.F. (1998) Increasing the performances of the coupled-dipole approximation: a spectral approach. *IEEE Trans. Antennas Propag.*, **AP-46**, 1126–1137.

489 Lakhtakia, A. (1992) General theory of the Purcell–Pennypacker scattering approach and its extension to bianisotropic scatterers. *Astrophys. J.*, **394**, 494–499.

490 Singham, S.B. (1986) Intrinsic optical activity in light scattering from an arbitrary particle. *Chem. Phys. Lett.*, **130**, 139–144.

491 Singham, S.B., and Salzman, G.C. (1986) Evaluation of the scattering matrix of an arbitrary particle using the coupled dipole approximation. *J. Chem. Phys.*, **84**, 2658–2667.

492 Singham, S.B., Patterson, C.W., and Salzman, G.C. (1986) Polarizabilities for light scattering from chiral particles. *J. Chem. Phys.*, **85**, 763–770.

493 Singham, S.B. (1987) Form and intrinsic optical activity in light scattering by chiral particles. *J. Chem. Phys.*, **87**, 1873–1881.

494 Taubenblatt, M.A., and Tran, T.K. (1993) Calculation of light scattering from particles and structures on a surface by the coupled-dipole method. *J. Opt. Soc. Am. A*, **10**, 912–919.

495 Schmehl, R., Nebeker, B.M., and Hirleman, E.D. (1997) Discrete-dipole approximation for scattering by features on surface by means of a two-dimensional fast Fourier transform technique. *J. Opt. Soc. Am. A*, **14**, 3026–3036.

496 Waterman, P.C. (1965) Matrix formulation of electromagnetic scattering. *IEE Proc.*, **53**, 803–812.

497 Waterman, P.C. (1971) Symmetry, unitarity, and geometry in electromagnetic scattering. *J. Phys. D*, **3**, 825–839.

498 Waterman, P.C. (1979) Matrix methods in potential theory and electromagnetic scattering. *J. Appl. Phys.*, **50**, 4550–4565.

499 Barber, P.W., and Yeh, C. (1975) Scattering of electromagnetic waves by arbitrarily shaped dielectric bodies. *Appl. Opt.*, **14**, 2864–2872.

500 Mishchenko, M.I., Travis, L.D., and Mackowski, D.W. (1996) T-matrix computations of light scattering by nonspherical particles: a review. *J. Quant. Spectrosc. Radiat. Transfer*, **55**, 535–575.

501 Mishchenko, M.I., Videen, G., Babenko, V.A., Khlebtsov, N.G., and Wriedt, T. (2004) T-matrix theory of electromagnetic scattering by particles and its applications: a comprehensive reference database. *J. Quant. Spectrosc. Radiat. Transfer*, **88**, 357–406.

502 Tsang, L., Kong, J.A., and Shin, R.T. (1985) *Theory of Microwave Remote Sensing*, John Wiley & Sons, Inc., New York.

503 Scheider, J.B., and Peden, I.C. (1988) Differential cross section of a dielectric ellipsoid by the T-matrix extended boundary condition method. *IEEE Trans. Antennas Propag.*, **AP-36**, 1317–1321.

504 Wriedt, T., and Doicu, A. (1998) Formulations of the extended boundary

condition method for three-dimensional scattering using the method of discrete sources. *J. Mod. Opt.*, **45**, 199–213.

505 Iskander, M.F., Lakhtakia, A., and Durney, C.H. (1982) A new iterative procedure to solve for scattering and absorption by dielectric objects. *IEE Proc.*, **71**, 1361–1362.

506 Iskander, M.F., Lakhtakia, A., and Durney, C.H. (1983) A new procedure for improving the solution stability, extending the frequency range of the EBCM. *IEEE Trans. Antennas Propag.*, **AP-31**, 317–324.

507 Lakhtakia, A., Varadan, V.K., and Varadan, V.V. (1984) Iterative extended boundary condition method for scattering by objects of high aspect ratios. *J. Acoust. Soc. Am.*, **76**, 906–912.

508 Iskander, M.F., and Lakhtakia, A. (1984) Extension of the iterative EBCM to calculate scattering by low-loss or lossless elongated dielectric objects. *Appl. Opt.*, **23**, 948–953.

509 Wriedt, T., and Doicu, A. (1997) Extended boundary condition method with multipole sources located in the complex plane. *Opt. Commun.*, **139**, 85–91.

510 Oguchi, T. (1960) Attenuation of electromagnetic wave due to rain with distorted raindrops. *J. Radio Res. Labs. (Tokyo)*, **7**, 467–485.

511 Morrison, J.A., and Cross, M.-J. (1974) Scattering of a plane electromagnetic wave by axisymmetric raindrops. *Bell Syst. Tech. J.*, **53**, 955–1019.

512 Rother, R., and Schmidt, K. (1996) The discretized Mie-formalism—a novel algorithm to treat scattering on axisymmetric particles. *J. Electromagn. Waves Appl.*, **10**, 273–297.

513 Rother, R., and Schmidt, K. (1996) The discretized Mie-formalism for plane wave scattering by dielectric cylinders. *J. Electromagn. Waves Appl.*, **10**, 697–717.

514 Rother, R., Schmidt, K., and Wauer, J. (1997) Plane wave scattering on hexagonal cylinders. *J. Quant. Spectrosc. Radiat. Transfer*, **57**, 669–681.

515 Ludwig, A.C. (1991) The generalized multipole technique. *Comput. Phys. Commun.*, **68**, 306–314.

516 Al-Rizzo, H.M., and Tranquilla, J.M. (1995) Electromagnetic wave scattering by highly elongated, geometrically composite objects of large size parameters: the generalized multipole technique. *Appl. Opt.*, **34**, 3502–3521.

517 Hafner, C. (1990) *The Generalized Multipole Technique for Computational Electromagnetics*, Artech House Books, Boston, MA.

518 Eremin, Y., and Sveshinikov, A.G. (1993) The discrete source method for investigating three-dimensional electromagnetic scattering problems. *Electromagnetics*, **13**, 1–22.

519 Eremin, Y., and Orlov, N.V. (1998) Modeling of light scattering by non-spherical particles based on discrete sources method. *J. Quant. Spectrosc. Radiat. Transfer*, **60**, 451–462.

520 Eremin, Y., Orlov, N.V., and Rozenberg, V.I. (1994) Scattering by non-spherical particles. *Comput. Phys. Commun.*, **79**, 201–214.

521 Leviatan, Y., Baharav, Z., and Heyman, E. (1995) Analysis of electromagnetic scattering using arrays of fictitious sources. *IEEE Trans. Antennas Propag.*, **AP-43**, 1091–1098.

522 Yee, K.S. (1966) Numerical solution of initial boundary value problems involving Maxwell's equations in isotropic media. *IEEE Trans. Antennas Propag.*, **AP-14**, 302–307.

523 Umashankar, K., and Taflove, A. (1982) A novel method to analyze electromagnetic scattering of complex objects. *IEEE Trans. Electromagn. Compat.*, **EMC-24**, 397–405.

524 Kunz, K.S., and Luebbers, R.J. (1993) *The Finite Difference Method for Electromagnetics*, CRC Press, Boca Raton, FL.

525 Taflove, A. (1995) *Computational Electrodynamics in the Finite-Difference Time Domain Method*, Artech House Books, Boston, MA.

526 Mur, G. (1982) Absorbing boundary conditions for the finite-difference approximation of the time-domain electromagnetic-field equations. *IEEE Trans. Electromagn. Compat.*, **EMC-23**, 377–382.

527 Sun, W., and Fu, Q. (2000) Finite-difference time-domain solution of light scattering by dielectric particles with large complex refractive indices. *Appl. Opt.*, **39**, 5569–5578.

528 Sun, W., Loeb, N.G., Tanev, S., and Videen, G. (2005) Finite-difference time-domain solution of light scattering by an infinite dielectric column immersed in an absorbing medium. *Appl. Opt.*, **44**, 1650–1656.

529 Yang, P., Kattawar, G.W., and Wiscombe, W.J. (2004) Effect of particle asphericity on single-scattering parameters: comparison between Platonic solids and spheres. *Appl. Opt.*, **43**, 4427–4434.

530 Chiappetta, P., and Torresani, B. (1988) Electromagnetic scattering from a dielectric helix. *Appl. Opt.*, **27**, 4856–4860.

531 Quirantes, A., and Delgado, A. (1998) Experimental size determination of spheroidal particles via the T-matrix method. *J. Quant. Spectrosc. Radiat. Transfer*, **60**, 463–474.

532 Fournier, G.R., and Evans, B.T.N. (1991) Approximation to extinction efficiency for randomly oriented spheroids. *Appl. Opt.*, **30**, 2042–2048.

533 Evans, B.T.N., and Fournier, G.R. (1994) Analytic approximation to randomly oriented spheroid extinction. *Appl. Opt.*, **33**, 5796–5804.

534 González, A.L., and Noguez, C. (2007) Influence of morphology on the optical properties of metal nanoparticles. *J. Comput. Theor. Nanosci.*, **4**, 231–238.

535 Kottmann, J.P., Martin, O.J.F., Smith, D.R., and Schultz, S. (2001) Dramatic localized electromagnetic enhancement in plasmon resonant nanowires. *Chem. Phys. Lett.*, **341**, 1–6.

536 Kottmann, J.P., Martin, O.J.F., Smith, D.R., and Schultz, S. (2001) Plasmon resonances of silver nanowires with a nonregular cross section. *Phys. Rev. B*, **64**, 235402–235411.

537 Kottmann, J.P., and Martin, O.J.F. (2000) Accurate solution of the volume integral equation for high permittivity scatterers. *IEEE Trans. Antennas Propag.*, **AP-48**, 1719–1726.

538 Yin, G., Wang, S.-Y., Xu, M., and Chen, L.-Y. (2006) Theoretical calculation of the optical properties of gold nanoparticles. *J. Korean Phys. Soc.*, **49**, 2108–2111.

539 Jin, R., Cao, Y., Mirkin, C.A., Kelly, K.L., Schatz, G.C., and Zheng, J.G. (2001) Photoinduced conversion of silver nanospheres to nanoprisms. *Science*, **294**, 1901–1903.

540 Aizpurua, J., Hanarp, P., Sutherland, D.S., Käll, M., Bryant, G.W., and García de Abajo, F.J. (2003) Optical properties of gold nanorings. *Phys. Rev. Lett.*, **90**, 057401.

541 Aizpurua, J., Blancob, L., Hanarp, P., Sutherland, D.S., Käll, M., Bryant, G.W., and García de Abajo, F.J. (2004) Light scattering in gold nanorings. *J. Quant. Spectrosc. Radiat. Transfer*, **89**, 11–16.

542 García de Abajo, F.J., and Howie, A. (1998) Relativistic electron energy loss and electron-induced photon emission in inhomogeneous dielectrics. *Phys. Rev. Lett.*, **80**, 5180–5183.

543 García de Abajo, F.J., and Howie, A. (2002) Retarded field calculation of electron energy loss in inhomogeneous dielectrics. *Phys. Rev. B*, **65**, 115418.

544 van der Zande, B.M.I., Böhmer, M.R., Fokkink, L.G.J., and Schönenberger, C. (1997) Aqueous gold sols of rod shaped particles. *J. Phys. Chem. B*, **101**, 852–854.

545 Hulteen, J.C., Patrissi, C.J., Miner, D.L., Crosthwait, E.R., Oberhauser, E.B., and Martin, C.R. (1997) Changes in the shape and optical properties of gold nanoparticles contained within alumina membranes due to low temperature annealing. *J. Phys. Chem. B*, **101**, 7727–7731.

546 Yu, Y.-Y., Chang, S.-S., Lee, C.-L., and Wang, C.R.C. (1997) Gold nanorods: electrochemical synthesis and optical properties. *J. Phys. Chem. B*, **101**, 6661–6664.

547 Mohamed, M.B., Ismael, K.Z., Link, S., and El-Sayed, M.A. (1998) Thermal reshaping of gold nanorods in micelles. *J. Phys. Chem. B*, **102**, 9370–9374.

548 Link, S., and El-Sayed, M.A. (1999) Spectral properties and relaxation

dynamics of surface plasmon electronic oscillations in gold and silver nanodots and nanorods. *J. Phys. Chem. B*, **103**, 8410–8426.

549 Mohamed, M.B., Volkov, V., Link, S., and El-Sayed, M.A. (2000) The "lightning" gold nanorods: fluorescence enhancement of over a million compared to the gold metal. *Chem. Phys. Lett.*, **317**, 517–523.

550 Link, S., Burda, C., Mohamed, M.B., Nikoobakht, B., and El-Sayed, M.A. (2000) Femtosecond transient absorption dynamics of colloidal gold nanorods: shape independence of the electron–phonon relaxation. *Phys. Rev. B*, **61**, 6086–6090.

551 Jana, N.R., Gearheart, L., and Murphy, C.J. (2001) Seed-mediated growth approach for shape-controlled synthesis of spheroidal and rod-like gold nanoparticles using a surfactant template. *Adv. Mater.*, **13**, 1389–1393.

552 Link, S., and El-Sayed, M.A. (2003) Optical properties and ultrafast dynamics of metallic nanocrystals. *Annu. Rev. Phys. Chem.*, **54**, 331–346.

553 McFarland, A.D., and van Duyne, R.P. (2003) Single silver nanoparticles as real-time and optical sensors with zeptomole sensitivity. *Nano Lett.*, **3**, 1057–1062.

554 Nikoobakht, B., and El-Sayed, M.A. (2003) Preparation and growth mechanism of gold nanorods (NRs) using seed-mediated growth method. *Chem. Mater.*, **15**, 1957–1962.

555 Kang, S.K., Chah, S., Yun, C.Y., and Yi, J. (2003) Aspect ratio controlled synthesis of gold nanorods. *Korean J. Chem. Eng.*, **20**, 1145–1148.

556 Liu, X., Yuan, H., Pang, D., and Cai, R. (2004) Resonance light scattering spectroscopy study of interaction between gold colloid and thiamazole and its analytical application. *Spectrochim. Acta Part A*, **60**, 385–389.

557 Bauer, L.A., Birenbaum, N.S., and Meyer, G.J. (2004) Biological applications of high aspect ratio nanoparticles. *Mater. Chem.*, **14**, 517–526.

558 Huang, C., Yang, Z., and Chang, H.-T. (2004) Synthesis of dumbbell-shaped Au–Ag core–shell nanorods by seed-mediated growth under alkaline conditions. *Langmuir*, **20**, 6089–6092.

559 Chang, J.-Y., Wu, H., Chen, H., Ling, Y.-C., and Tan, W. (2005) Oriented assembly of Au nanorods using biorecognition system. *Chem. Commun.*, **8**, 1092–1094.

560 Sönnichsen, C., and Alivisatos, A.P. (2005) Gold nanorods as novel nonbleaching plasmon-based orientation sensors for polarized single-particle microscopy. *Nano Lett.*, **5**, 301–304.

561 Ungureanu, C., Rayavarapu, R.G., Manohar, S., and van Leeuwen, T.G. (2009) Discrete dipole approximation simulations of gold nanorod optical properties: choice of input parameters and comparison with experiment. *J. Appl. Phys.*, **105**, 102032 (7 pages).

562 Pérez-Juste, J., Correa-Duarte, M.A., and Liz-Marzán, L.M. (2004) Silica gels with tailored, gold nanorod-driven optical functionalities. *Appl. Surf. Sci.*, **226**, 137–143.

563 Pérez-Juste, J., Pastoriza-Santos, I., Liz-Marzán, L.M., and Mulvaney, P. (2005) Gold nanorods: synthesis, characterization and applications. *Coord. Chem. Rev.*, **249**, 1870–1901.

564 Myroshnychenko, V., Rodríguez-Fernández, J., Pastoriza-Santos, I., Funston, A.M., Novo, C., Mulvaney, P., Liz-Marzán, L.M., and García de Abajo, F.J. (2008) Modelling the optical response of gold nanoparticles. *Chem. Soc. Rev.*, **37**, 1792–1805.

565 Chau, Y.-F., Tsai, D.P., Hu, G.-W., Shen, L.-F., and Yang, T.-J. (2007) Subwavelength optical imaging through a silver nanorod. *Opt. Eng.*, **46**, 039701.

566 Chau, Y.-F., Chen, M.W., and Tsai, D.P. (2009) Three-dimensional analysis of surface plasmon resonance modes on a gold nanorod. *Appl. Opt.*, **48**, 617–622.

567 Huang, H.J., Yu, C.P., Chang, H.C., Chiu, K.P., Chen, H.M., Liu, R.S., and Tsai, D.P. (2007) Plasmonic optical properties of a single gold nano-rod. *Opt. Express*, **15**, 7132–7139.

568 Alekseeva, A.V., Bogatyrev, V.A., Dykman, L.A., Khlebtsov, B.N., Trachuk, L.A., Melnikov, A.G., and

Khlebtsov, N.G. (2005) Preparation and optical scattering characterization of gold nanorods and their application to a dot-immunogold assay. *Appl. Opt.*, **44**, 6285–6295.

569 Brioude, A., Jiang, X.C., and Pileni, M.P. (2005) Optical properties of gold nanorods: DDA simulations supported by experiments. *J. Phys. Chem. B*, **109**, 13138–13142.

570 Schaich, W.L., Schider, G., Krenn, J.R., Leitner, A., Aussenegg, F.R., Puscasu, I., Monacelli, B., and Boreman, G. (2003) Optical resonances in periodic surface arrays of metallic patches. *Appl. Opt.*, **42**, 5714–5721.

571 Trinks, W. (1935) Zur Vielfachstreuung an kleinen Kugeln. *Ann. Phys. (Leipzig)*, **22**, 561–590.

572 Olaofe, G.O. (1965) Scattering of Electromagnetic Waves by Spherical Particles. PhD thesis. University of Manchester, Manchester.

573 Olaofe, G.O. (1970) Scattering by two Rayleigh–Debye spheres. *Appl. Opt.*, **9**, 429–437.

574 Liang, C., and Lo, Y.T. (1967) Scattering by two spheres. *Radio Sci.*, **2**, 1481–1495.

575 Bruning, J.H., and Lo, Y.T. (1971) Multiple scattering of EM waves by spheres. Part I. *IEEE Trans. Antennas Propag.*, **AP-19**, 378–390; Bruning, J.H., and Lo, Y.T. (1971) Multiple scattering of EM waves by spheres. Part II. *IEEE Trans. Antennas Propag.*, **AP-19**, 391–400.

576 Borghese, F., Denti, P., Toscano, G., and Sindoni, O.I. (1979) Electromagnetic scattering by a cluster of spheres. *Appl. Opt.*, **18**, 116–120.

577 Gérardy, J.M., and Ausloos, M. (1980) Absorption spectrum of clusters of spheres from the general solution of Maxwell's equations. The long-wavelength limit. *Phys. Rev. B*, **22**, 4950–4959.

578 Gérardy, J.M., and Ausloos, M. (1982) Absorption spectrum of clusters of spheres from the general solution of Maxwell's equations. II. Optical properties of aggregated metal spheres. *Phys. Rev. B*, **25**, 4204–4229.

579 Gérardy, J.M., and Ausloos, M. (1983) Absorption spectrum of clusters of spheres from the general solution of Maxwell's equations. IV. Proximity, bulk, surface, and shadow effects. *Phys. Rev. B*, **27**, 6446–6463.

580 Gérardy, J.M., and Ausloos, M. (1984) Absorption spectrum of clusters of spheres from the general solution of Maxwell's equations. III. Heterogeneous spheres. *Phys. Rev. B*, **30**, 2167–2181.

581 Jeffreys, B. (1965) Transformation of tesseral harmonics under rotation. *Geophys. J. R. Soc.*, **10**, 141–145.

582 Quinten, M., Kreibig, U., Schönauer, D., and Genzel, L. (1985) Optical absorption spectra of pairs of small metal particles. *Surf. Sci.*, **156**, 741–750.

583 Quinten, M., and Kreibig, U. (1986) Optical properties of aggregates of small metal particles. *Surf. Sci.*, **172**, 557–577.

584 Quinten, M., and Kreibig, U. (1988) Optical extinction spectra of systems of small metal particles with aggregates, in *Optical Particle Sizing* (eds G. Gouesbet and G. Gréhan), Plenum Press, New York, pp. 249–258.

585 Quinten, M., Schönauer, D., and Kreibig, U. (1989) Electronic excitations in many-particle systems: a quantitative analysis. *Z. Phys. D*, **12**, 521–525.

586 Fuller, K.A. (1987) Cooperative electromagnetic scattering by ensembles of spheres. PhD thesis. Stephen F. Austin State University, Nacogdoches, TX.

587 Fuller, K.A., Kattawar, G.W., and Wang, R.T. (1986) Electromagnetic scattering from two dielectric spheres: further comparisons between theory and experiments. *Appl. Opt.*, **25**, 2521–2529.

588 Fuller, K.A., and Kattawar, G.W. (1988) Consummate solution to the problem of classical electromagnetic scattering by an ensemble of spheres. Part I: linear chains. *Opt. Lett.*, **13**, 90–92; Fuller, K.A., and Kattawar, G.W. (1988) Consummate solution to the problem of classical electromagnetic scattering by an ensemble of spheres. Part II: clusters of arbitrary configuration. *Opt. Lett.*, **13**, 1063–1065.

589 Fuller, K.A. (1991) Optical resonances and two-sphere systems. *Appl. Opt.*, **30**, 4716–4731.

590 Fuller, K.A. (1993) Scattering and absorption by inhomogeneous spheres and sphere aggregates. *Proc. SPIE*, **1862**, 249–257.

591 Fuller, K.A. (1994) Scattering and absorption cross sections of compounded spheres. I. Theory for external aggregation. *J. Opt. Soc. Am. A*, **11**, 3251–3260.

592 Fuller, K.A. (1995) Scattering and absorption cross sections of compounded spheres. II. Calculations for external aggregation. *J. Opt. Soc. Am. A*, **12**, 881–892.

593 Hamid, A.-K., Ciric, I.R., and Hamid, M. (1990) Multiple scattering by a linear array of conducting spheres. *Can. J. Phys.*, **68**, 1157–1165.

594 Hamid, A.-K., Ciric, I.R., and Hamid, M. (1990) Electromagnetic scattering by an arbitrary configuration of dielectric spheres. *Can. J. Phys.*, **68**, 1419–1428.

595 Hamid, A.-K., Ciric, I.R., and Hamid, M. (1991) Iterative solution of the scattering by an arbitrary configuration of conducting or dielectric spheres. *IEE Proc. H*, **138**, 565–572.

596 Hamid, A.-K., Ciric, I.R., and Hamid, M. (1991) Scattering by systems of small conducting spheres, in Proceedings of Progress in Electromagnetic Research Symposium (PIERS), MIT, Cambridge, MA.

597 Hamid, A.-K., Ciric, I.R., and Hamid, M. (1991) Iterative technique for scattering by a linear array of spheres, in Proceedings of the IEEE AP-S International Symposium, University of Western Ontario, London, ON, pp. 30–33.

598 Mackowski, D.W. (1991) Analysis of radiative scattering for multiple sphere configurations. *Proc. R. Soc. London Ser. A*, **433**, 599–614.

599 Mackowski, D.W. (1994) Calculation of total cross sections of multiple-sphere clusters. *J. Opt. Soc. Am. A*, **11**, 2851–2861.

600 Xu, Y.L. (1995) Electromagnetic scattering by an aggregate of spheres. *Appl. Opt.*, **34**, 4573–4588.

601 Xu, Y.L., and Gustafson, B. (1997) Experimental and theoretical results of light scattering by aggregates of spheres. *Appl. Opt.*, **36**, 8026–8030.

602 Xu, Y.L. (1997) Electromagnetic scattering by an aggregate of spheres: far field. *Appl. Opt.*, **36**, 9496–9508.

603 Xu, Y.L. (1998) Electromagnetic scattering by an aggregate of spheres: errata. *Appl. Opt.*, **37**, 6494.

604 Xu, Y.L. (2001) Electromagnetic scattering by an aggregate of spheres: far field: errata. *Appl. Opt.*, **40**, 5508.

605 Rouleau, F., and Martin, P.G. (1993) Proximity effects in clusters of particles. *Astrophys. J.*, **416**, 707–718.

606 Rouleau, F. (1996) Electromagnetic scattering by compact clusters of spheres. *Astron. Astrophys.*, **310**, 686–698.

607 Ioannidou, M.P., Skaropoulos, N.C., and Chrissoulidis, D.P. (1995) Study of interactive scattering by clusters of spheres. *J. Opt. Soc. Am. A*, **12**, 1782–1789.

608 Vagov, A.V., Radchik, A., and Smith, G.B. (1994) Optical response of arrays of spheres from the theory of hypercomplex variables. *Phys. Rev. Lett.*, **73**, 1035–1038.

609 Stein, S. (1961) Addition theorems for spherical wave functions. *Q. Appl. Math.*, **19**, 15–24.

610 Quinten, M., and Kreibig, U. (1993) Absorption and elastic scattering of light by particle aggregates. *Appl. Opt.*, **32**, 6173–6182.

611 Claro, F. (1984) Theory of resonant modes in particulate matter. *Phys. Rev. B*, **30**, 4989–4999.

612 Claro, F., and Fuchs, R. (1986) Optical absorption by clusters of small metallic spheres. *Phys. Rev. B*, **33**, 7956–7960.

613 Clippe, P., Evrard, R., and Lucas, A.A. (1976) Aggregation effect on the infrared absorption spectrum of small ionic crystals. *Phys. Rev. B*, **14**, 1715–1721.

614 Ausloos, M., Clippe, P., and Lucas, A.A. (1978) Infrared active modes in large clusters of spheres. *Phys. Rev. B*, **18**, 7176–7185.

615 Rechberger, W., Hohenau, A., Leitner, A., Krenn, J.R., Lamprecht, B., and Aussenegg, F.R. (2003) Optical properties of two interacting gold nanoparticles. *Opt. Commun.*, **220**, 137–141.

616 Michel, B. (1995) A statistical method to calculate the extinction by small irregularly shaped particles. *J. Opt. Soc. Am. A*, **12**, 2471–2481.

617 Quinten, M. (1997) Enhanced optical response by soot agglomerates. *J. Aerosol Sci.*, **28** (S1), S769–S770.

618 Quinten, M. (1998) Enhanced light scattering and absorption by soot agglomerates, in Proceedings of PARTEC 98, Nürnberg, pp. 79–88.

619 Ruppin, R. (1999) Effects of high-order multipoles on the extinction spectra of dispersive bispheres. *Opt. Commun.*, **168**, 35–38.

620 Pustovit, V.N., Sotelo, J.A., and Niklasson, G.A. (2002) Coupled multipolar interactions in small-particle metallic clusters. *J. Opt. Soc. Am. A*, **19**, 513–518.

621 Smith, G.B., Vargas, W.E., Niklasson, G., Soleto, J.A., Paley, A., and Radchik, A. (1995) Optical properties of a pair of spheres: comparison of different theories. *Opt. Commun.*, **115**, 8–12.

622 Pack, A., Hietschold, M., and Wannemacher, R. (2001) Failure of local Mie theory: optical spectra of colloidal aggregates. *Opt. Commun.*, **194**, 277–287.

623 Andersen, A.C., Mutschke, H., Posch, T., Min, M., and Tamanai, A. (2006) Infrared extinction by homogeneous particle aggregates of SiC, FeO, SiO_2: comparison of different theoretical approaches. *J. Quant. Spectrosc. Radiat. Transfer*, **100**, 4–15.

624 Sotelo, J., and Niklasson, G.A. (1991) Optical properties of fractal clusters of small metallic particles. *Z. Phys. D*, **20**, 321–323.

625 Al-Nimr, M.A., and Arpaci, V.S. (1994) Optical properties of interfacing particles. *Appl. Opt.*, **33**, 8412–8416.

626 Lebedev, A.N., and Stenzel, O. (1999) Optical extinction of an assembly of spherical particles in an absorbing medium: application to silver clusters in absorbing organic materials. *Eur. Phys. J. D*, **7**, 83–88.

627 Barton, J.P., Ma, W., Schaub, S.A., and Alexander, D.R. (1991) Electromagnetic field for a beam incident on two adjacent spherical particles. *Appl. Opt.*, **30**, 4706–4715.

628 Cooray, M.F.R., and Ciric, I.R. (1989) Electromagnetic wave scattering by a system of two spheroids of arbitrary orientation. *IEEE Trans. Antennas Propag.*, **AP-37**, 608–618.

629 Cooray, M.F.R., and Ciric, I.R. (1991) Scattering of electromagnetic waves by a system of two dielectric spheroids of arbitrary orientation. *IEEE Trans. Antennas Propag.*, **AP-39**, 680–684.

630 Ciric, I.R., and Cooray, M.F.R. (1990) Admittance characteristics and far-field patterns for coupled spheroidal dipole antennas in arbitrary configuration. *IEE Proc. H*, **137**, 337–342.

631 Yousif, H.A., and Köhler, S. (1988) Scattering by two penetrable cylinders at oblique incidence. I. The analytical solution. *J. Opt. Soc. Am.*, **5**, 1085–1096.

632 Yousif, H.A., and Köhler, S. (1988) Scattering by two penetrable cylinders at oblique incidence. II. Numerical examples. *J. Opt. Soc. Am.*, **5**, 1097–1104.

633 Bever, S.J., and Allebach, J.P. (1992) Multiple scattering by a planar array of parallel dielectric cylinders. *Appl. Opt.*, **31**, 3524–3532.

634 Yousif, H.A., Mattis, R.E., and Kozminski, K. (1994) Light scattering at oblique incidence on two coaxial cylinders. *Appl. Opt.*, **33**, 4013–4024.

635 Felbacq, D., Tayeb, G., and Maystre, D. (1994) Scattering by a random set of parallel cylinders. *J. Opt. Soc. Am. A*, **11**, 2526–2538.

636 Lee, S.C., and Grzesik, J.A. (1998) Light scattering by closely spaced parallel cylinders embedded in a semi-infinite dielectric medium. *J. Opt. Soc. Am. A*, **15**, 163–173.

637 Au, W.C., Kong, J.A., and Tsang, L. (1994) Absorption enhancement of scattering of electromagnetic waves by dielectric cylinder cluster. *Microwave Opt. Tech. Lett.*, **7**, 454–457.

638 Ruppin, R. (1997) Infrared absorption of cube spheres. *Opt. Commun.*, **136**, 395–398.

639 Kottmann, J.P., and Martin, O.J.F. (2001) Plasmon resonant coupling in metallic nanowires. *Opt. Express*, **8**, 655–663.

640 Kottmann, J.P., and Martin, O.J.F. (2001) Retardation-induced plasmon resonances in coupled nanoparticles. *Opt. Lett.*, **26**, 1096–1098.

641 Byun, K.M., Kim, S.J., and Kim, D. (2006) Profile effect on the feasibility of extinction-based localized surface plasmon resonance biosensors with metallic nanowires. *Appl. Opt.*, **45**, 3382–3389.

642 Moharam, M.G., and Gaylord, T.K. (1982) Diffraction analysis of dielectric surface-relief gratings. *J. Opt. Soc. Am.*, **72**, 1385–1392.

643 Moharam, M.G., and Gaylord, T.K. (1986) Rigorous coupled-wave analysis of metallic surface-relief gratings. *J. Opt. Soc. Am. A*, **3**, 1780–1787.

644 Tsuei, T.-G., and Barber, P.W. (1988) Multiple scattering by two parallel dielectric cylinders. *Appl. Opt.*, **27**, 3375–3381.

645 Quinten, M. (1999) Optical effects associated with aggregates of clusters. *J. Cluster Sci.*, **284**, 319–358.

646 Kreibig, U., Quinten, M., and Schönauer, D. (1986) Optical properties of many-particle systems. *Phys. Scr. T*, **13**, 84–92.

647 Kreibig, U. (1986) Systems of small metal particles: optical properties and their structure dependence. *Z. Phys. D*, **3**, 239–249.

648 Kreibig, U., Quinten, M., and Schönauer, D. (1987) Optical investigations of aggregation processes in aqueous noble metal colloid systems, in *Time-Dependent Effects in Disordered Materials* (eds R. Pynn and T. Riste), Plenum Press, New York, pp. 103–115.

649 Kreibig, U., Quinten, M., and Schönauer, D. (1989) Many particle systems: models of inhomogeneous matter. *Physica A*, **157**, 244–261.

650 Kahlau, T. (1994) Aufbau eines Photogoniometers und Messung der winkelaufgelösten Lichtstreuung an kolloidaler Cluster-Materie. Diploma thesis. RWTH Aachen, Aachen.

651 Quinten, M., and Stier, J. (1995) Absorption of scattered light in colloidal systems of aggregated particles. *Colloid Polym. Sci.*, **237**, 233–241.

652 Kahlau, T., Quinten, M., and Kreibig, U. (1996) Extinction and angle-resolved light scattering from aggregated metal clusters. *Appl. Phys. A*, **62**, 19–27.

653 Reuter, G. (1994) Messung der optischen Absorption von Metall- und Fulleren-Clustern mittels photothermischer Laserstrahlablenkung. Diploma thesis. RWTH Aachen, Aachen,.

654 Boccara, A.C., Fournier, D., and Badoz, J. (1980) Thermo-optical spectroscopy: detection by the "mirage effect". *Appl. Phys. Lett.*, **36**, 130–132.

655 Rosencwaig, A., and Gersho, A. (1976) Theory of the photoacoustic effect with solids. *J. Appl. Phys.*, **47**, 64–69.

656 Murphy, J.C., and Aamodt, L.C. (1980) Photothermal spectroscopy using optical beam probing: mirage effect. *J. Appl. Phys.*, **51**, 4580–4588.

657 Jackson, W.B., Amer, N.M., Boccara, A.C., and Fournier, D. (1981) Photothermal deflection spectroscopy and detection. *Appl. Opt.*, **20**, 1333–1344.

658 Tam, A.C. (1986) Applications of photoacoustic sensing techniques. *Rev. Mod. Phys.*, **58**, 381–426.

659 McDonald, F.A., and Wetzel, G.C. (1988) Theory of photothermal and photoacoustic effects in condensed matter. *Phys. Acoust.*, **18**, 167–277.

660 Pacheco, J.M., and Ekardt, W. (1992) Microscopic calculation of the van der Waals interaction between small metal clusters. *Phys. Rev. Lett.*, **68**, 3694–3697.

661 Eckstein, H. (1992) Nichtthermische Einflüsse elektromagnetischer Strahlung auf die Aggregatbildung von Metallclustern. Diploma thesis. RWTH Aachen, Aachen.

662 Eckstein, H., and Kreibig, U. (1993) Light induced aggregation of metal clusters. *Z. Phys. D*, **26**, 239–241.

663 Hasegawa, H., Satoh, N., Tsujii, K., and Kimura, K. (1991) Fractal analysis of the coalescence process of Au nanometer particles dispersed in 2-propanol. *Z. Phys. D*, **20**, 325–327.

664 Satoh, N., Hasegawa, H., Tsujii, K., and Kimura, K. (1994) Photoinduced coagulation of Au nanocolloids. *J. Phys. Chem.*, **98**, 2143–2147.

665 Perminov, S.V., Drachev, V.P., and Rautian, S.G. (2007) Optics of metal nanoparticle aggregates with light induced motion. *Opt. Express*, **15**, 8639–8648.

666 Schönauer, D., Quinten, M., and Kreibig, U. (1989) Precursor-states of percolation in quasi-fractal many-particle systems. *Z. Phys. D*, **12**, 527–532.

667 Hellmers, J., Riefler, N., Wriedt, T., and Eremin, Y. (2008) Light scattering simulation for the characterization of sintered silver nanoparticles. *J. Quant. Spectrosc. Radiat. Transfer*, **109**, 1363–1373.

668 Karpov, S.V., Bas'ko, A.L., Popov, A.K., Slabko, V.V., and George, T.F. (2002) Optics of nanostructured fractal silver colloids, in *Recent Research Developments in Optics*, vol. **2** (ed. S.G. Pandalai), Research Signpost, Trivandrum, pp. 1–37.

669 Ou, D.R., Zhu, J., Zhao, J.H., Zhu, R.J., and Wang, J. (2006) Influence of the interaction between metal particles on optical properties of Ag–Si$_3$N$_4$ composite films. I. Experiment. *Appl. Opt.*, **45**, 1244–1248.

670 Ou, D.R., Zhu, J., Zhao, J.H., Zhu, R.J., and Wang, J. (2006) Influence of the interaction between metal particles on optical properties of Ag–Si$_3$N$_4$ composite films. II. Two-particle approximation. *Appl. Opt.*, **45**, 1249–1253.

671 Akamatsu, K., Takei, S., Mizuhara, M., Kajinami, A., Deki, S., Takeoka, S., Fujii, M., Hayashi, S., and Yamamoto, K. (2000) Preparation and characterization of polymer thin films containing silver and silver sulfide nanoparticles. *Thin Solid Films*, **359**, 55–60.

672 Akamatsu, K., and Deki, S. (1997) Characterization and optical properties of gold nanoparticles dispersed in nylon 11 thin films. *J. Mater. Chem.*, **7**, 1773–1777.

673 Bogatyrev, V.A., Medvedev, B.A., Dykman, L.A., and Khlebtsov, N.G. (2001) Light scattering spectra of colloidal gold aggregates: experimental measurements and theoretical simulations, *Proc. SPIE*, **4241**, 42–48.

674 Bogatyrev, V.A., Dykman, L.A., Krasnov, Y.M., Plotnikov, V.K., and Khlebtsov, N.G. (2002) Differential light spectroscopy for studying biospecific assembling of gold nanoparticles with protein or oligonucleotide probes. *Colloid J.*, **64**, 671–680.

675 Khlebtsov, N.G., Bogatyrev, V.A., Dykman, L.A., Khlebtsov, B.N., and Krasnov, Y.M. (2004) Differential light scattering spectroscopy: a new approach to studies of colloidal gold nanosensors. *J. Quant. Spectrosc. Radiat. Transfer*, **89**, 133–142.

676 Gartz, M., and Quinten, M. (2001) Broadening of resonances in yttrium nanoparticle optical spectra. *Appl. Phys. B*, **73**, 327–332.

677 Rouleau, F., Henning, Th., and Stognienko, R. (1997) Constraints on the properties of the 2175 ångström interstellar feature carrier. *Astron. Astrophys.*, **322**, 633–645.

678 Quinten, M., Kreibig, U., Henning, Th., and Mutschke, M. (2002) Wavelength-dependent optical extinction of carbonaceous particles in atmospheric aerosols and interstellar dust. *Appl. Opt.*, **41**, 7102–7113.

679 Michel, B., Henning, Th., Jäger, C., and Kreibig, U. (1999) Optical extinction by spherical carbonaceous particles. *Carbon*, **37**, 391–400.

680 Michel, B., Henning, Th., Stognienko, R., and Rouleau, F. (1996) Extinction properties of dust grains: a new computational technique. *Astrophys. J.*, **468**, 834–841.

681 Jäger, C., Henning, Th., Schlögl, R., and Spillecke, O. (1999) Spectral properties of carbon black, *J. Non-Cryst. Solids*, **258**, 161–179.

682 Fuchs, R. (1978) Infrared absorption in MgO microcrystals. *Phys. Rev. B*, **18**, 7160–7162.

683 Quinten, M. (1998) Optical response of aggregates of clusters with different dielectric functions. *Appl. Phys. B*, **67**, 101–106.

684 Friehmelt, R., and Heidenreich, S. (1999) Calibration of a white-light/90° optical particle counter for "aerodynamic" size measurements – experiments and calculations for

spherical particles and quartz dust. *J. Aerosol Sci.*, **30**, 1271–1280.

685 Umhauer, H. (1983) Particle size distribution analysis by scattered light measurements using an optically defined measuring volume. *J. Aerosol Sci.*, **14**, 765–770.

686 Quinten, M., Friehmelt, R., and Ebert, K.-F. (2000) Sizing of aggregates of spheres by a white-light optical particle counter with 90° scattering angle. *J. Aerosol Sci.*, **32**, 63–72.

687 Heyder, J., and Gebhart, J. (1979) Optimization of response functions of light scattering instruments for size evaluation of aerosol particles. *Appl. Opt.*, **18**, 705–711.

689 Lebedev, A.N., Stenzel, O., Quinten, M., Stendal, A., Röder, M., Schreiber, M., and Zahn, D.R.T. (1999) A statistical approach for interpreting the optical spectra of metal island films: effects of multiple scattering in a statistical assembly of spheres. *J. Opt. A*, **1**, 573–580.

690 Liu, Z., Wang, H., Li, H., and Wang, X. (1998) Red shift of plasmon resonance frequency due to the interacting Ag nanoparticles embedded in single crystal SiO_2 by implantation. *Appl. Phys. Lett.*, **72**, 1823–1825.

691 Liu, Z., Li, H., Feng, X., Ren, S., Wang, H., Liu, Z., and Lu, B. (1998) Formation effects and optical absorption of Ag nanocrystals embedded in single crystal SiO_2 by implantation. *J. Appl. Phys.*, **84**, 1913–1917.

692 Pivin, J.C., Garcia, M.A., Llopis, J., and Hofmeister, H. (2002) Interaction between clusters in ion implanted and ion beam mixed SiO_2:Ag films. *Nucl. Instrum. Methods Phys. Res. B*, **191**, 794–799.

693 Taleb, A., Russier, V., Courty, A., and Pileni, M.P. (1999) Collective optical properties of silver nanoparticles organized in two-dimensional superlattices. *Phys. Rev. B*, **59**, 13350–13358.

694 Pileni, M.P. (2000) Fabrication and physical properties of selforganized silver nanocrystals. *Pure Appl. Chem.*, **72**, 53–65.

695 Quinten, M., Stenzel, O., Stendal, A., and von Borczyskowski, C. (1997) Calculation of the optical properties of phthalocyanine thin films with incorporated noble metal clusters. *J. Opt.*, **28**, 245–251.

696 Denti, P., Borghese, F., Saija, R., Fucile, E., and Sindoni, O.I. (1999) Optical properties of aggregated spheres in the vicinity of a plane surface. *J. Opt. Soc. Am. A*, **16**, 167–175.

697 Denti, P., Borghese, F., Saija, R., Iati, M.A., and Sindoni, O.I. (1999) Optical properties of a dispersion of randomly oriented identical aggregates of spheres deposited on a plane surface. *Appl. Opt.*, **38**, 6421–6430.

698 Félidj, N., Aubard, J., and Levi, G. (1999) Discrete dipole approximation for ultraviolet–visible extinction spectra simulation of silver and gold colloids. *J. Chem. Phys.*, **111**, 1195–1208.

699 Xu, Y.L., and Gustafson, B. (1999) Comparison between multisphere light-scattering calculations: rigorous solution and discrete-dipole approximation. *Astrophys. J.*, **513**, 894–909.

700 Andersen, A.C., Sotelo, J.A., Pustovit, V.N., and Niklasson, G.A. (2002) Extinction calculations of multi-sphere polycrystalline graphitic clusters – a comparison with the 2175 ångström peak and between a rigorous solution and discrete-dipole approximations. *Astron. Astrophys.*, **386**, 296–307.

701 Iskander, M.F., Chen, H.Y., and Penner, J.E. (1989) Optical scattering and absorption by branched chains of aerosols. *Appl. Opt.*, **28**, 3083–3091.

702 Chen, H.Y., Iskander, M.F., and Penner, J.E. (1990) Light scattering and absorption by fractal agglomerates and coagulations of smoke aerosols. *J. Mod. Opt.*, **37**, 171–181.

703 Chen, H.Y., Iskander, M.F., and Penner, J.E. (1991) Empirical formula for optical absorption by fractal aerosol agglomerates. *Appl. Opt.*, **12**, 1547–1551.

704 Jain, P.K., Eustis, S., and El-Sayed, M.A. (2006) Plasmon coupling in nanorod assemblies: optical absorption, discrete dipole approximation simulation, and exciton-coupling model export. *J. Phys. Chem. B*, **110**, 18243–18253.

705 Mishchenko, M.I. (1991) Light scattering by randomly oriented axially symmetric particles. *J. Opt. Soc. Am. A*, **8**, 871–882.

706 Mishchenko, M.I., and Mackowski, D.W. (1994) Light scattering by randomly oriented bispheres. *Opt. Lett.*, **19**, 1604–1606.

707 Mishchenko, M.I., Mackowski, D.W., and Travis, L. (1995) Scattering of light by bispheres with touching and separated components. *Appl. Opt.*, **34**, 4589–4599.

708 Mishchenko, M.I. (1996) Coherent backscattering by two-sphere clusters. *Opt. Lett.*, **21**, 623–625.

709 Mackowski, D.W., and Mishchenko, M.I. (1996) Calculation of the T-matrix and the scattering matrix for ensembles of spheres. *J. Opt. Soc. Am. A*, **13**, 2266–2278.

710 Mishchenko, M.I., and Mackowski, D.W. (1996) Electromagnetic scattering by randomly oriented bispheres: comparison of theory and experiment and benchmark calculations. *J. Quant. Spectrosc. Radiat. Transfer*, **55**, 683–694.

711 Lou, W., and Charalampopoulos, T.T. (1994) On the electromagnetic scattering and absorption of agglomerated small spherical particles. *J. Phys. D*, **27**, 2258–2270.

712 Wang, Z.L., Hu, L., and Ren, W. (1994) Multiple scattering of waves by a half-space of distributed discrete scatters with modified T-matrix approach. *J. Phys. D*, **27**, 441–446.

713 Wang, Z.L., Hu, L., and Lin, W.G. (1994) A modified T-matrix formulation for multiple scattering of electromagnetic waves. *J. Phys. D*, **27**, 447–451.

714 Tamaru, H., Kuwata, H., Miyazaki, H.T., and Miyano, K. (2002) Resonant light scattering from individual Ag nanoparticles and particle pairs. *Appl. Phys. Lett.*, **80**, 1826–1828.

715 Drolen, B.L., and Tien, C.L. (1987) Absorption and scattering of agglomerated soot particulate. *J. Quant. Spectrosc. Radiat. Transfer*, **37**, 433–448.

716 Jones, A.R. (1979) Electromagnetic wave scattering by assemblies of particles in the Rayleigh approximation. *Proc. R. Soc. London Ser. A*, **336**, 11–27.

717 Jones, A.R. (1979) Scattering efficiency factors for agglomerates of small spheres. *J. Phys. D*, **12**, 1661–1672.

718 Stognienko, R., Henning, Th., and Ossenkopf, V. (1995) Optical properties of coagulated particles. *Astron. Astrophys.*, **296**, 797–809.

719 Ruppin, R. (1989) Optical absorption of two spheres. *J. Phys. Soc. Jpn.*, **58**, 1446–1451.

720 Shalaev, V., Botet, R., and Jullien, R. (1991) Resonant light scattering by fractal clusters. *Phys. Rev. B*, **44**, 12216–12225.

721 Mackowski, D.W. (1995) Electrostatics analysis of radiative absorption by sphere clusters in the Rayleigh limit: application to soot particles. *Appl. Opt.*, **34**, 3535–3545.

722 di Stasio, S., and Massoli, P. (1997) Morphology, monomer size and concentration of agglomerates constituted by Rayleigh particles as retrieved from scattering/extinction measurements. *Combust. Sci. Technol.*, **124**, 219–247.

723 Liu, L., Mishchenko, M.I., and Arnott, W.P. (2008) A study of radiative properties of fractal soot aggregates using the superposition T-matrix method. *J. Quant. Specrosct. Radiat. Transfer*, **109**, 2656–2663.

724 Vargas, W., Cruz, L., Fonseca, L., and Gomez, M. (1993) T-matrix approach for calculating local fields around clusters of rotated spheroids. *Appl. Opt.*, **32**, 2164–2170.

725 Ludwig, A.C. (1991) Scattering by two and three spheres computed by the generalized multipole technique. *IEEE Trans. Antennas Propag.*, **AP-39**, 703–705.

726 van de Hulst, H.C. (1980) *Multiple Light Scattering: Tables, Formulas and Applications*, Academic Press, New York.

727 Chandrasekhar, S. (1960) *Radiative Transfer*, Oxford University Press, London.

728 Mishchenko, M.I., Travis, L.D., and Lacis, A.A. (2006) *Multiple Scattering of Light by Particles: Radiative Transfer and*

Coherent Backscattering, Cambridge University Press, New York.

729 Kortüm, G. (1969) *Reflexionsspektroskopie: Grundlagen, Methodik, Anwendungen*, Springer, Heidelberg; translated by J. E. Lohr: *Reflectance Spectroscopy*, Springer, New York.

730 Murphy, A.B. (2006) Modified Kubelka–Munk model for calculation of the reflectance of coatings with optically rough surfaces. *J. Phys. D*, **39**, 3571–3581.

731 Brockes, A. (1960) Der Einfluß glänzender Oberflächen auf Remissionsmessungen. *Farbe*, **9**, 53–62.

732 Mudgett, P.S., and Richards, L.W. (1971) Multiple scattering calculations for technology. *Appl. Opt.*, **10**, 1485–1501.

733 Brinkworth, B.J. (1972) Interpretation of the Kubelka–Munk coefficients in reflection theory. *Appl. Opt.*, **11**, 1434–1435.

734 Reiss, H. (1988) *Radiative Transfer in Nontransparent Dispersed Media*, Springer, Berlin.

735 Curiel, F., Vargas, W.E., and Barrera, R.G. (2002) Visible spectral dependence of the scattering and absorption coefficients of pigmented coatings from inversion of diffuse reflectance spectra. *Appl. Opt.*, **41**, 5969–5978.

736 Saunderson, J.L. (1942) Calculation of the color of pigmented plastics. *J. Opt. Soc. Am.*, **32**, 727–729.

737 Vargas, W.E., and Niklasson, G.A. (1997) Forward average path-length parameter in four-flux radiative transfer models. *Appl. Opt.*, **36**, 3735–3738.

738 Ryde, J.W. (1931) The scattering of light by turbid media. Part I. *Proc. R. Soc. London Ser. A*, **131**, 451–464.

739 Reichman, J. (1973) Determination of absorption and scattering coefficients for nonhomogeneous media. I: theory. *Appl. Opt.*, **12**, 1811–1815.

740 Maheu, B., Letoulouzan, J.N., and Gouesbet, G. (1984) Four-flux models to solve the scattering transfer equation in terms of Lorenz–Mie parameters. *Appl. Opt.*, **23**, 3353–3362.

741 Maheu, B., and Gouesbet, G. (1986) Four-flux models to solve the scattering transfer equation: special cases. *Appl. Opt.*, **25**, 1122–1128.

742 Latimer, P., and Noh, S.J. (1987) Light propagation in moderately dense particle systems: a reexamination of the Kubelka–Munk theory. *Appl. Opt.*, **26**, 514–523.

743 Chu, C.M., and Churchill, S.W. (1955) Numerical solution of problems in multiple scattering of electromagnetic radiation. *J. Phys. Chem.*, **59**, 855–863.

744 Vargas, W.E., and Niklasson, G.A. (1997) Generalized method for evaluating scattering parameters used in radiative transfer models. *J. Opt. Soc. Am. A*, **14**, 2243–2252.

745 Vargas, W.E., and Niklasson, G.A. (1997) Intensity of diffuse radiation in particulate media. *J. Opt. Soc. Am. A*, **14**, 2253–2262.

746 Gate, L.F. (1971) The determination of light absorption in diffusing materials by a photon diffusion model. *J. Phys. D*, **4**, 1049–1056.

747 Gate, L.F. (1974) Comparison of the photon diffusion model and Kubelka–Munk equation with the exact solution of the radiative transport equation. *Appl. Opt.*, **13**, 236–238.

748 Ishimaru, A., Kuga, Y., Cheung, R.L.-T., and Shimizu, K. (1983) Scattering and diffusion of a beam wave in randomly distributed scatterers. *J. Opt. Soc. Am.*, **73**, 131–136.

749 Garg, R., Prud'homme, R.K., Aksay, I.A., Liu, F., and Alfano, R.R. (1998) Optical transmission in highly-concentrated dispersions. *J. Opt. Soc. Am. A*, **15**, 932–935.

750 Garg, R., Prud'homme, R.K., Aksay, I.A., Liu, F., and Alfano, R.R. (1998) Absorption length for photon propagation in highly dense colloidal dispersions. *J. Mater. Res.*, **13**, 3463–3467.

751 Niklasson, G.A. (1987) Comparison between four-flux theory and multiple scattering theory. *Appl. Opt.*, **26**, 4034–4036.

752 Maheu, B., Briton, J.P., and Gouesbet, G. (1989) Four-flux model and a Monte Carlo code: comparisons between two simple, complementary tools for multiple scattering calculations. *Appl. Opt.*, **28**, 22–24.

753 Briton, J.P., Maheu, B., Gréhan, G., and Gouesbet, G. (1992) Monte Carlo

simulations of multiple scattering in arbitrary 3-D geometry. *Part. Part. Syst. Charact.*, **9**, 52–58.

754 Vargas, W.E. (1998) Generalized four-flux radiative transfer model. *Appl. Opt.*, **37**, 2615–2623.

755 Elias, M., and Elias, G. (2002) New and fast calculation for incoherent multiple scattering. *J. Opt. Soc. Am. A*, **19**, 894–901.

756 Dusemund, B. (1991) Kollektive Effekte in den optischen Eigenschaften von Cluster-Materie. Diploma thesis. University of Saarland, Saarbrücken.

757 Kreibig, U. (1992) Optical properties of macroscopic many-cluster matter, in *Physics and Chemistry of Finite Systems: from Clusters to Crystals*, vol. II (eds P. Jena, S.N. Khanna, and B.K.N. Rao), Kluwer, Dordrecht, pp. 867–879.

758 Takahara, J., Yamagishi, S., Taki, H., Morimoto, A., and Kobayashi, T. (1997) Guiding of a one-dimensional optical beam with nanometer diameter. *Opt. Lett.*, **22**, 475–477.

759 Quinten, M., Leitner, A., Krenn, J.R., and Aussenegg, F.R. (1998) Electromagnetic energy transport via linear chains of silver nanoparticles. *Opt. Lett.*, **23**, 1331–1333.

760 Brongersma, M.L., Hartman, J.W., and Atwater, H.A. (2000) Electromagnetic energy transfer and switching in nanoparticle chain arrays below the diffraction limit. *Phys. Rev. B*, **62**, R16356–R16359.

761 Maier, S.A., Brongersma, M.L., and Atwater, H.A. (2001) Electromagnetic energy transport along arrays of closely spaced metal rods as an analogue to plasmonic devices. *Appl. Phys. Lett.*, **78**, 16–18.

762 Maier, S.A., Brongersma, M.L., Kik, P.G., Meltzer, S., Requicha, A.A.G., and Atwater, H.A. (2001) Plasmonics – a route to nanoscale optical devices. *Adv. Mater.*, **13**, 1501–1505.

763 Maier, S.A., Brongersma, M.L., and Atwater, H.A. (2002) Electromagnetic energy transport along Yagi arrays. *Mater. Sci. Eng. C*, **19**, 291–294.

764 Maier, S.A., Kik, P.G., and Atwater, H.A. (2002) Observation of coupled plasmon-polariton modes in Au nanoparticle chain waveguides of different lengths: estimation of waveguide loss. *Appl. Phys. Lett.*, **81**, 1714–1716.

765 Maier, S.A., Brongersma, M.L., Kik, P.G., and Atwater, H.A. (2002) Observation of near-field coupling in metal nanoparticle chains using far-field polarization spectroscopy. *Phys. Rev. B*, **65**, 193408, 4 pages.

766 Maier, S.A., Kik, P.G., Brongersma, M.L., Atwater, H.A., Meltzer, S., Requicha, A.A.G., and Koel, B.E. (2002) Observation of coupled plasmon-polariton modes of plasmon waveguides for electromagnetic energy transport below the diffraction limit. *MRS Proc.*, **722**, L6.2.1–L6.2.6.

767 Maier, S.A., Kik, P.G., Atwater, H.A., Meltzer, S., Harel, E., Koel, B.E., and Requicha, A.A.G. (2003) Local detection of electromagnetic energy transport below the diffraction limit in metal nanoparticle plasmon waveguides. *Nat. Mater.*, **2**, 229–232.

768 Maier, S.A., Kik, P.G., and Atwater, H.A. (2003) Optical pulse propagation in metal nanoparticle chain waveguides. *Phys. Rev. B*, **67**, 205402.

769 Salerno, M., Félidj, N., Krenn, J.R., Leitner, A., Aussenegg, F.R., and Weeber, J.C. (2001) Near-field optical response of a two-dimensional grating of gold nanoparticles. *Phys. Rev. B*, **63**, 165422.

770 Salerno, M., Krenn, J.R., Hohenau, A., Ditlbacher, H., Schider, G., Leitner, A., and Aussenegg, F.R. (2005) The optical near-field of gold nanoparticle chains. *Opt. Commun.*, **248**, 543–549.

771 Girard, C., and Quidant, R. (2004) Near-field optical transmittance of metal particle chain waveguides. *Opt. Express*, **12**, 6141–6146.

772 Gray, S.K., and Kupka, T. (2003) The propagation of light in metallic nanowire arrays: finite-difference time-domain studies of silver cylinders. *Phys. Rev. B*, **68**, 045415.

773 Pohl, D.W., Denk, W., and Lanz, M. (1984) Optical stethoscopy: image recording with resolution $\lambda/20$. *Appl. Phys. Lett.*, **44**, 651–653.

774 Pohl, D.W., Denk, W., and Dürig, U. (1986) Optical stethoscopy: imaging with λ/20. *Proc. Soc. Photo-Opt. Instrum. Eng.*, **565**, 56–61.

775 Dürig, U., Pohl, D.W., and Rohner, F. (1986) Near-field optical scanning microscopy. *J. Appl. Phys.*, **51**, 3318–3327.

776 Pohl, D.W. (1991) Scanning near-field optical microscopy (SNOM). *Adv. Opt. Electron Microsc.*, **12**, 243–312.

777 Heinzelmann, H., and Pohl, D.W. (1994) Scanning near-field optical microscopy. *Appl. Phys. A*, **59**, 89–101.

778 Girard, C., and Dereux, A. (1996) Near-field optics theories. *Rep. Prog. Phys.*, **59**, 657–699.

779 Synge, E.H. (1928) A suggested method for extending the microscopic resolution into the ultramicroscopic region. *Philos. Mag.*, **6**, 356–362.

780 Zenhausern, F., Martin, Y., and Wickramasinghe, H.K. (1995) Scanning interferometric apertureless microscopy: optical imaging at 10 ångstrom resolution. *Science*, **269**, 1083–1085.

781 Martin, Y., Zenhausern, F., and Wickramasinghe, H.K. (1996) Scattering spectroscopy of molecules at nanometer resolution. *Appl. Phys. Lett.*, **68**, 2475–2477.

782 Quinten, M. (2000) Evanescent wave scattering by aggregates of clusters – application to optical near-field microscopy. *Appl. Phys. B*, **70**, 579–586.

783 Doicu, A., Eremin, Y., and Wriedt, T. (2001) Scattering of evanescent waves by a sensor tip near a plane surface. *Opt. Commun.*, **190**, 5–12.

784 Wurtz, G.A., Dimitrijevic, N.M., and Wiederrecht, G.P. (2002) The spatial extension of the field scattered by silver nanoparticles excited near resonance as observed by apertureless near-field optical microscopy. *Jpn. J. Appl. Phys.*, **41**, L351–L354.

785 Xiao, M. (1997) Theoretical treatment for scattering scanning near-field optical microscopy. *J. Opt. Soc. Am. A*, **14**, 2977–2984.

786 Krenn, J.R., Wolf, R., Leitner, A., and Aussenegg, F.R. (1997) Near-field optical imaging the surface plasmon fields of lithographically designed nanostructures. *Opt. Commun.*, **137**, 46–50.

787 Fleischmann, M., Hendra, P.J., and McQuillan, A.J. (1974) Raman spectra of pyridine adsorbed at a silver electrode. *Chem. Phys. Lett.*, **26**, 163–166.

788 Jeanmaire, D.L., and van Duyne, R.P. (1977) Surface Raman electrochemistry. Part I. Heterocyclic, aromatic and aliphatic amines adsorbed on the anodized silver electrode. *J. Electroanal. Chem.*, **84**, 1–20.

789 Albrecht, M.G., and Creighton, J.A. (1977) Anomalously intense Raman spectra of pyridine at a silver electrode. *J. Am. Chem. Soc.*, **99**, 5215–5219.

790 Creighton, J.A., Albrecht, M.G., Hester, R., and Matthew, J. (1978) The dependence of the intensity of Raman bands of pyridine at a silver electrode on the wavelength of excitation. *Chem. Phys. Lett.*, **55**, 55–58.

791 Moskovits, M. (1978) Surface roughness and the enhanced intensity of Raman scattering by molecules adsorbed on metals. *J. Chem. Phys.*, **69**, 4159–4161.

792 Rowe, J.E., Shank, C.V., Zwemer, D.A., and Murray, C.A. (1980) Ultrahigh-vacuum studies of enhanced Raman scattering from pyridine on Ag surfaces. *Phys. Rev. Lett.*, **44**, 1770–1784.

793 McCall, S.L., Platzman, P.M., and Wolff, P.A. (1980) Surface enhanced Raman scattering. *Phys. Lett. A*, **77**, 381–383.

794 Murray, C.A., Allara, D.L., and Rhinewine, M. (1981) Silver-molecule separation dependence of surface-enhanced Raman scattering. *Phys. Rev. Lett.*, **46**, 57–60.

795 Gersten, J., and Nitzan, A. (1980) Electromagnetic theory of enhanced Raman scattering by molecules adsorbed on rough surfaces. *J. Chem. Phys.*, **73**, 3023–3037.

796 Furtak, T.E., and Reyes, J. (1980) A critical analysis of theoretical models for the giant Raman effect from adsorbed molecules. *Surf. Sci.*, **93**, 351–382.

797 Wetzel, H., and Gerischer, H. (1980) Surface enhanced Raman scattering

from pyridine and halide ions adsorbed on silver and gold sol particles. *Chem. Phys. Lett.*, **76**, 460–464.
798 Chen, C., and Burstein, E. (1980) Giant Raman scattering by molecules at metal-island films. *Phys. Rev. Lett.*, **45**, 1287–1291.
799 Creighton, J.A., Blatchford, C.G., and Albrecht, M.G. (1980) Plasma resonance enhancement of Raman scattering by pyridine adsorbed on silver or gold sol particles of size, comparable to the excitation wavelength. *J. Chem. Soc. Faraday Trans. II*, **75**, 790–798.
800 Dornhaus, R. (1982) *Surface Enhanced Raman Spectroscopy*, Festkörperprobleme: Advances in Solid State Physics, vol. 22, Springer, Berlin, pp. 201–228.
801 Chang, R.K., and Furtak, T.E. (1982) *Surface Enhanced Raman Scattering*, Plenum Press, New York.
802 Blatchford, C.G., Campbell, J.R., and Creighton, J.A. (1982) Plasma resonance–enhanced Raman scattering by adsorbates on gold colloids: the effects of aggregation. *Surf. Sci.*, **120**, 435–455.
803 Otto, A. (1984) Surface enhanced Raman scattering: "classical" and "chemical" origins, in *Light Scattering in Solids* (eds M. Cardona and G. Güntherodt), Springer Series Topics in Applied Physics, vol. **54**, Springer, Berlin, pp. 289–418.
804 Wang, D.-S., Chew, H., and Kerker, M. (1980) Enhanced Raman scattering at the surface (SERS) of a spherical particle. *Appl. Opt.*, **19**, 2256–2257.
805 Kerker, M., Wang, D.-S., and Chew, H. (1980) Surface enhanced Raman scattering (SERS) by molecules adsorbed at spherical particles. *Appl. Opt.*, **19**, 3373–3388.
806 Wang, D.-S., and Kerker, M. (1981) Enhanced Raman scattering by molecules adsorbed at the surface of colloidal spheroids. *Phys. Rev. B*, **24**, 1777–1790.
807 Barber, P.W., Chang, R.K., and Massoudi, H. (1983) Surface-enhanced electric intensities on large silver spheroids. *Phys. Rev. Lett.*, **50**, 997–1000.
808 Barber, P.W., Chang, R.K., and Massoudi, H. (1983) Electrodynamic calculations of the surface-enhanced electric intensities on large Ag spheroids. *Phys. Rev. B*, **27**, 7251–7261.
809 Meier, M., and Wokaun, A. (1983) Enhanced fields on large metal particles: dynamic depolarization. *Opt. Lett.*, **8**, 581–583.
810 Kreibig, U. (1984) Electromagnetic resonances in systems of small particles, S.E.R.S., in *Dynamics on Surfaces* (eds B. Pullman, J. Jortner, A. Nitzan, and B. Gerber), Reidel, Dordrecht, pp. 447–460.
811 Mrozek, I., and Otto, A. (1987) No evidence for local electromagnetic depolarisation effects in SERS from coldly deposited silver films. *J. Electron Spectrosc. Relat. Phenom.*, **45**, 143–152.
812 Otto, A. (1991) Surface enhanced Raman scattering of adsorbates. *J. Raman Spectrosc.*, **22**, 743–752.
813 Otto, A., Mrozek, I., Grabhorn, H., and Akeman, W. (1992) Surface-enhanced Raman scattering. *J. Phys. Condens. Matter*, **4**, 1143–1212.
814 Meier, M., Wokaun, A., and Liao, P.F. (1985) Enhanced fields on rough surfaces: dipolar interactions among particles of sizes exceeding the Rayleigh limit. *J. Opt. Soc. Am. B*, **2**, 931–949.
815 Carron, K.T., Fluhr, W., Meier, M., Wokaun, A., and Lehmann, H.W. (1986) Resonances of two-dimensional particle gratings in surface-enhanced Raman scattering. *J. Opt. Soc. Am. B*, **3**, 430–440.
816 Moskovits, M. (1985) Surface-enhanced spectroscopy. *Rev. Mod. Phys.*, **57**, 783–826.
817 Stockman, M.I., Shalaev, V.M., Moskovits, M., Botet, R., and George, T.F. (1992) Enhanced Raman scattering by fractal clusters: scale invariant theory. *Phys. Rev. B*, **46**, 2821–2830.
818 Kneipp, K., Wang, Y., Kneipp, H., Itzkan, I., Dasari, R.R., and Feld, M.S. (1996) Population pumping of excited vibrational states by surface-enhanced Raman scattering. *Phys. Rev. Lett.*, **76**, 2444–2448.
819 Bozhevolnyi, S.I., Markel, V.A., Coello, V., Kim, W., and Shalaev, V.M.

(1998) Direct observation of localized dipolar excitations on rough nanostructured surfaces. *Phys. Rev. B*, **58**, 11441–11448.

820 Markel, V.A., Shalaev, V.M., Zhang, P., Huynh, W., Tay, L., Haslett, T.L., and Moskovits, M. (1999) Near-field optical spectroscopy of individual surface-plasmon modes in colloid clusters. *Phys. Rev. B*, **59**, 10903–10909.

821 Lichtenecker, K. (1926) Die Dielektrizitätskonstante natürlicher und künstlicher Mischkörper. *Phys. Z.*, **27**, 115–158.

822 Beer, A. (1853) *Einleitung in die höhere Optik*, Vieweg, Braunschweig.

823 Gladstone, J.H., and Dale, T.P. (1863) Researches on the refraction and dispersion and sensitiveness of liquids. *Philos. Trans.*, **153**, 317–343.

824 Landau, L.D., and Lifshitz, E.M. (1974) *Lehrbuch der Theoretischen Physik VIII: Elektrodynamik der Kontinua*, Akademie Verlag, Berlin.

825 Looyenga, H. (1965) Dielectric constants of heterogeneous mixtures. *Physica*, **31**, 401–406.

826 Niklasson, G.A., Granqvist, C.G., and Hunderi, O. (1981) Effective medium models for the optical properties of inhomogeneous materials. *Appl. Opt.*, **20**, 26–30.

827 Garnett, J.C.M. (1904) Colours in metal glasses and in metallic films. *Philos. Trans. R. Soc. London Ser. A*, **203**, 385–420.

828 Bruggeman, D.A.G. (1935) Berechnung verschiedener physikalischer Konstanten von heterogenen Substanzen. I. Dielektrizitätskonstanten und Leitfähigkeiten der Mischkörper aus isotropen Substanzen. *Ann. Phys. (Leipzig)*, **24**, 636–679.

829 Bergman, D.J. (1978) The dielectric constant of a composite material – a problem in classical physics. *Phys. Rep.*, **43**, 377–407.

830 Bergman, D.J. (1978) Bulk physical properties of composite media, in *Les Méthodes de l'Homogénéisation: Théorie et Applications en Physique*, Editions Eyrolles, 57, pp. 1–128.

831 Bergman, D.J., and Stroud, D. (1980) Theory of resonances in the electromagnetic scattering by macroscopic bodies. *Phys. Rev. B*, **22**, 3527–3539.

832 Lamb, W., Wood, D.M., and Ashcroft, N.W. (1980) Long-wavelength electromagnetic propagation in heterogeneous media. *Phys. Rev. B*, **21**, 2248–2266.

833 Felderhof, B.U., and Jones, R.B. (1985) Spectral broadening in suspensions of metallic spheres. *Z. Phys. B*, **62**, 43–50.

834 Felderhof, B.U., and Jones, R.B. (1986) Multipole contributions to the effective dielectric constant of suspensions of spheres. *Z. Phys. B*, **62**, 215–224.

835 Felderhof, B.U., and Jones, R.B. (1986) Effective dielectric constant of dilute polydisperse suspensions of spheres. *Z. Phys. B*, **62**, 225–230.

836 Felderhof, B.U., and Jones, R.B. (1986) Multipolar corrections to the Clausius–Mossotti formula for the effective dielectric constant of a polydisperse suspension of spheres. *Z. Phys. B*, **62**, 231–237.

837 Gans, R., and Happel, H. (1909) Zur Optik kolloidaler Metallösungen. *Ann. Phys. (Leipzig)*, **29**, 277–300.

838 Doyle, W.T. (1989) Optical properties of a suspension of metal spheres. *Phys. Rev. B*, **39**, 9852–9858.

839 Ruppin, R. (2000) Evaluation of extended Maxwell-Garnett theories. *Opt. Commun.*, **182**, 273–279.

840 Chýlek, P., Srivastava, V., Pinnick, R.G., and Wang, R.T. (1988) Scattering of electromagnetic waves by composite spherical particles: experiment and effective medium approximations. *Appl. Opt.*, **27**, 2396–2404.

841 Videen, G., Ngo, D., and Chýlek, P. (1994) Effective-medium predictions of absorption by graphitic carbon in water droplets. *Opt. Lett.*, **19**, 1675–1677.

842 Borghese, F., Denti, P., Saija, R., and Sindoni, I.O. (1992) Optical properties of spheres containing a spherical eccentric inclusion. *J. Opt. Soc. Am. A*, **9**, 1327–1335.

843 Abelès, B., and Gittleman, J. (1976) Composite material films: optical properties and applications. *Appl. Opt.*, **15**, 2328–2332.

844 Granqvist, C.G., and Hunderi, O. (1977) Optical properties of ultrafine gold particles. *Phys. Rev. B*, **16**, 3513–3534.

845 Norrman, S., Andersson, T., Granqvist, C.G., and Hunderi, O. (1978) Optical properties of dicontinuous films. *Phys. Rev. B*, **18**, 674–695.

846 Bedeaux, D., and Vlieger, J. (1973) A phenomenological theory of the dielectric properties of thin films. *Physica*, **67**, 55–73.

847 Granqvist, C.G. (1981) Optical properties of cermet materials. *J. Phys. (Paris) Colloq. C1*, **42**, C1-247–C1-284.

848 Truong, V., Girouard, F.E., and Bosi, G. (1982) Optical properties of granular tin films from 0.22 to 1.0 μm. *Appl. Opt.*, **21**, 2508–2511.

849 Niklasson, G.A., and Craighead, H.G. (1983) Optical properties of codeposited aluminum–silicon composite films. *Appl. Opt.*, **22**, 1237–1240.

850 Thériault, J.-M., and Boivin, G. (1984) Maxwell-Garnett theory extended for Cu–PbI$_2$ cermets. *Appl. Opt.*, **23**, 4494–4498.

851 Chandonnet, A., and Boivin, G. (1989) Experimental study of Cu–PbCl$_2$, Cu–NaF, Ag–PbCl$_2$, and Ag–NaF cermet thin films. *Appl. Opt.*, **28**, 717–721.

852 Heilmann, A., Kampfrath, G., and Hopfe, V. (1988) Optical properties of plasma-polymer–silver composite films and their simulation by means of effective-medium theories. *J. Phys. D*, **21**, 986–994.

853 Heilmann, A., Hamann, C., Kampfrath, G., and Hopfe, V. (1990) Preparation and optical properties of plasma polymer silver composite films. *Vacuum*, **41**, 1472–1475.

854 Heilmann, A., Werner, J., Schwarzenberg, D., Henkel, S., Grosse, P., and Theiss, W. (1995) Microstructure and optical properties of plasma polymer thin films with embedded silver nanoparticles. *Thin Solid Films*, **270**, 103–108.

855 Leitner, A., Zhao, Z., Brunner, H., Aussenegg, F.R., and Wokaun, A. (1993) Optical properties of a metal island film close to a smooth metal surface. *Appl. Opt.*, **32**, 102–110.

856 Dobierzewska-Mozrzymas, E., and Bieganski, P. (1993) Optical properties of discontinuous copper films. *Appl. Opt.*, **32**, 2345–2350.

857 Nagendra, C.L., and Lamb, J.L. (1995) Optical properties of semiconductor–metal composite thin films in the infrared region. *Appl. Opt.*, **34**, 3702–3710.

858 Luo, R. (1997) Effective medium theories for the optical properties of three-component composite materials. *Appl. Opt.*, **36**, 8153–8158.

859 Wood, D.M., and Ashcroft, N.W. (1977) Effective medium theory of the optical properties of small particle composites. *Philos. Mag.*, **35**, 269–280.

860 Stroud, D., and Pan, F.P. (1978) Self-consistent approach to electromagnetic wave propagation in composite media: application to model granular metals. *Phys. Rev. B*, **17**, 1602–1610.

861 Kürbitz, S., Porstendorfer, J., Berg, K.-J., and Berg, G. (2001) Determination of size and concentration of copper nanoparticles dispersed in glasses using spectroscopic ellipsometry. *Appl. Phys. B*, **73**, 333–337.

862 Mandal, S.K., Roy, R.K., and Pal, A.K. (2003) Effect of particle shape distribution on the surface plasmon resonance of Ag–SiO$_2$ nanocomposite thin films. *J. Phys. D*, **36**, 261–265.

863 Takeda, Y., Lu, J., Plaksin, O.A., Amekura, H., Kono, K., and Kishimoto, N. (2004) Optical properties of dense Cu nanoparticle composites fabricated by negative ion implantation. *Nucl. Instrum. Methods Phys. Res. B*, **219–220**, 737–741.

Color Plates

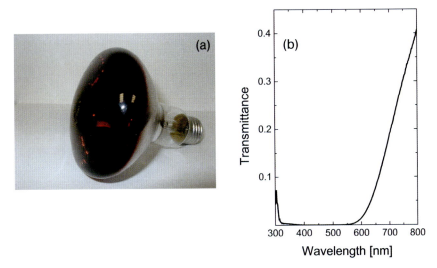

Figure 2.19 (a) Photograph of a red bulb for medical applications and (b) transmittance spectrum of such a red bulb. This figure also appears on page 35.

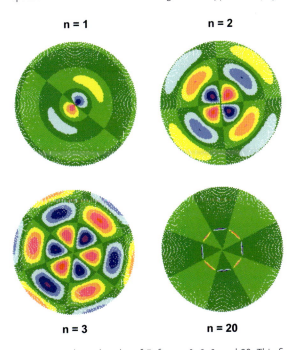

Figure 5.4 False color plot of E_θ for $n = 1, 2, 3$, and 20. This figure also appears on page 82.

Optical Properties of Nanoparticle Systems: Mie and Beyond. Michael Quinten
Copyright © 2011 WILEY-VCH Verlag GmbH & Co. KGaA, Weinheim
ISBN: 978-3-527-41043-9

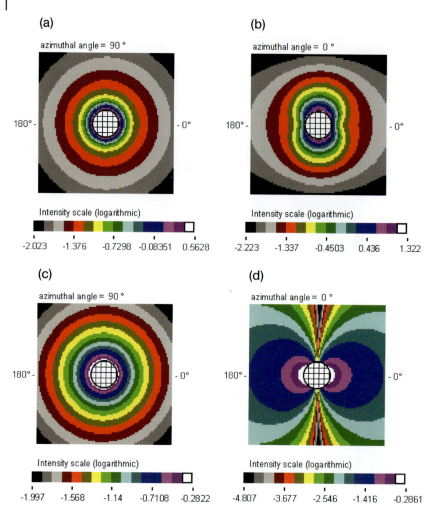

Figure 5.30 Angular distribution of the intensity of the light scattered by a spherical gold particle with diameter $2R = 100$ nm. (a) S_{per}; (b) S_{par}; (c) i_{per}; (d) i_{par}. This figure also appears on page 121.

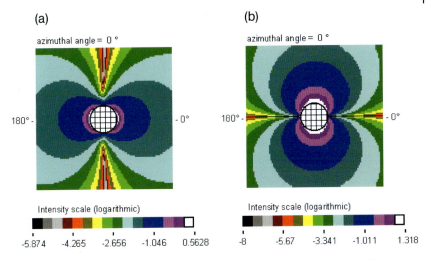

Figure 5.31 Contributions of (a) E_θ and (b) E_r to the modified angular distribution function S_{par} in the plane of incidence for the gold particle in Figure 5.29. This figure also appears on page 122.

Figure 6.22 Fluorescence of CdSe–ZnS nanoparticles. The size of the quantum dots increases from about 2 nm to about 6 nm (left to right). The emission spans across the visible part of the electromagnetic spectrum from the blue (2 nm nanocrystals) to the red (6 nm nanocrystals). This figure also appears on page 151. Reprinted from [155] with permission of Elsevier, copyright 2009.

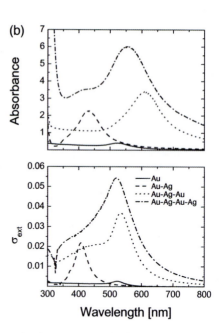

Figure 7.14 (a) Series of pictures for the multishell nanoparticle colloid Ag–Au–Ag–Au. Reproduced from [193] by permission of The Royal Society of Chemistry. (b) Measured absorbance spectra and calculated extinction efficiency spectra. This figure also appears on page 190. Data courtesy of Luis M. Liz-Marzan.

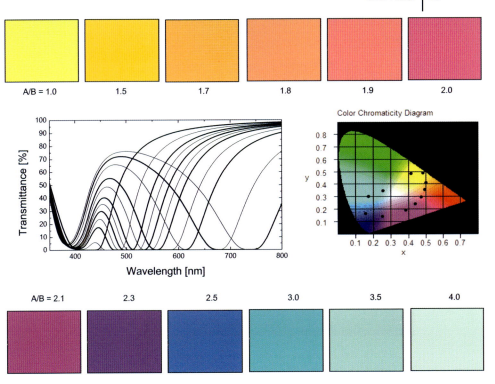

Figure 9.13 Transmittance spectra of ensembles of noninteracting prolate spheroidal particles in glass N-BK7 with varying ratio $r = A/B$. The corresponding colors and color coordinates are given in the rectangular boxes and the color chromaticity diagram. This figure also appears on page 263.

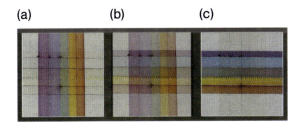

Figure 9.20 Five nanoparticle polarizers with different colors put perpendicular on five similar polarizers. (a) Incident light polarized vertically; (b) unpolarized incident light; (c) incident light polarized horizontally. This figure also appears on page 269. Image courtesy K.-J. Berg.

484 | *Color Plates*

Figure 11.10 Color pictures of samples with isolated and aggregated silver and gold particles. This figure also appears on page 351.

Figure 12.11 The Lycurgus cup. This figure also appears on page 406. Copyright: The Trustees of the British Museum.

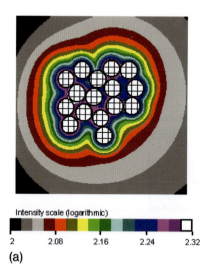
(a)

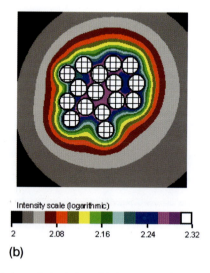
(b)

Figure 13.8 Near-field intensities in the equatorial plane (x–y plane) of a planar aggregate with $N = 16$ Au particles for an incident wave propagating along the z-axis, being polarized along (a) the x-axis and (b) the y-axis. This figure also appears on page 425.

Index

1/3–2/3 rule   210

a

absorbance   23
Absorbing Embedding Media   214
additional boundary condition   61
Aerosols   11
agglomerates   13
aggregates   13, 317, 341, 379, 422
aggregates of spheres   317, 422
alloy nanoparticles   99, 134
alloy particles   407
Alumina   168
amorphous carbon   156
angular distribution functions   117
anthropogenic air pollution   123
asymmetry in scattering   357
asymmetry parameter   86, 277

b

Backscattering   25
Bergman equation   430
blue moon   123
blue sun   123
Bouguer's law   24
boundary element method (BEM)   313
Brendel oscillator   40
Bruggeman ansatz   429

c

Carbonaceous Particles   156, 266, 291, 369
Cassius Gold Purple   131
Catalyst Metal Particles   139
ceria   170
chain-like aggregates   326
Charged Nanoparticles   206
chemical interface damping (CID)   241
clusters   388
coagulation aggregates   13

coalescence aggregates   13, 361
coated spheres   7, 177
Colloidal systems   1, 11
Color Pigments   398
convergence   332
coupled dipole method   302
cross-sections   63, 320
cubes   7, 246
Cubic Particles   296
curvilinear coordinate systems   60
cylinders   7
cylindrical vector harmonics   273

d

DDA   302, 387
Debye-Hückel   14
DGF   305
diamond   156
dielectric function   5, 25, 37, 59
dielectric nanoparticles   367
differential scattering cross-section   65, 86
diffuse reflectance   396
diffuse transmittance   396
digitized Green's function   305
dipole–dipole coupling   321, 324
discrete dipole approximation (DDA)   7, 246, 302, 387
dispersion integrals   52
DLVO theory   18
Drude dielectric function   41, 46
Drude Metal Particles   124
dynamic charge transfer   241

e

EBCM   305, 387
eccentricity   248
effective dielectric function   24, 427
effective medium   8, 427
effective-medium   24

effective medium theories 427
efficiencies 63
efficiency 277
Elastic light scattering 3
electromagnetic coupling 321
electromagnetic interactions 7, 341, 383
electronic resonances 88, 91, 92, 109, 142, 152, 252, 279, 280, 289, 297, 321, 333, 401
ellipsoids 7, 246
ellipsoids of revolution 256
enhancement factor 329
evanescent waves 226, 335, 417
evanescentwave scattering 417
exciton 151
extended boundary condition method 305, 387
extended boundary condition method (EBCM) 246
extinction 21
extinction cross-section 259, 342
extinction efficiency 124

f
FDTD 307, 387
field-based angular distribution functions 119
filling factor 13
finite cylinder 282
finite difference time domain 307, 387
fractal 388
fractal aggregates 14
fractal clusters 363, 390
fractals 14, 391
free path effect 235
Gamma distribution 28
Gaussian beam 68, 223

g
generalized Lorenz–Mie theory 224, 255
generalized Mie theory (GMT) 7, 36, 317, 341, 393
geometric resonances 88
geometry factors 255
GLMT 224, 282
GMT 317, 341, 408
graphite 156

h
harmonic oscillator 92, 147
harmonic oscillator model 38
haze 25
Helmholtz equation 58
hematite 163, 292
Hertz vectors 59

homogeneous electron gas 44
hot spots 425
Hubbard dielectric function 46

i
infinitely long cylinders 246, 273, 337
infrared absorbers 173
inhomogeneous incident waves 7, 68, 223
interband transitions 47, 102, 149, 152
interstellar dust 123
IR Absorbers 403
ITO 172, 267, 294, 301, 375

j
jellium model 18, 44, 233

k
Kramers–Kronig relations 52
Kubelka and Munk 8, 394
Kubelka–Munk theory 163, 394
Kubelka–Munk unit 395

l
Lambert–Beer law 24
lanthanum hexaboride 172, 267, 294, 375
layered spheres 145
Lindhard dielectric function 46
local density approximation 43
localization principle 224
localized beam approximation 224
log-normal distribution 27
longitudinal bulk plasmon modes 61, 104
longitudinal eigenmode 322
longitudinal plasmon resonances 281
Lorenz–Mie theory 75
Lycurgus cup 131, 404, 406

m
Magnetic Metal Particles 141
magnetic resonances 106
many-flux theories 36, 394
Mathiessen rule 42
Maxwell-Garnett equation 429
MDRs 109, 170, 229, 279, 289, 368, 401
mean free path 234
Metal Particles 259, 283, 299, 342
method 387
Mie scattering 55
Mie theory 75, 233
Mie's theory 6
morphology–dependent resonances 170
morphology-dependent resonances (MDRs) 88, 152, 159, 172, 187, 246
multilayer spheroids 254

multilayered cylinder   274
multiple resonances   101
multishell nanoparticles   189

n
nanocomposites   382
nanoparticle matter   1
nanoparticle system   10
nanoparticles   92
nanorods   313–315, 388
nanotechnology   1
near-field   8, 65, 411
near-field efficiencies   112
near-field optical microscopy   65
next-neighbor distance   327
noble metal nanoparticles   94
Noble Metal Particles   127
noble metals   310, 404
nonlinear optical effects   65
nonlinear optical properties   112
nonspherical nanoparticles   310
Nonspherical Particles   245
normal distribution   28

o
oblate   247
optical bistability   191
optical cross-sections   63, 83
optical density   23
optical particle counter   379
Optical particle sizing   379
optically active media   213
Ostwald ripening   19
Oxide Particles   162, 168, 292

p
Particles   265, 288, 299
peak splitting   327, 343
pertubation theory   246
Phonon Polaritons   170, 267, 293, 300, 372
Photothermal optical beam deflection   358
pigment particles   162
pigments   262
plasma frequency   209
plasma frequency ω_p 46
plasmons   46
polarizability   70
prolate   247
PTOBD   358

q
quantum dots   151
quantum size effects   151, 233, 235

r
radar backscattering cross-section   87
radiation pressure   86
radiative transfer   36, 393
Raman scattering   420
Rare Earth Metal Particles   142
Rayleigh approximation   69, 255, 296
Rayleigh scattering   55
Rayleigh–Debye–Gans approximation   71
refractive index   37, 59
relative depth   357
resonances   252, 342, 411

s
scalar wave equation   59
scanning near-field optical microscopy (SNOM)   112, 376, 415, 416
scattering   21
scattering intensities   84, 277, 320, 355
self-consistent field (SCF) *approximation*   45
Semiconductor Particles   151, 265, 288, 299, 364
Semimetal   265, 288, 299, 364
Semimetal Particles   148
SERS   112, 420
Silica   168
Silicon Nanoparticles   401
size effects   233
skin depth   53
small-angle scattering   71
SNOM   112, 376, 416
solar radiation budget   123
soot   391
soot agglomerates   391
spherical nanoparticles   123
spheroidal particle   246
spheroids   7, 247, 256
sp-hybridization   156
spill-out effect   234
SPP   252, 342, 411
SPP resonances   96, 108, 127, 142, 157, 172, 185, 194, 203, 229, 246, 259, 280, 283, 294, 297, 310, 326, 377, 428
stained glasses   406
Stochastically Distributed Spheres   382
Supported Nanoparticles   198
surface conductivity   207
surface plasmon polariton (SPP)   2, 88, 94, 95, 108, 125, 181, 245, 321, 342, 422
surface plasmon resonances   203
surface-enhanced Raman scattering   65, 112
surface-enhanced Raman spectroscopy   8

t

Tauc–Lorentz model 50
technique 307
Titania 170
T-matrix 387
T-matrix method 7, 246, 305
Transition Metal Particles 145
transmission loss 414
transverse eigenmodes 322

v

vector harmonics 60
vector spheroidal harmonics 247

vector wave equation 58
VIEF 305, 387
volume integral equation formulation 305, 387

w

Waveguiding 412
Weibull distribution 28
whispering gallery modes 88
White Pigments 399